Rethinking Bhopal

Rethinking Bhopal

A Definitive Guide to Investigating, Preventing, and Learning from Industrial Disasters

Kenneth Bloch

AMSTERDAM • BOSTON • HEIDELBERG • LONDON
NEW YORK • OXFORD • PARIS • SAN DIEGO
SAN FRANCISCO • SINGAPORE • SYDNEY • TOKYO

Elsevier
Radarweg 29, PO Box 211, 1000 AE Amsterdam, Netherlands
The Boulevard, Langford Lane, Kidlington, Oxford OX5 1GB, UK
50 Hampshire Street, 5th Floor, Cambridge, MA 02139, USA

British Library Cataloguing-in-Publication Data
A catalogue record for this book is available from the British Library

Library of Congress Cataloging-in-Publication Data
A catalog record for this book is available from the Library of Congress

ISBN: 978-0-12-803778-2

Publisher: Joe Hayton
Acquisition Editor: Fiona Geraghty
Editorial Project Manager: Maria Convey
Production Project Manager: Julie-Ann Stansfield
Designer: Matthew Limbert

Typeset by TNQ Books and Journals

Cover photo credit: David Graham/Bhopal Medical Appeal

To my father, for encouraging me to write this book and for making my professional goal to solve one problem more than I create.

Contents

About the Author

Kenneth Bloch is a senior HES professional who specializes in petrochemical industry incident investigation and failure analysis. His experience includes 30 years of downstream service in maintenance, PSM, technical, and operations roles. He speaks regularly at AFPM, API, and AIChE process safety symposiums about experiences that help prevent recurring process safety failures throughout the manufacturing industry. Mr. Bloch graduated with honors from Lamar University in Beaumont, Texas, with an Environmental Science Degree in 1988.

Foreword

When the Bhopal disaster happened in December 1984, I had been employed by a multinational petrochemical company for over two decades. I recall that Bhopal floated to the top of the news as the enormity of the event became ever more graphic. I was then a mechanical engineer immersed in machinery reliability improvement and had learned not to form a premature opinion on matters. Premature judgment would have been especially inappropriate in what appeared to be a complex incident at a facility owned by a legacy corporation headquartered in the United Sates. To withhold judgment was the correct notion, because it did take years to more fully comprehend the underlying causes of the Bhopal incident. Underlying causes needed to be studied in addition to the sequence of mistakes that caused the loss of thousands of lives. We did realize, though, that nothing in the incident matched the legal profession's generally misguided definition of "Act of God." It was clear that this was an artificial chain of events. Actions and inactions, omissions and commissions were involved; we knew it in 1984 with absolute certainty.

Years later we read around the fringes of the various commission reports and could observe the predictable thinking processes of litigators and counselors. Their overriding concern was to lessen the financial or career impact of whatever parties had hired them. While it would not be correct to claim that the truth was deliberately suppressed by anyone, we only now have full access to documents unlocked by the Indian legal system beginning in 2010. Although books on Bhopal fill the library shelves, we only now have a book that makes use of many of these latest documents.

In addition to this recent access to previously sealed documents, we now have, for the very first time, an insightful book written by a highly competent author. The author of this latest book is trained in TapRooT® and other advanced troubleshooting and incident investigation approaches. He applies these proven techniques to all facts in evidence and meticulously lists the hundreds of references that led him away from guesses, suppositions, and theories. He inserts actual on-the-record Bhopal events into a superbly structured troubleshooting and failure analysis approach, which has been taught on all six continents with great success for the past 20 years. Combining TapRooT® with facts in evidence makes this book totally unique and fascinating reading. Here, for the very first time, is an author who applies a time-tested failure analysis method to all the relevant data and does so without bias or predisposition toward any particular outcome of the investigation.

A perceptive reader now has access to, and will be absorbing, lessons that were previously shrouded in the fog of opinion and the concessions made to corporate allegiance. Also, until this book was compiled and published, the real lessons were often subsumed into agendas or reports ranging from sketchy and fragmented to downright erroneous. Other previously enunciated lessons likely became inaccessible or were shelved indefinitely because of the organizational dynamics and reward systems in place in 1984. Sadly, many companies are still encumbered by lagging organizational dynamics and wholly counterproductive reward systems. As of today and despite its importance, structured failure analysis has not been implemented or updated in all segments of modern industry. Yet, structured incident analysis must be adopted or, in some cases, a current failure analysis method must be modified or adjusted. The implementation and updating issue is linked to professional ethics, by those who are serious about asset protection and by those who place the value of human life above anything else. This is a text that should be placed in the hands of equipment and process reliability specialists who desire to understand and then fully accept their professional and ethical responsibilities. In other words, this latest Bhopal book speaks to those who need a wake-up call; it also speaks to those who genuinely want to hear, and then act on, the unvarnished truth. We simply believe that it is the equipment and process reliability specialists' responsibility to detect, analyze, and resolve repeat failure events. Human lives may be at stake—and thousands of them at that.

The book is profusely illustrated and one photo in particular (Figure 1) should serve as a fitting reminder that process pumps and mechanical seals were involved in this incomprehensible tragedy. There were obvious and documented equipment reliability issues. Erroneous decisions were made in the interest of coping with, and circumventing, a chronic process pump problem. As the plant operators tried to get around the pump problem, they went down a path that led to misguided process-related shortcuts. Logical steps taken to devise alternative process operating strategies introduced unknown and potentially unacceptable hazards. The book describes these steps in detail and demonstrates how equipment reliability, process reliability, industrial productivity, and process safety are inextricably linked. Creative alternative operating strategies ("process optimization steps") may seem justified in the interest of confronting legitimate personal safety concerns. However, their impact on process safety, their unintended consequences, their "closing of escape routes" can be disastrous.

FIGURE 1

A reminder of the most tragic loss of life in the history of process plants.

Bhopal Medical Appeal, www.bhopal.org, used by permission.

Bhopal serves as a supreme lesson about why the focus must be on resolving equipment reliability defects. Doing so often makes it possible to eliminate risk-inducing components or elements. Unswerving focus, early remedial action, and zero tolerance for unexplained repeat failures of mechanical assets are the most valuable safeguards against a potentially devastating sequence of events. In the case of process reliability events, preemptive action may be needed. But all of this drives home a singular message: Facts are far more important than opinions. Arriving at facts becomes an ethical obligation in process incident investigations.

That obligation was not lost on the book's author. He places before us the exceedingly well-documented and ultimately fatal sequence of events that preceded a disaster. He describes with uncanny clarity and precision the equipment reliability and operational concerns that funneled process-related decisions into a dead end. With a misdirected focus, each remaining opportunity to redirect events vanished. As a consequence, events were inexorably moving toward disaster; each corrective step became a misstep that took the plant closer to an irreversible tragedy.

Finally, this reviewer believes that engineering colleges and their industries training or technical conference adjuncts must accept an obligation to explain the sequence of events that caused the Bhopal disaster. The events must also be traced back to business schools. Some of these institutions of higher learning are on paths that convey the notion that success is rooted in money-saving and/or schedule-improving strategies. But we now know the sequence of precursor events, which resulted in the worst industrial disaster in human history. Many of these precursor events were driven by a form of "management think" which did not nurture and groom and reward what might be called "safety and reliability think." We should never push safety into a subordinated role. We need to share our knowledge of the fact that blind faith in misdirected management can cause unmitigated disaster. In other words, subject matter experts with experience and wisdom may have to nudge management toward treading on safe pathways only.

Not teaching or not passing on the crucial explanations in this latest text will impoverish future generations of operators, technicians, and engineers. Not teaching about Bhopal will deprive these employees and their managers from fully grasping the importance of sound science, good engineering judgment, and good management. Lack of engineering judgment led to an unprecedented catastrophe in Bhopal. No compassionate human being would ever want to see a repetition of the sequence of events leading up to the disaster. Accepting this "no repeat" premise, we must pay close attention to the wise words uttered by W. Edwards Deming in the late 1940s: "It's not enough to do your best. You must first know what to do, and then do your best." To this reviewer and to the author of this book, Deming's words translate into an insistence on pleading with those who have ears to listen. It also means that now is the time to tell them.

Heinz P. Bloch

Acknowledgments

I would like to extend my personal gratitude toward specific individuals, whose direct or indirect contributions made this book possible:

To my wife Theresa and also Gerritt, Callan, and Jillian for your infinite patience, support, and the sacrifices you made in my interest.

Sincere thanks to my friend and colleague Briana Jung for helping me to develop the thesis and for demonstrating extraordinary courage, dignity, and wisdom in applying process safety principles to the difficult work that you do.

I thank the truly gifted mentors I have worked with, for taking a personal interest in my development as an industry professional, including Paula McCain, Art Brunn, Rick Gilmore, Jim Simon, Phil Gaarder, and Jeff Caudill. You continue adding value through the work of those you directed.

I am indebted to Jeff Wilkes for selecting me to be on his team and for the countless experiences that resulted in tangible process improvements.

Never could I have imagined the privilege of working with Ray Brooks, Joe Marra, and Don McCord who integrate Process Safety Management into all aspects of the manufacturing business.

Thank you, Dr. Tony Sofronas, Paul Barringer, and Norm Lieberman for providing the technical oversight that is woven deeply into this analysis. Also to Allen Budris for proofreading and editing Chapter 15.

Personal thanks to Doug Dunmire for everything you have done to relate asset reliability to process safety. I am so grateful to have worked with you.

I credit Dr. Shyam Shukla and Dr. Keith Hansen for preparing me for, and directing me to, a career in the manufacturing industry.

Many thanks to Kristie Young and Karen Nichols at the Lamar University Foundation.

I respect and acknowledge the operational discipline demonstrated by Gary Ista, Tracy Clem, Nicole Birchall, Tom Church, Geoff Glasrud, Rochelle LeVasseur, Larry Patterson, Mike Brose, Don Johnson, Joe Fronczek, Dawn Wurst, and Helen Rikard. Recalling your examples made certain sections of this book easier for me to write. Thank you.

Thank you, Dennis Hendershot for sharing your photos of the Bhopal factory with me and also for the work you have done to promote Inherently Safer Design concepts.

Thank you, Fiona Geraghty, Maria Convey, and Julie-Ann Stansfield for your help and support at Elsevier.

Special recognition to Ingrid Eckerman, Colin Toogood, IChemE, and the Bhopal Medical Appeal.

Amanda Quinn, thank you for understanding and we will work together soon.

Introduction

Rethinking Bhopal started out as a trade book named *Iron and Clay*. Its original title was symbolic of the dysfunctional relationship that often develops between man and machine. The major theme for me involved recognizing how good intentions can transform into big problems. Avoiding a path that ends in self-destruction involves taking the time to fully understand our mistakes before it is too late. Only then it is possible to correct a relatively minor issue which, if not managed properly, could result in a disaster. The most difficult part about this is admitting our mistake. It requires humility to address the situation. Our natural inclination is to apply another "solution" and instead of solving one problem, we create another. This is where things get messy.

The theme of well-intentioned actions or inactions transforming into big problems applies to all areas of life. In other words, a technology background is not needed to get the point. The book originally communicated the sequence of events behind history's worst industrial disaster as a self-improvement topic. Its purpose was to assist people in making courageous decisions, to avoid misguided solutions that complicate life.

But industrial disasters deserve much more coverage than can be written into a self-improvement book. Disasters deserve meticulous analysis to improve productivity throughout the manufacturing industry. And so, after a considerable amount of personal thought, input from trusted coworkers, and patience from my family, I decided to rewrite and reshape the manuscript into its present, much more technical format. By taking this approach, the book targets those best equipped to act responsibly with the weighty information it contains. Indeed, the worst problems faced by industry originate with how relatively small issues are being managed. While taking action to address a process problem, special care must be taken to avoid inadvertently making matters worse. Provisions must also be made to prevent exchanging one problem for another, potentially a worse situation. Understanding the details behind history's worst industrial disaster brings a vivid sense of awareness to these facts. Through the process of imparting this knowledge we illustrate what must be done to achieve a satisfactory level of process safety performance in the manufacturing industry.

Process safety management (PSM) as we know it today got started in the 1980s. Contrary to popular belief, process safety was not a primitive concept back then. From industry's start, the importance of designing and operating with safety in mind was appreciated by all involved in a manufacturing process. The sudden release of process energy in the form of an explosion or fire had proven deadly on previous occasions. The goal of PSM was to exercise a level of control sufficient to avoid the accidental release of process materials. Everybody was aligned on protecting their mutual interests by safely managing their industrial processes. Safety was the foremost commitment in everybody's mind.

That prevailing attitude has not changed very much over the years. Industry professionals at all levels within a typical manufacturing organization are still devoted to designing and operating industrial processes safely. Due to these ongoing efforts, the total recordable injury and illness rate in the United States (US) petroleum refining sector has dropped significantly since 1988 (Fig. 1) [1]. However, industry still

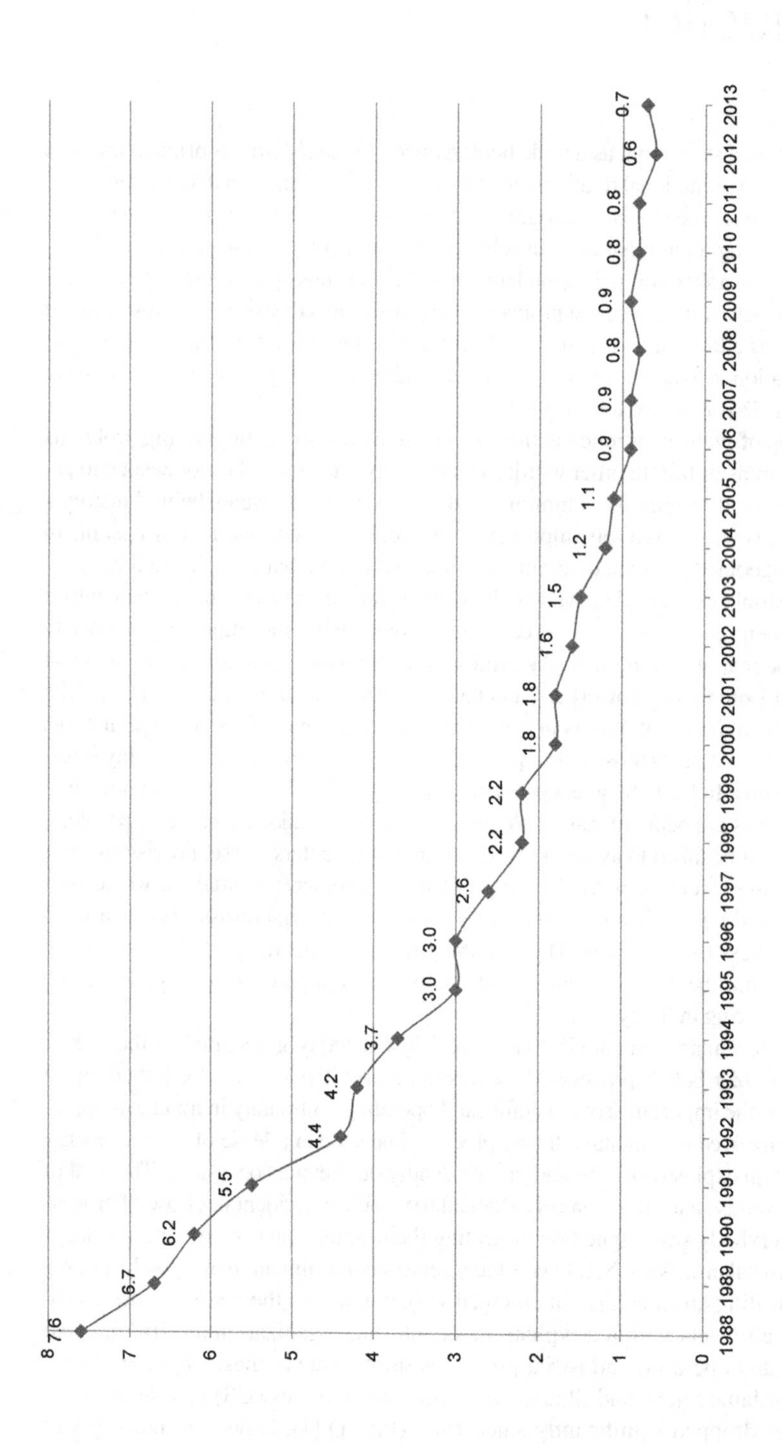

FIGURE 1
Total recordable incident rate per 100 full-time employees—American Fuel and Petrochemical Manufacturers (AFPM) petroleum refining members.

experiences devastating setbacks resulting from periodic and catastrophic process releases. These incidents make it difficult for consumers to appreciate the difference between industry's past and present performance. More concerning, however, is the fact that industrial disasters continue to damage industry's reputation and perceived ability to govern itself. Quite understandably, ongoing process spills, leaks, and disasters are viewed as proof.

Cooperation and communication are two major areas within the manufacturing industry that have dramatically improved since process safety's popularity surged in the 1980s. Before the PSM standard was formalized, individual parties responsible for maintaining industrial process control acted independently. For the most part, everyone kept silent over how they were managing process safety within their facilities. All of that changed almost instantly on December 3, 1984. On that date the Bhopal disaster unified industry's approach to process safety. Soon afterward, a group of committed industry volunteers collaborated to formalize process safety practices ultimately on a worldwide scale. Their goal was to simply eliminate catastrophic process incidents. This led to the official establishment of the Center for Chemical Process Safety (CCPS) on March 23, 1985 [2]. Ironically, the devastating refinery explosion and fire in Texas City that propelled process safety back into the spotlight occurred on the CCPS's 20th anniversary. Although separated by exactly two decades, facility-siting practices and atmospheric relief systems were two of the many common issues shared between the two industrial disasters. That disaster, along with others that have occurred since then, are a painful reminder that industry has yet to reach the process safety vision that represents our universal goal.

Looking back at the history of the manufacturing industry, the Bhopal disaster is the worst industrial disaster to have ever occurred [3]. It was unique in many ways. Industry has never recovered from the toll and consequences of the disaster. Never before, nor ever since, has a single industrial incident caused so much damage in the course of only a few hours. From an outsider's perspective, the incident came without warning. It occurred on a night no different from any other. The dense population living on the outskirts of the factory involved seemed oblivious to the hazard that literally descended upon them in their sleep. What followed for them was an evening of unprecedented chaos, terror, and tragedy. Unfortunately, the nightmare never ended for many of the survivors. No industrial disaster in the history of the world has managed to persist for so long. Not even the Chernobyl nuclear disaster compares in terms of its enduring impact on the way industry operates today. Looking back, the Bhopal disaster was a pivotal development in the history of modern industry. It settled any debate over the importance of process safety.

Industry learned much from the Bhopal disaster. Foremost, perhaps, is our understanding that all the money in the world cannot undo the pain that a single industrial disaster can inflict. An indifferent approach to managing process safety based on the severity of an incident is unacceptable. The focus, therefore, must be on incident prevention. There are certain lessons that we simply cannot afford to learn again. In that aspect, the price for the Bhopal disaster was much too high.

In recognition of the high cost of process safety failures, the Bhopal disaster changed how industry operates and is governed today. Regulations introduced in response to the Bhopal disaster made PSM an integral part of our manufacturing processes. But despite these advancements, tragic incidents like the explosion in Pasadena, Texas (1989), the Gulf of Mexico oil spill (2010), and the previously mentioned Texas City Refinery explosion (2005) raise questions about what would have happened if the PSM practices industry embraced following the Bhopal disaster had been in place when the factory was designed. Would these practices have prevented the Bhopal disaster? Would we have identified hazards that were only acknowledged after the incident made them blatantly clear? If not then what else should we be doing today to make our processes safer? If so, then what are we not currently doing well enough or incorrectly? On one point all must agree: something is missing. Otherwise, we would not be experiencing the kinds of disasters that have continued since the Bhopal disaster changed the way that industry prevents incidents.

As time goes by, we seem to forget at least as much as we learn about industrial disasters. This is unfortunate because the CCPS and other industrial organizations typically increase their activity immediately following an incident. The ultimate purpose of these activities is to document and preserve a record about what happened as soon as possible. Sharing this information throughout industry is the first step to avoiding similar incidents. There is, indeed, a genuine interest in preventing others from personally experiencing the same type of breakdowns. Yet, personal experience is the most effective teacher. It stimulates internal protective action. Maybe it originates with an instinctive will to survive.

Perhaps this explains why the frequency of industrial disasters has not yet been reduced to an acceptable level. Those targeted by information made available about how similar incidents can be avoided may not be motivated to take action until the issue directly involves them. It's not that they have not learned about how to avoid an incident, but rather that it just has not involved them personally. Thus, there is no triggering of an instinctive will to survive. This is where gaining insight into the Bhopal disaster has far greater relevance than might have been expected earlier. A modern, independent analysis of the incident draws in the unaffected. It puts them in the middle of an operation trying to cope with the same types of problems which, to this day, are regularly encountered within our own industrial processes. In a very compelling sense a modern and independent analysis creates a virtual experience to which all of us can relate. For the first time we are able to experience the Bhopal disaster in a very personal and technical way without having to relive the experience personally. This virtual experience will motivate industry professionals to intervene in a dangerous situation before it is too late. By using a systematic investigation framework to examine the facts behind this disaster, we can finally grasp how we are all involved.

As hard as it is to make industry professionals part of an incident that they were not personally involved in, it is even more difficult to make the experience real to those who have never walked into a factory. The problem is magnified for many graduating students entering the industrial workforce. Most have only limited hands-on experience. Granted, they may start their career with a basic understanding of their assigned role. Yet, how that specific role influences process safety is an entirely different matter.

Like some of my classmates, I took a part-time job in a factory before I graduated. This experience gave me some idea about the type of control each individual member of a manufacturing operation can have over the process. Once inside the factory, I was introduced to many important principles of responsible process control. But speaking personally, my most memorable lessons came long after that. Only when I had direct responsibilities over a manufacturing process as a full-time employee did I recognize how my personal actions or inactions made the difference between success and failure. Some of what I learned came about as a result of trial and error. When something I did worked, I kept doing it. When something did not work, I stopped doing it. Many times I did what I thought was right—only to find out that it was wrong after many repetitions. I was fortunate enough to work for a progressive manufacturing company that exercised patience in me while I developed and grew into the safety professional I ultimately became. Those who can relate to my experience will agree that as we advance toward retirement we become more useful and less dangerous to ourselves and others. It seems like at the point where we have the most to offer our employers, we are obligated to hand over our responsibilities to the incoming generation. The moment that we reach maturity is when we must pass our responsibilities down to the incoming rookies. So as to truly appreciate their involvement in PSM, the new team must successfully navigate through the same obstacles that we did. They too will ultimately find themselves in the same embarrassing situations at some point during their careers. And in due time, they too will be faced with surrendering their responsibilities to the next generation of replacements; a generation that also believes that they are prepared to take control. Breaking the cycle of each generation reliving the same failures requires making the past real enough to any person who might that might contribute to the design, construction, or operation of a manufacturing process. All of us must be equally affected by the previous incidents to avoid rediscovering them over and over. Industrial disaster education therefore must begin in the classroom so that the graduating student knows what to expect or, even more importantly, what to look for upon reporting to work the very first day.

There is no question that anyone qualifying for an industrial career will be exposed to the practices and principles needed for safe operation. Process operation, design, thermodynamics, fluid mechanics, chemical reactions, and PSM are only some of the relevant topics that are covered in the classroom to prepare students for a full-time industrial career. However, regardless of what specific position we are interested in, we can easily miss how these practices interact to either promote or defeat process safety. Only a definitive analysis of the Bhopal disaster can help students understand how these concepts influence critical decisions within the factories where they will be employed. This background prevents them from making similar mistakes based on limited, incomplete, or erroneous process safety knowledge. Without this knowledge, students are left to figure out potentially dangerous relationships between process design, maintenance, and operations on their own terms upon entering the workforce. This information should therefore be considered an essential part of industrial career development. It is just as important as any of the other information that a student might be expected to cover to prepare for a job in the manufacturing industry.

It is with the students' best interest in mind that all royalties from this book are being donated to the Lamar University Process Safety Heritage Trust. This scholarship has been established to recognize science and engineering students who will assume control over the industrial processes that we operate today. The hope is to effectively promote the operating qualities requisite for PSM success within successive generations before they enter the factory gates as full-time employees. Only through their leadership can we realistically expect to achieve the goal for which the PSM standard was introduced. We show our respect to the many touched and affected personally by industrial disasters by helping others make constructive use of reality. On an individual basis we show this respect by vowing to do everything that we possibly can to avoid repeating the same mistakes that led, or will lead, to disaster. This includes preserving a historical record that represents extraordinary value for industry. You are holding the objective historical record in your hands even as you read this text. I, and many others, beg you to fully absorb its meaning.

Kenneth Bloch

REFERENCES

[1] American Fuel & Petrochemical Manufacturers, Refinery Safety at a Glance. Available from: http://www.afpm.org/Refinery_Safety_at_a_Glance/, April 30, 2015.
[2] L. Horowitz (Ed.), Process Safety in Action: 2012 Annual Report, 2012.
[3] M.S. Mannan, A.Y. Chowdhury, O.J. Reyes-Valdes, A portrait of process safety: from its start to its present day, Hydrocarbon Processing. 91 (7) (2012) 60.

A Note for Readers

Throughout the book, I have used the term “supervisor” and variations of that term to denote the management level in charge of the line organization (workers). Depending on any respective site’s location and composition, this level collectively includes the Plant Manager, Division Manager, or Works Manager, and their direct reports.

Also, the book conforms to the principles of professional industrial incident reporting practices. In proper application of these principles, case histories are described without using proper names of companies, individuals, or trademarked products unless doing so would create ambiguity. Lessons learned from the analyses contained within are not limited to a specific company, business sector, individual, or product line. The selected reporting approach will assist you in determining where similar circumstances apply in unrelated organizations, including your own.

Kenneth Bloch

PART

Process Reliability

1

Process reliability is the industrial science that defines the relationship between chemical and mechanical interactions. Developing this knowledge is crucial to controlling damage mechanisms that affect manufacturing processes. Because this science relates directly to operations and maintenance activities, the following chapters impart considerable focus to establishing the principles of process reliability that apply in all industries.

It could be argued that equipment reliability and process reliability are governed by the same principles, or that there would be no process reliability without equipment reliability. While these definitions and their respective subcategories are entirely true, we would not want readers to miss the point made in this text. Process reliability is an important science and overlooking its individual importance or using sweeping generalizations can be perilous.

Therefore, this section of our text explores and explains process reliability as a discipline capable of resolving persistent process phenomena that can lead to a disaster. Manufacturing processes are fertile territory for failure mechanisms related to process reliability issues. Process reliability professionals take the lead in maintaining operating stability by diagnosing minor issues before they can contribute to more serious consequences.

The methods described in this section assist with maintaining operating simplicity. Simpler processes are easier to control and thus less likely to fail than their more complex counterparts. Additionally, simpler processes embody the transparency needed to preclude the pursuit of misguided solutions. Although "misguidance" may not be obvious in certain expedient and seemingly acceptable solutions, applying the principles of process reliability prevents decisions and actions that can have counterproductive results.

CHAPTER

Industrial Learning Processes

1

The manufacturing industry is the backbone of the world's commercial system. Its purpose is to provide an adequate supply of useful products that consumers are inclined to purchase. Industry thrives by generating economic wealth for investors while improving the standard of living for people around the world. Generating this wealth and making its fruitage both available and affordable is the perfect example of a win–win situation that rewards both consumers and those who are willing to accept manufacturing responsibilities.

If industry's goal is to offer security and comfort to the public through the development of useful products, then nothing could have defeated that mission more completely than the Bhopal disaster. This major incident occurred in a relatively small factory that had operated rather anonymously in central India up until December 3, 1984. The Bhopal disaster made an unparalleled impression and left questions for modern industry that have persisted for decades. This is one of the reasons why the Bhopal disaster is considered history's worst industrial accident. A brief overview of the circumstances behind the Bhopal disaster is appropriate for a number of reasons; however, recognizing the full value of thoroughly and authoritatively analyzing the event ranks first and foremost among these reasons. Knowing and fully understanding the circumstances will serve as an indelible case study for people with the responsibility to protect the lives of others both directly and indirectly.

THE BHOPAL DISASTER

The international subsidiary of an American multinational chemical company commissioned a full-conversion pesticide manufacturing process in Bhopal, India, in 1979 [1]. The manufacturing process consisted of using raw materials to generate intermediate products that were then fed to a reactor to make the final product. One of the intermediate products was a reactive, toxic, volatile, and flammable derivative of phosgene [2]. Its flammability and toxicity hazards made it unsuitable for discharge into the environment [3]. A lethal dose could realistically develop upon its release into the atmosphere. Process containment, therefore, was highly important under both normal and emergency operating conditions. Accordingly, safe design and operating practices were initially incorporated and implemented; they embodied the commendable goal of maintaining adequate process control at all times.

Rethinking Bhopal.

The exposure hazards associated with the toxic intermediate were well known. In its vaporized form, the toxic intermediate was a serious health hazard. Breathing it in could result in a fatal lung injury [4]. However, contact by any means was not recommended.

With the unforgiving process in mind, it was somewhat comforting to know that the business was in good hands. The parent company was one of the most trusted, innovative, experienced, successful, and respected organizations in the chemical manufacturing industry at the time. For over a decade the same manufacturing process had been safely generating the toxic intermediate compound in a domestic factory in the United States. That process served as the design template for the overseas factory in India. On the basis of accrued design and operating experience, it was reasonable to expect that the overseas manufacturing operation (the Bhopal factory) would be at least as safe as the existing domestic production site in the United States [14].

Routine maintenance was taking place in the Bhopal factory on the evening of December 2, 1984. The toxic intermediate compound was contained in three stainless-steel storage tanks (Fig. 1.1) according to normal procedures. An exothermic reaction could result if the tank contents were contaminated by an incompatible substance, which included water. If not detected, this adverse operating condition could propagate into a thermal runaway reaction which could realistically result in a catastrophic process release. That is why special provisions were included by design to keep contaminants out of the storage tanks. Safety systems were also provided to manage the chemical reactivity hazard if the tank contents should somehow become contaminated.

Shortly before midnight, factory workers noticed a stinging sensation in their eyes and throats [8]. Under suspicion that the process might be leaking somewhere, they started checking the pipes. The workers soon located gas and water escaping from an overhead vapor line and immediately took steps to address the problem [9–12]. It was unclear to them whether or not the leak signified another problem elsewhere in the factory. Process leaks were a common occurrence within the factory site [13]. Up to this point, all of the gauges that would indicate a problem inside the storage tanks were reading within their normal ranges.

Following standard protocol, the workers reported what they had found to the control room [14]. This prompted a worker at the control panel to take another look at the storage tank pressure gauges. This time, he noticed that one of the three storage tanks' pressure gauges was now reading higher than normal [15]. Although this immediately drew the workers' attention to the contaminated tank, it was already too late. Water had been contaminating the tank for about 2 h. An exothermic chemical reaction was already in full progress and the tank contents were hot [16]. The workers took immediate steps to control the unstable condition, but the situation continued to deteriorate rapidly. Not long after that, the pressure inside the tank exceeded the pressure relief system's set point. At that very moment hot, boiling, lethal process vapor and liquid charged into the tank's vapor management system [17].

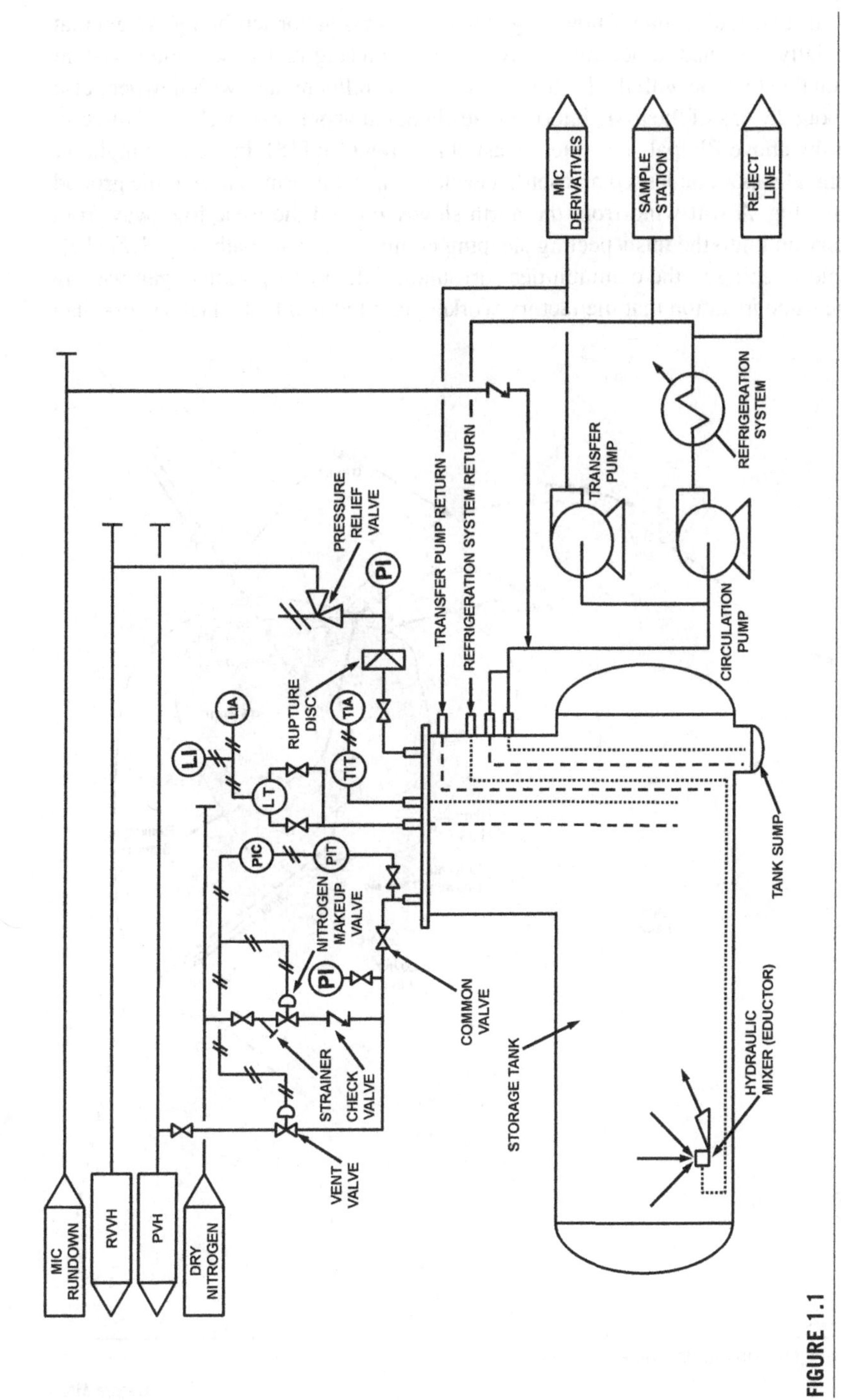

FIGURE 1.1
Intermediate storage tank [5–7].

The unstable tank contents now surged into a process vapor scrubbing system that was specially designed to neutralize any material entering it. The scrubbing system proved unable to cope with the high-volume process influent and, with nowhere else to go, about 28 tons of the toxic intermediate chemical vapor was discharged directly into the sky above Bhopal. The release lasted for about 2 h [18]. In the cool night air the chemical vapor condensed and settled back to earth, hovering close to the ground as a moist fog. A soft wind from the north slowly pushed the toxic fog away from the factory and into the unsuspecting sleeping community in its path (Fig. 1.2) [19].

People sleeping in the communities surrounding the factory soon began waking up to the same irritation that the factory workers had felt inside the factory less than

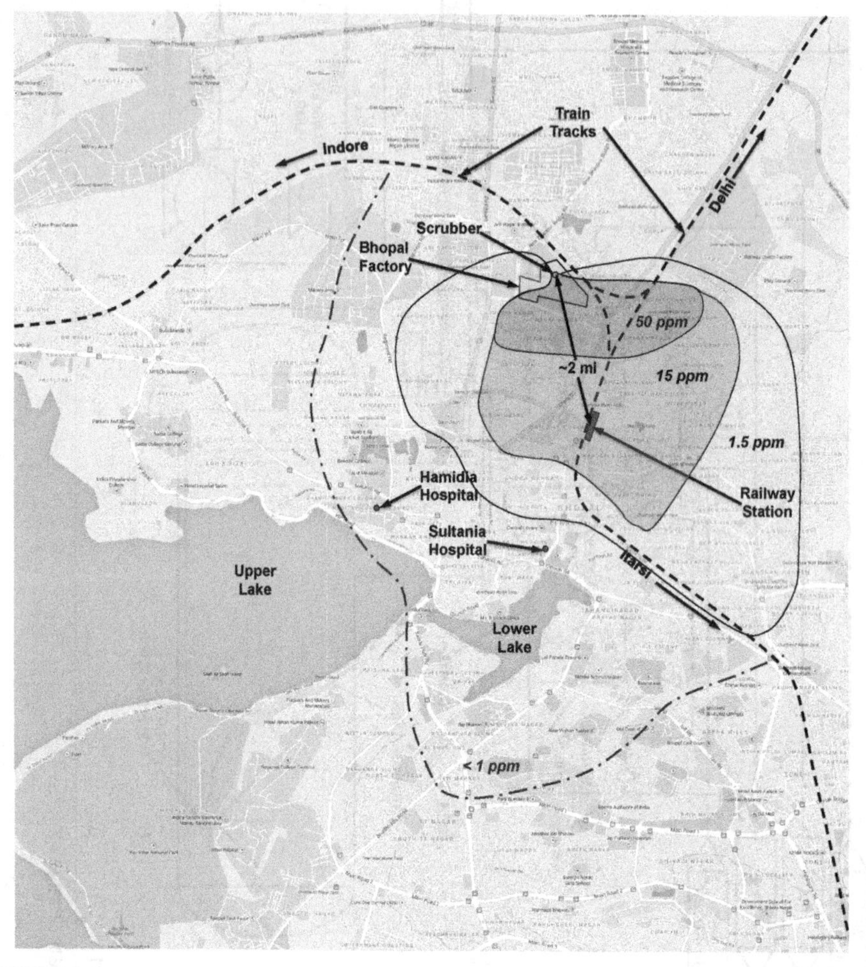

FIGURE 1.2

Bhopal disaster dispersion analysis [20].

Google Maps

an hour earlier. Before long they were gasping for air. At this point many of them fled outside their homes hoping to find comfort, only to find that the situation was even worse in the streets.

Those in the direct path of the vapor cloud were blinded and unable to breathe. Overcome with confusion, the crowds in the streets started to panic. Adults and children alike started running in every direction. In the darkness, nobody could see what was happening. A safe haven could be found nowhere. Many died in the stampede that followed as chaos spilled into the streets surrounding the factory [21].

The intermediate storage tank emptied and the release stopped at about 2:15 am [22]. More than 100,000 partially blinded survivors arrived at local hospitals during the first night [23]. Many of them were severely injured and suffering from acute breathing disorders. However, the hospitals were in no way prepared to handle the unexpected influx of patients requiring treatment and quickly ran out of supplies [24]. By the end of the disaster over 200,000 people were injured [25,26]. Sadly, many of these injuries were permanent.

At daybreak, thousands of people and animals were found dead or barely clinging to life in the streets of Bhopal [27]. Human bodies were piled up everywhere the eye could see [28,29]. According to the first incident report 2850 victims died on the night of the disaster [30]. Updated official fatality statistics later pushed the count up to 3828 deaths [31]. The unofficial cumulative death toll since the disaster took place now numbers into the tens of thousands [28]. More than 2000 animals also died and the environmental impact was not only severe, but persistent [32].

The Bhopal factory was shut down on that very night and never reopened for commercial production. Operations in the US factory were suspended pending further regulatory review [33,34]. The parent company resorted to taking drastic measures to improve their image, but never managed to recover from the damage that the disaster had on its reputation [35]. Slowly, the company was parted out as its multiple operating divisions were sold in a desperate attempt to remain financially stable. However, the sell-off ended in 1999 when the last functional business unit was sold to a competitor in the chemical manufacturing industry [36]. This quietly sealed the fate of a once flourishing and influential company known for its consistent introduction of progressive technology. Indeed, to this very day we would not even think of living without much of the advanced chemistry introduced by this technological pioneer.

POLICIES AND REGULATIONS

As this example demonstrates, the consequences of a catastrophic process release can be severe. Its repercussions can extend far beyond the health and environmental damages that usually receive the most attention immediately following an incident. Jobs can be lost, an economic crisis might develop, and stable institutions can collapse. Industry well knows that there is no business proposition that supports a tolerant attitude for catastrophic process releases. It is for this reason that industry takes aggressive measures to prevent them.

Regulatory agencies also take careful note of the potential harm that can result from a catastrophic process release. Consequently, these agencies are directly engaged in the advocacy, stewardship, and implementation of programs intended to improve process safety performance. Neither industry nor regulatory agencies take industrial disasters lightly. Responsible and responsive organizations and their associated entities actively and consistently develop strategies to prevent industrial disasters.

After a serious incident, industry typically confers over how the existing system broke down and what needs to change. This knowledge is then typically shared in a manner that promotes the development and unification of best operating practices to prevent similar failures throughout industry. Adjustments are usually recommended in cases where lessons from an incident increase industry's knowledge about how to improve process control.

Because industry does not have regulatory power, it can only advocate *voluntary* participation in programs designed to further the application of best operating practices at local sites. No matter how effective these practices might actually be, industry does not have the power to enforce them. Voluntary-participation programs are useful only to those who appreciate their value and are committed to applying them. This is accomplished at local sites by making applicable safety practices, a matter of policy. This is where regulatory agencies step in.

Regulatory agencies are in control of industry's mandatory compliance programs. Industrial safety and environmental protection bills that become laws are implemented by the appropriate regulatory agency. Regulations interpret the law. They explain the technical, operational, and legal details needed to implement various laws that are passed. Regulatory agencies also have the power and authority to impose punitive actions upon companies that either willingly or unwillingly demonstrate noncompliance.

Regulatory agencies are primarily concerned about serving the needs of citizens who could be adversely affected by industrial disasters. Take, for example, the Gulf oil spill (2010). This disaster not only resulted in multiple casualties and significant environmental damage, but the incident also affected other industries as well. Tourism and the fishing industry are just two examples of other industries that were directly affected by the disaster. Interconnected commercial relationships such as this serve as the basis for new laws that restrict industry's ability to choose what best operating practices are appropriate, and how to practice them. Sometimes, the influence that one sector of industry has over multiple other sectors of industry is not quantified until a disaster puts the entire economic system to the test.

The different degrees of actual compliance with policies and regulations come down to strictness of enforcement. At some locations, strictness of enforcement is a huge variable swayed by all kinds of considerations. Suffice it to say that enforcement often varies from place to place and for a very wide range of reasons. Regulatory agencies can apply *mandatory* rules that industry is legally required to comply with whereas industry can only encourage the *discretionary* use of best operating practices. However, regulatory agencies use auditing practices to verify that local sites are in compliance with their instituted policies, whether or not local site policies reflect a legal obligation. Organizations can be cited for not managing their business in accordance with documented policies that define how they expect to conduct business.

The Bhopal disaster brought about sweeping changes intended to eliminate process safety failures. Many of the regulations and operating principles commonly practiced throughout the industry today were introduced in response to sharp awareness over how a catastrophic process release could negatively impact community health and the environment. The foremost of these regulations is the process safety management (PSM) standard, which dictates the requirements for managing catastrophic process release hazards. Collectively, the various policies and regulations implemented globally after the Bhopal disaster can be grouped together as "PSM."

THE PROCESS SAFETY MANAGEMENT MOVEMENT

The PSM standard was adopted by industry to control potential catastrophic process release hazards after the Bhopal disaster. It was published into the Code of Federal Regulations (CFR) on February 24, 1992. This development made PSM a legal requirement within the United States. Criminal charges can now be brought against companies or individuals that fail to establish or practice the PSM standard as prescribed by law within their respective entities or organizations.

Although the PSM standard (29 CFR 1910.119) refers to laws designed to promote industrial safety specifically in the United States, most industrialized countries have parallel regulations in place. For example, COMAH (Control of Major Accident Hazards) regulations are imposed on industry to protect public health and safety and the environment in the United Kingdom. Although the specific requirements for legal compliance may differ between continents, there is a general acceptance that the PSM standard's dictated principles are needed to adequately control catastrophic process release hazards. Indeed, the Center for Chemical Process Safety's (CCPS) effort to unify process safety practices throughout the world has evolved into a "universal language" that is now spoken by industry around the world. It specifies the minimum requirements for managing process safety hazards to a tolerable level.

In reality, the PSM standard originated with a collection of 14 different essential components (elements), all of which are necessary to realistically exercise an acceptable level of hazard control (Table 1.1). Its stated purpose is to prevent or minimize the consequences of catastrophic releases of toxic, reactive, flammable, or explosive substances that may result in toxic, fire, or explosion hazards [37]. Regulatory agencies in the United States (OSHA (Occupational Safety and Health Administration)) and abroad (such as HSE (Health and Safety Executive) in the United Kingdom) enforce the laws created under their respective PSM policies to eliminate catastrophic process releases. However, responsible organizations worldwide appreciate

Table 1.1 The Process Safety Standard's Fourteen Elements

Subpart	Process Safety Management Standard Element
(c)	Employee participation
(d)	Process safety information
(e)	Process hazard analysis
(f)	Operating procedures
(g)	Training
(h)	Contractors
(i)	Prestartup safety review
(j)	Mechanical integrity
(k)	Hot work permit
(l)	Management of change
(m)	Incident investigation
(n)	Emergency planning and response
(o)	Compliance audits
(p)	Trade secrets

how PSM practices can enhance their production value. To them, the embracing of PSM concepts creates a competitive advantage that would not otherwise be available. Indeed, some had similar standards in place well before they were mandated or legislated more broadly.

Collectively, significant progress has been made in unifying PSM practices throughout industry on a global scale. However, despite industry's voluntary and compulsory indoctrination of PSM around the world since the mid-1980s, catastrophic process releases from local production sites continue to be a recurring issue (Table 1.2). Unfortunately, a catastrophic loss of primary containment incident brings attention not only upon those directly involved, but also upon industry in general. In the eyes of the community, industry is at fault when one of its representative members fails to adequately contain the process.

REPEAT FAILURES

Repeat failures are a common ingredient shared between industry's worst catastrophic process releases. Repeat failures are tied together with the normalization of deviance. Both subjects will be examined more closely in later chapters. However, seemingly subtle similarities between the multiple incidents that have occurred since the Bhopal disaster consist of repeat failures. It could also be said that in the majority of cases, several discernible deviations combined into a very major process release event. Again, our text will elaborate and explain in later chapters.

Table 1.2 Consequential Manufacturing Industry Catastrophic Process Releases

Industry	Description	Date	Country	Fatalities
Oil and gas	BLEVE	November 19, 1984	Mexico	500
Chemical	Toxic chemical release	December 3, 1984	India	3828
Nuclear	Core meltdown	April 26, 1986	Ukraine	31[a]
Oil and gas	Oil platform explosion	July 6, 1988	UK	167
Chemical	Vapor cloud explosion	October 23, 1989	USA	23
Oil and gas	Exchanger rupture	September 25, 1998	Australia	2
Oil and gas	Acid tank explosion	July 17, 2001	USA	1
Chemical	Reactor explosion	April 24, 2004	USA	5
Oil and gas	Refinery explosion	March 23, 2005	USA	15
Chemical	Reactor explosion	December 19, 2007	USA	4
Chemical	Vessel explosion	August 28, 2008	USA	2
Oil and gas	Exchanger rupture	April 2, 2010	USA	7
Oil and gas	Gulf oil spill	April 20, 2010	USA	11
Oil and gas	Refinery explosion	August 25, 2012	Venezuela	48
Oil and gas	Refinery explosion	September 19, 2012	Mexico	26
Chemical	Chemical explosion	April 17, 2013	USA	14
Chemical	Toxic chemical release	November 15, 2014	USA	4

[a]*Reports explain that the incident caused 237 cases of acute radiation sickness. Of these cases, 31 died in the first 3 months.*

Simply stated, a repeat failure occurs when equipment breaks down on a recurring basis. This assumes that the equipment is not meeting performance expectations. A repeat failure can occur if:

- A previously applied solution did not work,
- Recommended actions from an investigation are pending (an example might be for a scheduled turnaround), or
- The decision was made not to apply a known solution.

Most of industry's catastrophic process release incidents result either directly or indirectly from repeat failures. Notice that the Bhopal disaster stands out as an extreme outlier event in terms of how many fatalities were involved. In less than 24 h it created more damage than the others did over the next 30 years combined. Indeed, it left a permanent mark upon industry's history. However, the unforgiving nature of the process is what sets the Bhopal disaster apart. If the severity of the process is not considered, then the incident becomes very similar to the others that have continued to occur since then (Table 1.2). In fact, the volume of process material released in the Bhopal disaster is dwarfed by some of the catastrophic process releases that have been experienced since 1984. Yet, the consequences have not been nearly as severe as the Bhopal disaster. Many industrial incidents and near misses that have occurred since PSM was embraced globally could have been just as bad or perhaps even worse. They were not as bad because the process materials were not as dangerous as those involved in the Bhopal disaster. When the severity of the process is held constant, the incidents are similar. Their most distinguishable quality is the type of process that was released.

Another similarity can be observed in the specific sequence of events behind some of the better documented incidents. When equipment reliability is taken into consideration, it appears that none of the better documented incidents occurred without warning. Although the incidents themselves surely caught many people by surprise, upon further review it becomes abundantly clear that each of them could have been avoided. In all cases, ample time existed to analyze and correct the persistent equipment performance issue that later became an integral part of a disaster. For example, the refinery explosion in Texas City, Texas (2005), was determined to be a disaster whose probability increased and which "brewed" over the course of many years, perhaps even decades. The equipment involved in this catastrophic process release was part of a series of less consequential incidents (including fatalities) in the years leading up to the disaster.

Considering these relationships, repeat failures exist in two dimensions. On a minor level, repeat failures occur on equipment that individual production sites use to support the manufacturing process. We see this in the sequence of events leading up to a catastrophic process release at a local production site. For that matter, this category includes all of the individual production sites that collectively represent the global manufacturing industry. Repeat equipment failures within local production sites can result in a significant release of process materials and should not be ignored. Repeat failures occurring at individual sites are representative of an *acute* problem probably originating with investigation practices, or a lack thereof.

On a major level, industry encounters repeat failures with similar circumstances between multiple independent locations and often among different business sectors. Repeat failures occurring on a much wider or even worldwide scale throughout the manufacturing industry can be considered a more *chronic* problem. The chronic nature of repeat failures results in recommendations to change industry's approach to managing process safety hazards.

THE INCENTIVE TO CHANGE

Examples of more recent industrial incidents provide further proof that the consequences of future catastrophic process releases can be just as deadly as they were in previous incidents. Nothing good can happen if the process is involuntarily released due to an equipment failure or process upset. Industry recognizes that there is no business justification for a tolerable attitude with respect to a potentially catastrophic process release. In response to these types of incidents, industry introduces or modifies policies to offer better prevention. Likewise, regulatory agencies attempt to lower the incident rate by imposing more regulations. These mutual reactions to adversity explain the growth of voluntary and mandatory programs designed to lower the frequency and consequence of catastrophic process releases since the Bhopal disaster (Table 1.3).

Table 1.3 Voluntary and Mandatory Actions Taken to Prevent Incidents

Incident	Year	Voluntary Industry Program	Mandatory Regulatory Program
The Bhopal disaster	1984	1985: The American Institution of Chemical Engineers (AIChE) establishes the Center for Chemical Process Safety's (CCPS) as a global community dedicated to preventing and mitigating catastrophic releases of hazardous materials through effective engineering and management practices.	1986: The Emergency Planning and Community Right to Know Act (EPCRA) is implemented to help communities plan for emergencies involving hazardous substances. The law covers hazardous chemical emergency planning by federal, state, and local governments; Indian tribes; and industry. It also specifies industry reporting requirements on the storage, use, and release of hazardous chemicals to federal, state, and local governments.
Chemical plant explosion in Pasadena, Texas	1989	1992: The Safety and Chemical Engineering Education (SAChE) program is introduced as a cooperative effort with the CCPS to educate undergraduate and graduate students studying chemical and biochemical products on the elements of process safety, to promote safer operations.	1990: The Clean Air Act of 1970 is amended to specify requirements for accidental release prevention so that facilities are required to reduce the likelihood and severity of accidental chemical releases. It also requires OSHA to adopt a process safety management (PSM) standard to protect factory workers. This led to the adoption of the PSM standard on February 24, 1992.
Refinery explosion and fire in Texas City, Texas	2005	2010: The American Petroleum Institute (API) issues API RP-754 to standardize the reporting of leading and lagging performance indicators throughout the refining and petrochemical industries. 2012: The American Fuels and Petrochemical Manufacturers (AFPM) event sharing database goes live for users to access information about the causes of actual incidents and lessons taken from them.	2007: OSHA launches the Petroleum Refinery Process Safety Management National Emphasis Program (CPL 03-00-004) to reduce or eliminate workplace hazards associated with the catastrophic release of highly hazardous chemicals at petroleum refineries. This involves verifying compliance with OSHA's PSM standard as defined in 29 CFR 1910.119.
Fertilizer factory explosion in West, Texas	2013	2014: The CCPS introduces "Vision 20/20" as a call to action for "broad industrial and societal change [43]."	2013: Executive Order (EO) 13650 is issued to prioritize "Modernizing OSHA's PSM standard to improve safety and enforcement [44]."

More changes can be expected in response to recent incidents; these incidents demonstrate that industrial safety improvement is not progressing at a satisfactory rate. Over time, stagnating performance encourages even more aggressive changes. For example, one of the more significant appeals for industrial change since the Bhopal disaster occurred soon after an explosion in a relatively small factory in central Texas (United States) on April 17, 2013. Fourteen people perished in the disaster and over 200 more were injured. This incident was followed by Executive Order (EO) 13650, which was issued on August 1, 2013, in response to "recent catastrophic chemical facility incidents in the United States" [38]. The stated purpose for EO 13650 is to, "...enhance the safety and security of chemical facilities and reduce risks associated with hazardous chemicals to owners and operators, workers, and communities." On December 3, 2013, a request for information was formally processed for potential revisions to the PSM standard [39].

About a year later, the US Chemical Safety Board (CSB) expressed a growing concern over the manufacturing industry's process safety performance by calling attention to "...stagnating risk levels, even as industry-recommended best practices and technology continue to advance in the U.S. and overseas." In response to this concern, the CSB has added the "Modernization of Process Safety Regulations" to their "Most Wanted" list [40]. This announcement was made shortly after the California Department of Industrial Relations called for "...a sweeping rewrite of the PSM standard for oil refineries as well as several new management system elements [41]."

These developments are expected to fuel discussions over significant changes in the way that industry manages process release hazards at local production sites. The requested course of action is consistent with the anticipated response from industry and regulatory agencies following an industrial disaster. A more drastic approach to PSM changes seems to be developing as the number of catastrophic process release incidents grows.

Change requests are not a new development for the manufacturing industry. What is different about this particular request for reform is that it targets the fundamental system that industry uses to manage process safety. This may be the first time since the Bhopal disaster that industry could be required to implement significant changes directly to the PSM standard.

FROM SIMPLICITY TO COMPLEXITY

Complexity is the enemy of safety [42]. As changes are layered upon industry in an effort to make it safer, process design, maintenance, and operations tend to become more complex. This can have a counterproductive impact on the system upon which the changes are being applied to make it safer. With time, the processes we operate can become more dangerous and more likely to fail.

The continuing occurrence of catastrophic process releases following the Bhopal disaster indicates either a problem with our policies and regulations or that we are missing something important. The imbalance between industry's investment in PSM and its actual PSM performance as documented or inferred by the recent appeal for PSM reform tends to favor the notion that something critical, perhaps essential, is missing.

To illustrate this point, imagine that the "80-20 rule" applies to PSM. In this case, it might be possible to achieve a satisfactory level of industrial safety performance by implementing only the basic principles. In other words, applying only the basic principles would result in an 80% success rate. It is not perfect, but it gets you within the reach of your target and, at least, it stops the bleeding. From this point, enhancements in the form of policies and regulations can be added if needed to capture the remaining available opportunity for improvement.

If the basics are not adequately covered, then the 80% success rate will not be possible. The highest possible score might be in the range of only 40–60%. While this is better than nothing, it will likely not be considered acceptable, taking into account how much effort is needed to apply volume of enhancements you would now need to get you to your target. The worst part about not having the basics adequately covered is that the additional work it creates for you pushes you nearer to a disaster. Instead of getting what you expect from all the hard work, you get exactly the opposite. This is the peril involved when complexity is added as a result of not adequately covering the basics. Without firmly establishing the basic requirements, there is no hope in achieving success through hard work.

That said, a basic purpose or target achievement of process reliability is to roll back complexity in response to a catastrophic process release. Achieving this target begins with an objective and well-structured, repeatable analysis of the sequence of events that preceded an incident. Well-structured and repeatable approaches are the exact opposite of guesswork. This text deals extensively with structure, because structure immensely facilitates the learning process that achieves a satisfactory level of incident prevention. A systematic approach of analyzing the Bhopal disaster develops a level of transparency that can prevent recurring process releases. This "thread of thought" stitches together a thick protective blanket around PSM.

INCIDENT PREVENTION

Effective incident prevention requires learning from experience. The regular occurrence of industrial disasters since 1984 suggests a serious difficulty with preventing incidents by incorporating lessons from previous experience. Perhaps any one of the incidents that followed the Bhopal disaster could have been avoided if industry's learning process had functioned properly. Reversing this course involves using information readily available through multiple resources to diagnose the causes for unacceptable process and equipment performance.

Failure to properly diagnose the causes for a catastrophic process release can spread learning defects throughout the manufacturing industry and undermines the learning process. The learning process begins with an effective investigation according to the basic principles of PSM. If this process breaks down, then repeat failures like the ones observed in recurring incidents can be expected. The causes of repeat failures since the Bhopal disaster might be rooted in a "solution" used to address a specific problem that simply did not work. Instead of a solution it may have been a measure to take care of a symptom. Or perhaps the problem was not given sufficient attention because action was taken according to preliminary, superficial information about the sequence of events. Either situation might lead to disappointing results that encourage the piling on of additional "enhancements" that do not strike deep enough into the core of the problem. In the end, the failure to address a relatively minor issue can lead to a disaster. At some point it becomes necessary to sort out the cause and effect and unravel the chain of events behind a disaster, to resolve interconnections and relationships that might be hidden far below the visible surface.

In a sense, an incident investigation immediately becomes overdue at the point of a disaster. The best time to resolve unexplained process and equipment performance issues is *before* a catastrophic process release occurs. Based on the consistent precursor experience of repeat failures in history's most noteworthy industrial incidents, there is sufficient time to diagnose active failure mechanisms that could potentially result in more severe consequences. Taking advantage of this opportunity requires identifying repeat failures and viewing them as warning signals. The time to start an investigation is when warning signals, which include repeat minor failures, are noticed. Using this approach makes it possible to maintain process simplicity and to avoid more policies and regulations that otherwise might favor adding more complexity to an aging industrial infrastructure. Regrettably, some of this infrastructure is already in a weakened condition.

The importance of analyzing unacceptable system performance at all levels, from the smallest to the largest of possible malfunctions, cannot be overemphasized. The objective should always be to trigger an investigation with enough time left over to apply effective recommendations. The objective should also include sharing the investigation report throughout the enterprise at risk. Indeed, depending on its importance, relevant information should also be exchanged formally through professional industry networks including conferences organized by the CCPS, Institution of Chemical Engineers (IChemE), American Petroleum Institute (API), and American Fuel and Petrochemical Manufacturers (AFPM) in addition to relevant industrial trade journals.

A systematic and repeatable approach to investigating the causes for process upsets, equipment failures, and excessive maintenance promotes the most simple and straightforward solution to a problem. Properly documented investigations make it possible to retain important lessons for future generations of industry employees. They can use this information to avoid the mistakes made by others earlier and often at devastating cost. Better yet, they can learn an effective and repeatable approach to process reliability instead of having to unlearn unstructured approaches which, in some instances, have left very much to be desired.

INVESTIGATION REPORTS

An incident investigation is at the heart of industry's continual learning process. Some incidents are too tragic to experience. Among these are some that only happened once. It should be considered unacceptable for a serious problem to repeat over and over because of a failure to incorporate lessons learned from previous incidents.

An objective investigation is without bias. Objectivity comports with science and fact-based findings. Their application prevents recurring incidents by incorporating lessons from previous events.

A well-documented investigation report makes its readers a part of the experience that took place. The report should be written simple enough for the reader to clearly understand how it applies directly to process under their control as operators, engineers, project managers, executives, and site supervision. Upon reviewing the information it contains, there should be no question about how they are involved.

Regardless of how PSM rules might change in the future, the principles of process troubleshooting and incident investigation will remain the same. The influence that these principles have on stable operation and disaster prevention is as significant now as it will be into the future. Industry's learning process begins with a comprehensive incident investigation followed by a formal report. Unless these sequential requirements are satisfied, any recommendations directed at preventing recurring incidents will not be effective or will have temporary value at best. With temporary fixes, industry ultimately misses a shot at global improvement. The learning process is undermined and an incident of equal or greater magnitude will likely develop. In response, industry and agencies invariably take additional action to curtail an unacceptable incident rate. These actions then tend to add complexity that can actually hasten the approach of more serious incidents. Avoiding this progression of events begins with a systematic, effective investigation at local sites, where incidents are experienced.

Understanding how a minor equipment malfunction can escalate into a major industrial catastrophe requires a basic understanding of equipment reliability principles. Human response to equipment reliability issues is a multidimensional process that concerns:

- Equipment design
- Operating procedures
- A failure mechanism (mode)
- Mean Time Between Failures (MTBF)
- Mean Time to Repair (MTTR)
- Availability
- Utilization

Each of these factors will be examined in the next chapter. Their integration explains the motivation for human action and reaction to an unexpected operating phenomenon or a string of occurrences.

LESSONS FOR US

In this chapter we introduced the basic process reliability concept of making factory operation less complicated. We also described how changes in policies and regulations are guided by industry's process safety performance. Other important messages to remember include the following:

- Some incidents are too harmful to experience, if only even once.
- Repeat failures are a recurring theme among industrial disasters.
- A single catastrophic process release can remain with industry or the affected public forever.
- Complexity degrades process safety.
- Industrial processes become progressively more complicated as more policies and regulations are applied.
- Industrial disasters trigger changes in policies and regulations.
- Different industrial disasters share many similarities.
- Catastrophic process releases are separated by the severity of their respective processes.
- The success of industry's learning process is dependent on effective incident investigation practices at local production sites.
- Prevention is the most effective form of protection for all involved, and should therefore be the primary goal of a process safety program.

REFERENCES

[1] P. Shrivastava, Bhopal: Anatomy of a Crisis, 41, 1987, ISBN: 088730-084-7.
[2] Union Carbide Corporation, Review of MIC Production at the Union Carbide Corporation Facility Institute West Virginia, April 15, 1985, 1-1, 1985.
[3] Union Carbide Corporation, Review of MIC Production at the Union Carbide Corporation Facility Institute West Virginia, April 15, 1985, 2-2, 1985.
[4] Union Carbide Corporation, Review of MIC Production at the Union Carbide Corporation Facility Institute West Virginia, April 15, 1985, 1-3, 1985.
[5] T.R. Chouhan, et al., Bhopal: The Inside Story, Carbide Workers Speak Out on the World's Worst Industrial Disaster, 53, 1994, ISBN: 0-945257-22-8.
[6] A.S. Kalelkar, Investigation of large-magnitude incidents: Bhopal as a case study, in: IChemE: Symposium Series No. 110, Preventing Major Chemical and Related Process Accidents, May 10–12, 1988, 576, 1988.
[7] Union Carbide Corporation, Bhopal Methyl Isocyanate Incident Investigation Team Report, March 1985, 7.
[8] S. Diamond, The Bhopal Disaster: How It Happened, The New York Times, January 28, 1985.
[9] T. D'Silva, The Black Box of Bhopal, 126, 2006, ISBN: 978-1-4120-8412-3.
[10] I. Eckerman, The Bhopal Saga: Causes and Consequences of the World's Largest Industrial Disaster, 48, 2005, ISBN: 81-7371-515-7.
[11] A.S. Kalelkar, Investigation of large-magnitude incidents: Bhopal as a case study, in: IChemE: Symposium Series No. 110, Preventing Major Chemical and Related Process Accidents, May 10–12, 1988, 567, 1988.
[12] A. Agarwal, S. Narain, The Bhopal Disaster, State of India's Environment 1984–85: The Second Citizens' Report, 207, 1985.

[13] Supreme Court of India Criminal Appellate Jurisdiction, Application for Directions to Institute Charges U/S 302 (For Offence U/S 300(4)) Read With S. 35 of the Indian Penal Code, 1860 Against the Respondents Herein, Curative Petition (Criminal) No. 39-42 of 2010 in Criminal Appeal No. 1672-75 of 1996 in the Matter of Central Bureau of Investigation Versus Keshub Mahindra & Ors., 50, April 2011.
[14] District Court of Bhopal, India, State of Madhya Pradesh Through CBI vs. Warren Anderson & Others, Criminal Case No. 8460 of 1996, 47, June 7, 2010.
[15] Union Carbide Corporation, Bhopal Methyl Isocyanate Incident Investigation Team Report, March 1985, 11.
[16] A. Agarwal, S. Narain, The Bhopal Disaster, State of India's Environment 1984–85: The Second Citizens' Report, 207, 1985.
[17] S. Berger, Status of AIChE Initiatives to promote effective management of chemical reactivity hazards, in: 3rd International Symposium on Runaway Reactions, Pressure Relief Design, and Effluent Handling, Cincinnati, Ohio, USA, November 1–3, 2005, 2, 2005.
[18] Union Carbide Corporation, Bhopal Methyl Isocyanate Incident Investigation Team Report, March 1985, 24.
[19] I. Eckerman, The Bhopal Saga: Causes and Consequences of the World's Largest Industrial Disaster, 83, 2005, ISBN: 81-7371-515-7.
[20] M. Sharan, S.G. Gopalakrishnan, R.T. McNider, M.P. Singh, Bhopal gas leak: a numerical simulation of episodic dispersion, Atmos. Environ. 29 (16) (1995) 2062.
[21] District Court of Bhopal, India, State of Madhya Pradesh Through CBI vs. Warren Anderson & Others, Criminal Case No. 8460 of 1996, 12, June 7, 2010.
[22] W. Morehouse, M.A. Subramaniam, The Bhopal Tragedy: What Really Happened and What It Means for American Workers and Communities at Risk, 21, 1986, ISBN: 0-936876-47-6.
[23] I. Eckerman, The Bhopal Saga: Causes and Consequences of the World's Largest Industrial Disaster, 87, 2005, ISBN: 81-7371-515-7.
[24] T. D'Silva, The Black Box of Bhopal, 13, 2006, ISBN: 978-1-4120-8412-3.
[25] W. Morehouse, M.A. Subramaniam, The Bhopal Tragedy: What Really Happened and What It Means for American Workers and Communities at Risk, 1, 1986, ISBN: 0-936876-47-6.
[26] United States District Court for the Southern District of New York. In Re: Union Carbide Corporation Gas Plant Disaster at Bhopal, India in December, 1984, MDL No. 626; Misc. No. 21-38, 634 F. Suppl. 842, May 12, 1986, 1.
[27] District Court of Bhopal, India, State of Madhya Pradesh Through CBI vs. Warren Anderson & Others, Criminal Case No. 8460 of 1996, 3, June 7, 2010.
[28] Supreme Court of India Criminal Appellate Jurisdiction, Application for Directions to Institute Charges U/S 302 (For Offence U/S 300(4)) Read With S. 35 of the Indian Penal Code, 1860 Against the Respondents Herein, Curative Petition (Criminal) No. 39-42 of 2010 in Criminal Appeal No. 1672-75 of 1996 in the Matter of Central Bureau of Investigation Versus Keshub Mahindra & Ors., 4, April 2011.
[29] W. Morehouse, M.A. Subramaniam, The Bhopal Tragedy: What Really Happened and What It Means for American Workers and Communities at Risk, 23, 1986, ISBN: 0-936876-47-6.
[30] Supreme Court of India Criminal Appellate Jurisdiction, Application for Directions to Institute Charges U/S 302 (For Offence U/S 300(4)) Read With S. 35 of the Indian Penal Code, 1860 Against the Respondents Herein, Curative Petition (Criminal) No. 39-42 of 2010 in Criminal Appeal No. 1672-75 of 1996 in the Matter of Central Bureau of Investigation Versus Keshub Mahindra & Ors., 9, April 2011.

[31] District Court of Bhopal, India, State of Madhya Pradesh Through CBI vs. Warren Anderson & Others, Criminal Case No. 8460 of 1996, 6, June 7, 2010.

[32] P. Shrivastava, Bhopal: Anatomy of a Crisis, 48, 1987, ISBN: 088730-084-7.

[33] W. Morehouse, M.A. Subramaniam, The Bhopal Tragedy: What Really Happened and What It Means for American Workers and Communities at Risk, 11, 1986, ISBN: 0-936876-47-6.

[34] Union Carbide Corporation, Review of MIC Production at the Union Carbide Corporation Facility Institute West Virginia April 15, 1985, 1985.

[35] T. D'Silva, The Black Box of Bhopal, 27, 2006, ISBN: 978-1-4120-8412-3.

[36] R. Willey, D. Hendershot, S. Berger, The accident in Bhopal: observations 20 years later, in: 40th Loss Prevention Symposium, Orlando, Florida, USA, April 23–27, 2006, 13, 2006.

[37] Office of the Federal Register, 29 CFR Chapter XVII (July 1, 2012 Edition), 354, 2012.

[38] C. Durkovich, D. Michaels, M. Stanislaus, Executive Order 13650 Actions to Improve Chemical Safety and Security – A Shared Commitment: Report for the President, iii, 2014.

[39] U.S. Department of Labor, News Release: US Labor Department Seeks Public Comment on Agency Standards to Improve Chemical Safety, December 3, 2013.

[40] U.S. Chemical Safety Board, News Release: CSB Board Members Identify Modernization of Process Safety Management Regulations as the Agency's Second "Most Wanted Safety Improvement", December 1, 2014.

[41] M. Farley, California Proposes Major Changes to Refinery PSM Standard, Environmental Advisory, September 19, 2014.

[42] B. Storrow, Sinclair Officials on Refinery Working Conditions: 'We've Made Mistakes', Casper Star-Tribune, May 9, 2014.

[43] J. McCavit, S. Berger, C. Grounds, L. Nara, A Call to Action: Next Steps for Vision 20/20, in: 10th Global Conference on Process Safety, New Orleans, Louisiana, USA, March 30-April 2, 2014, 10, 2014.

[44] C. Durkovich, D. Michaels, M. Stanislaus, Executive Order 13650 Actions to Improve Chemical Safety and Security – A Shared Commitment: Report for the President, x, 2014.

CHAPTER

Equipment Reliability Principles

2

Thousands of books have been written about asset maintenance and reliability. You could attend perhaps more than 100 conferences, conventions, and symposia dealing with maintenance and reliability topics each year. Still, the goal of improved maintenance and reliability equaling zero catastrophic incidents has yet to be realized. Even as you read these lines, unforeseen events seem to happen at random, albeit with clearly undesirable regularity. Whenever the unforeseen happens, catastrophic equipment failures create process safety hazards and drain a very significant amount of revenue from the manufacturing industry each year. In a variety of ways, factory maintenance is directly involved with process safety and manufacturing productivity.

A local site's production rate (also called its "output") is governed by the type and capacity of the equipment operating inside the factory. In a perfect world, this equipment never breaks down and has no capacity limit. In reality, equipment failures are an anticipated consequence of factory operation. With very few exceptions (such as gears that "wear in" before they "wear out") equipment performance begins to deteriorate from the very moment we place it in service. Progressive deterioration is an unavoidable fact of life. It is therefore far more productive for local sites to properly define their specific type of process equipment and set operating limits accordingly. While a site aims to preserve equipment life to the fullest extent possible, it is even more important to appreciate the intrinsic value of using operating limits to prevent catastrophic equipment failures.

It is at this point where our understanding has to become a continuum. We join those who define it as a continuous sequence in which adjacent elements are not perceptibly different from each other, although the extremes are quite distinct and this is where we must close the loop: Catastrophic equipment failures are process failures. They represent intolerable risks to our process equipment. Catastrophic equipment failures and the degree of hazard acceptance that led us there started with our not appreciating the extent to which flawed thinking processes were involved. The whole notion of "stuff happens" deserves to be replaced by an absolute belief that all failures have causes. Stuff happens only because we let it happen. And when we come to this realization, we begin asking ourselves the most important question:

What can I do to prevent failures from happening?

BEING IN CONTROL

Back to operating limits and the objective for setting operating limits. The objective for establishing operating limits is to exercise full control over equipment performance so that a prescribed maintenance program can be effective. Never should the equipment be allowed to dictate the plan. To remain in control, factory personnel must fully understand the causes and consequences for equipment breakdowns. While maintenance would (occasionally) include precautionary replacement of worn or time-limited components, breakdowns are failure events that occur between scheduled or prescribed maintenance intervals.

A simple "break-even analysis" demonstrates the strength that various equipment reliability principles exert over local site production and maintenance plans. Such an analysis sets the stage for a discussion about the importance of being in complete control over operations and maintenance at all times. If we are not in control, factory performance can be limited by unstable equipment. Accepting unstable equipment behavior can lead to process upsets which, in turn, can result in a catastrophic process release. Allowing assets to dictate control over you and your plans will lead to trouble. Disregarding fixed costs or not understanding variable costs can create production difficulties. For now, and as a very important prelude to several later chapters, we need a simple but solid example. We find it in the concept of a break-even analysis, or the question of what break-even quantities are needed to cover the site's operating costs.

BREAK-EVEN ANALYSIS

Recall, please, our statement that equipment capacity is one of the major parameters that needs to be selected by and for an enterprise. Common sense tells us that we must select equipment of adequate capacity to achieve a specific production target that generates spendable income (profit). The break-even quantity (BEQ) is useful for two different, but equally important reasons:

1. Reliability engineers are encouraged to quantify the likely savings which result from improving equipment or components. Understanding and calculating the BEQ may provide all the proof needed to move in a certain direction. Also,
2. BEQ calculations can be powerful screening tools toward justifying the construction of new factories, perhaps to expand a customer base or reduce distribution costs.

Along these lines, the minimum number of units, in pounds, for example, which a factory must produce to cover its operating costs is defined by Eq. [2.1] [1]

$$\mathrm{BEQ} = \frac{C_\mathrm{f}}{(P_\mathrm{s} - C_\mathrm{v})} \qquad [2.1]$$

where, C_f is the fixed cost; C_v is the variable cost per unit; P_s is the sale price per unit; BEQ is the break-even quantity (ie, number of units).

Simply stated, as soon as the manufacturer breaks even, any remaining capacity that can be utilized for production represents a profit. It is therefore in the producer's best interest to minimize the BEQ. Doing so helps increase the process rate, or production capacity, without having to seek approval for a capital project. Generally speaking, capital projects are initiated to expand production or develop alternative operating methods to obtain output that exceeds the process' existing capacity. Seeking a safe alternative is called debottlenecking and process optimization.

Based on the BEQ formula and targeting the lowest break-even quantity, there are three possible ways to minimize the BEQ:

1. *Keep the fixed costs low.* Fixed costs represent necessary costs that are not likely to change based on factory output. Costs in this category include items big and small—like building rent and maintenance, copier toner, utilities, technology licensing fees, and marketing. These and other fixed costs can be difficult to reduce; they tend to be very consistent and predictable. It is safe to assume that fixed costs are always being held to a minimum. Lowering fixed costs in response to the onset of financial hardships always leads to uncomfortable questions about the late response. In other words, the question that is normally raised is, "Why did not we take advantage of the savings opportunity earlier?"
2. *Increase the unit price.* The unit (sales) price establishes how much the consumer must pay for the product after it reaches the market. The producer can raise the unit price with relative ease, but doing so might disrupt the balance of supply and demand. It might convince customers to consider alternative products or endure the hardship of doing without consuming or utilizing the preferred product. Contested price increases and reduced sales volume are prone to create an unfavorable climate for the producer. They might actually cause more losses than if no action was taken.
3. *Reduce the variable cost.* The final option involves improving control over the variable cost, which relates to charges that increase and decrease according to production. Items in this category include machinery maintenance and raw material purchases. Reducing the variable cost essentially increases the profit margin without having to raise the unit price. Indiscriminately cutting variable costs, however, is a potentially dangerous practice. It cannot realistically be done unless the factors responsible for the variable costs are somehow controlled with better precision. Machinery maintenance is one of the most significant variable costs within the manufacturing industry.

Since the variable cost is a function of machinery maintenance spending, reliability penalties directly impact the BEQ. As this relationship demonstrates, the BEQ will rise as the variable cost increases. In a case where equipment reliability meets expectations, the variable cost for machinery maintenance can be held constant and in agreement with the production plan, similar to what might be expected from a fixed cost. However, a higher-than-predicted failure rate leads to additional machinery maintenance requirements, which directly impacts a factory's variable costs.

Getting back to Eq. [2.1], dividing the total maintenance cost by the total number of units produced elevates the variable cost and drives up the BEQ. Suppose the company's objective is to maintain a target production rate so as to generate no less than a certain "must have" amount of hard currency. In that case, offsetting the additional variable cost imposed by poor equipment reliability will require either raising the unit price or lowering the fixed cost. As was stated earlier, neither of these actions comes without consequences. Again, and before changing anything, determining why the equipment is not performing according to life cycle expectations would be the best choice of action. Correcting the problem at its source may allow implementation of options including a one-time upgrading of weak or troublesome components. A compelling justification would explain how the proposed solution would reduce the BEQ; thereby increasing the production value.

A simple and entirely hypothetical example demonstrates the importance of how equipment reliability, variable costs, the BEQ, and factory operation are all intertwined (Table 2.1). First, consider the BEQ in Case 1 that could be presented as a basis to justify a factory construction or expansion project with projected annual variable costs of $240,000. The variable costs are split equally between maintenance and nonmaintenance necessities. Maintenance costs are based on a projected machinery overhaul frequency of once every 4 years, assuming that there would be no need to shut down for corrective maintenance in between.

Through a supply and demand study, the manufacturer decides to set the unit price at $2.00 per pound. This creates a high profit margin (45%) that justifies the spending of $200,000 in annual fixed costs. In this case the BEQ is 142,857 pounds, provided that variable costs can remain under control. Based on this assessment the factory is constructed with equipment properly sized and licensed to manufacture a maximum of 400,000 pounds a year. When the factory is commissioned the company expects the facility to operate "lean and mean," thus using up all of its remaining capacity to generate an annual profit of $360,000.

Let us say, however, that equipment reliability does not meet expectations. After the factory is commissioned, a reliability issue surfaces and causes recurring catastrophic

Table 2.1 Example of a Break-Even Analysis

	Case 1 (Design)	Case 2 (Actual)
Production capacity	400,000 pounds	400,000 pounds
Fixed cost	$200,000	$200,000
Total nonmaintenance cost	$120,000	$120,000
Total maintenance cost	$120,000	$480,000
Unit price	$2.00	$2.00
Break-even quantity	142,857 pounds	400,000 pounds
Remaining capacity	257,143 pounds	0 pounds
Profit margin	45%	0%
Potential profit	$360,000	$0

equipment failures. Instead of overhauling equipment once every 4 years during scheduled maintenance, major equipment failures occur and overhauls are now being performed every 12 months (Case 2). This quadruples the total annual maintenance costs to $480,000, which generates a mere $0.50 per unit sold. Notice how the variable cost increase under these specific circumstances raises the BEQ to 400,000 pounds, thus consuming all available capacity. Despite the fact that revenue is still being generated with every pound of product sold, there is no longer any capacity remaining that can be utilized to generate a profit. No amount of hard work can overcome the capacity limitation set forth by design. In reality, customer orders cannot be filled either, since the factory cannot manufacture product during shutdowns and these are now occurring annually. Under the circumstances, the factory loses a significant amount of money and is faced with the prospect of closing or being put up for sale. Good employees typically seek greener pastures with companies that can run their business more responsibly. This further increases the probability of a business failure.

This example is an appropriate introduction to the principles of equipment reliability. Several important and highly relevant points can be gleaned from these principles and the following points rank among the more noteworthy:

1. Process design sets the hard limit on factory output. As variable costs rise, the process is less capable of generating the revenue needed to support continued business operation. Every additional monetary outlay contributing to variable costs above the design basis decreases the factory's remaining capacity after the capacity reserved for the BEQ has been filled (Fig. 2.1).
2. If the goal is to keep the process "full," then catastrophic equipment failures contradict this objective by causing it to run empty. Factory output ceases when the production line shuts down for major unplanned maintenance work. On the other hand, scheduled plans for major maintenance take into consideration customer supply needs. Plans would therefore stock up inventory before shutting down, to satisfy projected sales volume for a predetermined length of downtime.

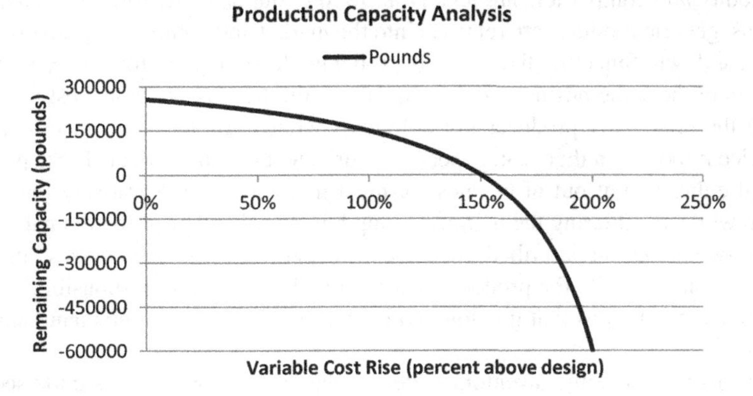

FIGURE 2.1

Production capacity remaining after satisfying the example break-even quantity (BEQ).

3. Revenue generation alone does not make a business profitable. This example perfectly demonstrates how a competitive profit can disappear although technically money is being generated with the sale of every unit. In this case, a potential profit margin of 45% was of no value to a business dominated by an equipment reliability problem.
4. As variable costs increase in response to maintenance requirements, it becomes increasingly difficult to compete for the resources needed to address the problem. Notice that the decreasing profit margin causes less cash generation for factory investments. In our example, increased maintenance expenses are elevating the variable cost. A formal investigation might provide the basis for a capital project that can remedy the problem responsible for excessive maintenance expenses (variable costs). However, the problem's impact on the BEQ can make funding that capital project difficult. Regardless of what options might be available to solve the problem, it is always safe to stick to the core belief that a good justification always takes into account the BEQ impact. A valid project is justified by showing the return created by lowering the BEQ. In other words, the justification process prevents the company from collapsing over a reliability issue that drives up the BEQ. As we consider case histories in equipment failures, observe how ineffective solutions drive up the BEQ.

PATENTS AND TRADE SECRETS

Companies that hold patents for the products they invent have a competitive advantage, at least for a time. If the product is good enough and the public wants it bad enough, then the manufacturer can raise the price with little consequence. In effect, inflating the price adds capacity relative to what the design can provide. Essentially, the consumer pays for the producer's inefficiency. All of this, however, changes when the patent expires. At that point, the specialty product instantly becomes a commodity and competitors are legally allowed to duplicate its formula. When this happens, generic products are released into the market and competition can drive the unit price down. Since the fixed cost is probably close to equal for different manufacturers in the same business, the competitor with the least variable cost stands to benefit the most. Any producer not able to generate sufficient hard income at the unit price imposed on them can expect to be driven out of the market. These producers will either be put out of business or must live with a comparatively low profit margin while shouldering the manufacturing burden. This then shifts the balance of the win–win situation described in chapter "Industrial Learning Processes" in favor of the consumer, while the producer is left with all of the legal responsibilities and compliance obligations that go along with operating a factory in the manufacturing industry.

For this reason, some inventors prefer to market their products as trade secrets. Trade secrets have proven to be very successful for many companies including IBM, Kentucky Fried Chicken, and Coca-Cola. In cases where a product is manufactured under a trade secret, it might never become a commodity. The company in possession of the intellectual property is its protector and guardian. Although they have entitlement to a specific formula, they are not offered any legal protection. The longevity of the product's specialty classification is limited only by the producer's ability to keep a secret. Once the secret becomes public, other companies can duplicate it. At this point, it too becomes a commodity.

Patents, on the other hand, are awarded only after full disclosure of the intellectual property in the patent application. In exchange for this disclosure, the inventor enjoys exclusive manufacturing privileges protected by law for a limited amount of time. When the patent expires the product becomes a commodity. When this time arrives the inventor loses exclusive manufacturing rights, but can perhaps retain most of the market share by enjoying the benefits of name recognition [2]. Indeed, some customers prefer to purchase the original over generic products even though the original might cost a bit more. In the end, however, the competitor is probably offering the exact same molecule, part, or product. Conceivably, customer loyalty is determined by product distribution and availability within a certain geographic region. Customers will invariably switch brands based on supply issues. Depending on their need, if the consumer cannot obtain a product they want from a preferred supplier, then they will buy it from a competitor that has the ability to deliver, perhaps for a lower unit price.

PROCESS FLEXIBILITY

Manufacturers are keenly aware of the relationship between product availability and customer loyalty, because it is simply a normal way to react. If the product you need is unavailable from one supplier you will likely contact alternative suppliers instead of suffering the pain of going without. It is for this reason that manufacturers go to excessive lengths to meet production targets regardless of what obstacles they might encounter. In most cases, they would prefer to retain or grow their customer base even at the expense of some profit. As our example plainly illustrates, a recurring reliability issue can make this option infeasible.

As discussed earlier, variable costs can rise due to chronic reliability issues. Such issues will increase the recurring maintenance costs; this cost increase may prompt a manufacturer to raise the price of a product in order to maintain an acceptable profit margin. This approach also lowers the BEQ, perhaps as a defense mechanism to keep it constant. But again, this option becomes much less practical after the product becomes a commodity. If an identical product can be obtained for less, the consumer will likely switch brands.

Another way to address the situation is by ramping-up production in an effort to generate revenue and to thus counteract BEQ inflation. However, since the production capacity remaining after meeting the BEQ is much reduced, it becomes necessary to modify the system in a manner which creates additional capacity. This translates into an additional construction project to augment an existing design; such plans would address process inefficiencies that were not recognized in the original project's proposal. But this pursuit would amount to an admission of failure by the project management team responsible for the original factory design.

Board members are typically not favorably receptive to these types of projects. Therefore, attention usually shifts to creative solutions that take advantage of process flexibility incorporated in the original design. In such cases, system operating and maintenance procedures are modified to tap into stagnating availability that is either reserved for a purpose other than production, or exploiting underutilized process capacity. When a solution is found, the existing process is modified to increase production volume. This type of activity encourages the development of workaround solutions, which include shortcuts. If successful, the selling of additional products generates cash flow that can keep the business operating without changing the unit price. Never forget, though, that workaround solutions are fraught with potential problems. It is much safer to operate equipment according to procedures that take into account specific design requirements. Even if a workaround solution is supported by scientific principles, the process might not be able to accommodate it. Process changes must correspond with the prevailing design intent to be successful. In a process plant, innovation and its unchecked pursuit can be dangerous.

Generating cash flow by increasing production volume only works if there are consumers in need of the additional inventory. If the market does not support the additional production volume, then raising production rates does not solve the problem and can create a glut. Gluts can damage market stability and once again do not favor the win–win situation that is needed to satisfy the mutual interests of both business owners and consumers.

As the BEQ increases, it will be hard to resist the temptation to access unused capacity that can readily be found within a process. But notice how we now come back to emphasize the importance of reliability concerns and the often underappreciated need for reliability that corresponds with anticipated design performance. To some project managers, reliability is a *future concern* whereas budgetary constraints—regardless of how unreasonable they may be—are a *present concern*. Present concerns receive most of the attention while future concerns are often ignored or simply transferred to others upon commissioning a new process. But here is the point again: Unless the equipment is designed to allow modifying its service conditions, changing its operation to obtain greater capacity raises the likelihood for a catastrophic equipment failure. Operating limits, in the form of temperature, pressure, flow, and composition are needed to prevent repeat failures. These limits are usually documented in original equipment manufacturer literature and through other accessible resources. They must, however, be formalized inside the local production sites to be effective. Failure to clearly define, embed, and enforce (with compliance audits) operating limits is a certain prescription for a catastrophic equipment failure. There is no question that operating equipment outside of its capabilities is a recurring theme in catastrophic process releases.

Further insight on specific reliability principles is needed to fully appreciate the relationship between equipment operation and failure. A satisfying level of equipment performance can be obtained by applying these principles throughout the manufacturing industry. It also helps to recognize specific equipment reliability (or failure) patterns and to look for these patterns while investigating a catastrophic process release.

ASSET PROTECTION

Without exception, every industrial disaster involves a catastrophic equipment failure. Examples from the past include the Hindenburg airship, the Titanic ocean liner, the Concorde supersonic aircraft, and the nuclear reactor involved in the Chernobyl nuclear disaster. Even the two disastrous space shuttle events support this fact. In every case, an equipment malfunction had a far-reaching impact on the industry it represented. In hindsight, none of these disasters would have occurred if proper attention had been given to understanding how and why the equipment was prone to failure. All failures are explained by previous experience in managing or handling an asset's reliability. This theme is further substantiated by the case studies described in this text.

Equipment reliability principles dictate how equipment can be expected to perform under certain process and mechanical operating conditions. No matter what service the equipment operates under or how the process is configured, process equipment behaves in a predictable and controllable manner upon being placed into service. The purpose for understanding basic equipment reliability principles is to exercise a satisfactory level of process control. This can only be done by preventing the types of equipment failures which then make it necessary to respond to a corresponding process upset. Never would we want to increase the potential for a catastrophic equipment failure due to not having applied the basic principles of equipment reliability. Instead, we would want to show our equipment the level of respect it deserves by making sure it operates according to our expectations at all times.

The global manufacturing industry uses a wide range of different equipment types; all are in service within various processes. Process operating conditions typically exert a heavy demand on this equipment. In many cases, the process operates nonstop, 24 h a day and 7 days a week to manufacture products that we regularly buy and consider essential to maintaining a reasonable quality of life. It is understandable, then, why we refer to any piece of process equipment that contributes to the overall success of a manufacturing operation as a "physical asset" (or simply "asset"). We fully realize that this term shows our respect and appreciation for personal belongings that either create or represent some value to us. The term also implies our genuine interest in protecting our valuables from damage that could reduce its usefulness. Indeed, our manufacturing processes would be nothing without the assets that do much of the work for us.

Every asset has a predefined function that delineates its specific role in an integrated manufacturing process. An asset's function can be expressed in logical generic terms or in more specific terms that apply to its required overall performance criteria (Table 2.2). Both descriptions are extremely useful for explaining how reliability principles apply to a manufacturing operation. An asset's design function is

Table 2.2 Generic Asset Functions and Example Specific Functions

Location	Generic Function	Design Function
Compressor	Move process gas from a point of lower pressure to a point of higher pressure	Maintain up to 2.4 mmscfd of recycle hydrogen gas flow
Exchanger	Transfer thermal energy across a boundary	Preheat reactor feed to 450°F
Flare	Destroy any unstable material escaping the process	Provide safe disposal of up to 250 mmscfd of process gas flow
Furnace	Directly raise the temperature of a process liquid or gas	Heat unit charge entering the tower to 700°F
Motor	Convert electrical energy into mechanical energy	Drive a 1024 horsepower centrifugal pump
Pump	Move process liquid from a point of lower pressure to a point of higher pressure	Process 30 MBPD of raw crude into distillation unit
Pipe	Route liquid or gas from one place to another	Carry 5 GPM of stripped sour water into overhead heat exchanger outlet
Tank	Store liquid or gas	Supply product pumps with up to 50,000 gallons per day of feed
Turbine	Convert kinetic and heat energy into mechanical energy	Drive a 1024 horsepower centrifugal compressor
Vessel	Manage liquid and gas distribution in a continuous process	Separate entrained liquids to produce 99.8% pure overhead gas
Valve	Control liquid or gas flow through a pipe	Maintain at least 10% level in tank to avoid pump cavitation

determined by its exact physical location in a process system. The generic function makes it easier to understand how assets are designed to support multiple functions within a single manufacturing process.

For example, the generic function of a shell and tube heat exchanger location is to transfer thermal energy across a boundary. This description accurately defines what the asset, a heat exchanger, is designed to do. Based on these nonspecific terms, the owner could probably envision using it for multiple purposes, including:

- A waste heat boiler to generate steam while cooling down the process,
- A process cooler/heater, as in a feed-effluent exchanger that heats up unit charge while cooling the rundown stream, or
- A steam condenser to increase turbine efficiency by allowing more heat content to be extracted from a pound of steam.

But the owner-operator could be making a serious mistake thinking that an asset with the generic name "heat exchanger" will give the same reliability in any of the above bulleted services or functional locations. This is a hugely important point. An automobile tire is a generic component that allows a vehicle to get from point "A" to point "B." Quite evidently, a particular tire will be designed and built for a low-power passenger vehicle, while a very different tire construction will be needed for a high-performance racing car. Again, each is rightly called a tire.

Hundreds of similar analogies pertain to safety factors and other ratings. In the case of automobile tires speed ratings only apply to tires that have not been damaged, altered, underinflated, or overloaded. Additionally, most tire manufacturers maintain that a tire that has been cut or punctured no longer retains the tire manufacturer's original speed rating after being repaired, because the tire manufacturer can not control the quality of the repair. While a tire with a speed rating of 115 mph might be able to do 130 mph for short duration on a flawlessly maintained highway, it could never perform as well on an unpaved road with potholes, or for 4 h on a flawlessly maintained highway heated to 130 °F.

It is no different with the mechanical assets in a factory. If a fluid machine at design conditions has a safety factor of 1.5 at 100% output, any increase in rate-of-flow beyond 100% will reduce its safety margin. Allowing continuous operation beyond 100% will compromise the intended margin of safety. Combining and piling on unintended fluid properties (eg, higher temperatures, pressures, pH-values, solids content, etc.) may even drive the safety factor into negative territory. Catastrophic failures and process releases are nearly certain when that happens. The message is clear: Purposely and deliberately allowing operation outside of set limits deprives us of the safety margins reserved for events that are neither purposeful, deliberate, nor foreseeable. Operation outside of design limits is contrary to the principles of asset reliability.

Keeping in mind all of the above, a particular machine (or any asset, for that matter) may be designed for a particular duty, or with parameters which impart a defined set of safe operating limits. Hopefully, these limits are never willfully exceeded. Even in the case where this machine is inserted into a factory for temporary use in one of a number of different services—and quite obviously one service at a time—human factors are involved. Human factors induce risk and these must be taken into account. As just one example, pipes connected to and from isolating valves and switching valves might allow the same pump to be used in different processes within a single production unit. The same pump and its connected electric motor driver could possibly be used in a product rundown service to storage, as the side draw or pump-around product pump from a vessel such as a packed column, or as a charge pump sending raw material into a reaction process. The more versatile we consider the asset to be, the more receptive we become to using it for purposes contrary to its original design intent.

Getting some idea of an asset's basic function helps us to envision some alternative use for it, so that as industry professionals we can live up to our expectation of maximizing its value. That sounds quite logical, but it really emphasizes resourcefulness and innovation to the detriment of reliability principles. Without verifying fitness for service and ascertaining that operational function will remain within set limits, "alternative use" may actually and regrettably refer to the asset's generic function only. In that case, its planned-for alternative use may expose it to conditions contrary to its designated, location-specific functional requirements and operating limits. Carefully note that this would be a violation of reliability principles.

Admittedly, under ideal circumstances there is rarely a need to consider how relocating or repurposing the assets can cause problems. Most of the time, at least based on its generic description, the relocated or repurposed asset will meet performance expectations or resolve a particular operating issue. But sometimes things go wrong. Repurposing an asset might have very unexpected consequences on the process. Moreover, we lose sight of the fact that actual process conditions in the manufacturing industry are rarely ideal. We might be misled by thoughts that, through our own personal experience, we have become adequately familiar with the way a particular process actually operates. And so we think about ways to modify the process configuration so that we can make it do what we *want* it to do, which many times goes far beyond what it is *supposed* to do. When there is a substantial difference between actual operation and intended operation, we think of ways to address the problem with resources under our control. This involves taking advantage of an asset's generic function, which might result in expanding or eliminating its design function. Please rethink, because that expansion or elimination can become a serious problem, as you will learn in one of the later chapters. Meanwhile, recall that in all processes, assets are provided to perform a specific function. Of course, the design intent was for the asset to do something very specific. However, we are in complete control over our assets and can therefore use them for what we *want* them to do. The two do not always match. As you will see, what an asset does and what you want the asset to do may not be the same. All failures have causes, always.

FUNCTIONAL COMBINATIONS

It is not unusual for the function of one asset to directly influence another asset's function. As a result, a problem in one area of the process can interfere with stable performance in another area. An exothermic reaction process (Fig. 2.2) can be used to demonstrate how different asset functions are integrated in a manufacturing process.

In this fictitious example, a heat exchanger location serves as a waste heat boiler that uses thermal energy as a by-product of the reaction process to generate saturated steam. The saturated steam then passes through a superheater, which is another heat exchanger designed to boost the steam temperature. The superheated steam is then fed into a turbine that drives a compressor. The compressor boosts the low-pressure (LP) feed gas pressure high enough to overcome the high-pressure (HP) reactor pressure, which sends gas flowing into the reactor. Inside the reactor, an exothermic reaction takes place. The hot product leaving the reactor is cooled down as it passes through the superheater, followed by the waste heat boiler. The waste heat boiler quenches the product temperature down to an acceptable level for storage as it generates more saturated steam. This final process exits the system as cool product rundown and is stored in an appropriate container.

Situations like this where one asset is designed to manage multiple process functions are not uncommon. In fact, engineers are trained to look for ways to combine multiple functions where practical. This is by far the most efficient way to operate

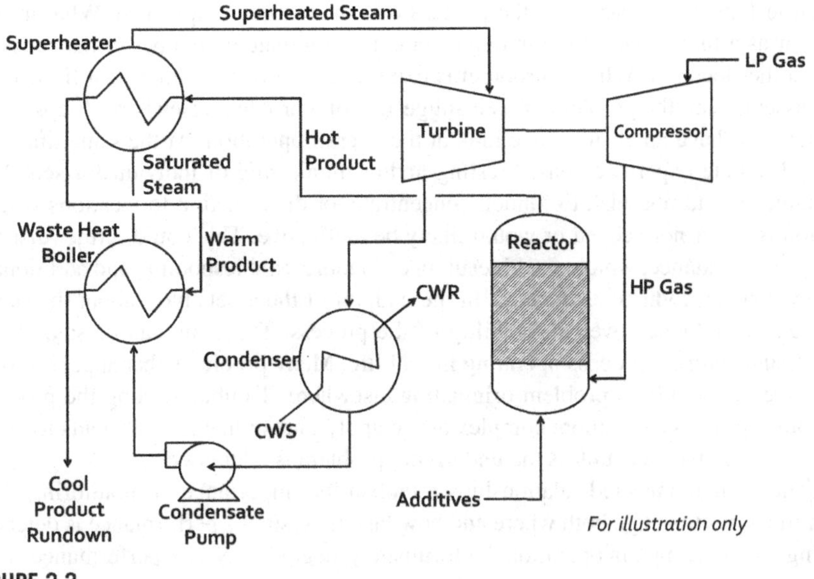

FIGURE 2.2

Example exothermic reaction process.

the process. It eliminates extra assets that might be included when a fundamentally different design approach is used. The goal in this multifunctional approach is to recover the maximum amount of energy and materials that would otherwise be lost through process inefficiencies. In exchange for this conservative approach to process design, individual dependencies are created that cannot easily be broken without disrupting the stability of the entire process.

For example, notice that the superheater is a multifunction asset that influences the performance of every other asset in the process configuration. The function of the superheater is to both generate steam and drop the temperature of the product leaving the reactors. Both functions are critical to the system's overall performance. The superheater's performance not only affects the temperature of the product rundown to storage but also the turbine efficiency. In the turbine it affects the speed and power delivered to a connected compressor that delivers HP feed gas into the reactor. The turbine's discharge pressure is dependent on the performance of the surface condenser, while also affecting the amount of steam needed to maintain target production rates controlled by the compressor's performance. In this example, the energy produced inside the reactor governs the performance of the superheater, waste heat boiler, turbine, and even the condenser. When the various assets meet performance expectations, the system functions are all kept in balance.

Which single asset is most important to the system's productivity? It is difficult to say in this example. There are many ways for a performance issue with any one asset to interfere with stable operation in multiple other assets. Due to these interwoven relationships, troubleshooting the process becomes more complicated. What might appear as a functional failure in one asset could originate with a performance issue in another location. When a problem is experienced in one location, is it limited to the asset having the problem, or is it suggestive of a problem elsewhere? Steps must be taken to have an all-inclusive look at the overall operation. At the same time we would initiate experience-based testing and/or monitoring of individual assets. We certainly would not wish to funnel, concentrate, or divert action to locations where action is either not needed or would likely be ineffective. This could further disrupt the process balance, which would result in even more troubleshooting and additional, likely counterproductive, changes. In the end, all of the assets have about the same amount of influence over the stability of the process. The same can be said about any manufacturing process operating in real life. Many problems that appear in one area are initiated by a problem originating elsewhere. Troubleshooting the process becomes progressively more complex as we apply unrestrained adjustments to keep the system in balance, unless the underlying problem is addressed.

These interconnected relationships establish the importance of monitoring system functions to gauge both where and how fast the system's performance is deteriorating after placing it in operation. Unfortunately, degraded system performance is an unavoidable consequence of standard process operation. Performance deterioration commences from the moment that charge is introduced to a system. Performance can be expected to deteriorate even when things appear to be perfectly under control and progressing according to plan. It is therefore appropriate to design an asset in a way that allows a certain amount of performance loss during its useful life. Think again of the automobile tire analogy. After a few months of use, careful measurements would indicate tread wear. Nevertheless, the tire is considered perfectly fit for service and remains useful for a long time to come.

MANAGING PERFORMANCE LOSS

A certain tolerance for asset performance deterioration must be exercised when the process is being designed. This is done by adding extra capability that can be sacrificed without consequence during the asset's useful life (Fig. 2.3). As long as the performance deterioration does not cut into the asset's desired performance, there is no unacceptable penalty for its degraded performance. The degraded performance is nothing more than an anticipated consequence of process operation that has been planned for in advance. A repair is needed when the cumulative amount of performance loss exceeds that which can reasonably be allowed without introducing a production constraint, in the form of a safety, environmental, reliability, or economic hazard. At this point a decision would have to be made about how to address the problem.

While remembering that it is outside the scope of this book to teach mechanical engineering subjects, we wish to convey knowledge via practical examples to which our readers can relate. In this instance a basic design principle can be demonstrated by calculating the minimum metal thickness of a vessel to withstand internal

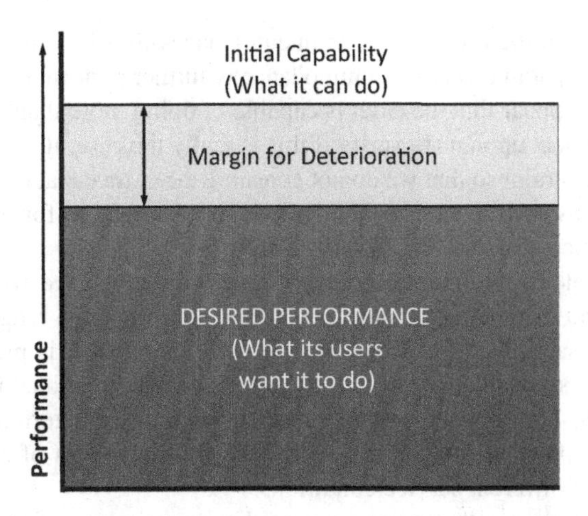

FIGURE 2.3

Allowing for deteriorating asset performance [3].

pressure. The minimum thickness of a pressure vessel needed to contain a process operating under certain conditions is given by Eq. [2.2]:

$$t = \frac{P \times R}{S \times E - 0.6 \times P} \quad [2.2]$$

where, t is the minimum required thickness of shell (inches); P is the internal design pressure (PSI); R is the inside shell radius (inches); S is the maximum allowable stress value (dependent on the material of choice); E is the joint efficiency (confidence in the integrity of weld seams).

This calculation provides the minimum acceptable metal thickness that must be preserved to responsibly operate a cylinder with longitudinal joints up to a certain maximum pressure limit [9]. Let us say that a modest amount of corrosion is expected in this particular process service. If a vessel was built strictly with nothing more than the metal thickness prescribed by the above equation, then there would be no margin for deterioration. Thus, there would be no tolerance for the anticipated service conditions—conditions where some corrosion can be expected. As a result, the performance loss would immediately begin consuming the metal thickness needed to safely contain the process. It is therefore practical to design the process with the forethought of managing a controllable rate of performance deterioration after the system is brought online by adding excess capacity.

FUNCTIONAL FAILURES

A functional failure occurs when the asset's deterioration margin has been completely consumed. After this, the asset's *desired* (its originally expected or stipulated) performance is no longer in play. With its deterioration margin consumed, we must now focus on what the deteriorated asset is *capable* of doing. It can no longer be used in full accordance with the original design intent. Concessions have to be made on one or more operating parameters to accommodate any further performance deterioration.

While it may appear that the asset is capable of doing more than what we need it to do, it is incumbent upon us to verify if this is really the case. It is therefore important to exercise restraint so that we do not consume the extra capacity that is reserved for other functions such as safety. Giving up capacity reserved for other designated functions to support production can be disastrous.

In our example, a minimum amount of corrosion allowance would have to be restored to extend the useful life of the asset into an acceptable range. This equates to making extra sacrificial metal available in addition to what is prescribed by the minimum thickness calculation (Eq. [2.2]). A foreseeable functional failure can only be avoided by designing assets with a sufficient margin for deterioration. This is a universal design practice that applies across all different types of assets operating under a variety of different service conditions.

For example, let us say that we are interested in operating an asset (pressure vessel) for at least 40 years before having to replace it. Based on practical experience, we predict that an average corrosion rate of 0.030 inches per year, or 30 mils per year (MPY) will be observed during the asset's useful life. In this case, a 40-year life expectancy would result by adding an extra 1.2 inches of metal thickness (40 years × 0.030 inches per year). If our assumptions are correct then it will take about 40 years at a 30 MPY uniform corrosion rate to reach the minimum wall thickness, at which time the asset would have to be replaced. We consider the cost difference between adding an extra layer of metal in the asset as corrosion allowance; alternatively we consider replacing a less expensive thinner-walled asset more frequently. We would probably find investing in the corrosion allowance the more economical choice. At any given time, the remaining life can be determined by dividing the remaining corrosion allowance by the active corrosion rate (Eq. [2.3]):

$$L_r = \frac{CA_r}{CR} \quad [2.3]$$

where, L_r is the remaining life; CA_r is the remaining corrosion allowance; CR is the corrosion rate (inches per year).

However, things are not always perfect. Let us say that during the life of the asset a valve begins leaking acidic process fluid into the vessel. During this time the corrosion leaps to 800 MPY. Perhaps it takes us 6 months to detect and correct this unintended operating condition in the asset's 25th year of operation. The remaining life then comes to 2.2 years instead of the 15 years that would have been possible if the acid leak had not occurred:

$$L_r = \frac{1.2\text{ inches} - (24.5\text{ years} \times 0.030\text{ inches per year} + 0.5\text{ year} \times 0.800\text{ inches per year})}{0.030\text{ inches per year}}$$

$$L_r = 2.2$$

At this point a decision must be made. Let us take this problem a step further by saying that the next time slot or "window" available to replace the damaged asset without suffering a devastating production penalty is at least 3 years off into the future. Based on the remaining life with no further upsets, the asset's current condition will not accommodate operating for this length of time. Therefore, we are now faced with the prospect of shutting down the process between maintenance cycles for an unscheduled repair at considerable cost. Otherwise, a way must be found to increase the corrosion allowance without shutting down. Doing this would essentially require somehow lowering the minimum allowable thickness in exchange for additional corrosion allowance. Rearranging the thickness calculation, (Eq. [2.2]) makes it possible to determine how far the asset's maximum allowable internal pressure would have to be dropped (rerated) to make an adequate amount of metal available as extra corrosion allowance (Eq. [2.4]):

$$P = \frac{S \times E \times t}{R + 0.6 \times t} \quad [2.4]$$

where, P is the internal design pressure, psi; S is the maximum allowable stress value, ksi (dependent on the material of choice); E is the joint efficiency, percent (confidence in the integrity of weld seams); t is the minimum required thickness of shell, inches; R is the inside shell radius, inches.

Let us say in our example we have a 28-inch diameter carbon steel vessel designed for 560 psi pressure. Under these conditions the minimum thickness comes out to be 0.5 inches (Eq. [2.2]). In order for us to achieve our target run length, we will still need to recover an additional 15 years of useful life at the projected deterioration rate of 30 MPY. This corrosion rate assumes that we learned something important about how to better control the process from the investigation into the accelerated corrosion incident. It also requires that we have taken appropriate action based on this information. Extending the vessel's life for another 15 years at this point amounts to reserving an *additional* 0.385 inches of extra metal from the original 0.5-inch base metal thickness as corrosion allowance. This would leave only 0.115 inches of metal at the projected end of run 15 years in the distance. Using the thickness calculation formula, this drops the asset's maximum pressure rating to 130 psi at the end of the run:

$$P = \frac{20,000 \text{ ksi} \times 0.8 \times 0.115 \text{ inches}}{14 \text{ inches} + 0.6 \times 0.115 \text{ inches}}$$

$$P = 130 \text{ psi}$$

This option would therefore call for operating the vessel at a pressure of only one-fifth of its initial design value in order to recover a sufficient amount of metal thickness to extend its useful life to the original target (40 years between replacements). Unfortunately, this option proves too restrictive. Attention therefore shifts back to shutting down the process early, and at great expense, for an unscheduled repair.

But we have not completely run out of options; it makes sense to examine if the goal could be changed to simply reaching the next convenient "shutdown window" in 3 years. The presently available corrosion allowance is 65 mils. At the current corrosion rate (30 MPY) only 25 mils (65 minus 90) would be taken away from the base metal thickness at the end of the third year. In this case the vessel's maximum pressure rating would only have to be lowered to 500 psi (Eq. [2.4]). This might be a more practical option. It could probably be done by making a few minor operating adjustments without much cost. A plan could then be developed including increased monitoring frequency to verify that no further surprises develop in the interim before the asset can be replaced or repaired. The plan would also include provisions to stock up inventories over the remaining operating window; the obvious purpose being to minimize any potential supply chain interruption upon shutting down for the repair. Supply chain interruptions could alienate loyal customers who might then defect and seek alternative suppliers. The plan would also incorporate the appropriate change management process, to maintain overall process stability upon making an adjustment in one part of the system.

RELIABILITY PLANS

A reliability plan is the roadmap to successful asset maintenance that maximizes production value. The reliability plan's effectiveness is impacted directly by the quality of information used to generate it. A considerable amount of thought is invested into knowing when the asset is most likely to fail. This reliability plan makes it possible to prepare adequately in advance for major maintenance followed by a successful restart. The primary goal upon restart after major maintenance is to enter another extended run without having to interrupt production for corrective maintenance. Success in this aspect depends on the quality of the reliability plan and the information used in its development.

The relationship between deteriorating asset performance and the asset's useful life can be converted into a functional failure probability curve (Fig. 2.4). In actuality, these curves or charts are based on empirical test data that simulate operating conditions on an asset introduced to the market. These statistics can also be derived from actual operating experience; operating experience that is commonly based on demonstrated system performance after a process is commissioned. The chart displays the cumulative total percent of failures in a population of test samples over a set period of time.

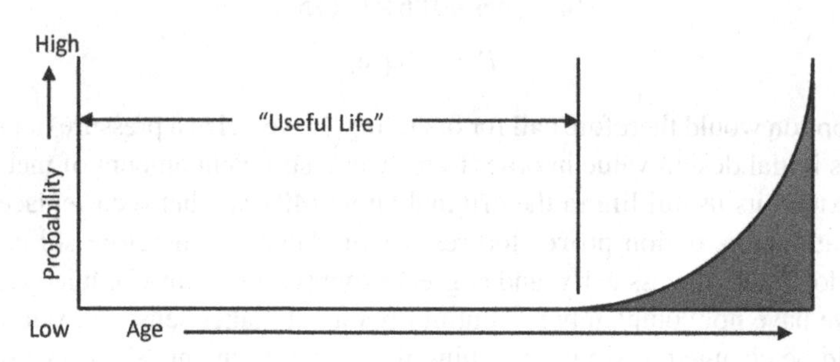

FIGURE 2.4

Functional failure probability curve [4].

Recognizing when a functional failure is most likely to occur makes it possible to control the outcome. In all instances the objective of *staying in control* is to prevent process upsets that could lead to involuntary shutdowns. The relationship between an asset's age and the probability for a functional failure translates into a mandatory, but practical, maintenance plan that avoids such incidents. Normally, these maintenance plans are executed during a turnaround (TAR), also known as inspection and repair downtime (IRD). An IRD requires that the process is completely de-energized, drained, depressured (and sometimes even inert gas-purged) in anticipation of major invasive maintenance. Needless to say, such maintenance is thoroughly planned; the intent of thorough planning is to complete the work without a safety incident. Working on a completely de-energized, drained, and depressured process unit avoids having to manage the elevated hazards involved with doing similar work on a partially or fully operating process.

RELIABILITY ANALYSIS

A complex mechanical system such as an automobile's transmission demonstrates how reliability statistics are factored into the reliability plan. A typical transmission consists of many different components. Each individual component could be considered an asset by itself. If any one of the transmission's critical components fails while in use, then the consumer might have to call a tow truck for assistance. Other inconveniences would certainly follow before the vehicle is repaired and returned to the consumer. The same is true with process operation.

To illustrate, let us say that through a series of performance tests an automobile manufacturer develops a transmission failure probability curve that anticipates trouble-free operation for 50,000 miles. After reaching that mileage the transmission is increasingly likely to experience a functional failure that could seriously, and negatively, affect its performance. The manufacturer therefore sets the warranty at 50,000 miles and recommends a precautionary service visit every 15,000 miles.

You, the consumer, find yourself calling a tow truck when the transmission fails at 50,001 miles. Is the timing of your failure coincidental, meaning that you were simply the victim of nothing more than bad luck? Or is there a more scientific explanation for why failure occurred only one mile after the warranty expired? Did you remember your service interval at 45,000 miles? Rest assured there was an anonymous reliability professional in the background who set the warranty period as part of a reliability plan that protected the manufacturer's best financial interests. The truth is that your failure occurred almost exactly as (or when) predicted, which is why the manufacturer made the bet that your transmission would not fail before 50,000 miles. In a similar sense, the manufacturing industry schedules preventive (PM) or corrective maintenance interventions according to a reliability plan. The intent is to prevent inconvenient business interruptions or downtime resulting in production losses based on the failure probability of critical assets.

Reliability professionals are responsible for analyzing deteriorating system performance; this analysis should predict how long a process can operate before a functional failure will involuntarily force an immediate shutdown. A controlled shutdown can then be scheduled before such a time by implementing a maintenance plan that includes the amount of time, resources, and replacement or repair parts that must be budgeted to restore an acceptable length of time, or distance traveled, or some other margin for deterioration. It is also during these major maintenance intervals or preplanned outage events that system upgrades and special projects can be completed. It should be noted that reliability professionals who work for reliability-minded organizations confine their work effort to data analysis, problem identification, incident investigations, and authoritative recommendations for permanent solutions. They are not involved in routine data collection. Such data collection is normally done by operators, operating technicians, and process technicians who are in direct control of the process. The line organization must be fully engaged in the manufacturing process at all times.

JUSTIFYING DESIGN CHANGES

Performance often deteriorates when an attempt is made to operate an asset beyond its design function. A typical example of this involves operating two centrifugal pumps in parallel. Suppose the two pumps have identical head-versus-flow (H/Q) performance curves and that normal process operation would require one pump to operate (or be online), while the second pump would serve as a normally nonoperating standby, called an "installed spare." To reiterate, while the primary (running) and spare (nonrunning) pumps would be *installed* in parallel, only one of the two is designed to be operating at any given time. Operating only a single pump being the design intent, many pumps in modern industry are supplied with impellers that exhibit a characteristic curve that flattens somewhere to the left of the "best efficiency point," or BEP (Fig. 2.5). The operational H/Q intersect is in the "knee" of the curve. If, contrary to the design intent, both of the pumps depicted on Fig. 2.5 are *deliberately* operated in parallel, even minor differences in system resistance (metal roughness, length, size, elevation, geometry/configuration of piping, etc.) might cause the operating point of one of the two pumps to move to the left and into the flat part of the curve. Operation at that intersecting H/Q point may now be too close to zero flow; operation at or near zero flow tends to accelerate deterioration; and can even destroy centrifugal process pumps.

As we then continue with our pump example, we wish to remind the reader that reliability thinking and reliability principles are the main topics. Reliability thinking and its underlying principles must migrate to every one of the thousands of different assets we find in a modern factory. Although among the most frequently used machines in modern process plants, pumps take no special place among the thousands of assets to which identical reliability principles apply. We therefore discuss pumps as an easy example. A facility with insufficient reliability focus may allow pumps to operate contrary to design specifications. Say, the factory desires to overcome high pressure

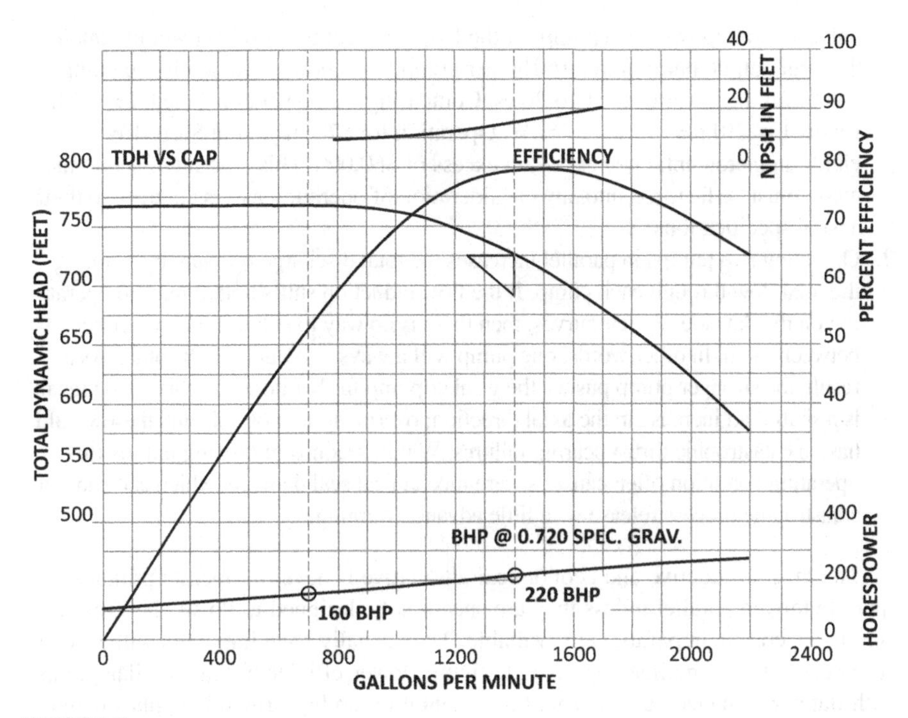

FIGURE 2.5

Typical "flat" curve found in many high-speed modern primary and spare pump locations.

drop in its discharge piping network or to add flow by running two pumps simultaneously. Note, again, how these seemingly innocent and low-cost process configuration changes (ie, running both flat-curved centrifugal pumps simultaneously) tend to increase the rate of asset performance deterioration.

We again acknowledge that short-term gains might be obtained by increasing total flow to overcome a production constraint. But there are at least two important drawbacks to this operating-in-parallel approach that can offset any short-term gains.

1. Operating a pump to the left of its best efficiency point increases the potential for cavitation and wastes energy. Notice how the pump's efficiency drops off rapidly below its design flow rate (marked by a right triangle). There are incremental variable costs associated with this approach, in both reliability and energy penalties. The objective should always be to run assets as close to their highest efficiency as possible, which is normally as per design specifications. As described earlier in this chapter, increasing variable costs shrinks process capacity by devaluing production. As a result, more product must be sold to break-even on production costs.

   As indicated in Fig. 2.5, a single pump with its impeller designed for 1400 GPM will absorb 220 hp. In an approximation (for demonstration purposes only), operating two such pumps at the H/Q intersect of 700 GPM would double the total output, but requires 160 hp per pump. Note how inefficiently the pumps operate at the lower of the two flows. Comparing the power consumed: 2 × 160 hp versus 1 × 220 hp = 100 hp or 75 kW. Operation for 8700 h/year at $0.10/kWhr yields an incremental power cost in excess of $60,000. This amount of economic waste must be factored into any consideration of operating assets contrary to their design specifications.

2. Operating two pumps in parallel increases the total discharge pressure by reducing the total flow through each pump. If the flow reduction shifts both pumps to operating on the flat side of their curves, then there is no way to balance the hydraulics between them. In other words, one pump will always win out over the other. As a result, the stronger pump pushes the weaker pump further down its curve. The result is a shaft load increase in the axial direction; overloads are possible and these would hasten catastrophic thrust bearing failures. Within seconds of bearing failure, this operating condition often causes severe mechanical seal damage, which can cause a catastrophic process release with little advance warning.

The most productive and economical alternative is never to operate primary and spare pumps in parallel unless they are specifically designed to do so. Designs suitable for operation in parallel are available. They usually have impellers with steeper H/Q curves but sometimes operate at slightly lower efficiency than similar pumps with flat performance curves. This fundamental reliability principle applies equally to all types of process assets in the manufacturing industry.

The choice of installing impellers with flat H/Q curves, as well as similar choices governing all instances of responsible asset management, requires operational discipline. The pressure to do things as we have always done them is extraordinarily high when a production incentive favors operating an asset outside of its limits. In this example, when the production constraint could be overcome by running (unsuitable) pumps in parallel, restraint must be exercised. Sticking to reliability principles and maximizing asset preservation means operating the pumps according to their design. We should not subscribe to fallacious reasoning in this case. The thought might be that the consequences in this instance would affect only discretionary maintenance and reliability costs. But in reality, safety and environmental concerns are also involved. In the end, strict adherence to operational discipline, which in this case translates into running both the process and assets within their respective design specifications, should drive the decision over what changes, if any, should be made. It is possible that production volume is curtailed by following a conservative operating approach that favors the denial of exceeding operating limits. If there are economic (profit-driven) incentives for design changes to capture the losses, these should be pursued. Redesigns and upgrades, even the addition of a new factory may be cost-justified in some instances. Never would overcoming a production constraint be worth running an asset into a premature, perhaps catastrophic, failure. In other words, the rule should be to never alter design performance for the purpose of gaining more output than the asset was designed for. Running primary and spare pumps together in disregard of impeller-related constraints is a common, but unfortunate, practice. Our narrative is again and again making it known that operating any asset outside of set limits is contrary to reliability principles. It also sets a bad example and must therefore be avoided.

An unfortunate reality in modern industry is that very few functional failures are governed strictly by age [5]. Many functional failures are random and unexpected. Relatively few failures are predicted ahead of time and occur when expected. Certain predictive maintenance (PdM) instruments are available as tools to observe relevant performance excursions in time to safely effect shutdown. Recall, however, that all maintenance execution routines constitute a variable cost and, as such, affect the factory's BEQ. Random deviations from normal operation and unforeseen departures from a safe operating envelope are most often caused by one or more of the four different functional failure "agents" [6]. Age (time) is simply one of them. Time, however, is the only agent guaranteed to cause the inevitable functional failure. The three other agents that can (and often will) cause premature failure are:

- Force (steady, transient, or cyclic—but out-of-range)
- Reactive environment (chemical or nuclear—out-of-range)
- Temperature (low, room, or elevated, and steady, transient, or cyclic—out-of-range)

However, force, reactive environment, and temperature might not cause assets to experience uniform performance deterioration over time. When these agents are present and out of their normal design ranges, the predictable process performance deterioration relationships (or rates of progress toward serious loss of containment) no longer apply. Unpredicted timing now enters the picture and the failure of an asset in use can cause the loss of primary containment (LOPC).

Process containment is a secondary function that applies to all physical assets configured in a process. In other words, all assets placed into service are designed to prevent the fugitive loss of process materials. Therefore, any leak that occurs in any conceivable asset that is part of a manufacturing process can be defined as a functional failure. However insignificant it may seem, no valve, vessel, pipe, furnace, exchanger, pump, compressor, turbine, or tank should ever be expected (or willfully predetermined) to ultimately leak massive amounts of process into the atmosphere. Minor leaks should always be considered a telltale sign of a greater problem worthy of analyzing more fully. All functional failures deserve to be addressed.

Whereas the consequences for progressive asset performance deterioration might appear to be manageable, the consequences of a catastrophic process release can be quite severe. Catastrophic process releases can result in irreversible health, safety, environmental, property, and economic damages. It is therefore essential to monitor, test, inspect, or analyze asset performance in an effort to capture the onset of excessive degradation before a catastrophic failure is imminent. Again, the emphasis is on developing a plan that offers the most amount of time to acceptably control a situation that could develop into a catastrophic asset failure. The notion that one can run all assets until they fail catastrophically is inconsistent with fundamental asset reliability principles and operational discipline.

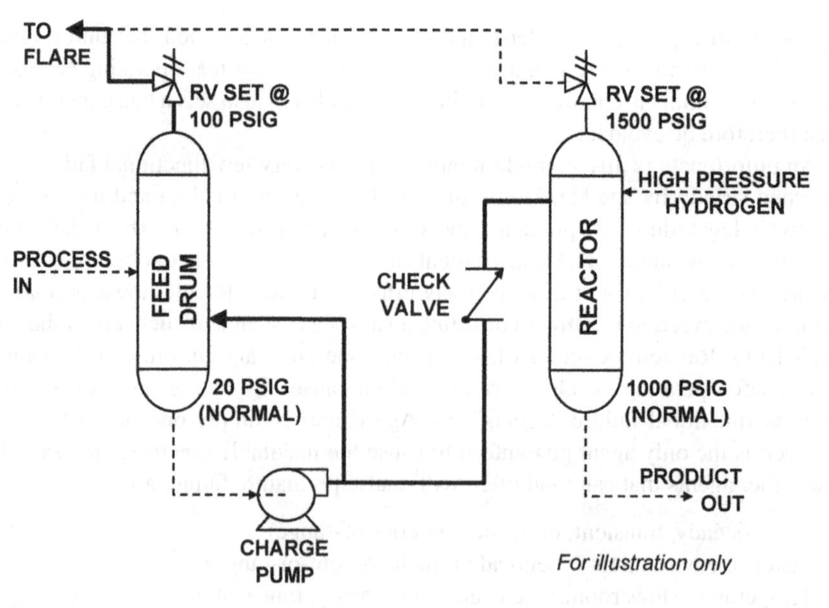

FIGURE 2.6

Process flow reversal upon pump trip.

Premature failures are often the impetus for adding redundancy into a system that does not perform well. However, redundant assets do not solve reliability problems. A common mistake is to install multiple assets in the same service in the hope that doing so will raise performance to an acceptable level. For example, a check valve failure can result in reverse flow through a pipe from a HP location to a LP location (Fig. 2.6). Let us say that a pump trip causes the sudden loss of flow from a LP feed drum into a HP reactor. This, in turn, results in a reverse-flow scenario that elevates the feed drum pressure above its maximum allowable working pressure. In this instance, the automatic pressure relief valve would be expected to open, thereby protecting the feed drum by directing its pressurized contents into the flare header (solid line). This automatic safety provision prevents the feed drum from rupturing. However, let us say that the process contains sulfur, and the resultant combustion at the flare releases sulfur compounds above the permissible reportable quantity limit. The system worked as designed; it prevented a safety issue by relieving HP in the feed drum to the flare. In exchange there was an environmental incident, but only because the check valve failed. Naturally, an investigation would be triggered to diagnose the cause of the incident and determine what steps can be taken to prevent recurrence.

Before taking this hypothetical example any further it is important to point out three reliability principles that translate into process safety concerns:

- *Hazard exchange.* Notice that the demand placed on the feed drum's automatic pressure relief system is what created an environmental hazard. The argument could be made that if the process had not vented to the flare header, then there would have been no environmental incident. Of course, this alternative scenario would have been completely unacceptable. There might have been an even more consequential safety incident. The lesson that applies to all industrial processes is that mitigating one hazard creates at least one other hazard that must then be mitigated as well; thus a check valve was inserted into the process line. Its designated function was to prevent the reverse flow scenario. The check valve would have prevented both incidents if it had functioned correctly. The check valve, however, failed.

- *Hidden failures.* The incident involved a check valve failure. It is almost certain that the check valve failed long before the actual reverse flow scenario created a demand for it. Hidden failures are failures that are revealed only when there is a demand for a specific function. Of course, this is not the best time to find out that the asset, and therefore your plan, has failed. These types of failures can only be detected through periodic inspection and testing programs. Hidden failures are best mitigated through the application of inherently safe design practices. For example, consideration might be given to designing the feed drum to withstand the pressure that would be expected in a reverse-flow scenario.
- *Reliability focus.* Industrial processes are designed with forethought about how adequate control can be maintained during an upset. Some system functions are provided for reacting to an event caused by an unrelated functional failure, such as a check valve that is added to stabilize the process in the event of a pump trip. In order for there to be complete satisfaction when a functional failure occurs, all system functions provided must be available on demand. However, this "perfect" process condition is difficult to establish and even more difficult to maintain, especially when hidden failures might be involved. Moreover, depending on how much thought is given to the installation of independent safeguards, a malfunction in one location might involuntarily disable multiple other system functions, which might include the loss of safety provisions (see chapter: Process Hazard Awareness and Analysis). The problem is better solved by preventing any incident that might create a periodic demand for various system functions. Understand that these system functions may or may not be available when needed. This solution simply serves as one more reminder that incident prevention makes much more sense than having to respond to an incident. Thus, reliability-minded organizations pursue incident prevention as their first priority.

These are the deeper issues that an incident investigation would seek to resolve, so that other systems impacted by hazard exchange, hidden failures, and our entire reliability focus can be adequately protected from similar consequences. In a practical sense, all manufacturing processes are potentially affected by these concepts. Realize, please, that a superficial investigation into this reverse flow incident would concentrate only on the check valve failure. In this case the recommended action might be to install another check valve in series to achieve a satisfactory level of incident prevention through redundancy. That way, both check valves would have to fail for a similar reverse flow incident to happen—highly improbable if we apply conventional wisdom. Along the lines of customary reasoning, the faulty check valve would be replaced with a new one, and a second check valve added in the same line as a precaution. But suppose that 3 months later the exact same reverse flow scenario and incident occurs, this time with both check valves in place. Would it then make sense to add yet another check valve in series?

If the original failure involved a manufacturing defect (a rather rare occurrence) then there would be a basis to replace it with one that is verified to be free of similar imperfections. Otherwise, any additional check valves inserted in the same line can be *expected* to also fail at about the same time. After all, they are of the same design and exposed to identical operating conditions as the one that initially failed. The same concept applies to any type of spare asset that is exposed to the same operating conditions as its corresponding primary asset. We will discuss more about the *responsible* use of multiple assets to increase reliable redundancy in chapter "Process Hazard Awareness and Analysis."

PREVENTING PREMATURE FAILURE

The Process Safety Management (PSM) standard outlines the basic requirements for asset performance monitoring. This approach has been adopted within the manufacturing industry to avoid being caught off guard by an undetected failure. Since industrial disasters consistently involve the sudden and massive release of hydraulic, pneumatic, chemical, nuclear, or electrical energy, there is never enough time to plan for a decision in the heat of the moment, when a catastrophic event is underway. Routine inspections and operator surveillance provide the mechanisms to take appropriate action before it is too late. The time available to develop an appropriate response to a functional failure in progress is defined by the failure's P-F interval (Fig. 2.7).

The P-F interval is the amount of time between when an approaching functional failure can be detected (point "P") and when it actually occurs (point "F"). The P-F interval will vary in duration; it will be influenced by the features inherent to a specific asset, its operating conditions, and the active failure mode. For

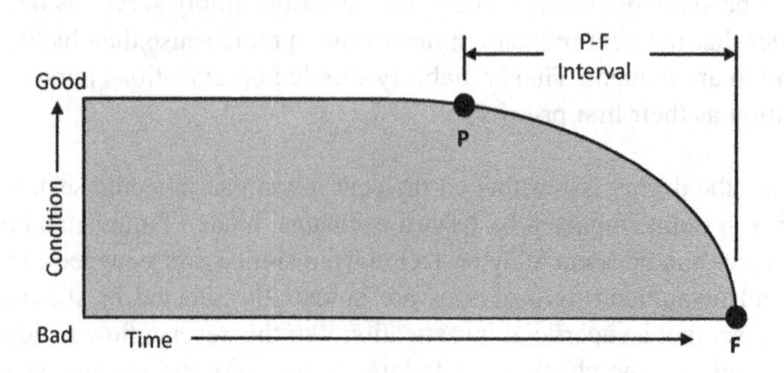

FIGURE 2.7

P-F interval [7].

example, the P-F interval for a catastrophic reciprocating compressor failure due to high suction knockout drum level is extremely low. In this specific case, and unless a properly positioned high-level shutdown switch is activated, the time between when the failure can be detected and when it actually occurs is near-instantaneous. There are other situations involving high-speed machinery where only perhaps seconds at the most are available to respond to a detectable hazardous condition before a functional failure occurs. In those instances, too, it would be appropriate to automate the shutdown feature and to not depend solely on a heroic response by a speedy operator.

On the other hand, the P-F interval for a catastrophic centrifugal pump thrust bearing failure may be days to weeks, depending on the failure mechanism involved. Characteristic vibration signatures can usually be detected on PdM instruments before the damage becomes so extensive that unusual sounds and smoke are providing even more direct warning signals. Still, the practical use of real-time instrumented methods of machinery performance monitoring will often allow us to extend the elapsed time between diagnosing problems and having to take appropriate preventive action. Continuous monitoring practices provide the most immediate form of detection when performance deterioration begins. This must never be used as an excuse, however, to abandon routine surveillance and monitoring methods, which are both essential for operating a stable process.

Routine surveillance, performance monitoring, and functional testing are among the programs that industry uses to detect a failure in progress (point "P"). Nondestructive examination for fixed assets and vibration monitoring for rotating assets are two specific programs that are implemented on a recurring basis for early detection of an impending functional failure. These programs are instrumental for exercising good engineering judgment over a maintenance situation which, if not addressed, could lead to a catastrophic process release. The reliability plan therefore tunes system monitoring and response according to the prevailing failure mode that could impact asset reliability. In cases like the one described above where the P-F interval is less than a second an automated shutdown (interlock) is appropriate whereas a weekly vibration monitoring schedule might be acceptable for managing potential thrust bearing failures. Detecting a failure before it occurs makes it possible to take advantage of all available time as defined by the P-F interval; it allows us to plan how to manage the failure on acceptable terms.

Process performance monitoring often detects an operating condition that can only be addressed by shutting down the asset for an offline repair. For example, a pump's mechanical seal cannot be changed while the asset is operating. Similarly, a repair that involves disassembling process assets while performing invasive maintenance can only be done safely under isolation after removing the asset from service. Spare assets are normally installed to cover the types of maintenance situations we have discussed up to this point. The intent for installing a spare asset is to continue production without interruption while a defective asset is unavailable. The manufacturing process continues without consequence to the consumer while the installed spare is being utilized so that the main (primary) asset can be repaired or replaced.

Companies that wish to indoctrinate asset reliability principles into the workforce incorporate installed spares into normal process operation. Here is why we want to again resort to our automobile analogy. Most owners use the spare tire quite rarely and only when needed. When it is used, it usually stays on the car for only a short time. They sell or otherwise dispose of the car with its spare tire often in virtually pristine condition. A set of four tires might last the car's owner 24 months. However, a few thoughtful users include the spare tire in their time-based maintenance, which includes tire rotation. A set of five instead of four tires now lasts them 30 months. Similarly, the most thoughtful users call their two pumps the "A" pump and the "B" pump. "A" pumps operate in January, March, May, etc. "B" pumps operate in February, April, June, etc. Each pump has the same probability of experiencing a random outage. The probability of both becoming unserviceable due to equal wear is statistically low. On centrifugal pumps, monthly switchovers keep operators well trained. Assets sitting there but not being used are still subject to atmospheric dust and water vapor ingestion. Unless blanketed with nitrogen or oil mist, nonoperating assets are still subject to corrosion. Accordingly, the advantages of periodic switchovers usually outweigh the disadvantages.

As so often, there is a finer point to be made here: Reliability thinking must reach into areas previously left to tradition or areas where it (seemingly) made little or no difference. Nevertheless, some operating practices deserve to be reevaluated in the context of fundamental asset reliability principles. Over the course of time, it is not uncommon for traditions and "urban legends" to displace more reliable explanations for process malfunctions. If the operating mentality is one of hazard acceptance and tolerance for repeat failures, then it absolutely must be replaced by a far more preventive approach. Inculcating and adhering to reliability principles is hugely important.

SPARE ASSETS

The potential for downtime or a serious production rate reduction during major unplanned maintenance leads to evaluating design options that allow taking part of the system offline without limiting output. Economics usually support purchasing and installing spare assets to offset the excessive business interruption losses that can result from an *unplanned* critical asset failure (if all maintenance events could be planned then there would be little need for spare assets). Service requirements and life cycle expectations will determine if an installed spare is appropriate. For example, it would likely not be practical for a process vessel with a 40-year life expectancy to have an installed spare, whereas a centrifugal pump with a 4-year life expectancy might be a good candidate for an installed spare, depending on its criticality. Compared to a fixed (stationary) asset like a vessel, rotating machinery is typically exposed to more failure mechanisms during its useful life. Additionally, the P-F interval for rotating machinery is substantially less than what might be expected for fixed assets such as vessels, storage tanks, and heat exchangers. This gives less time to develop a comprehensive maintenance plan before a failure on rotating machinery would be expected to occur.

When considering the installation of spare assets it is important to remember that any additional unused capacity provided by an installed spare in standby mode is reserved for maintenance purposes only. In other words, operational discipline prohibits using an installed spare to supplement production above what the primary asset can provide. More importantly, the installed spare cannot be used to compensate for degraded performance experienced with the primary asset. The installed spare can only take the place of a corresponding asset that has been isolated for maintenance.

In reference to the parallel pump example we discussed earlier, continuously operating a spare pump while also operating the primary asset violates this reliability principle. The spare's sole purpose is to provide uninterrupted service upon removing the other pump from operation. Nevertheless, an available spare must be run periodically to prevent its mechanical componentry from deteriorating to a point where both primary and spare assets must be repaired at the same time. However, our wear-out concerns apply mainly to the hydraulic end of the pump. As to the mechanical end, think of a sandblasting crew preparing both pumps for repainting. Imagine what could happen if a careless crew blasted highly abrasive grit into all bearing housings in sight. You will soon have multiple bearing failures on the same day.

Remember again our discussion of pump switching matters. We do advocate speaking of (and so labeling) the "A" pump set and the "B" pump set, as described earlier. Only in case of both pumps failing would production cease and the spare's purpose would be defeated. Again, abiding by this principle requires demonstrating operational discipline when the temptation to consume unused availability arises. Changing process conditions will always provide an incentive to operate spare assets contrary to the design purpose for which they are intended. In such cases we must again recognize the reliability principles that apply to installed spares, and exercise operational discipline knowing that this is the safest approach to long-term productivity.

Many of these concepts are integrated into the early phases of process design to provide the most reliable and economical process configuration option. Since spare assets are installed to maintain production when a major asset is shut down and isolated for maintenance, it is most logical to provide an installed spare where an asset failure would give no choice other than to interrupt the manufacturing process. The most conservative approach would be to equip all assets with an available spare, but doing so is unrealistic. Neither is it necessary when proper attention is given to how the process can be designed to offer the most flexibility if faced with a major unscheduled repair. A practical example (Fig. 2.8) demonstrates the thought process behind installing spare assets where it makes the most sense.

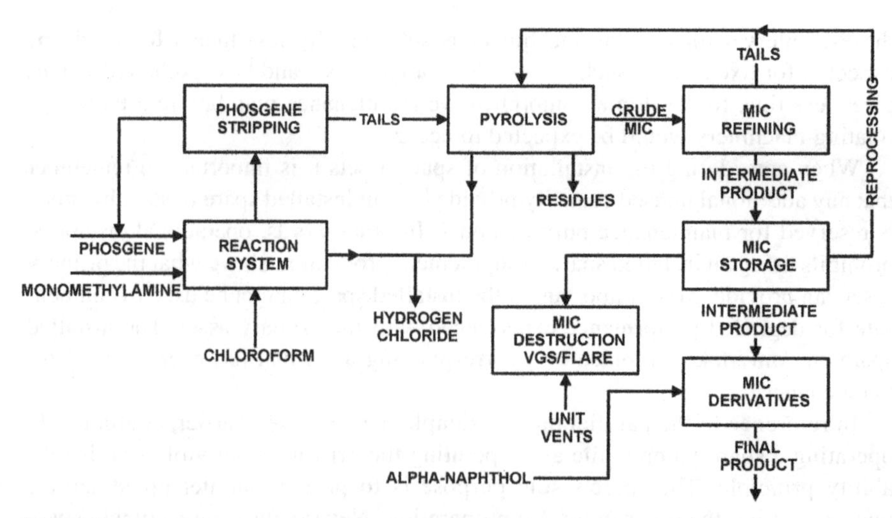

FIGURE 2.8

Bhopal factory process configuration [8].

This brief introduction to asset reliability principles provides a basis to assess the functional layout of the process involved in the Bhopal disaster. The three distinct process sections in the Bhopal factory that apply to our analysis include:

1. *Synthesis.* The synthesis section contained various assets specialized in the production of an intermediate product (methyl isocyanate, or MIC) used in the reaction to manufacture the final pesticide product beginning with indigenous raw materials.
2. *Storage.* The storage area consisted of two rundown tanks (Fig. 1.1) that fed the derivatives section with purified intermediate product collected from the synthesis section. There was also a temporary reserve tank that could be used to process off-spec (impure) intermediate product and provide additional process cooling in the event of an emergency [10].
3. *Derivatives.* Intermediate product from the storage area was transferred into the derivatives section where it was reacted with another essential ingredient to manufacture the final product, which could then be diluted (formulated) into different grades and sold to customers.

Examining this process more closely provides insight into the thought processes behind preparing a system for major maintenance with the least production consequences. Notice that the storage area both sends and receives intermediate product without contributing to its synthesis from raw materials or conversion into final product in any way. It is simply a collection spot where the intermediate product temporarily resides before being transferred into the final reaction process in the derivatives section. In this specific location, the storage area separates the intermediate production assets in the synthesis section from the manufacturing assets in the derivatives section. There is no way to get intermediate product from the synthesis section into the derivatives section without passing through the storage area.

Connecting the synthesis section directly to the derivatives section would significantly raise the reliability requirements for synthesis section assets. Depending on actual reliability performance, this process configuration could seriously limit factory production, possibly on a rather frequent basis. A more flexible design option was selected. Extra storage volume was inserted to provide a function whereby intermediate product inventory could be built up to satisfy customer orders without having to operate the synthesis section. This made it possible for the synthesis section to be taken out of service, if necessary, for major maintenance and with little consequence to factory production. This design option provided a degree of tolerance for unplanned maintenance in the synthesis section, much like an installed spare would. Perhaps operating experience factored into the decision to design the process with rundown tanks between the synthesis and derivatives sections. This would make sense since this specific design detail was copied directly from the original domestic factory in the United States [11].

The placement of a storage area between the synthesis and derivatives sections reflects careful thought about the best way to manage asset reliability in a factory operating in a remote part of the industrialized world. Collectively, it would be fair to say that lead times for certain specialty components (replacement parts, etc.) in a remote a geographic location could exceed lead times in more densely industrialized parts of the world. In this process configuration, the synthesis section could be taken out of service for major maintenance for an extended period of time without curtailing production. However, a sufficient volume of product would have to remain in the rundown tanks for the duration of any maintenance event that required an involuntary synthesis section shutdown. If the rundown tanks in the storage area ran empty while the synthesis section was isolated for maintenance, then there would likely be a business interruption. At this point there would be no intermediate product available to feed the derivatives section, and customer orders might not be filled. This could cause substantial financial losses as well as unhappy customers who might start looking for an alternative supplier.

The foregoing describes a preferable design option considering the differences between assets located in the synthesis section and storage area. Compared to the specialized, complex assets in the synthesis section, maintenance requirements in the storage area were confined to common, simple industry assets and their related failure mechanisms. Therefore, installing spare assets in the storage area represented a practical design choice; the selection served three different purposes. In the 1980s, this design choice incorporated considerations which are still relevant to modern industry today and will be in the future:

- Process design simplification and optimization,
- Business interruption prevention, and
- Allowing major maintenance to take place safely and with minor consequence to factory productivity.

However, the selected design option significantly increased the criticality of reliable asset performance in the storage area. In this configuration, the assets in the storage area would have to be adequately covered by installed spares to maintain continuous production in the derivatives section. This shifts the primary reliability focus onto a relatively simple process where maintenance could be managed with more ease. Should either rundown tank or one of its auxiliary assets need to be shut down unexpectedly, production could be easily switched over to the spare (standby) rundown tank and its auxiliary assets. Necessary maintenance could then be completed without interrupting final product manufacturing operations. The storage area was the critical link for continuous production in this process configuration. Its designated function shifted a considerable amount of the reliability concern away from the more sophisticated and proprietary assets in the synthesis section.

Based on this process configuration, it is reasonable to conclude that the Bhopal factory was intended for continuous production. Indeed, the factory was designed with the forethought of preventing an upset in the synthesis section that could curtail production rates. As long as adequate inventory remained inside the rundown tanks, operating the synthesis section was unnecessary. The factory would still be in a position to maintain targeted production rates. In this respect, the factory design embodied operating flexibility.

This strategy corresponds to the reliability plan discussed earlier in this chapter. When downtime for maintenance is needed, a good plan provides for maintaining product distribution by storing up excess inventory in advance. The function provided by storage area assets satisfied this purpose. For this plan to be effective, however, both rundown tanks could not run empty. Neither could major maintenance on both rundown tanks be overlapping at the same time. There was no way to bypass the rundown tanks and send product from the synthesis section directly into the derivatives section. All the manufacturing availability in the world does no good unless the assets are there to support utilization. Otherwise, a completely capable system (the synthesis section) could sit idle while production ceased due to loss of product feed from the storage area.

RELIABILITY PERFORMANCE METRICS

Asset maintenance requirements will determine whether or not an asset will be available when needed for production and if an available asset can be utilized. These two variables depend mostly on an asset's failure rate and the time it takes to complete a repair. Reliability performance metrics not only provide insight into how well a factory is utilizing its assets, but also how much human intervention is required to sustain the manufacturing process (Eq. [2.5]).

$$\text{Asset utilization rate (percent)} = \frac{\text{Actual Output}}{\text{Designated Output}} \times 100 \qquad [2.5]$$

Notice that the asset utilization rate takes into account the system's *designated* capacity. A system's designated capacity can be different from its design capacity. This is an important distinction to keep in mind when gauging a system's performance.

Table 2.3 Bhopal Factory Asset Utilization Rate

Year	Actual Output			
	Derivatives (MT)	**Utilization**	**Synthesis (MT)**	**Utilization**
1979	1468	29%	–	–
1980	1534	30%	374	19%
1981	2658	53%	864	43%
1982	2271	45%	623	31%
1983	1727	34%	535	27%
1984	1101	22%	313	16%

For example, the Bhopal factory's designated output capacity was lower than its design capacity for both final product (derivatives section) and intermediate product (synthesis section). The Bhopal factory was scaled to produce up to 6,500 MT of final product and 3,200 MT of intermediate products annually [12]. However, the designated capacity in this case was set by the factory's operating license, which limited production to 5000 metric tons (MT) of final product and 2000 MT of the intermediate product, annually [13]. A trending analysis of the Bhopal factory's asset utilization rate raises questions about asset reliability issues that limited production from the time that factory commissioning process began in 1979 (Table 2.3) [14].

Likewise, asset utilization trends are an important indicator for modern industry. Since the most economical outfit is best positioned to survive the behavior of a commodity market when a downturn is experienced, utilization is critical for the company that intends to remain on top. Utilization is solely dependent on the time it takes to repair an asset. Assets that are under repair cannot be utilized and therefore contribute nothing to production. This takes us back to our earlier discussion about the BEQ, by demonstrating how difficult it becomes to make up for production losses when asset reliability issues are hijacking the maintenance plan.

MEAN TIME TO REPAIR

When an asset is taken offline for major maintenance, a series of events must take place before it can be returned to service (Fig. 2.9) [15]. Each of these events requires time, and they are all human-dependent processes. Among them are:

- *Run-down.* This step involves coordinating the safe, stable shutdown of a major asset and possibly an entire operating section. This is a human-intensive period of operation that requires precise communication and a solid plan from which there can be no deviation. It is no surprise, then, that the potential to experience a catastrophic incident runs higher during the run-down period compared to normal operation [16]. Compare it to the aviation industry where accidents are more likely on take-offs and landings as compared to issues arising when in level flight at a prescribed altitude.

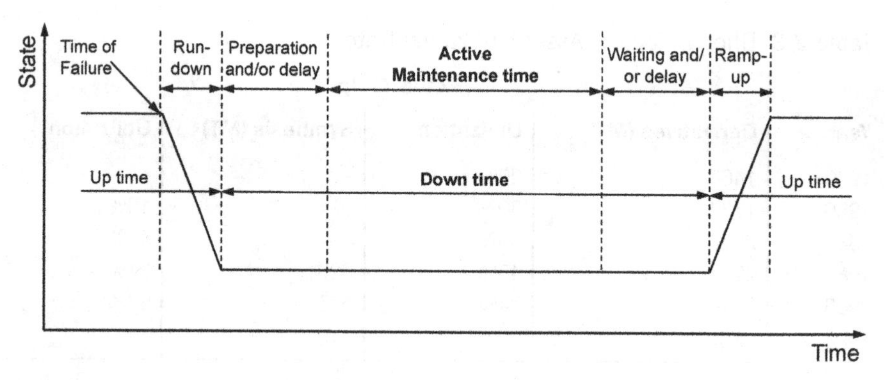

FIGURE 2.9

Different phases of a major repair cycle.

- *Preparation and/or delay.* Process isolation using conventional methods of lockout-tagout (LOTO) and blinding are applied during this phase. Depending on the complexity of the process, a considerable amount of time might be necessary to adequately isolate the process from the area where a repair is needed. Considering the complications that can arise from inadequate process isolation, this is time well spent. Asset purging, cleaning, draining, and depressuring all take place during this period. The objective for these safe work practices is to provide fellow workers with safe conditions while performing major invasive maintenance. Because this phase of operation requires opening flanges and bleeder valves, the potential for personal exposure to process hazards is elevated. Depending on the availability of replacement parts, skilled labor, and the details of needed repairs, this period can be drawn out over days to weeks, or longer.
- *Active maintenance time.* Repairing the asset can become quite time-consuming, especially if rework is required due to quality defects discovered only after the restart attempt fails.
- *Waiting and/or delay.* Once again, hazards are elevated at this time. In most plants, this is the time shortly before restarting the process. Isolation relative to adjoining units is being removed and system checks are in progress; these checks verify that maintenance is complete and has been performed correctly. Any fugitive process that has not been effectively isolated can escape upon opening flanges to remove blinds, leading to personal injury and perhaps even a major process release.
- *Ramp-up.* After the green light is given to restart the process, there is normally a period of time that must pass before the asset is fully functional. This applies to both fixed assets and rotating machinery. Some fluid machines need to transition through a progressive warm-up period where a certain temperature rise over time must occur. Many drivers and some driven machines, if brought on too quickly, could experience spikes or "surging." Severe events of this nature can cause asset damage and shut the process down for further repairs. Again an automobile

analogy: Even a modern automobile could not be expected to survive a cold start followed by instantly redlining the engine and driving at maximum speed. The same truth applies to industrial processes.

The average time it takes to complete the repair cycle represents the mean time to repair, or MTTR [17] (Eq. [2.6]):

$$\text{MTTR} = \frac{\text{Total Maintenance Time}}{\text{Number of Repairs}} \quad [2.6]$$

Although installing spare assets might provide some tolerance for a less reliable process, the emphasis on a "timely repair" does not diminish. When a primary or spare asset is out of service for maintenance, system operability becomes limited to the available asset, which is likely being utilized to keep up with production. The longer a spare remains unavailable, the more probability there is for an unpleasant business interruption. All it takes for such interruptions is a failure of the now nonspared asset to occur. Therefore, time is of the essence. The goal is to complete the repair cycle as efficiently as possible. This is to avoid an embarrassing situation if a spared asset should fail while the corresponding asset is down for maintenance. For this reason, spared assets really have no influence over the MTTR or failure rate. A spare asset simply prevents an unplanned business interruption if a major offline repair is needed.

FAILURE RATE

Despite the fact that long, drawn-out repair cycles can be a major inconvenience, they provide learning opportunities that progressive manufacturing organizations both cherish and use to their advantage. In a sense, the price paid for the failure and repair can be offset by using the experience to prevent similar failures on other assets. The purpose is to reduce the failure rate (Eq. [2.7]) to an acceptable level. An "acceptable level" could be defined as one that minimizes variable costs and prevents a repeat failure that could produce even worse consequences:

$$\text{Failure Rate } (\lambda) = \frac{\text{Number of Failures}}{\text{Period of Time}} \quad [2.7]$$

Reliability (R) is a function of an asset's failure rate (λ) and the time needed for it to remain functional (t). Having this information is essential for maintenance planning that depends on managing assets into a prearranged TAR cycle. The prearranged cycle is aimed at preventing business interruptions or production losses that can raise the BEQ. The probability for an asset to remain in service over a specified length of time can be estimated by Eq. [2.8]:

$$R_t = \mathrm{e}^{-\lambda t} \quad [2.8]$$

where, e is the natural logarithm base (2.718281828); t is the length of time that the asset would be expected to provide reliable service; λ is the asset's actual failure rate.

So, in a factory with a 4-year TAR interval and a pump that fails once every 7 years ($\lambda=0.1429$), the probability to work it into the scheduled TAR period is 56%. This result means that the odds for reliable service until the scheduled maintenance event are in our favor. One hundred percent confidence is never possible. The factory's bet would likely be that the maintenance schedule is realistic and there would be no reason to expect a premature shutdown as long as the required PM and routine surveillance programs are properly applied.

Compare that to a pump that fails maybe once every 9 months ($\lambda=0.75$). The probability for that pump to last long enough for scheduled maintenance drops to 5%. These odds are against operating the pump for a sufficient length of time to plan for a repair during a TAR running on a 4-year cycle. There are several ways to address this situation:

1. Modify the TAR interval to accommodate the pump's actual performance. This would involve shutting the factory down for major maintenance regardless if other more reliable assets need it or not.
2. Diagnose and correct the cause for asset reliability that does not correspond with planned maintenance on a more normal frequency. This would require conducting an investigation, applying a solution, and monitoring performance to validate success.
3. Accept the performance of the asset as it exists. This approach is surprisingly common when an installed spare is available, although its impact on the BEQ can be enormous. However, there are more serious concerns that can develop when this approach is selected.

MEAN TIME BETWEEN FAILURE

Keeping in mind that repeat failures are commonly found in history's worst industrial disasters (see chapter: Industrial Learning Processes), a case study in machinery reliability demonstrates the seriousness of accepting poor asset reliability. On April 8, 2004 a fire and explosion occurred in an oil refinery operating in the south central part of the United States [18]. The release occurred during routine maintenance to replace a spare pump's mechanical seal. Two of the six injuries that resulted were critical, and involved serious burns. Property damage exceeded $13 million. Production losses were excessive due to the extended unplanned outage that followed the incident, which was needed to restore processing capabilities to the damaged assets.

The incident followed a series of maintenance events involving a history of pump seal failures (Table 2.4). The cause for these failures had not been addressed. Instead, the problem was being managed by using the installed spare to perform frequent offline repairs (breakdown maintenance) [19]. In the 1-year period leading up to the incident, 23 work orders had been written in response to seal leaks at this pump location.

Table 2.4 Work Order History for Process Pumps

Date	Pump	Problem
April 17, 2003	P-5A (electric)	Seal leak
May 9, 2003	P-5A (steam)	Pump spraying from seal
May 23, 2003	P-5A (electric)	Repair seal
June 9, 2003	P-5A (steam)	Repair seal
June 9, 2003	P-5A (electric)	Repair seal
June 18, 2003	P-5A (electric)	Repair seal
June 20, 2003	P-5A (electric)	Replaced seal
July 31, 2003	P-5A (electric)	Replaced seal
August 22, 2003	P-5A (steam)	Seal leak
August 25, 2003	P-5A (steam)	Replaced seal
September 26, 2003	P-5A (steam)	Replaced seal
September 26, 2003	P-5A (electric)	Replaced seal
October 14, 2003	P-5A (electric)	Seal leak
December 6, 2003	P-5A (electric)	Replaced seal
December 9, 2003	P-5A (steam)	Seal leak
December 9, 2003	P-5A (electric)	Seal leak
December 15, 2003	P-5A (steam)	Replaced seal
December 15, 2003	P-5A (electric)	Seal leak
January 28, 2004	P-5A (electric)	Seal leak
March 22, 2004	P-5A (electric)	Seal leak
April 1, 2004	P-5A (electric)	Pump seal leaking
April 3, 2004	P-5A (electric)	Pump seal leaking
April 7, 2004	P-5A (electric)	Repair pump seal

Based on these reliability statistics, the mean time between failure (MTBF) for seal failures at this pump location alone was 0.087 years (Eq. [2.9]) compared to a modest but acceptable industry average of 5.5 years [20]. This reliability management plan proved nearly fatal after an isolation defect resulted in the catastrophic release of hot, flammable process material. It is important to note that the incident occurred not during the course of normal operation, but when mechanics attempted to remove the pump from its field position for maintenance. Here is the equation:

$$\text{MTBF}\ (\theta) = \frac{T}{r} \qquad [2.9]$$

where, T is the total time in service (sum for all assets in a group); r is the total number of failures (sum for all assets in a group).

The isolation defect that led to the release involved the improper operation of a block valve that was supposed to be locked in the *closed* position in preparation for safe invasive maintenance (Fig. 2.10). Instead, the valve was found locked *open* after

FIGURE 2.10

Pump isolation valve handle and position indicator as found after incident.

Photo credit: U.S. Chemical Safety Board.

the incident. A gear-operated actuator had been originally provided to open and close the valve. However, the design was modified by replacing the actuator with a 2-ft long removable pipe, which was inserted into holes drilled into the valve stem. In a position perpendicular to the flow through the pump suction line, the valve appeared to be closed. However, looking more closely at Fig. 2.10, a position indicator on the valve itself showed that the valve was still open. In this position, it was unable to block process flow that was stagnating in the line under 150psig pressure at a temperature of 350°F. The release occurred when the mechanics separated the pump case flange to remove the pump rear assembly, thinking that the process was completely isolated.

THE HUMAN CONDITION

The diversity of opinions that can be expressed over the causes for this incident is amazing. Many would argue that inadequate training is what led to the incident. The operator who made the mistake should have been familiar with the proper way to operate the valve and should have used the flow indicator instead of the valve handle, which had been added some time earlier upon the original actuator's removal. Unfortunately, this training works only until you run across a valve position indicator that defies this sensible rule (Fig. 2.11). At that point, your training sets you up for your next mistake.

FIGURE 2.11

Valve position indicator perpendicular to actual flow through pipe (note embossing).

Others might find fault with a modification involved in the valve actuator's design, or the improvised approach to operating the pump suction isolation valve. Indeed, this shifts the blame from the operator who isolated the process over to anonymous people who had not considered the hazards of someone mistaking the valve for being closed on the basis of the handle's orientation relative to the direction of process flow through the pipe. Perhaps it was a management of change issue, which again implicates a group of individuals who should have done a better job at some previous time when the change was made. However, it is safe to conclude that problems with the actuator are what likely led to its removal, and thus the observed change. The original gear-operated actuator may have seemed to take "forever" to rotate only a quarter turn on the valve, depending on the actual gear ratio that would make the valve easier to operate. Turning it twice (closed then opened) every 4 years is one thing, but having to turn it 46 times a year, which was the frequency when the incident occurred, would represent a major inconvenience. This situation would have provided the incentive to modify the valve actuator design to something more practical. The design modification could simply have been a matter of convenience. After all, why tolerate the inconvenience of constantly turning a valve wheel hundreds of times, perhaps for 20 min or more, when all that is really needed is a partial (quarter) turn of the valve stem? Indeed, the suction block valve should not need to be manipulated at all except for pump maintenance, which had become quite frequent. Therefore, there are possibly *two* incidents that are directly related in this one event:

1. Valve operability issues that led to an improvised design solution, and
2. Improper valve operation by the person confused by the improvised design modification.

In going about addressing this situation you might recommend:

- Conducting more training on the proper use of modified valves (pointing out that exceptions apply), and
- Disciplining those responsible for modifying the original actuator design (perhaps after having initially rewarded them for coming up with a creative, zero-cost workaround solution to a major maintenance inconvenience).

There is, however, a deeper issue that ties the entire incident together. It is obvious to many that the system's maintenance frequency directly contributed to the potential for the release [21]. To understand the human factors involved in the incident, consideration must be given to the functional capacity of industry's most versatile, adaptable, and multifunctional asset: The worker who is responsible for safe maintenance job execution.

There is a direct relationship between MTBF, MTTR, and individual workload (Fig. 2.12). As discussed earlier, every maintenance event is a multistage, completely human-dependent process. Factories are designed with an assumed level of reliability that might be falsely higher than its actual reliability performance. The example we just discussed may be more common than you think. Indeed, you will frequently encounter situations where the process is designed with the forethought of how to manage an unexpected failure with installed spares, to lessen the severity of business consequences. In addition, some failures are considered minor, or inconsequential, as long as they can be addressed without interrupting factory output. All of this adds up to excessive time that the asset is unavailable.

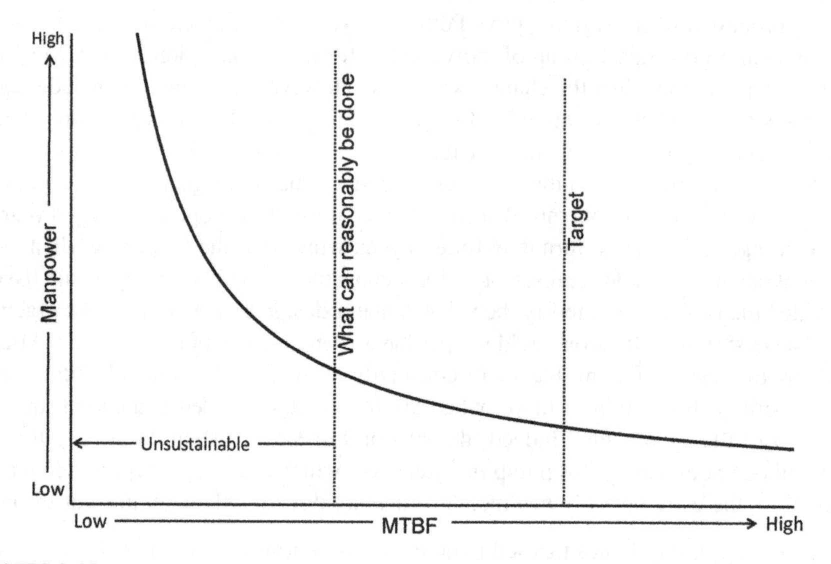

FIGURE 2.12

Human output needed to support production. *MTBF*, mean time between failure.

Notice that the combination of MTBF and MTTR becomes exponentially large as the MTBF decreases. Since maintenance cycles are human dependent, this translates into rising reliance on human intervention to manage the process. Like any other asset, there is a limit to what the worker is capable of doing. When that limit is exceeded, something will break. Just like the assets we encounter throughout the factory where we work, it may not happen instantly but it *will* happen if not corrected. When it does happen, it is usually in the form of the massive release of process energy—an industrial disaster.

Improvising an asset "solution" in the manner described in the foregoing case study represents a consistent pattern in industrial disasters. When the limits of human utilization capacity are exceeded, the work still gets done but by alternative and "more efficient" means. Such instances create an incentive to implement shortcuts and workaround solutions. It is for that reason, after a catastrophic process release, an improvised approach to doing routine work can usually be found, and the system has likely been operating in a manner completely contrary to its design, perhaps for an extended period of time. The source for such regrettable changes always involves the reliability of the process where the incident occurred. The incentive to modify original asset design to make its operation more reliable and efficient comes down to how fast the assets around you are failing (MTBF) and how long it takes for you to get the assets working again (MTTR). With these thoughts in mind it becomes possible to identify the most probable sequence of events that led to the catastrophic process release, including the source of the problem:

1. Frequent maintenance created a major inconvenience in the work involved with opening and closing a quarter-turn process block valve.
2. A major maintenance inconvenience led to an improvised design solution.
3. The improvised design solution led to confusion over how to operate the valve.
4. Confusion over how to operate the valve led to an operator error.
5. Operator error led to a catastrophic process release.
6. The catastrophic process release led to six injuries.

In light of this explanation, there is no substitute for diagnosing the causes for asset failures that represent a production constraint, and maintaining the asset's design basis with respect to its MTBF. Repeat failures are the root cause for industrial disasters. Asset reliability is the substance of industrial disaster prevention.

LESSONS FOR US

In this chapter we examined how consistently applying the basic principles of asset reliability provides the most satisfying approach to continuous process control. Consistently applying the basic principles of asset reliability means that we must diagnose the causes for deteriorating asset performance. It also requires demonstrating that we are aware of deviations from a maintenance plan that can result in safety, environmental, asset, and economic damage. Other important messages to retain from this discussion are:

- All assets, regardless of their designated function and service that they operate in, perform according to basic reliability principles. They are all equal in this respect.
- Overlapping functions between multiple assets can make it difficult to modify an asset's designated function without affecting the balance of the entire process.
- Factory production capacity is a function of variable costs.
- The most reliable manufacturer is also the most productive manufacturer.
- The specific function of every asset is limited by its design, but all assets are expected to contain the process.
- Capacity limits do not necessarily represent the most that an asset can do.
- Offline maintenance can take place without shutting down production when an installed spare is available.
- Installed spares make it easier to tolerate repeat failures.
- Running assets within their designated limits requires exercising operational discipline.
- Storing up product inventory for a production interruption allows a business to retain its customer base when the process is shut down for maintenance.
- Major releases are perhaps more likely to occur during maintenance, when the process is believed to be isolated.
- Improvised design solutions end with tragic human error.

REFERENCES

[1] B.S. Dhillon, H. Reiche, Reliability and Maintainability Management, 72, 1985, ISBN: 0-442-27637-0.

[2] T. D'Silva, The Black Box of Bhopal, 195, 2006, ISBN: 978-1-4120-8412-3.

[3] J. Moubray, Reliability-Centered Maintenance, second ed., 1997. ISBN:978-0-8311-3164-3, 23.

[4] J. Moubray, Reliability-Centered Maintenance, second ed., 1997. ISBN:978-0-8311-3164-3, 132.

[5] J. Moubray, Reliability-Centered Maintenance, second ed., 1997. ISBN:978-0-8311-3164-3, 140.

[6] H. Bloch, F. Geitner, Machinery Failure Analysis and Troubleshooting, second ed., 1994, ISBN: 0-87201-232-8, 4.

[7] J. Moubray, Reliability-Centered Maintenance, second ed., 1997. ISBN:978-0-8311-3164-3, 145.

[8] Union Carbide Corporation, Bhopal MIC Incident Investigation Team Report, March 1985. 4.

[9] American Society of Mechanical Engineers (ASME), ASME Boiler and Pressure Vessel Code Section VIII, Division I, Paragraph UG-27, 18, 2004.

[10] Union Carbide Corporation, Bhopal MIC Incident Investigation Team Report, March 1985. 5, 9.

[11] W. Worthy, Methyl isocyanate: the chemistry of a hazard, Chem. Eng. News 63 (6) (1985) 29.
[12] T. D'Silva, The Black Box of Bhopal, 205, 2006, ISBN: 978-1-4120-8412-3.
[13] T. D'Silva, The Black Box of Bhopal, 193, 2006, ISBN: 978-1-4120-8412-3.
[14] National Environmental Engineering Research Institute (NEERI), Assessment and Remediation of Hazardous Waste Contaminated Areas in and around M/s Union Carbide India Ltd, 4, 2010.
[15] The International Organization for Standardization, Petroleum, Petrochemical and Natural Gas Industries – Collection and Exchange of Reliability and Maintenance Data for Equipment, 22, 2006.
[16] The American Oil Company, Safe Ups and Downs for Refinery Units, revised second ed., 1963, 4.
[17] Department of Defense, MIL-STD-721-C: Definitions of Terms for Reliability and Maintainability, 7, 1981.
[18] U.S. Chemical Safety and Hazard Investigation Board, Case Study 2004-08-I-NM: Oil Refinery Fire and Explosion, 2005.
[19] U.S. Chemical Safety and Hazard Investigation Board, Case Study 2004-08-I-NM: Oil Refinery Fire and Explosion, 5, 2005.
[20] H. Bloch, Pump Wisdom: Problem Solving for Operators and Specialists, 184, 2011, ISBN: 978-1-118-04123-9.
[21] U.S. Chemical Safety and Hazard Investigation Board, Case Study 2004-08-I-NM: Oil Refinery Fire and Explosion, 11, 2005.

CHAPTER

Incident Investigation

3

An incident investigation is a reverse-engineering practice triggered by an event that does not meet our asset or process performance expectations. By definition, an incident investigation would not be needed unless all process hazards had received proper attention. When an incident occurs, there is no question that something is not working the way it was intended. The problem can be expected to persist until appropriate action is taken. The purpose for an incident investigation is to implement a practical solution that will effectively remove or control the hazards responsible for unacceptable process performance. The amount of protection that results from an incident investigation depends largely on how investigation triggers are defined. The level of protection we acquire from investigation triggers relates to our tolerance for unidentified process hazards.

Most people would agree that the actual or potential consequences of a catastrophic process release are sufficient to trigger an investigation. While some might consider investigating anything less than a catastrophic incident of questionable value, all genuine incident investigations have merit. After a serious incident has occurred, a diligent researcher will likely find a long history of less consequential incidents that preceded it [1]. It follows that demonstrating a tolerant attitude for these less consequential failures is a sign of the normalization of deviance.

THE NORMALIZATION OF DEVIANCE

In a pure sense, the normalization of deviance involves widening an asset's operating envelope through an unrestrained process of trial and error. An example in chapter "Equipment Reliability Principles" involved calculating the minimum thickness of a metal necessary to contain a process at a certain operating pressure. The example explained that industry's operating limits are set conservatively to include a safety margin sufficient to avoid immediate failure. It would not be appropriate to set operating limits so tight that a catastrophic failure would instantly occur upon reaching the predefined limit. For a limit to be effective and deserving of consideration for a *safe* operating limit, there must be sufficient time available to detect the start of a failure (point "P") before it actually occurs (point "F"). Increasing the P–F interval makes more time available to coordinate an effective plan to maintain full control over the process. The P–F interval may be so short that only an automatic safety instrumented system can realistically provide a fast enough response. For the outcome to be satisfactory, an appropriate action must be taken *before* the functional failure occurs. This "before action" is mandatory in all instances.

Figuratively speaking, the normalization of deviance represents an interlock bypass for the human conscience, allowing an abundance of warning signs to be dismissed until finally there is no time to react. Warning signs that could have been recognized and acted upon in advance offer no value. Everything leading up to a catastrophic failure appears to be normal operation to anyone with jurisdiction over the process. When this normalization of deviance is implanted, only a catastrophic failure of extraordinary magnitude can offer sufficient motivation to reset the operating limits to more protective values. Only with this reset can the P–F interval be expected to provide sufficient reaction time to avert a disaster.

Unfortunately, this trial and error approach to failure avoidance does not correspond to, or align with, responsible process operation. To make matters worse, the normalization of deviance occurs over time and quite gradually. With time, normalization of deviance is virtually impossible to detect by those who are affected by it. It is very difficult to admit that the normalization of deviance is obscuring hazards that really cannot be tolerated. Only an incident of extraordinary significance can expose the error of unconditional hazard acceptance produced by the normalization of deviance.

The normalization of deviance concept was introduced by a technological failure that started with a conservative limit. As time went on, the limit kept getting progressively less restrictive as operating experience was applied [2]. The absolute limit was found when a catastrophic failure occurred, which resulted in seven fatalities and the destruction of a multibillion dollar asset. The event contributed to industry becoming ever more appreciative of pursuing ways to eliminate, and not just control, hazards. In retrospect many, if not all, other technological (industrial) disasters involve normalization of deviance to some degree; perhaps to the extreme. Moving forward from this realization, detecting and preventing such normalization became a priority.

As we ponder an industrial incident of historic significance, we acknowledge that limits can be adjusted to operate assets more responsibly in the future. However, it is never possible to get back what has been lost. For good reason, industry therefore advocates establishing conservative limits before it is too late. The practice of setting conservative limits provides an opportunity for principled decision-making based on numerous inputs that affect safety, the environment, asset value, and economics. The ultimate purpose for setting operating limits is to prevent any incident that might suddenly shut down the process for mandatory unplanned maintenance, or worse. Setting operating limits in advance is perhaps the most effective way to resist losing control through the normalization of deviance.

The normalization of deviance is a human-dependent factor that typically adds complexity to an incident investigation. Still, there are many other nonmechanical causes or contributing factors that can interact to make an incident investigation difficult. Rarely do any of these circumstances originate as a deliberate attempt to distort the facts or mislead an investigation. However, a disorderly incident investigation approach makes it virtually impossible to distinguish fact from fiction.

This results in confusion that only fans the flames of controversy that can lead to ineffective recommendations or diverging theories to explain what might have happened. Much of this controversy can be avoided by implementing a methodical and repeatable incident investigation procedure. Deviating from the investigation procedure makes it impossible to separate important from useless information. Straying away from the plan can make a relatively simple stepwise approach infinitely more difficult, confusing, and frustrating. It is no wonder, then, why many dread the prospect of an incident investigation after a serious incident occurs. Never would we want to participate in unproductive investigation practices that defeat the objectives of process safety management (PSM). In most cases, the facts and evidence needed to piece together a credible account of what really happened are there to be found. These facts and evidence are not always readily apparent. A measure of hidden or obscured information can always be expected at the beginning of an investigation. Therefore, a systematic approach is needed to bring the important information to the top. Only a systematic pursuit makes an investigation both efficient and productive.

An industrial incident investigation cannot become an unrestrained brainstorming project. Imagine the time that can be wasted if someone contends that accepted science is not relevant in diagnosing the root causes for process failures, or that our assets fail for reasons deeply buried in ancient mythology—reasons that defy anything we have ever learned in our technical courses. Therefore, it is vitally important to agree that in our industrial world, scientific cause and effect relationships govern both asset and process performance.

INVESTIGATION TRIGGERS

An effective set of investigation triggers is needed for local sites to initiate an effective investigation. Therefore, a successful investigation starts by defining exactly what type of event represents an incident. Surely a fire or an explosion or even a "near miss" involving either of these could be counted as an incident. Maybe an incident that generates business interruption losses in excess of a specified amount would qualify for an investigation. One might also consider a relatively inconsequential failure to be an incident if it occurred twice in a 12-month period. Such recurring incidents would create a distraction, to say the least. With quite obviously a multitude of different options for setting investigation triggers, where should the line be drawn?

How we define an incident worthy of a purposeful examination represents a balance between the learning opportunity and the amount of effort that is typically involved in driving an unresolved technical problem to a satisfactory solution. A cohesive, systematic, reliable, and structured approach to problem solving improves both the efficiency and accuracy of an incident investigation. Consider again, please, the undesirable alternative to this. Although an event might have triggered an incident investigation, it is not productive to wander aimlessly through an investigation while wasting precious time and confusing matters even further. The preferred and obvious choice is to implement a reliable, repeatable, and structured investigation method. A properly conducted, comprehensive incident investigation greatly facilitates deriving the most appropriate future failure avoidance strategy. Such value-adding incident investigations provide the confidence needed to assign effective recommendations. Upon completion or implementation, a conscientious and structured approach will start paying back; it will unfailingly produce measurable performance improvements that prevent recurring failures.

Risk is a function of likelihood and consequence (Eq. [3.1]). Incident triggers that are based on the *likelihood* for a process or asset malfunction are infinitely more protective (risk-reducing) than their *consequence* counterparts. For example, most companies in the manufacturing industry define their incident tolerance levels according to consequence thresholds. They might assign a maximum amount of safety, environmental, asset, and economic damages as their formal investigation triggers. This common practice makes it extremely difficult to find the right balance between the learning opportunity and the effort needed to investigate an incident. Setting the threshold too high favors normalization of deviance, where being aware of an actual hazard, or understanding its true significance, is delayed until the pain becomes severe enough to make us take action. Think this through, however; deviance having been normalized makes it now too late to *prevent* the losses that could have been saved if we had been more sensitive to the hazard. By the same token, setting the tolerance threshold too low also favors the normalization of deviance. In this case, the distinction between incidents of great and much lesser significance is blurred by distractions that result from "investigation overload." An overload situation involves devoting the majority of available investigation time and resources to less valuable learning opportunities.

$$\text{Risk} = \text{Likelihood} \times \text{Consequence} \quad [3.1]$$

Progressive manufacturing organizations quite evidently channel their time and resources into detecting, evaluating, and addressing hazards that could potentially lead to a catastrophic process release. We note that these failure-resistant organizations limit their attention to three major performance areas that contain the most commonly missed warning signals for incidents of major consequences (Table 3.1). One of these investigation triggers (repeat failures) has already been covered in fine detail in the previous chapters. The two remaining investigation triggers are less familiar and therefore deserve further explanation:

1. **Production constraint**. Every process and asset operating within that process has its limits. A functional failure can be anticipated when operating outside of these limits. We formalize those limits to avoid operating the process outside of mandatory safety boundaries, where a failure can be expected. A limit is a type of policy that represents our balance between risk (Eq. [3.1]) and reward. We honor operating limits because we know that without them, we would damage our assets beyond the incremental production value that could be gained by running them irresponsibly. It would be a bad economic business decision. There is nothing wrong with voluntarily establishing operating limits to protect us

Table 3.1 Preventive Investigation Triggers

Performance Category	Investigation Trigger	Examples (Not All Inclusive)
Production	Production constraint	• Any unplanned business interruption caused by a major process upset or asset failure • Process or utility system foaming • Maximum or minimum automatic control valve output (loss of process control) • Carryover or contamination
Maintenance and operations	Workaround solution	• Acid injection into cooling water inlet of heat exchangers • Water washing turbine blades • Adding steam to compensate for loss of heater function • Bypassing a malfunctioning control valve
Reliability	Repeat failure	• Pump or compressor mechanical seal replacements • Pump or compressor bearing replacements • Asset replacement due to aggressive and unrestrained corrosion or cracking mechanisms • Hot spots (with or without refractory failures)

from making an uneconomical and potentially unsafe business decision. As we discussed earlier, operating limits are not only acceptable, but desirable.

Production constraints, in contrast, are not to be confused with operating limits. Production constraints are involuntary limits that result from premature failures or accelerated asset performance losses beyond what should reasonably be expected. The protection we establish by purposefully implementing operating limits can be defeated by a prevailing production constraint. These involuntary, and extraordinarily penalizing, production restrictions are a source of great frustration that create an incentive for improvised design solutions, as will be discussed in the next few pages. Production constraints differ from limits, in that they frequently originate with hazards that are potentially dangerous, expensive, and therefore completely unacceptable. Limits, on the other hand, are always protective.

2. **Workaround solution**. This is a special category of unauthorized procedures or practices that are neither formally recognized nor recommended by professional reliability organizations or maintenance institutions. In every case involving a workaround solution, a specific maintenance or operating function is provided that contradicts approved industry standards, specifications, codes, policies, or recognized and generally accepted good engineering practices (RAGAGEP) related to the process. Each case involves a persistent failure mechanism that can be managed without shutting down the process for safe maintenance. At first, there is heightened sensitivity to what all recognize as being a direct violation of operating mentality that has been reinforced perhaps over a significant amount of time. There may even be a cultural resistance to performing this type

of procedure, which is the preferred impulse to recognize and preserve. But over time a dependency develops over these types of unconventional maintenance or operating practices to maintain production while avoiding the inconvenience of recovering lost process functions by shutting down manufacturing operations. In all cases, however, there is a damage mechanism that usually decreases the mean time between failure (chapter: Equipment Reliability Principles), which results in more frequent asset and component replacements at a minimum. As a workaround solution becomes integrated with normal operation, there might even come a time when permanent spares are installed, so that damaged assets can be routinely swapped out or repaired without incurring lost production. When this happens, permanent system modifications are incorporated into factory design to accommodate unauthorized, and potentially seriously dangerous, maintenance and operating practices.

For many good reasons, failure-restrictive organizations resist triggering an investigation only according to an arbitrary level of penalty that might fluctuate between business sectors, local sites, and production units. They do not wish to discover that any of the precursors to more serious process instabilities were limiting system performance prior to investigating an incident that meets a certain consequence threshold value. Neither are they receptive to possibly promoting a fictitious impression of performance that meets life cycle expectations in exchange for implementing unconventional maintenance or operating practices that could insert an unknown defect. Failure-restrictive organizations support their internal learning process by triggering an incident investigation based on answering one simple question:

Is my process functioning according to its design expectations?

If the answer is "no" then an investigation is warranted. Triggering an investigation based on performance criteria obviously requires documenting those performance criteria in advance. Besides effectively preventing a disaster, activating an investigation based on process and asset performance criteria serves multiple other purposes. Reaching or consistently reestablishing design expectations brings many benefits (recall chapter: Equipment Reliability Principles). A good investigation procedure provides the accuracy needed to improve the value, efficiency, safety, and profitability of a factory or enterprise by the following:

1. **Reducing variable costs.** Performance-based warning signals (preventive investigation triggers) decrease variable costs. Therefore, the time and resources devoted to diagnosing and addressing the causes of warning signals are usually easy to justify. The resulting performance improvement essentially decreases the break even quantity (see chapter: Equipment Reliability Principles), which increases operating capacity and production value. When commodity competition is involved, controlling variable costs favors staying on top of the commodity market after the patents expire [3].
2. **Improving witness cooperation.** Witnesses or others who can support an investigation are usually more willing to participate before something catastrophic occurs.

They are less reluctant to take ownership when they are asked to support an investigation to *prevent* an incident rather than becoming involved only *after* an incident. There are many different incident investigation methods available to industry. Good investigations have in common that they are fact-finding, not fault-finding. However, the only practical way to convince your witness of not trying to find fault or place blame is by not having an incident where people might sense any personal involvement. This, of course, means that we must prevent an incident from happening in the first place. Once an incident has occurred, we will again realize what we have always suspected: people, by nature, are skeptical of the purpose for an investigation and are therefore less willing to cooperate.

3. **Protecting business autonomy and privacy**. To prevent facts from being distorted or misused for unproductive pursuits, companies operating in the manufacturing industry are generally advised to be selective with information shared publicly about incidents. Sadly, privacy in these matters interferes with industry's responsibility to protect others who might unknowingly have to cope with similar process upsets in the future. In law enforcement, if you are in possession of information that can prevent others from calamity but fail to act upon it, you are considered an "accessory." With this thought in mind it should be noted again that the best option for all involved is to investigate, evaluate, and communicate. The freedom to do this only results when an investigation is completed with sufficient time remaining to avoid a more serious incident that could be publicized for nefarious reasons. By definition, an investigation triggered by *exceeding* what one might consider to be an acceptable limit of pain (a consequence threshold value) cannot possibly prevent an extraordinarily painful experience. Under ideal circumstances, there would be no need for distributing lessons learned from industrial incidents throughout the global manufacturing industry. The incidents that create unacceptable pain for both industry and the community would simply not exist. Likewise, there would be no "accessories." Triggers would activate an investigation to resolve the cause for any warning signal detected under normal operating conditions.

CULTURE AND CONTEXT

The word "culture" means different things to different people. We must therefore be specific when using the term to describe principles and practices that could perhaps be conveyed in more precise language. Generalizations and/or all-encompassing terminologies are the opposite of "specific" or "detailed." Speaking about these matters with sweeping generalizations can inadvertently defer personal actions. When all is said and done, personal actions are needed to prevent industrial incidents. Generalizations are subject to interpretation. Generalizations might be confused, misinterpreted, or not implemented uniformly; thereby reinforcing cultural defects (such as the normalization of deviance) that promote repeat failures.

However, the use of the term "organizational culture" cannot be more specific when topics like the normalization of deviance and investigation triggers are being discussed. An organization infected by the normalization of deviance cannot detect hazards that could lead to catastrophic failures. In such cases, the true significance of the normalization of deviance is rarely appreciated before an industrial disaster occurs. In sharp contrast an organizational culture which values doing things right will use an incident investigation program to prevent the normalization of deviance.

We want to be sure to strongly make this point: A culture influenced by the normalization of deviance works (sometimes tacitly or ignorantly) in a direction that deemphasizes or even negates the reliability principles described in the two preceding chapters. Unfortunately, organizational culture varies between different companies that collectively operate within the manufacturing industry. Cultural differences often penetrate deep into the local sites operated by those companies, and even down to the different units operating within those local sites. Indeed, organizational culture has proven to be a complex and inconsistent subject with very few absolutes. For this reason, we will define organizational culture according to seven common attributes that apply universally to all organizations throughout the manufacturing industry (Table 3.2). This approach makes it possible to understand how an organization would likely respond to specific operating difficulties that it might encounter, including:

- Production constraints,
- Process upsets, and
- Unpredicted asset failures.

As we examine information collected as part of an investigation, we would want to be sensitive to what it tells us about a specific organizational culture. Does the information reflect a failure-restrictive or failure-permissive culture? Our ability to predict how an organization might respond to any operating difficulty depends on the prevailing attitude demonstrated by actions that were taken prior to the catastrophic incident. This cultural assessment can be made by matching observed actions to the following criteria:

1. **Uniformity in message**. Is safety truly "everybody's business" (failure-restrictive), or is safety only the business of the safety department (failure-permissive)?
2. **Hazard awareness**. Is interest shown in diagnosing, correcting, and keeping a distance from the causes for unexplained, chronic process phenomena (failure-restrictive), or are heroic recoveries from process upsets glorified, recognized, and rewarded (failure-permissive)?
3. **Operating limits**. Are limits set as conservatively as possible with long-term process stability and profit generation in mind (failure-restrictive), or is the objective to set limits as close to failure as possible to capture the short-term incentive (failure-permissive)?

Table 3.2 Attributes That Define an Organizational Culture

Attribute	Failure-Restrictive	Failure-Permissive
Uniformity in message	Constantly setting the example for what you expect from others while striving to eliminate any ambiguity.	Holding others accountable for demonstrating a higher standard of performance than you are able or willing to provide.
Hazard awareness	Detecting contradictory process performance and implementing an incident investigation to initiate the steps to address the specific hazard(s) responsible for it.	Acquiring valuable experience only by responding to failures worthy of examining more closely based on exceeding a consequence threshold.
Operating limits	Identifying, documenting, and controlling the process in accordance with process and mechanical limits implemented as policy to preserve asset reliability and process performance.	Generating formal or informal exceptions to operating limits, ignoring chronic deviations from established limits, or operating the process without any limits.
Improvisation	Refraining from using tools, assets, maintenance, and processing capacity for purposes contrary to their design functions.	Willingness to resort to unconventional solutions to provide a substitute function for tools, assets, and maintenance in process service.
Casual compliance	Complying with policies, procedures, and regulations at all times, regardless if you believe they are necessary or not.	Complying with policies, procedures, and regulations, and best-operating practices only when being watched or when it is convenient.
Tribal knowledge	Documenting and validating expectations, instructions, and operating requirements in policies, procedures, and best-operating practices.	Leaving certain critical operations up to a limited number of individuals who through experience have learned unconventional ways to make the process work.
Transparency	Freely sharing information you have with all individuals who might realistically be affected by it.	Maintaining confidentiality over information that others might be entitled to know about.

4. **Improvisation**. Are incident investigations launched after something unexpected is detected to correct the problem with a scientific solution (failure-restrictive), or are process problems addressed without restraint by modifying system configuration, design, operating limits, and/or procedures (failure-permissive)?
5. **Casual compliance**. Is rigid adherence to documented policies demonstrated (failure-restrictive), or has flexibility toward policies developed to adapt to temperamental process conditions (failure-permissive)?

6. **Tribal knowledge**. Are all those who might operate a process able to do so according to documented design expectations (failure-restrictive), or do only a handful of people "have the gift" of being able to make the process work (failure-permissive)?
7. **Transparency**. Is important information being shared with affected people even when they do not ask for it (failure-restrictive), or do people get information that they might be entitled to only when they specifically ask for it (failure-permissive)?

The secret to solving a problem based on facts and evidence is to know what to look for and where to find it. Using the above screening method to define organizational culture provides valuable insight into the basis for decisions or acts of omission and commission that facilitate a disaster. The purpose for making this assessment is to establish a precedent, or finding a consistent cultural basis for specific actions that appear in the sequence of events leading up to an industrial disaster. This prevents us from distorting or misinterpreting facts and evidence that could easily be taken out of context. By following this approach, a reasonable explanation for meaningless actions can usually be determined with considerable accuracy. While the technical details concerning various industrial incidents are rarely similar, the cultural causes for incidents are extraordinarily consistent. Even within other industries, identical cultural patterns or shortcomings are quite often found [4]. Experience proves that once an organizational culture is firmly established it becomes very difficult to change. A failure-permissive organizational culture cannot change unless the true causes for industrial incidents are diagnosed and eliminated. More often than not, making this change requires addressing at least one of the seven attributes of organizational culture as directed by investigation results.

Whatever the template might be for a prevailing culture, the incidents it produces seem to replicate as if they were cast in a mold. We see the replication even in instances after an incident investigation assigns recommendations to address specific cultural defects. The same defects are inherited by recurring failures of significant consequence.

To illustrate this investigation principle, let us first consider the history of cultural issues that interfered with the mechanical integrity of the Space Shuttle Program. On January 28, 1986 the space shuttle Challenger exploded 73 s into its 10th mission. It was a very public disaster and a presidential commission was formed to determine its causes. The final report listed technological and cultural defects among the primary causes for the disaster and assigned specific recommendations to correct them. In the final report's documentation of the space shuttle Challenger disaster we find [5]:

> *The decision to launch the Challenger was flawed. Those who made that decision were unaware of the recent history of problems concerning the O-rings and the joint and were unaware of the initial written recommendation of the contractor advising against the launch at temperatures below 53 degrees Fahrenheit and the continuing opposition of the engineers at Thiokol after the management reversed its position. They did not have a clear understanding of Rockwell's concern that it was not safe to launch because of ice on the pad. If the decision makers had known all of the facts, it is highly unlikely that they would have decided to launch 51-L on January 28, 1986.*

Within this information we can recognize at least three of the seven cultural attributes that contributed to the disaster:

1. *Hazard awareness* was defective in this case. It appears that those who approved the launch were not familiar with the hazards involved with a cold temperature launch nor were they aware of the worst thing that could happen if the O-ring failed to adequately seal the joint.
2. *Operating limits* were not formally established. The summary presents information about a low temperature limit of 53°F that existed in some, but not all, minds of the technical staff. The information also indicates that a recommendation to establish an official minimum temperature limit was rejected.
3. *Tribal knowledge* interfered with the communication of important historical information up the chain of command. Notice that the decision-makers were not aware of all the facts involved, which included a history of problems concerning the O-rings and the mechanical joint that the report suggests would likely have prevented the launch.

Compare these events and findings to the cultural factors that were involved in the space shuttle Columbia disaster 17 years later. The investigation report documenting this disaster [6] concluded:

> *Risk, uncertainty, and history came together when unprecedented circumstances arose prior to both accidents. For Challenger, the weather prediction for launch time the next day was for cold temperatures that were out of the engineering experience base. For Columbia, a large foam hit – also outside the experience base – was discovered after launch. For the first case, all the discussion was pre-launch; for the second, it was post-launch. This initial difference determined the shape these two decision sequences took, the number of people who had information about the problem, and the locations of the involved parties.*

Once again, problems with hazard awareness, operating limits, and tribal knowledge are described as they pertain to "a large foam hit" on the leading edge of a space shuttle's wing and the number of people who had information about the problem. The discovery process occurred in response to a failure of extraordinary magnitude. By considering the circumstances behind the first incident, we are able to accurately predict their involvement in the second. In all cases when an investigation is taking place, the same truth applies. Different pieces of information and evidence can be made more understandable when the context is brought into the picture.

Highly important parallel accounts were also observed in two industrial process settings. The similarities between the 2005 Refinery Explosion in Texas City, Texas, USA, and the 2010 Oil Spill (Deepwater Horizon offshore drilling platform disaster) in the Gulf of Mexico can also be explained according to their organizational culture attributes. In both cases, a worst case scenario was experienced within the same organization. Likewise, that same organization was in control of the process. In a study report completed for the Gulf oil spill [7] we read:

> *This disaster also has eerie similarities to the BP Texas City refinery disaster. These similarities include: a) multiple system operator malfunctions during a critical*

period in operations, b) not following required or accepted operations guidelines ("casual compliance"), c) neglected maintenance, d) instrumentation that either did not work properly or whose data interpretation gave false positives, e) inappropriate assessment and management of operations risks, f) multiple operations conducted at critical times with unanticipated interactions, g) inadequate communications between members of the operations groups, h) unawareness of risks, i) diversion of attention at critical times, j) a culture with incentives that provided increases in productivity without commensurate increases in protection, k) inappropriate cost and corner cutting, l) lack of appropriate selection and training of personnel, and m) improper management of change.

This description references perhaps all seven of the cultural attributes to which we would want to be alert when performing an investigation. Once again, our purpose is to define the deep-rooted connections that explain how an entire sequence of events unfolded from start to finish. Seeing these connections facilitates understanding how our actions play a vital role in preventing industrial incidents. But getting to that point involves much more than simply documenting the cultural defects that are involved in an incident. Placing the information into its proper context requires that we take the next step, which is finding the evidence we need to explain how those decisions were reached. Defining a problem's cultural attributes and ingredients simply helps us to remain neutral and open-minded with respect to assessing information we collect during an investigation. We need this kind of definition to fully understand what the collected information really means once the investigation is over so that universal lessons can be applied.

NEUTRALITY

Independence must envelop the entire investigation procedure from start to finish. Independence makes it possible for the investigation team to be objective. In practice, being objective means that premature judgment is withheld while facts and evidence are being collected. While we may have access to individual facts, we do not yet knew how these relate to adjoining facts. Therefore, formulating opinions in the initial phase of an investigation only fuels speculation and controversy. Clearly, such "fueling" serves no useful purpose in an incident investigation and only complicates matters as the sequence of events becomes more transparent. For an investigation to be successful, it cannot be directed by opinions, assumptions, theories, or conjectures. It must be directed by the facts and evidence alone as they come in, without any external interference. It also must never be allowed to contradict science.

By the same token, an objective investigation requires *acting* with neutrality toward people and toward bits of evidence. Special attention must be given to avoiding an activity that could even remotely be considered a conflict of interest. This is done by rejecting any relationship with external forces that could possibly influence, control, manipulate, or interfere with the investigation. Even traces of a conflict of interest can damage an investigation's credibility. At a minimum, only the perception of a conflict of interest is needed to create an unnecessary distraction that could undermine the integrity of a cohesive incident investigation.

Remaining neutral toward evidence during the initial stages of an investigation can be accomplished by treating each different group of evidence like a single "pixel" that when combined with the other forms of evidence creates a complete and very compelling picture. This pursuit requires isolating different types of evidence at the beginning of the investigation. For example, witness statements, physical evidence, and the timing of events are three distinct categories that must not be allowed to overlap during the initial part of an investigation. Different segments of the investigation will come together later, at the proper time. However, remaining neutral toward people-related as well as asset-related information requires separating different types of information during the initial period of evidence collection.

Without this separation, it becomes very difficult to avoid forming unsubstantiated opinions about the sequence of events that was involved in an incident. With this separation and by adhering strictly to the investigation procedure the sequence of events will, in due time, become remarkably clear. The entire picture will be developed by allowing the investigation to be directed by facts and evidence, and properly handling those facts and evidence as they are obtained. Acting contrary to this advice in an effort to save time would amount to taking a shortcut. Shortcuts lead to regrettable failures. With that in mind, ask a few introspective questions: What does the way you manage an incident investigation reveal about your organizational culture? In the areas of "uniformity in message" and "casual compliance" does your attitude represent failure-restrictive or failure-permissive conduct? Remember, your actions during the investigation and the results of your investigation always speak louder than words. Your goal is for the investigation to echo in value long after your departure.

THE INVESTIGATION PROCEDURE

A cohesive, comprehensive, repeatable, and structured incident investigation takes place according to a clearly defined procedure. The investigation team (consisting of two or more people) must agree to adhere to the procedure before the investigation gets underway. Interestingly, the selected investigation procedure is far less important than the investigation team's commitment to adhere to the selected approach. A well-rounded investigation procedure incorporates the basic elements required to accurately determine and resolve the probable cause for a failure. These basic elements include the following:

- Collecting information,
- Interacting with witnesses,
- Constructing a timeline,
- Resolving contradictions and inconsistencies, and
- Developing practical recommendations.

Experienced incident investigators recognize the value of being competent in a wide range of different investigation practices [8]. Being fluent in multiple investigation practices prevents stagnation in any part of the procedure where steady progress is always expected. It is not realistic to successfully troubleshoot an ailing process or diagnose the root causes for an industrial failure by simply wandering around aimlessly; hoping that with patience everything will eventually make perfect sense. Again: It is completely acceptable to incorporate multiple investigation practices in the investigation procedure. In that manner we stay on track and continue making steady progress toward solving the problem. The investigation procedure simply defines the basic steps that are followed to perform an objective, independent analysis of facts and evidence related to an incident. Success ultimately depends on adherence to the investigation procedure, specific requirements of the tools you incorporate in that procedure, and to what extent you have not been directed by your own opinions but have allowed the facts and evidence to guide you through the entire procedural sequence.

As mentioned above, the investigation procedure you select is far less important than rigid adherence to a selected procedure. The steps described in our discussion relate to a generic, transferable investigation procedure (Fig. 3.1) that is applicable across a wide range of investigation methods marketed to the industry. Depending on the complexity of the incident, any one of these investigation methods may or may not be adequate by itself to deliver a credible explanation for probable cause at the end of the investigation. The more complex the incident and industrial process involved, the more likely it will be that weaving complementary methods into the procedure will lead to a timely and satisfactory conclusion. Still, the discussion in this chapter draws heavily upon the practices specific to the TapRooT method [9]. The TapRooT method of investigation has been implemented with remarkable success through virtually all sectors of the manufacturing industry.

We will demonstrate the application of a generic investigation procedure that takes place in three distinct phases:

1. Front-end loading,
2. Pattern matching [10],
3. Closure.

PHASE I: FRONT-END LOADING

Front-end loading refers to and involves the initial gathering of information and conducting an analysis of facts and evidence related to the failure. It is also during this phase that we begin to identify patterns that can be used to define organizational culture; finding patterns is helpful in obtaining a proper sense of the context that envelops the incident. It must be remembered that information and evidence gathering does not end after the front-end loading phase is finished. The investigation continues to establish a factual basis for conclusions through the discovery of information from start to finish.

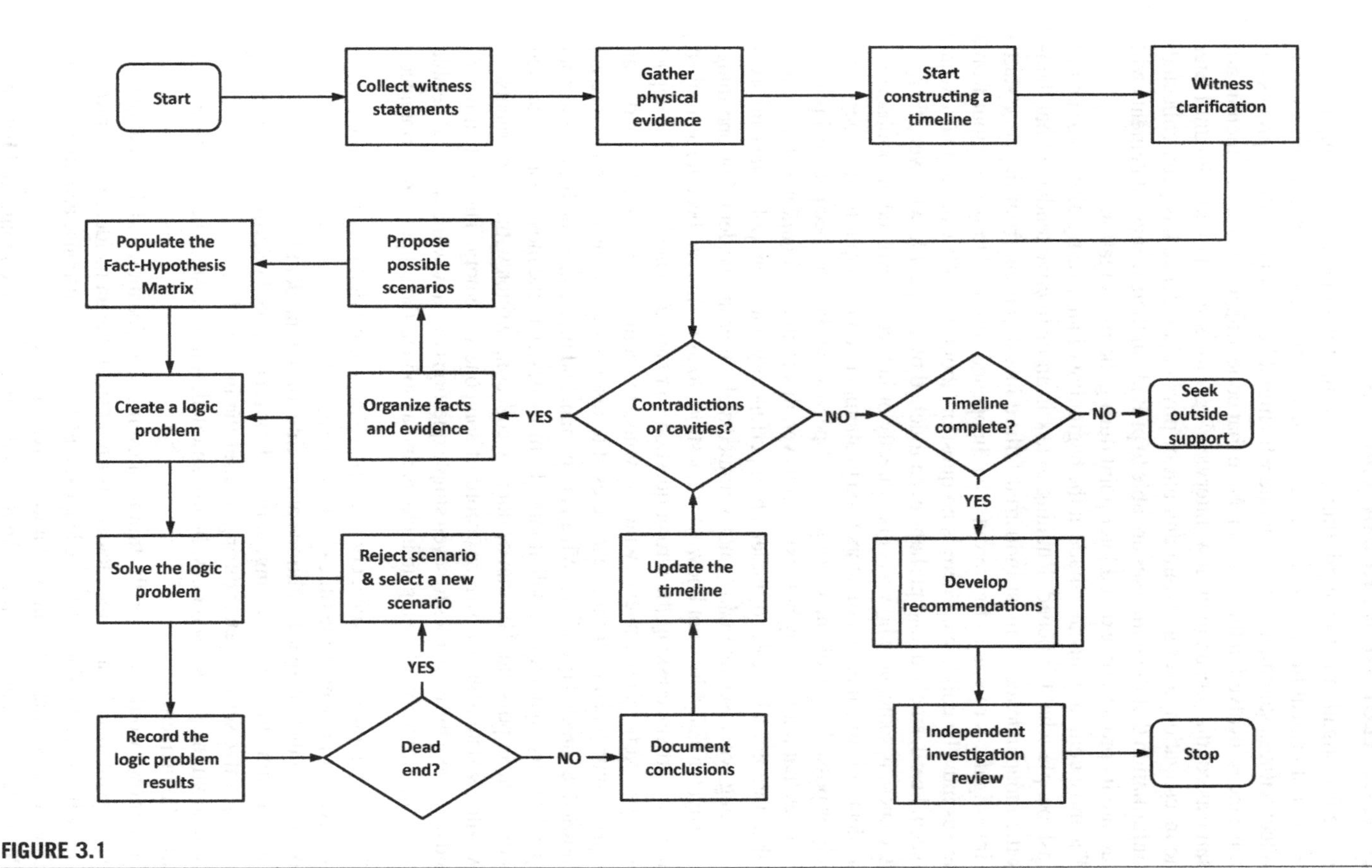

FIGURE 3.1
Comprehensive industrial failure investigation procedure (generic).

STEP 1: COLLECT WITNESS STATEMENTS

After the investigation is triggered, immediate contact with people who might be in possession of critical information is needed. Therefore, the first order of business is to collect witness statements from anyone who might be familiar with the process or circumstances involved in the incident. Note that collecting a witness statement is not the same as conducting an interview. Interviewing witnesses is not advised until later in the investigation, after a certain amount of physical evidence has been collected. The information that the witnesses are able to provide in their witness statements will be put into its proper context at a later point during the investigation.

The most common mistake made at the beginning of an investigation is to run out and ask people what happened. Running to ask is an instinctive tendency that damages the independence of most industrial failure investigations from the very start. The investigator's role is to figure out what happened. For that to occur, input from witnesses must be collected. There is no question about that. We want to know what was seen, heard, smelled, and perhaps even touched by the people who were there.

A witness is someone who was either directly or indirectly involved in, or observed, an incident. It is true that investigators need to attempt to collect input from witnesses as quickly as possible to avoid memory loss and speculation. People speculate and ponder over causes that may, or may not, exist. However, this is an *investigation*, not *litigation*. It is therefore not advisable to implement a perfunctory interview policy that involves interrogating witnesses and taking notes immediately after an incident. For one thing, the investigator really does not know what questions to ask. This can become a problem at the very start since asking the wrong question can have a counterproductive impact on employee participation. Interviewing the witnesses comes later if needed, and at an appropriate time when their input can be used to resolve specific inconsistencies based on knowledge-directed questions. What is important and practical at the beginning of an independent investigation is to solicit input from witnesses in the most nonpersonal and objective manner possible. This can be done by collecting written witness statements.

A witness statement is a written record of someone's memory. The written record should document the responses to seven simple questions that any witness should be able to answer immediately following an incident and as soon as their input can be obtained:

1. Who are people (employees, contractors, visitors, etc.) that may have been involved or present at the time of the incident?
2. What were the weather conditions at the time of the incident?
3. List anything that was rearranged, moved, or repositioned (equipment, debris, manual valve settings, etc.) during or after the incident?
4. Note anything you sensed (saw, heard, smelled, felt, or tasted) before, during, or after the incident?
5. In your own words, describe the timing of events, your location and the location of others prior to and during the incident, warning signs that an incident was approaching, actions that you or other people took, and emergency response activities.
6. Other than the incident, did you notice anything unusual?
7. Please describe any situations like this that you may have encountered before.

When asking your witness to answer these questions, advise the witness that names are less important than generic job functions. Job titles by role (like foreman, spotter, operator, and pipe fitter) are appropriate and preferred in place of individual names. Affirm that the intent for any investigation is to find or define the causes of an event so that action can be taken to prevent a recurrence. There is really no need for the names of individuals when fact-finding and not fault-finding is our primary concern. Most importantly, the witnesses must be *supervised* while documenting their witness statements.

Supervised Witness Statements

For witness statements to be "supervised," the investigator or an appointed designee observes how the witness actually commits his or her statement to paper. We want to be sure that the statements are truly authentic and have not been influenced by others. Supervising witness statements involves having an investigation team member or other designated stand-in representing the investigation team present. The supervising person monitors the witness as he or she writes down their own personal statement. This level of oversight is needed to make certain that witnesses will provide input without coaching.

Say, eg, there is a crane incident involving a team of workers. A crane operator, rigger, and supervisor witnessed the event. At least three witness statements are therefore to be collected. However, the input form cannot be handed out with a friendly, "Please get these back to me before you leave today." Neither should the witnesses be allowed to discuss their responses as a group before they answer or as they are answering the questions. Each witness must provide their own input individually and without interference from others, as soon as possible. Completed witness statements are to be filed away and not reviewed or accessed until the initial collection of physical evidence is complete. Be determined to follow this method to maintain independence during the investigation procedure and start with the people who are most likely to provide vital information.

STEP 2: GATHER PHYSICAL EVIDENCE

Conventional forms of physical evidence include paper, electronic, and location (position) evidence. It is important to capture as much of this information in all conceivable and available formats immediately after an incident occurs. Clearly, some of this information will be available or accessible for the shortest of times. Information degrades or starts to deteriorate immediately following an incident; it therefore becomes progressively less available as time passes.

Some incidents involving the release of process liquids or gases can result in a fire or an explosion. In all cases where a process release has occurred, do not attempt to enter the immediate area until the "all-clear" signal has been received. During the interim period when physical entry into the immediate area is not possible, there is already an abundance of relevant information that can be captured. Good use of the time available involves immediately concentrating on the information that can be captured; do not allow unproductive concerns about (as yet) inaccessible information to delay action. Some data collection must wait until the site is safe for physical entry.

Drawings are an important part of any industrial process incident investigation. Drawings help to create a working knowledge about the process involved in the incident. Working knowledge makes it possible for the investigator to become "a part" of the investigation. Investigation drawings consist of both personal, hand-drawn sketches, and professional electronic computer-aided design files.

Piping and instrumentation diagrams (P&IDs) are an essential form of paper evidence that can often be accessed immediately after an incident occurs. Aerial view diagrams that show the layout and positioning of equipment are also very useful in establishing spatial orientation and an awareness over where things (objects, assets, process materials, etc.) belong. Keep up with technology and consider hiring a qualified firm that can send drones over the facility. Aerial photographic mapping routines at the conclusion of a construction project can be helpful if an incident happens years later. In some cases, observation drones can be activated within hours of an incident.

Together, these approaches and documentation methods start to promote a general awareness about how the process flows in and out of various units or processing sections. These inputs should then be used to sketch an *unofficial* process flow diagram (PFD) that the investigator can take into the unit after receiving the "all-clear" signal.

Transferring your perceptions about the process configuration onto paper is the first step to developing an objective working knowledge about the process involved in the incident. For example, the simplified process block diagram that was discussed in chapter "Equipment Reliability Principles" can be converted to an aerial view of the Bhopal factory (Fig. 3.2). The aerial view drawing shows the relative location

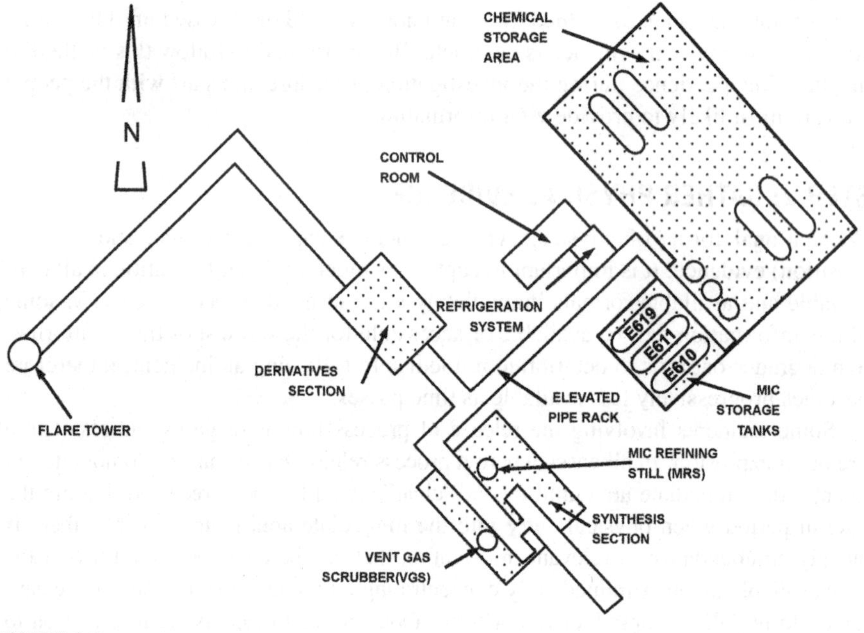

FIGURE 3.2

Aerial view drawing of the Bhopal factory. *MIC*, methyl isocyanate.

of different assets involved in the manufacturing process that took place within the factory. This makes it a bit easier to envision how different process components or materials were moved about through operating areas on their way to becoming technical grade (final) product.

There is nothing too extraordinary about the way that the Bhopal factory process was configured, based on its physical layout and geometry. In many aspects it closely resembles the type of design that can be typically found in a modern factory. In terms of today's industry standard with an emphasis on facility siting, some discussion would likely be generated over the location of the control room. Its proximity to the chemical storage area would be noted. Factory employees would then be working in areas where toxic chemicals—including chlorine gas—were being processed. Of course, the Bhopal disaster and industrial disasters including the 2005 refinery explosion in Texas City, Texas, USA, created a sense of awareness about the hazards involved with locating personnel close to the process. As a result, many responsible organizations in the manufacturing industry now take more definitive steps to manage their process exposure hazards. In particular, they design and modify their facilities in accordance with API Recommended Practice (RP) 752, *Management of Hazards Associated with Location of Process Plant Permanent Buildings*.

Electronic evidence is another form of physical evidence that must usually be collected in an incident investigation. In modern industry, databases are widely used to store a variety of useful investigation information, including equipment work order histories and historical process control information such as pressure, temperature, and flow measurements. Management of change (MOC) information and previous failure data including inspection and investigation reports are other forms of information that can often be accessed electronically within most factories today. E-mails and control board (console) reports are also vital forms of electronic information that may be needed to establish the complete timeline and sequence of events that explains how a failure developed over time.

Position evidence is very important, especially in cases where the sudden release of energy interrupts what appears to be normal process operation. This type of evidence can only be obtained by entering the production area to conduct an inspection. At locations where an incident occurred, investigators would enter and commence their evidence gathering immediately after clearance is given.

In all industrial processes involved in an incident there will always be pressure to begin cleaning up and repairing the impacted area immediately. This type of activity is necessary to prepare for an aggressive restart attempt. Still and all—the investigation team must be allowed to preserve any evidence that might begin to disappear once the cleanup effort is initiated. Area access needs to be highly restricted until the initial sweep for position evidence has been performed. Strictly doing things in proper sequence prevents anyone from contaminating, altering, removing, or doing anything else that could later be construed as tampering with physical evidence.

Collecting position evidence is aimed at recording exactly how valves, equipment, hoses, switches, objects, or anything that could possibly be moved or manipulated had been set when the incident took place. This especially applies to manual bypass valves around automatic control valves and temporary hoses. Position evidence should be documented directly on the PFD sketch that was created at the start of the investigation.

In many cases, the emergency response will lead to unavoidable direct contact with the physical evidence. A certain amount of this type of involuntary "tampering" is expected while bringing an unstable situation under control. This makes it necessary to interview emergency response personnel after the "all clear" signal is received. They might be able to verify the as-found position data when the responders arrived at the incident location. Under no circumstances should the investigation team pass judgment on position data while it is being collected. The investigation team's responsibility is to simply document everything the way it was found upon arrival, regardless if it looks suspicious or was likely manipulated during the emergency response.

Since the quality of evidence is reasonably expected to deteriorate progressively with the passage of time, it is necessary to immediately capture, preserve, or "freeze" as much evidence as possible. With this purpose in mind, digital photography has become an indispensable tool for preserving position (location) evidence during the initial stages of an investigation. The photographer's goal is to capture anything that might increase in significance after the investigation begins making progress. Remember, at the start of an investigation the line between important and useless information can be blurred. Once the investigation team begins looking for specific evidence that might have been missed on the initial pass, digital photography becomes very important.

Start by photographing anything that obviously looks suspicious. Then photograph everything else regardless of its perceived significance. Continue taking pictures until the entire process area can be reconstructed in photographs on a table. Having an ability to reconstruct the post-incident scene avoids many problems and possible complexities after the area is cleaned up and back in service. Remember there is a dual purpose for taking photographs during the initial inspection:

- Recording the condition of equipment and the area directly involved in the incident and
- Providing a reference to return to after the process is restarted. Concern yourself with recording things that may not seem significant in the early stages of the investigation.

STEP 3: START CONSTRUCTING A TIMELINE

At this point in the investigation process, the different types of information collected must come together so that a timeline can be developed. A timeline, chronology, or sequence of events is a fundamental component in every industrial failure investigation. The timeline serves as the anchor point that hooks together events and conditions. Timelines also help the investigation team to keep track of their own progress. Although timelines are chronologies of events, they allow the team to see where they are in the investigation procedure and how much work remains to be done.

A SnapCharT is a dynamic timeline specific to the TapRooT investigation method. It demonstrates how events and conditions interact to form or create an incident (Fig. 3.3). Regardless of what investigation method is utilized or being conducted by the team, all events and sequences relating to an incident must be shown on a timeline. The timeline's main function is to constantly direct the investigation team's attention to contradictions or missing information ("cavities") that must be resolved so that no missing links or inconsistencies remain in the final analysis. Without a timeline, there is a good chance that important facts will be misrepresented or misunderstood, and that events will not be recorded in their proper sequence from beginning to end.

A SnapCharT organizes events and conditions that collectively represent the entire incident. Events are the major occurrences that take place according to a specific chronology. On the other hand, conditions cover the details concerning the "who, what, why," and "severity" of an event. Therefore, conditions are always a subset of an event. Events and conditions appearing on the SnapCharT or timeline will be collected, added, and modified throughout the investigation process as inconsistencies are resolved to bring about a cohesive timeline. Because a timeline keeps things in chronological order, it tells the story from start to finish. A cohesive timeline leaves no holes or questions about how or why things happened the way that they did. This method then points to solutions that can be pursued to prevent recurring incidents in other applicable processes.

One or more of the conditions that appear in an incident's timeline will be a causal factor. On a SnapCharT, causal factors are marked by a triangle that contains the abbreviation, "CF." Causal factors are conditions that either caused or allowed an incident to develop. In other words, any causal factor, if reversed or done differently, could have stopped the sequence of events that propagated into an actual incident. Other conditions will have no influence over the failure at all, but will be listed simply to provide clarity over the sequence of events.

An incident timeline will always contain at least one causal factor, but industrial incidents of a catastrophic nature will normally have multiple causal factors. This is because industrial processes normally contain multiple protective barriers designed to adequately manage known process hazards. Somehow, each of these multiple barriers must fail for an incident to occur. One of the primary goals for the timeline is to identify each of these protective barriers so that attention can be given to how they failed. It is also important to determine if those failures are independent or instead somehow related to each other. Our text will elaborate on this topic in chapter: Process Hazard Awareness and Analysis.

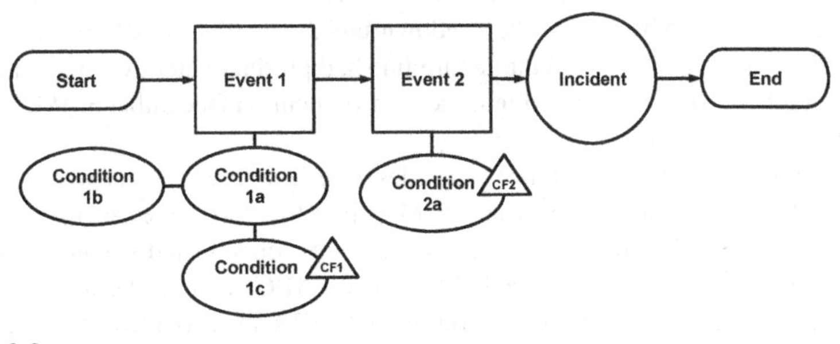

FIGURE 3.3

A SnapCharT defines the interaction between events and conditions.

Causal factors are addressed by recommendations that reduce the probability for a recurrence. Causal factors should be corrected at the greatest depth known, based on the investigation results. As a simple and arbitrary example, attributing a failure to "bearing failed" would be shallow and superficial. Bearings fail for a reason and discovering that a bearing failed because the fitting-to-shaft method was flawed in a particular way adds depth to the investigation. As the investigation probes deeper into the causes for an incident, the recommendations supported by the analysis become more universal in their application. Properly addressing causal factors makes it possible to avoid depending on unpleasant experiences in other parts of the system to provide you with a basis to take action. In other words, we exercise a "failure-restrictive" attitude by applying recommendations at the greatest depth possible. Superficial recommendations (closer to the failure) favor a "failure-permissive" attitude. At every step in the investigation procedure, you set the example of organizational culture. Do your actions indicate that you operate along the lines of an organizational culture that restricts or permits process failures? Taking early and appropriate steps to apply a universal approach to distributing investigation recommendations is by far the most efficient and practical way to prevent incidents. It not only sets the right example, but also accelerates the rate of process safety and asset reliability improvement.

Managing Inconsistent Information

During an investigation you will detect at least one inconsistency in the facts and evidence you receive. In the final analysis, an investigation is a contradiction of its own, considering the fact that system malfunctions are not expected and are definitely not intentionally designed into the process. The amount of effort needed to resolve a contradiction depends on your diligence with regard to the freezing of evidence at the very start of the investigation. The time invested in freezing evidence pays tremendous dividends as you then follow-up and look for additional input needed to resolve an inconsistency. The more effectively you have frozen the evidence, the easier it becomes to reconcile contradictions before resorting to more sophisticated methods of problem solving, which we will soon discuss.

For example, basic raw materials were reacted in the Bhopal factory to make methyl isocyanate (MIC) (refer back to Fig. 2.8). MIC manufactured in the synthesis section (Fig. 3.2) was routed to a pair of rundown tanks (E610 and E611) for temporary storage. The MIC inside the rundown tanks was then transferred as needed into the derivatives section to manufacture finished product batches. The contents of rundown tank E610 were released into the environment on December 3, 1984, which resulted in the Bhopal disaster.

According to credible references that describe how the process was designed, differential pressure was used to transfer MIC into the derivatives section [11–14]. This transfer method required feeding nitrogen into isolated rundown tanks to overcome the downstream pressure needed to transfer MIC without the use of pumps. However, a process and instrumentation diagram (P&ID) provided by the process designer [15] (Fig. 3.4) shows that transfer pumps were used to send MIC into the derivatives section [16–19]. This process configuration would mean that the system

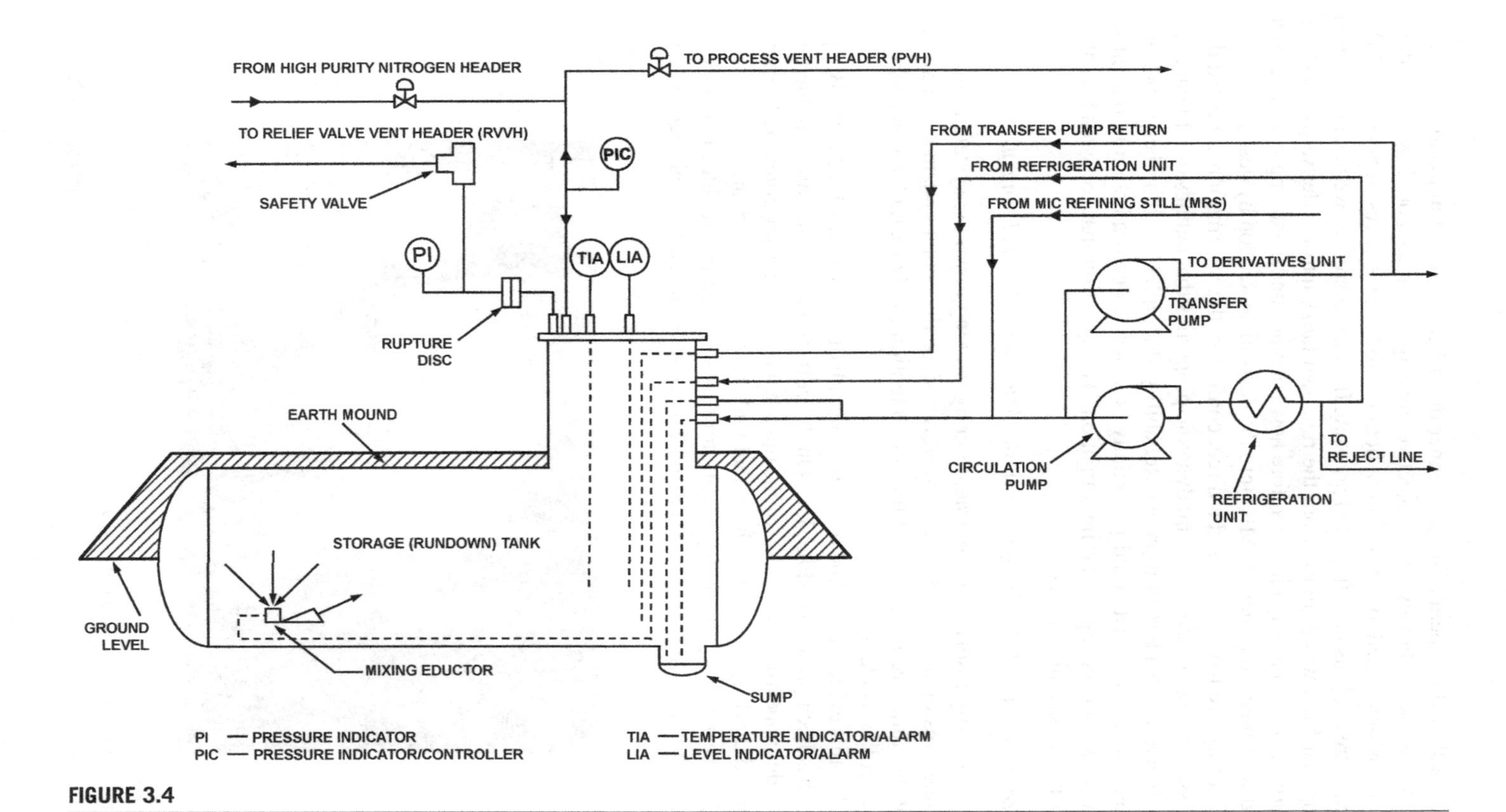

FIGURE 3.4

Original methyl isocyanate rundown tank drawing showing a transfer pump location.

was not designed to operate in the manner described by the aforementioned references. At this point, however, an objective investigation does not allow us to draw any conclusions based on this contradiction. Some references report that MIC transfers were made using differential pressure. In other references we find information showing that MIC was pumped from the rundown tanks into the derivatives section. The purpose of the investigation therefore now becomes understanding which transfer method was intended to deliver MIC into the derivatives section by design.

Obviously, we have uncovered a direct conflict in the information collected thus far. Pressure is one of several thermodynamic properties that defines or influences an asset's function (chapter: Equipment Reliability Principles). It is rather important, therefore, to resolve what could potentially be a serious contradiction in our information before advancing in the investigation. The question that we would want to answer in this specific case is:

Were pumps included by design to transfer MIC out of the rundown tanks?

We can easily resolve this contradiction by walking down the system to put our hands and eyes on the actual assets. However, since the actual system is not accessible at this time, we must instead turn our attention to photographic evidence collected after the incident.

Within this evidence we find that a photograph of the tank involved in the Bhopal disaster has been preserved (Fig. 3.5). On the corresponding tank drawing there are four nozzles appearing on the side of the rundown tank, which are connected to pump suction and discharge lines. On the actual tank photograph, there are also four nozzles. Comparing these two independent information sources provides the evidence we need to comprehend a stated fact that the purpose for feeding dry nitrogen into the

FIGURE 3.5

Excavated rundown tank (E-610) that was involved in the Bhopal disaster.

Photo by Paul Cochrane.

tank was to create "slight" (low) pressure [20]. Operating the tank at a slight pressure would make sense only if a transfer pump was provided to move refrigerated MIC into the derivatives section.

Now that we have an idea of how the tank was designed to operate, other sources of information begin increasing in value. For example, we now understand the significance of information from credible sources stating that the rundown tanks were designed for operation at a maximum pressure of 2 psig [21]. Yet we also know that by the end of the process' life cycle the rundown tanks were supplied with at least 14 psig to send MIC into the derivatives section without the transfer pump [12].

Our conclusions are further validated in corroborating facts from another independent reference. It shows corresponding transfer pumps in the domestic (USA) factory that served as the Bhopal factory's design template [22]. Therefore, there is consistency in the original design between the two similar processes. Finding corroborating evidence in the answer to particular questions is common to many investigations. In all cases after causes have been properly identified there will be no further contradictions; only confirmations.

However, notice how resolving the one inconsistency examined in the preceding paragraph raises more questions. The investigation team is now responsible for answering these additional questions before completing the investigation:

- What was the purpose for changing the design from a pumped system to a differential pressure transfer system?
- Does the observed design change represent an improvised design solution?
- If an improvised solution is represented by this design change, then what precedent might have made it okay? In other words, what does this observation tell us about the Bhopal factory's organizational culture?
- What information can we find referencing the reliability of the MIC pumps shown on the P&ID (Fig. 3.4)?
- What hazards might have been created by implementing the design change and how were these hazards mitigated?
- What specific operating practices were needed to develop sufficient differential pressure inside the rundown tanks to abandon the transfer pump?

For now, these questions must be put aside. The investigation team would have confidence that these and other questions will eventually be answered with facts and evidence by adhering to the investigation procedure. At this point, we have satisfactorily resolved a contradiction, which provides essential knowledge about how the process was designed to operate. The priority now is to simply move ahead with the investigation.

Note, again, that the SnapCharT (timeline) is a tool that makes it easier to identify contradictions. After an asset fails there will generally be at least one contradiction waiting to be resolved. By evaluating contradictions, an investigator adds superior value to the entire incident investigation. Although we might never have expected one of our assets to have failed, always focus on identifying the defect that did, in fact, make it fail. Failures occur due to improper operation or tolerating design defects under which the asset simply cannot be expected to provide reliable service.

To reemphasize: resolving contradictions very often releases a torrent of information that otherwise would have been overlooked or would have remained unexplored. A contradiction is, potentially, a direct link to critical information. That is why resolving contradictions generally opens a door to understanding what happened and what can be done to prevent incidents from happening again.

STEP 4: WITNESS CLARIFICATION

After assembling the initial timeline, it is important to review the sequence of events with the witnesses whose input was originally solicited for witness statements. Their cooperation is now needed to verify the investigation's accuracy before moving on to more sophisticated analysis methods. Most often, the witnesses are met with individually in an informal discussion with the investigator.

A poorly executed meeting will certainly complicate the investigation process. That makes it critically important to remember three important points when working with witnesses who observed, or might know something about, an incident. These experience-based tips and reminders will make the discussion more productive:

1. The witness probably believes that your purpose is fault-finding, not fact-finding,
2. The amount of cooperation you receive from the witness depends entirely on the types of questions that you ask, and
3. Never ask anyone a question to which you already know the answer. It comes across as being insincere and reinforces the witness' impression that you are on a fault-finding mission.

It is also important to remember that reviewing information about an incident can be an intimidating or stressful experience for both the witness and the investigator. The interaction that you facilitate must make the witness feel comfortable to speak openly about what he or she knows. Sensitivities are more likely to flare up at this point during the investigation. A mutual level of trust must be established for the witnesses to fully cooperate.

You, as the investigator, must come into the discussion knowing the right questions to ask. Chances are that you will be discussing a sensitive issue. There will be times when the witness will cooperate only when asked the right question. Alleviate the potential stress by working backward in time from when the failure occurred. It is more likely to recognize a transition point by walking backward through the sequence of events. Unresolved transition points are usually safe topics for conversation between the investigator and witness.

Transition Point

Simply stated, a transition point represents a marker in the timeline where a step change and possibly a permanent change is identified. Does the investigation resolve the conditions responsible for the change? Keep in mind that there are no random, meaningless changes in the manufacturing industry. Process changes occur for specific reasons. Keep searching for the explanation and you will find it. An investigation should start with this search. Finishing the investigation may require going further back in time to answer questions about precedents that justify later actions. But for now, the starting point is an event that represents a change of some sort in the timeline leading up to the incident. This usually involves a change in thermodynamic properties, operations, mechanical integrity, or asset functionality. In our example involving a possible change in tank operation, we would begin the discussion by asking the witnesses to confirm how the tank was being operated at the time of the incident and would there have been any reason to discontinue using the transfer pumps. Depending on the answer, we can determine if we have sufficient clarity to move ahead in the timeline, or if we have to go back further in time to gather more information that might have been missed on the first pass.

There are times when group interviews can be more advantageous than speaking to witnesses individually. A group interview can be conducted with all of the witnesses present at a convenient time either before or after their normal work shift. In such cases, the investigator is often able to observe cooperation between individual team members who are able to sort out discrete events in sequentially correct chronology. This can be an especially productive experience when the console operator is invited to participate in the discussion with field operators. Again, it is the investigator's responsibility to ask the types of questions that promote group cooperation. The interviewer does this by examining the facts and evidence prior to the group meeting so that effective questions can be asked. Going into a group interview unprepared is a certain recipe for a communication failure. Group interviews should therefore be approached with both preparation and caution.

Gaining the witness' trust begins with populating the timeline with as much information as possible. The witnesses have to be convinced of your sincere desire to create transparency, and that their honest input will not be used against them. In most cases, the investigation process can be greatly simplified by performing due diligence in the preliminary phases of information collection. Try to resolve as many contradictions as you possibly can before seeking assistance from the witnesses. If you have resolved all of the inconsistencies then the purpose for meeting with individuals or groups is to gain their alignment before moving on to the closing phase.

Many times, however, the witness clarification step creates more questions than it answers. Accept that this seemingly inconvenient fact can be a turning point where an investigation begins making serious advancement. Any remaining contradictions or cavities (missing links in the sequence of events documented on the timeline) send the investigation directly into the important pattern matching exercise. This is the phase where information tends to suddenly make more sense and where disjointed globs of information get shaped into meaningful and matching blocks of understanding. Continuing with the metaphor, shapeless islands of information now undergo an objective, logical evaluation process where their meaning is coming out of the fog. The logical evaluation process moves us forward in big steps. Meaningful information previously trapped in shapeless islands of witness recollections and investigative work now begins to be revealed.

STEP 5: ASSESS CONTRADICTIONS AND CAVITIES

This step involves answering a simple question:

Are there any remaining contradictions or cavities?

A contradiction is an inconsistency within your timeline or discontinuity in the sequence of events. A cavity is a missing piece of information, such as an event that does not clearly link together the event immediately ahead and the event immediately following. If there are no more contradictions or cavities, then your investigation is complete and you move on to the closing phase that we will talk about later. If, however, contradictions or cavities remain then you must continue the problem-solving exercise until the investigation is complete. The investigation is complete only when all contradictions, cavities, and inconsistencies have been resolved on the basis of available facts and evidence. Facts are not conjecture, facts are scientifically accurate.

PHASE II: PATTERN MATCHING

At this point, most incident investigations move into the second phase, which is called "pattern matching." Pattern matching is a systematic process of elimination that defines the most probable cause for an incident purely on the basis of facts and evidence. Following the process creates a rationale for why certain scenarios apply while others do not apply. During an incident investigation this method is essential to develop clear understanding of cause-and-effect that explains how an incident happened.

Pattern matching is based on the premise that certain scenarios, if they occurred, will produce a specific pattern of evidence (or "markers") that can then be verified through a logical process of elimination. If the evidence needed to validate the scenario is found, then the scenario is considered valid. If, however, the evidence is not found, then the scenario is rejected and a different explanation must apply. The process is repeated until all contradictions and cavities have been resolved on the basis of available evidence.

Pattern matching is a more sophisticated problem-solving approach than the method used to resolve the contradiction described earlier in this chapter. Recall that the earlier method called for photographic (frozen) evidence to be captured and compared for purposes of verifying asset design features. The investigation sought physical evidence to ascertain the chemical intermediate transfer method that was commissioned by design. Whenever there is insufficient information to decisively resolve a contradiction through physical evidence, pattern matching is used as the next step. While not prejudging anything, pattern matching simulates the evidence you need to resolve an inconsistency. Once you envision the evidence without prejudice, you know what you are looking for and where it can likely be found. From that point on you can resample your sources, knowing exactly what you need to find. If the evidence is there, then you have solved the problem.

As an investigation approach, pattern matching is vital to avoid any speculation over what may or may not have occurred. All incidents that are under investigation are historical events. Each of them occurred sometime in the past. You can only be expected to recreate past history on the basis of available evidence. As was discussed previously, the effort you invest at the beginning of an investigation in terms of recording, documenting, and freezing evidence will determine how much work is required to find the evidence needed to complete the pattern matching exercise. A passionate viewpoint not backed by evidence only causes a loss of credibility and stirs controversy. Your job as an investigator is to allow the facts and evidence to direct you to a credible explanation about what happened and why it happened. To do this, your major contribution in the investigation process is to set the best possible example by following the procedure, by not taking shortcuts, not forcing results, and by holding others accountable for doing the same.

Pattern matching begins with defining the problem. What do you wish to accomplish by performing the pattern matching exercise? Your purpose must be clearly defined. The problem can either be a difference between expected and actual process or asset performance [23] or any inconsistency found in the as-designed versus as-implemented process configuration. There is no reason to be obsessive or to elaborate excessively in your problem statement. Simply state what it is that you expect to accomplish by performing a pattern matching exercise.

Let us say that you have detected another inconsistency in the investigation. Perhaps the witnesses are finding it difficult to remember significant details about an exposure event that occurred years prior to a catastrophic process failure. Some of your sources are saying that a previous release was caused by a valve failure while others are telling you that a pump seal leak was the cause. A word of advice in this situation: Never play referee. Once again, only facts and evidence are allowed to direct the thought processes. Your rationale for determining a most likely scenario must be based on an objective assessment of the information you have collected. Making sense of that information can require a bit more sophistication than simply looking at photographic evidence. It begins with organizing your facts and evidence in a manner that makes it possible for you to perform a logical assessment.

STEP 6: ORGANIZE FACTS AND EVIDENCE

There is an inconsistency related to a serious incident that occurred inside the Bhopal factory on January 9, 1982, where about 24 workers had to be hospitalized to recover from severe respiratory illnesses. One witness recalls that the incident resulted when thick clouds of phosgene were released after a valve on a phosgene line broke off [24]. However, other witnesses recorded that the incident on January 9, 1982, occurred after MIC was released from a mechanical pump seal failure [25]. So, within this one contradiction we actually find *two* inconsistencies that must now be resolved:

1. Was phosgene or MIC released during the incident on January 9, 1982?
2. Was a valve or a pump failure involved in the catastrophic process release on January 9, 1982?

Our conclusions about whether or not the process release involved a pump or valve failure will be directed by a pattern matching exercise. By assessing the initial information that we received in the first phase of the investigation we can demonstrate how the most reasonable answers to these questions can be derived (Table 3.3). The exercise begins by defining its purpose in a practical problem statement. Our purpose in this case can be stated as:

Determine the most probable scenario involved in the catastrophic process release on January 9, 1982.

Keep in mind that quality is much more important than quantity when organizing facts and evidence for a pattern matching exercise. There is no minimum amount of evidence upon which an accurate assessment can be made (aside from none, of course). For example, regarding the inconsistency surrounding the incident in the Bhopal factory on January 9, 1982, there are initially only seven points of factual evidence directly related to the process release:

- The incident occurred on January 9, 1982.
- The incident occurred about 2 weeks after a fatal incident involving exposure to phosgene.
- The incident occurred about 3 days before the refrigeration system was shut down.
- The incident occurred in the morning hours, during night shift.
- About two dozen workers were sent to the hospital to recover from exposure symptoms after the release.
- The workers were treated for phosgene exposure.
- There were no fatalities.

At times it may be useful to think about facts and evidence in four dimensions to extract as much useful data from what little information might be available [23].

1. **Identity (I)**. This category defines the item that is (or has) a problem (usually a process or an asset) and what is wrong with it.
2. **Location (L)**. This category relates to the specific or general location where the incident occurred. It can also detail what specific part or component in a system was involved.
3. **Time (T)**. Items falling into this category have to do with the timing of an incident, as to when exactly it occurred. Timing is very important when ruling scenarios in or out. Defining when specific events happened can be as important as when they did not happen.
4. **Size (S)**. This category describes the severity of the incident. It could contain information about the distribution pattern of a process release, the number of injuries, or how much of the total process was involved. If a repeat failure is involved, were the consequences more severe or less severe than on previous occasions?

Table 3.3 Pattern Matching Worksheet

Define the Purpose										
Determine the most probable scenario involved in the catastrophic process release on January 9, 1982.										
List the Facts and Evidence (I, L, T, and S)										
ID	Facts and Evidence									
1	The incident occurred on January 9, 1982.									
2	The incident occurred about two weeks after a fatal incident involving exposure to phosgene.									
3	The incident occurred three days before the refrigeration system was shut down.									
4	The incident occurred in the morning hours, during night shift.									
5	About 24 workers were sent to the hospital to recover from exposure symptoms after the release.									
6	The workers were treated for phosgene exposure.									
7	There were no fatalities.									
8										
9										
10										
Propose Possible Scenarios										
ID	Possible Scenarios									
A	A valve broke off the phosgene pipe.									
B	A mechanical MIC pump seal failed.									
C										
D										
E										
Fact-Hypothesis Matrix										
ID	1	2	3	4	5	6	7	8	9	10
A	NA	NA	NA	NA	NA	+	-			
B	NA	NA	?	NA	NA	-	+			
C										
D										
E										
Create a Logic Problem										
ID	If-then statement									
A7	If there were no fatalities, then the workers were not directly exposed to phosgene gas.									
B6	If the workers were exposed to phosgene, then phosgene was in the MIC storage tanks.									
Solve the Logic Problem										
ID	Results									
A7	Fail: Workers were sent into the hot zone without wearing fresh air. [Chouhan, 34]									
B6	Pass: MIC contains 200-300 PPM phosgene as an inhibitor to prevent polymerizing. [Agarwal, 219]									
Document conclusions										
"Scenario B" has been selected as our working hypothesis. References solve the logic problem (B6) and also provide a marker that supports "B3" in the Fact-Hypothesis Matrix regarding circumstances that might have led to the Refrigeration System shutdown. Solving the logic problem on this basis also resolves a discrepancy discovered in the reported incident date, which was not previously noticed.										

Organizing facts and evidence into different categories is extensively practiced in highly disciplined and effective problem-solving approaches such as the Kepner-Tregoe method [26]. Even if it is not a direct part of the problem-solving process, thinking in these collective terms still serves a useful purpose. In particular, it helps to define what is known about a problem may have left behind very few clues as to what might have happened [10].

STEP 7: PROPOSE POSSIBLE SCENARIOS

After all of the facts and evidence have been written down, it is time to list any possible scenarios that might explain what happened. Many pattern matching exercises will require brainstorming to come up with possible scenarios that can then be evaluated. Here, we have a different situation. Our purpose for this pattern matching exercise is to demonstrate the basic principles involved in a neutral and objective problem-solving process. In many cases we would be performing this exercise to follow a trail of evidence that directs us to the cause for an asset failure. In this case we will be using the same principles to resolve an inconsistency in the details about a specific incident that occurred inside the Bhopal factory on January 9, 1982. There are two scenarios provided by our sources that must be screened to determine which one (and only one) is the most consistent with the facts and evidence:

1. A valve broke off the phosgene pipe or
2. A mechanical MIC pump seal failed.

STEP 8: POPULATE THE FACT–HYPOTHESIS MATRIX [27]

At this point we begin incorporating the principles of falsifiability into the investigation [28,29]. This concept involves putting at least as much effort into disproving a hypothetical situation as is invested into proving it. Falsifiability involves using inconsistent information patterns to validate that a scenario fails, instead of only searching for information that reinforces a favored scenario. In all cases, the information found upon returning to the actual evidence serves as the basis to accept or reject a specific scenario [10]. An objective investigation results from applying this practice.

This step in the procedure involves the actual assessment of the information that we have prepared in the previous steps (Table 3.3). More precisely, we perform a systematic analysis of our proposed scenarios according to the facts and evidence listed. A table is constructed with letters to represent the proposed scenarios (vertical, left side) and numbers for the facts and evidence (horizontal, top). We then proceed to rank each item of evidence according to how it validates the scenario we are assessing:

- **Plus sign ("+")**: The facts and evidence proves the scenario.
- **Minus Sign ("–")**: The facts and evidence disproves the scenario.
- **"NA"**: The fact does not appear to have any relationship to the scenario. It neither proves nor disproves it.
- **Question mark ("?")**: A mental connection is possible with this scenario, although more information is needed to either prove or disprove it. If no supporting information can be found, then the scenario cannot be validated by this fact and we consider it to carry no significance.

For example, the first item of evidence (the fact that the incident occurred on January 9, 1982) really does nothing for us in terms of validating or invalidating either scenario. In this case, knowing the specific date of the incident does not help us in any way to determine if a broken valve or mechanical seal failure is what caused the catastrophic process release. Therefore, we mark each box under the first column of evidence with an "NA" and move on to the second item.

Eventually, we come across item number 3, which lists the fact that the incident occurred 3 days before the refrigeration system was shut down (January 12, 1982 [25]). This is a very interesting clue from a "time" (T) perspective. Notice that once again it means nothing to us if the release was caused by a broken valve on a phosgene line. The two events are random and coincidental. There is no way to join them together. Again, we mark box "A3" as "NA."

However, when considered in the context of an MIC pump seal failure, the relationship could be significant. Referring back to the rundown tank P&ID (Fig. 3.4) we find that an MIC pump (the circulation pump) continuously processed the tank contents through a refrigeration system consisting of a heat exchanger. A mechanical pump seal failure at this location would correspondingly have shut down refrigeration. Having previously established the fact that the transfer pump's use was discontinued at some point, we are left with only the circulation pump as an option. Could the precedent that we observed with regard to abandoning the transfer pump provide a clue about how similar problems with the circulation pump might have been managed? This is just one of the questions that would begin swirling in our head at this time.

Unfortunately, this is a mental stretch at this point. We simply do not have enough information to rule this scenario in or out on the basis of this specific fact. But the scenario at least deserves to be marked as "?" since more information is needed. We do not pass judgment at this time on this fact–hypothesis relationship. Therefore, we keep it in mind as we continue to evaluate facts and evidence to ascertain its possible contribution to the timeline.

Since the concept of falsifiability involves making a special effort to disprove a fact–hypothesis relationship, what we are really looking for specifically is the evidence that disproves a proposed scenario. Really, there would be no reason to perform a pattern matching exercise if all of the facts and evidence are consistent with a specific explanation or cause for an incident. The facts and evidence that disprove a scenario are what we will use in the next step, which requires creating a logic problem.

We come across our first "invalid" combination at "B6," which does not support an MIC release to have caused phosgene exposure. You could at this point rationalize that if MIC and phosgene produce the same health impacts then the treatment for MIC exposure might be the same as phosgene exposure. But rationalizing is illegal when filling in the fact–hypothesis matrix. We rationalize only during the next step, in order to create a logic problem. We let the facts and evidence speak clearly about the truth and mark the combination appropriately. So in this case ("B6") we mark the relationship invalid ("–") while marking the phosgene release from a broken valve ("A6") as valid ("+").

The last item of evidence produces an invalid result on "A7" while supporting "B7." Based on the fact that a fatality occurred upon being exposed to phosgene (fact number 2) less than 2 weeks before the incident we are assessing, it would be reasonable to expect that at least one phosgene exposure (out of about 24) might have resulted in a fatality. But knowing that the purpose for refrigerating MIC was to control its vapor pressure [30,31], a low volatility release would have produced injuries, but not necessarily fatalities. So, we mark "A7" with a "–" while marking "B7" with a "+." Having completely populated the fact–hypothesis matrix, we now move to the next step.

STEP 9: CREATE A LOGIC PROBLEM

In this part of the exercise we are tasked with creating a test question (logic problem) for each hypothesis, using the output from the fact–hypothesis matrix. We do this by observing where any combination between a scenario and a fact produces an invalid relationship in the fact–hypothesis matrix. Our first invalid combination occurs in cell "A7," which does not explain why there were no fatalities if the workers were directly exposed to phosgene gas. Were they wearing respirators? Were they holding their breath? Were 24 workers just "lucky" that day compared to the one poor soul who lost his life only 2 weeks earlier for being exposed to the same material? Something does not make sense. We must now phrase a logic problem in a way so that the invalid relationship *does* make sense.

This is done by creating a hypothetical "if–then" statement. Simply stated, a question is phrased in a way making it so that *if* some reported observation is true, *then* this (something else) would need to have happened. Based on the invalid relationship that exists in cell "A7" a logic problem can be phrased in the following sense:

If there were no fatalities, then the workers were not directly exposed to phosgene gas.

Solving a logic problem forces us to seek additional sources of information to confirm our rationale for disproving a particular scenario. In this instance, if solving the logic problem directs us to information that explains why the workers were not directly in harm's path, then we have substantiated our basis for accepting the scenario. Essentially we will have found the answer that we were looking for, so that anyone who raises the same question later can possess the facts and evidence they need to understand why the exposure to phosgene resulted in multiple injuries, but no fatalities and why the consequences were different than what was experienced 2 weeks before. If, however, we find additional evidence that the workers were inadequately protected from the release whereby fatalities would have resulted as they did earlier, then we must reject that scenario and turn our attention to the other scenario, hoping that the same process will reconcile the inconsistency that was identified in cell "B6." In either case, we must now solve the logic problem by finding at least one additional fact.

STEP 10: SOLVE THE LOGIC PROBLEM

Now that we know what we are looking for and have a good idea where to find it, we turn our attention back to our references to perhaps find a fact that might have been missed earlier, before the logic problem pointed us in a certain direction. Sure enough, in our references we find a note that states that the workers purposefully entered the immediate area without wearing fresh air breathing masks [32]. In this case we have *not* found the evidence we need to accept the proposed scenario. Workers entering the immediate area of a phosgene release without respiratory protection as the new fact indicates would probably be the best way for them to be directly exposed to a deadly gas release. The puzzle becomes more complex with this piece of evidence. The evidence we found actually falsifies the if–then statement. Therefore, the logic problem has failed, and we must preserve a record of the information that conflicts with our rationale that we created to solve this logic problem.

STEP 11: RECORD LOGIC PROBLEM RESULTS

It is important to keep a running record of whether or not a logic problem passes or fails on the basis of acquired evidence. The question captured in a logic problem is likely to be asked again by anyone who is not part of the investigation team. Consistency in explaining the technical basis for accepting or rejecting a specific scenario establishes credibility that cannot be achieved through lip service alone.

In the case we just discussed we hit a dead end, which means that the evidence we collected after defining exactly what we needed to solve logic problem did not support the scenario, but made it less likely. We might phrase our results something like this, to retain a permanent record of the evidence that was used to disprove the scenario that we evaluated, while linking it to a specific reference:

Workers were sent into the hot zone without wearing fresh air [Chouhan, 34].

At this point, we have hit a dead end with this particular scenario. Hitting a "dead end" by no means interrupts the investigation, though.

STEP 12: DEAD-END ASSESSMENT

All but one of the proposed scenarios should fail the logic problem. If the logic problem fails, we have reached what is known as a "dead end" in following a specific trail of evidence. Instead of digging any further into the matter by attempting to resolve any new inconsistencies that develop from the evidence we found, we must answer "yes" to the question about reaching a dead end and reject the scenario.

STEP 13: REJECT SCENARIO AND SELECT A NEW SCENARIO

At this point, we have discovered a basis to reject the scenario that involves a valve breaking off of a phosgene line. After we reject a scenario, all is not lost. It is actually a desired output from an investigation procedure. If we are truly objective with respect to a specific hypothesis that can resolve the sequence of events responsible for an incident, then rejecting a scenario on the basis of facts and evidence frees us from the frustration we would otherwise experience at the end of a long, unproductive path that leads to nowhere. If the evidence we obtained in Step 10 does not solve the logic problem, then we must remain neutral with respect to any emotional outbreak or demands for further clarification. We must simply accept the results of a failed attempt to solve the logic problem and move on to other scenarios until the logic problem directs us to evidence that supports it. No matter how much we might want to believe a specific scenario on the list, as independent investigators we must really have no attachment to any of them.

Notice in our procedure, that rejecting a scenario puts us back in the loop to evaluate the next proposed scenario on our list (Step 9). Using the pattern matching worksheet, the facts and evidence are already listed for the next scenario on the list. At this point, we could possibly update the list of facts and evidence with the new information we obtained while trying to solve the first logic problem, but this is optional. It is perfectly acceptable to assess all proposed scenarios with the same information. For demonstration purposes, we will move on to assessing the second proposed scenario using the same information that was used to assess the first scenario. Remember, we are looking for quality facts and evidence. Increasing the quantity at this point does nothing to improve the initial screening of plausible explanations.

Moving on to the next and only other possible scenario, we create a new logic problem that resolves the inconsistency found in cell "B6" of the fact–hypothesis matrix. This particular combination does not explain why the 24 victims who were exposed to the process were treated for phosgene exposure if an MIC pump seal failed. Was phosgene in the MIC rundown tank for some reason? Are the health impacts between MIC and phosgene similar? If so, then why would a refrigerated release of MIC liquid (boiling point 39.1°C [33]) have created what appears to be a significant inhalation hazard during the winter? Again, something does not add up here. We could rationalize by making up a few mental networks to authoritatively resolve these questions and probably sound like we know what we are talking about, but instead we hold ourselves accountable for following the procedure whereby we can always expect the end result to be more satisfactory than making things up as we go along. Therefore, we create a new logic problem in the form of an "if–then" statement that can help us locate the missing evidence:

> *If the workers were exposed to phosgene, then phosgene was in the MIC rundown tanks.*

Turning our attention once again to our technical references, we find a note in a report that MIC contains 200–300 ppm phosgene to inhibit polymerization [34]. This note is confirmed by multiple corresponding references that describe MIC as a molecule that is stored with a small amount of phosgene to inhibit polymerization reactions [35,36]. Recalling Le Chatelier's principle from college chemistry [37], there appears to be a scientific explanation for adding phosgene to the MIC rundown tank contents. In the Bhopal factory, MIC was manufactured by reacting phosgene ($COCl_2$) with monomethylamine (CH_3NH_2) according to the following reaction [38] (Eq. [3.2]):

$$COCl_2 + CH_3NH_2 \rightarrow CH_3NCO + 2HCl \quad [3.2]$$

According to Le Chatelier's principle, extra reactant (phosgene) pushes the equilibrium to the product side of the reaction, which includes MIC. This adjustment can improve the stability of the product inside the MIC rundown tanks by resisting potentially dangerous side reactions. Thus, the relationship in cell "B6" not only passes on the basis of the new evidence, but also establishes a scientific basis for the presence of phosgene in the MIC rundown tanks, just as described in the literature. We would then document how the evidence we were directed to this time solved the logic problem (Step 11) by adding the following statement to the worksheet:

MIC contains 200-300ppm phosgene as an inhibitor to prevent polymerizing [Agarwal, 219].

While looking for this information, we also notice three additional independent references that consistently corroborate a pump seal failure's involvement in the catastrophic process release that sent about 24 people to the hospital [32,39,40]. Perhaps we missed this detail earlier, because these particular references list the incident date on either February 9, 1982 [39], or February 10, 1982 [32,40], instead of January 9, 1982. Of particular interest is one of the three references that quotes a witness with intimate knowledge about the Bhopal factory's history saying that there were only *two* (not three) catastrophic process release incidents prior to disaster on December 3, 1984 [39]. As we have discussed, the first of these two incidents occurred on December 24, 1981 when a factory worker was exposed to phosgene gas during a maintenance activity and later died [24,32]. The *second* (and only other) incident referenced is the one that we just assessed through a pattern matching exercise, which occurred on January 9, 1982, in response to the catastrophic pump seal failure. References to what might be a *third* incident, on or around February 9, 1982, suspiciously involve the exact same circumstances that were embedded in references to the second incident on January 9, 1982 (a catastrophic pump seal failure that led to the hospitalization of about two dozen workers). It is also worth mentioning that the source documents that speak of the incident in January 1982 [24,25,41] contain no information about a third incident in February 1982, and vice versa [32,39,40]. Using our facts and evidence, the disparity between what might appear to be two separate incidents in January 1982 and February 1982 can be resolved by combining the two events into one. In other words, the evidence serves as a basis to conclude that any references made to what might appear to be a *third* catastrophic process release in February 1982 are incorrect and likely resulted from a misunderstanding about specific events on certain dates that are all closely related.

To clarify the point we are making, notice the specific date involved with the *second* incident corresponds with the timing of a significant and related event that occurred on February 9, 1982 [25]. On that date, exactly 1 month after the catastrophic process release that severely injured about 24 workers [41], the Bhopal factory's trade union filed a formal grievance with regard to hazard awareness inside the factory. The very next day (February 10, 1982) a formal, independent investigation was commissioned by the State Labor Department [40,42].

Labor unions might consider the 1-month anniversary of a catastrophic process release that seriously injured 24 workers to be an appropriate date to launch a propaganda campaign aimed at directing community awareness to a series of incidents that had occurred in rapid succession inside the factory [43]. Along those lines, we can see how facts and evidence can inadvertently be distorted when a logical thought process is not used to organize conflicting facts obtained during an investigation. Most of the time, these kinds of differences are not intentional. Personal accounts of the same event can differ between witnesses. Not everybody sees the information the same way as it is being received. As time passes by, facts are skewed and embellished as they are communicated verbally down through generations. In addition, memories fade and untrained note takers may not always write down exactly what people are saying. Others may simply not take the time to verify that they understood what they think they heard. There are many reasons, none of them deceitful, for contradictory information to pass into an investigation. This reinforces the importance of using a systematic process of elimination to reconcile any conflicting information we might come across while investigating an industrial incident. Only through a structured and repeatable process can we objectively discern the most probable sequence of events based on the totality of available facts and evidence, while assembling the entire picture from multiple inputs as was demonstrated here.

Back to the scenario that solves the logic problem, as the investigation progresses the credibility of our assessment is raised even further by three more pieces of corroborating evidence that attribute the catastrophic process release that seriously injured about two dozen workers to a mechanical pump seal failure, not a broken valve on a phosgene pipe [32,39,40]. One of those references provides valuable insight into the pump's operating history by recording that:

> *...leaks resulting from this type of mechanical failure never exceeded the toxicity level above which such incidents were likely to be fatal [32].*

In other words, this was not the first time that the MIC pump's seal failed. The catastrophic MIC pump seal failure that occurred on January 9, 1982, was a repeat failure. Prior to this incident, however, the consequences of the same failure were much less severe. A near-fatal incident involving about two dozen serious injuries would likely be expected to generate a sobering discussion in the factory by daybreak concerning what could be done *immediately* to stop the persistent cycle of repeat seal failures. Thus, we have now established that the normalization of deviance was involved in a chronic asset reliability issue that created an actual process safety hazard in the Bhopal factory about 3 years before the disaster. In this case, shutting down the circulation pumps would definitely eliminate any pump failure mechanism that interfered with process safety. It would also follow the precedent (improvisation) set forth by abandoning the parallel transfer pumps. Looking closely at our P&ID (Fig. 3.4), we see that shutting down the circulation pumps would also disable the refrigeration system.

We now have enough information to reconstruct a probable sequence of events (pattern) strictly on the basis of our available facts and evidence. From the information that we have collected, validated, and recorded at this point in the investigation, we can develop a relatively comprehensive narrative of the incident on January 9, 1982:

> *As some had predicted, the Circulation Pump's mechanical seal failed again on January 9, 1982. Per normal routine, the workers entered the area wearing their typical Personal Protective Equipment (PPE) to isolate the MIC leak and prepare the pump for maintenance as they had successfully done many times before. Upon their arrival, they immediately realized that this particular leak was worse than they had anticipated. Already committed, and witnesses to catastrophic process release getting worse by the moment, they had no choice other than to manage the situation as they had done previously many times by manually shutting down the pump and isolating the process without wearing fresh air protection. By the time that the initial responders and reinforcements had isolated the leak, at least 24 workers had been exposed to low levels of phosgene. The fact that the refrigerated process (boiling point 39 °C) was in its liquid form at the time of its release prevented more serious injuries. But the phosgene inhibitor (boiling point 8 °C) in the MIC vaporized more readily as the process escaped from the cooling circuit, thereby creating an immediate inhalation hazard that sent the victims to the hospital for recovery from phosgene exposure. In the aftermath of this incident, supervisors recognized that the circulation pump repeat failures could no longer be tolerated and therefore issued a technical instruction note authorizing their immediate shutdown, which disabled the refrigeration system as well. This series of events agitated workers who flexed their Union muscle by filing a formal grievance on February 9, 1982, exactly one month after the catastrophic process release that injured 24 workers, only two weeks after the death of a coworker who was directly exposed to phosgene while performing maintenance inside the factory. An independent investigation was organized the very next day by the State Labor Department in response to the union's concern.*

At the end of this exercise we have a much greater depth of understanding about the incident on January 9, 1982, than we might have ever thought possible before applying a systematic evaluation approach to the information in our possession. Those of us with process experience can easily relate to how an unstable process condition like the one just described could quickly unravel. However there are now additional questions that must also be answered before we are done with our investigation:

1. What was different about the failure this time? What made it more severe than any of the ones that preceded it?

2. Aside from disabling the refrigeration system, what other impact might shutting down the circulation pumps have on the balance of process functionality?
3. What do the actions referenced in this sequence of events say about organizational culture, and is there any consistency demonstrated with other events in our timeline?
   - Uniformity in message?
   - Hazard awareness?
   - Improvisation?
   - Casual compliance?

Having gotten past the dead end by solving the logic problem in the above write-up relating to the Bhopal disaster, we can now document our conclusions.

STEP 14: DOCUMENT CONCLUSIONS

We finish this pattern matching exercise by documenting our conclusions on the same worksheet that contains all the information used in the assessment. This step is needed to preserve the technical basis for solving one of the more complex problems we can expect to encounter during this, or any other, investigation. Documenting the rationale for rejecting any scenario is just as important as preserving the rationale for the favored explanation.

As the investigation continues, we would want to be receptive to the discovery of any new information that either supports or appears inconsistent with the output from a pattern matching exercise. Any new inconsistencies, contradictions, or cavities would return the investigation back to Step 6 (Organize Facts and Evidence) for another pattern-matching exercise to validate the possible scenario. Recall, however, that when the actual scenario has been positively identified, there will be no other inconsistencies. Any additional information sources will only help to clarify the exact sequence of events, or shape our working knowledge of the event into perfection.

After documenting our conclusions about a specific scenario at the end of the pattern-matching exercise, we update the timeline so that it captures what we have learned.

STEP 15: UPDATE THE TIMELINE

The timeline must be updated to keep track of where we are in the investigation process. Any validated event needs to be added in its proper location, so that it links the two events on either side of it in a logical manner (Fig. 3.6). What we end up with is quite literally the sequence of events that links the incident's entire life cycle together from start to finish.

It is not only possible, but likely, for the details to change as the entire picture comes into focus. The timeline is not finished until the investigation is closed. The time for closure comes when no inconsistencies, contradictions, or cavities remain open and all logical questions have been answered by the evidence preserved in the timeline. This is a very good reason to remain somewhat flexible with the timeline

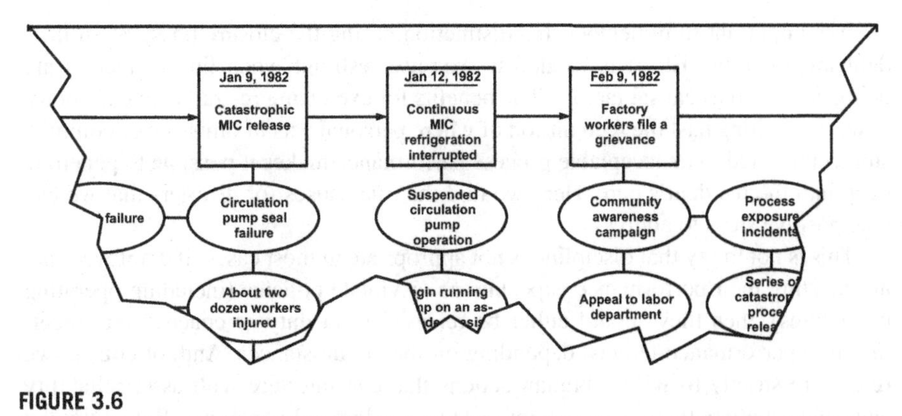

FIGURE 3.6

Single timeline events are linked together in sequential order.

until sufficient transparency pushes the investigation into the third and final phase of the procedure which will then close the investigation.

PHASE III: CLOSURE

In the final phase of the investigation, we make use of the sequence of events established from facts and evidence. We also perform necessary quality checks to verify that the most probable cause has been defined. Facts and tangible evidence will be cited in support of realistic and achievable recommendations. Once implemented, realistic and achievable recommendations are fully expected to prevent the next failure. This implies paying close attention to the most effective option available to us. With every investigation, we have the opportunity to "strike while the iron is hot." If complacency was a factor in the incident, then we have a chance to address our dull sensitivities by establishing effective operating limits to remain sharp in the future. We also examine the attributes of organizational culture that contributed to the incident as we close an investigation.

DISCIPLINE

Depending on their own personal experience and observations, factory workers may find it difficult to express confidence in an incident investigation program. They may consider the incident investigation program an extension of the site's disciplinary program. With this thought in mind, please note that disciplinary action is not one of PSM's elements while an incident investigation is. The two must therefore, remain separate. It is easy to understand the principle behind this fact. After an incident has occurred, no discipline can erase it. The multiple chances to prevent it have been lost. The most that can be hoped for after an incident is for the investigation to resolve lessons learned that can be used to prevent future incidents.

It is important to remember this distinction during the closure phase of an incident investigation. Effort is needed to exercise restraint when human factors are being described as causal factors. The benefits for exercising restraint are enormous, however. Getting past the recognition of where personal acts of omission or commission contributed to unacceptable process performance makes it possible to penetrate deep into the incident's core. Here we find the true causes for disasters that we can take positive steps to address.

This is not to say that discipline is not appropriate in most cases. If employees are not meeting their performance expectations or violate policies (including operating procedures), then they should either be released from duty or coached into meeting their performance targets, depending on the circumstances. And, of course, we refer here strictly to willful human actions that can interfere with asset reliability and safety matters that have been entrusted to authorized personnel. But again, the time for coaching, correction, guidance, and termination is *before* an incident investigation documents the sequence of events that resulted in an incident. Disciplining employees after an incident is not a reasonable or defensible approach to maintaining a satisfactory level of process control. Any discipline administered in response to an investigation report will be perceived as retaliation that makes the workforce less willing to cooperate in the future. It also reinforces the myth that an incident investigation is a tool that belongs to "human resources" (HR) rather than PSM. Instead of making good use of lessons learned, the company loses in the end, with more serious failures to contend with.

When applied to prevent an incident that has not yet happened, disciplinary action demonstrates an appropriate level of respect and concern for an individual's health. It is a visible form of active (not passive) protection. Discipline translates into warnings that progress up to termination in order to protect the individual and other potential victims. Not everyone is fit for the job of managing and operating a process. Even those with years of experience could be dangerous, and might even be comparatively more dangerous, if they are unwilling to address performance issues brought to their attention. By firing people who consistently demonstrate a willingness to do things "their way," which is not the "right" or prescribed way, you are essentially telling them, "I care about you enough to release you from your duties here that might allow you to do something that could interfere with the quality of your life, your family's life, or the lives of others" [44]. The opportunity to show this amount of respect for human life is possible only through discipline—discipline that is implemented *before* the incident. In this regard, discipline is an essential aspect of incident *prevention.*

STEP 16: ASSESS TIMELINE COMPLETION

Addressing all of the inconsistencies detected in our investigation allows us to review the timeline for completion. The timeline is complete when all of the causal factors are clearly defined and the sequence of events bridges or weaves the incident together from start to finish. At the end of the investigation all of the logic problems have been solved. If any logic problems are still unsolved, then the investigation should be turned over to an independent outside expert, a specialist in the technology involved in the incident.

Still, seeking outside support to resolve how the incident occurred is the least preferred option for closing an investigation. The goal of the internal investigation team should be to finish the investigation with a complete understanding of what happened. Nevertheless, in those instances where a satisfactory explanation does not become transparent, it will be appropriate to seek outside support from a qualified subject matter expert (SME) who can push the investigation in the right direction.

The right time to call in outside reinforcements, however, is only *after* the internal team has exhausted all routes to make sense of the problem being investigated. Only *after* the investigation procedure enters the closure phase is it advantageous for the internal team to call in an external SME. At this point the internal team will have amassed a sizable volume of relevant information: drawings, procedures, process history, documentation of pattern-matching exercises, etc. Any and all of this information can be handed over to the SME in whatever form makes it possible for that person to efficiently determine what is being missed. The handover, if needed, sets up the internal investigation team to either accept or reject the SME's independent conclusions. Again, this transition would take place only after the team came to an impasse and had reached a dead end to their understanding. Once more this explains why it is not advisable to engage an outside SME at the beginning of an investigation, which is to say before the internal team has established at least some sense of direction. Either way, the request for outside support will be countered by a request for information. The value of the information delivered to an external SME increases exponentially if the internal team has digested it beforehand.

In practice it is rarely necessary to call in an outside resource to essentially lead an investigation or take over an investigation that is not making satisfying progress. What you will find through experience is that following an investigation procedure always leads you to a satisfactory conclusion. Indeed, it does not matter what procedure you follow; what matters is that you follow the procedure. Back to points made at the beginning of this chapter, complications are created by skipping steps in an investigation procedure or by "doing what comes naturally" without much organization.

Speculation and improvisation are what come naturally. Speculation and improvisation are unacceptable approaches to solving a technical problem in the manufacturing industry. These substitute investigation practices increase the potential for an industrial disaster as we shall see in later chapters of this book. Only a systematic approach to investigating industrial incidents can prevent a minor upset from becoming a disaster.

STEP 17: DEVELOP RECOMMENDATIONS

Assuming that the investigation timeline is complete and the entire story can be told, you now enter the most important step in the procedure. Your goal is to avoid any chance for recurring incidents at the location being investigated and at locations elsewhere. You avoid recurrence by recommending a practical set of corrective actions.

Developing recommendations is a twofold process that involves first addressing what might be considered the more direct causes for an incident. Following this initial step, you proceed to examine the underlying issues related to an organizational culture. An entire book could likely be written on the topic of establishing effective investigation recommendations. For our purpose, however, we will limit our discussion to the most important principles that govern the creation of effective recommendations at the end of any investigation:

1. **Use precise language**. After diligently following the precursor steps we have outlined, we are ready to formulate effective recommendations. An effective recommendation clearly describes what must be done to solve a problem and to avoid its repetition. How you write that recommendation influences the degree of performance you can reasonably expect. It is essential to state both *what* must be done and *why* it has to be done. Many recommendations fail simply because they state what must be done, but not why it must be done. Excluding *why* the recommendation applies can and most often does lead to confusion over the circumstances behind a specific recommendation. Perhaps the person who receives the action item believes that the intention is to take advantage of a production incentive when, in fact, a safety issue is involved. Do not take for granted that the person you assign to take action is as informed as you are about the importance and purpose of the recommendation. More often than not, you must spell out the action item in precise detail before becoming reasonably confident that it will be carried out or implemented correctly. An example of a recommendation written in acceptable format might look something like this:

   Install a minimum flow spillback line on the discharge of the splitter bottoms pumps to avoid pressure spikes in the rundown line if the pumps trip offline due to low flow during startup.
2. **Be assertive**. The recommendation should be assertive, in the sense that it identifies the problem and what actions are needed to correct it. Never should an investigation produce a recommendation to evaluate a hazardous situation further, or to perform another investigation sometime in the future. The investigation team's responsibility is to "strike while the iron is hot." At this stage the captive audience that commissioned the investigation must be told what they need to do or else the opportunity to effect necessary change will be lost. The investigation team's effort would have been wasted and a repeat failure could occur. Using assertive language requires confidence in recommending a specific course of action. In many instances the team will have spoken directly to an appropriate SME. As discussed earlier, engaging SMEs during the first two phases of an investigation should be approached with caution. But when the investigation enters its last phase, speaking to SMEs has a decidedly more specific purpose. Conversing on the feasibility of a recommendation might provide additional input; for a certainty it validates that the recommended path is both practical and appropriate. After communicating (with SMEs) the investigation team can complete and close the investigation with full confidence that satisfactory performance will result.
3. **Be conservative**. Being conservative in an investigation has nothing to do with politics. The number of corrective actions you assign directly reflects

how well you understand the problem. It might sound counterintuitive, but the more corrective actions you assign, the less certainty you have about the root cause. And the less certain you are about the root of the problem, the more likely it is for the issue to surface again; perhaps sooner than you think. Recall the principles elaborated on in chapters "Industrial Learning Processes" and "Equipment Reliability Principles." Giving people more work to do does not automatically make a process safer. Process safety results from addressing the problem at its source. This can very often be accomplished in a single corrective action.

Turning our attention to addressing the physical hazards or process defects that must be addressed, it is important to express recommendations in singularity. In the context of report writing, singularity means that separate recommendations relate to, contain, or capture a single, independent cause. Most investigation reports go way overboard in assigning too many corrective actions. In many cases this fault originates with failing to differentiate independent causes. Let us return to our case study involving repeat mechanical seal failures that we discussed in chapter "Equipment Reliability Principles" and see how many independent causes apply.
Recall that causal factors on a timeline are denoted in a triangle containing "CF." Looking at the timeline for the incident on April 8, 2004 (Fig. 3.7), there appear to be four causal factors. Each of these causal factors deserves an independent recommendation:

- **CF1: Repeat mechanical seal failures ($\lambda = 0.065$)**. The actual process configuration made it possible to switch over to spare pump operation when a mechanical seal failed. Unanticipated maintenance (breakdown maintenance) had to be performed to continue uninterrupted production. However, the repair rate (its frequency of occurrence) was abnormal and extreme. The incident occurred due to a maintenance procedure defect involving lockout/tagout. Had the pump not required maintenance, then the incident would not have happened. Preventing the incident would require performing an investigation to fully understand the problem and confront it with a practical recommendation. This was not done in this instance. Instead the process was allowed to remain in operation by swapping back and forth between the installed spare.
- **CF2: Valve handle was mounted perpendicular to flow in pipe**. An MOC error was definitely reflected in this causal factor. The workaround solution to address an inconvenience that likely resulted from frequent process isolation at this valve involved an improvised homemade device that could be disconnected to allow better clearance for people walking around the unit when the handle was not being used. In exchange for this versatility that no doubt prevented many bruises, bumps, scrapes, and other minor personal injuries, the handle could be repositioned such that it could lead to human error when process isolation was necessary. The hazard was not recognized before the incident. Why? There was no MOC [45]. Nobody involved in the design modification recognized that this was *not* a "replacement in kind." Therefore, more MOC training would likely be considered appropriate to address this knowledge deficiency.

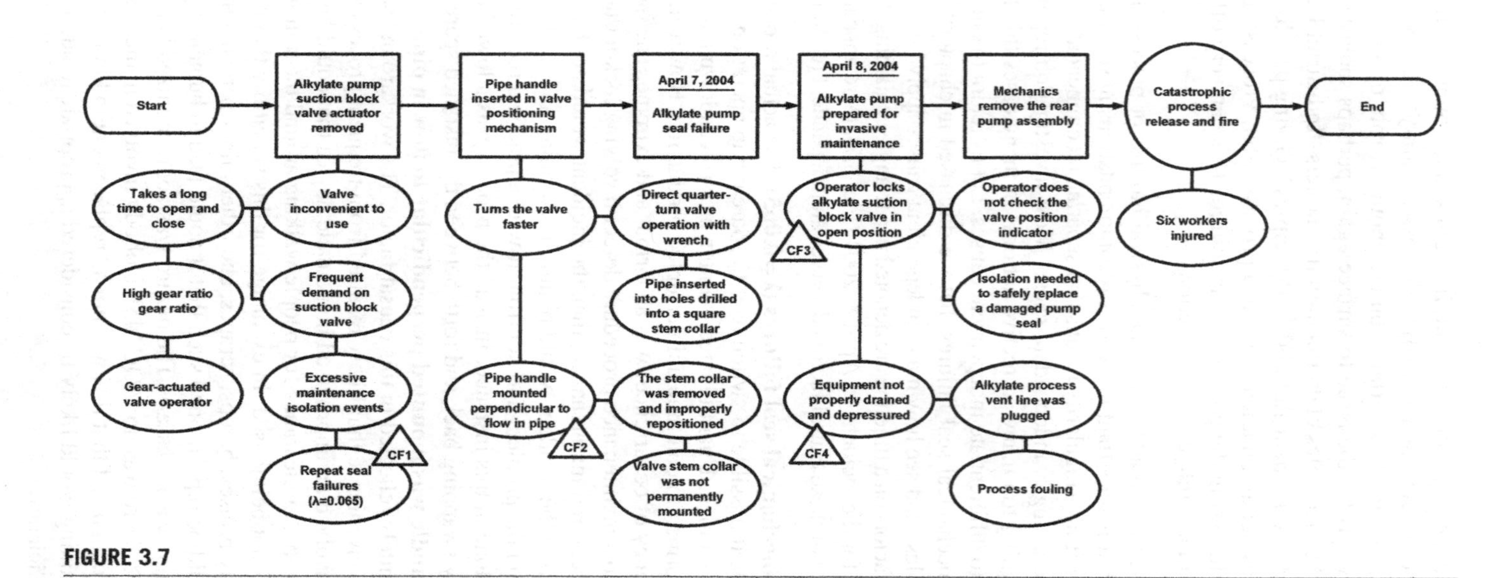

FIGURE 3.7

Refinery alkylate release incident SnapCharT summary.

- **CF3: Operator locked pump suction block valve in open position**. This again is a training issue. Apparently not all of the operators received proper training over using the valve position indicator to determine the direction of flow through the pipe. The training department would therefore get an action item to promote awareness throughout the organization and would need to keep updating the training plan appropriately to catch all new hires and sustain existing staff performance.
- **CF4: Equipment was not properly drained and depressured**. Initially the process appeared to be adequately prepared for invasive maintenance. Unfortunately, the line was choked by process solids. Unknown to the operators that checked the bleeder valves, the process block valve was still open. Perhaps a more sophisticated approach to verifying effective process isolation was warranted, especially since the process was known to contain excessive solids. Again, these are good lessons that can effect a positive change in the form of performing additional cross-checks before issuing permits for invasive maintenance. Options might include using a rod-out tool to verify that lines are clear, and perhaps adding isolation blinds every time the pump seal must be replaced in the future. There are many things that can be done. Nothing additional was done before the incident in this case.

But back to the points made earlier: There is really only one independent causal factor in this incident. The maintenance requirement at this pump location exceeded the limits of human capability. How do we know this? For one thing, the use of an improvised tool to actuate the block valve. Although the Chemical Safety Board (CSB) report does not provide information on previous unrelated incidents inside the refinery, it is likely that this was not the first time that improvisation was used to make the process more manageable. Another indicator is human error. Repeat failures result when improvisation is incorporated into the process. Its prolonged use could lead to misunderstandings. We therefore only have one issue to address that solves the entire problem: The unacceptable maintenance frequency. Causal factor 1 (CF1), repeat mechanical seal failures ($\lambda = 0.065$), is the only independent causal factor in the entire chain of events. Solving it gets rid of all the other problems. The other issues that all relate to organizational culture are to be addressed separately. Assigning only one corrective action in this case promotes the basic principles of process safety by removing the distractions that result from addressing dependent causes in addition to the direct cause. After the investigation is finished, carefully evaluate the timeline to be sure that *dependent* causes do not receive separate recommendations.

4. **Apply the hierarchy of controls** [46] **to address the direct causes**. Industry has known for a long time that not all solutions are created equal. There are usually multiple options available to address a single direct cause. These options are grouped into the following categories, in order of effectiveness from most effective to least effective (Fig. 3.8):
   - **Elimination**. In some cases, inherent safety is made possible by using an alternative substance or process that removes the hazard. When the hazard is

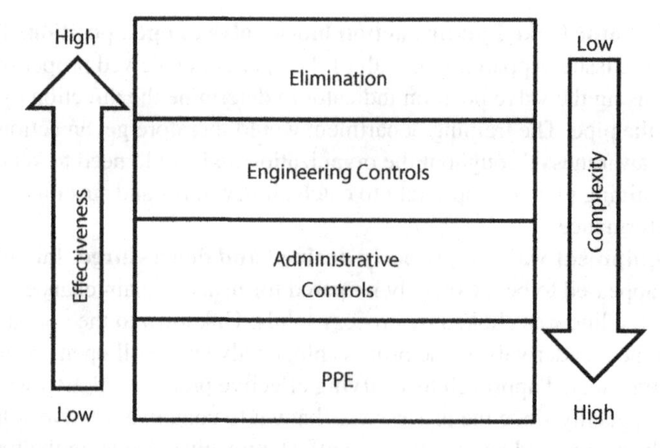

FIGURE 3.8

The hierarchy of controls.

gone, there is no longer a basis for concern over possible harm. For example, when considering ways to control the hazard resulting from loss of process containment from a chemical storage tank, the thought might turn to ways to eliminate the tank rather than ways to control the hazard inside or outside the tank (see chapter: Spent Caustic Tank Explosion: A Case Study in Inherently Safer Design).

- **Engineering controls**. Asset design considerations are brought into the picture here. Most failure mechanisms are properly addressed through engineering controls; such controls usually take advantage of advancing technology. Upgrades are often appropriate and safety systems most certainly belong here. Interlocks, relief valves, coalescers, atmospheric scrubber towers, and flare systems are among many examples commonly found throughout industry.
- **Administrative controls**. Alarms, procedures, and policies all fit into this category. Administrative controls are human-dependent processes. They all require human awareness and interaction to be successful.
- **Personal protective equipment (PPE)**. PPE does not prevent the scenario that creates harm in any way; it simply prevents people from being the hazard receptor. This category includes safety glasses and goggles, ear plugs, chemical-resistant boots, hard hats, fire retardant clothing, protective gloves, and fresh air breathing equipment, to name just a few.

In all cases, it is best to recommend the most effective option. However, many factors will determine which option is both practical and feasible. For example, in a scenario involving water ingress into a hot oil system, the seemingly best option would be to install a relief valve or rupture disc capable of handling the overpressure situation. Unfortunately, the instantaneous expansion of water to steam in a 1700:1 volumetric ratio does not favor addressing the hazard with an engineering control.

Relief devices do not respond well to sudden jolts of energy and are likely to fail under these conditions. We must therefore turn our attention to administrative control options. These options would include adjusting process pressures in our favor on both sides of a barrier and configuring alarms to alert us if the pressure balance becomes unstable. Without getting overly specific, the point being made is that careful thought must be given to strictly recommend only actions that can be expected to work as required. Start with high expectations, but be prepared to accept a less cost-effective but more realistic option based on feasibility that tends to be inconsistent between process systems. In other words, there is never a blanket solution that applies universally. You must always "do the math" to be certain that a solution will work under specific inherent operating and design conditions.

5. **Apply the PSM standard to address organizational culture**. Many companies operating in the manufacturing industry are under obligation to implement the PSM standard. As we have discussed, some other companies voluntarily use the PSM standard to create a competitive advantage in a commodity market. The application of the PSM standard as described here (during the closing phase of an investigation) involves a more interactive use of PSM principles to address organizational culture issues. An effective industrial failure investigation requires using PSM to shape a failure-restrictive organizational culture.

Let us assume that the investigation has revealed flaws in one or more of the seven organizational culture attributes. Depending on the consistency among the flaws detected, these should be addressed with a recommendation to improve the effectiveness of its corresponding PSM standard element (Table 3.4). For example, the use of a homemade tool (improvisation) that resulted in an incident would indicate a problem with the factory's auditing function. Specific recommendations in this case would be made for addressing audit program gaps that allowed the use of an improvised tool to go undetected and without discipline.

Homemade tools are a violation of company policies in most factories. Health, environmental, and safety policies usually clearly state this and employees routinely acknowledge their familiarity with this knowledge. Nevertheless, an audit program is necessary to enforce the policy. We are reminded of the truism: "You get what you *ins*pect, not what you *ex*pect." Appropriate supervision would be made aware of any policy deviations for applicable disciplinary action to be taken before an incident involving a homemade tool triggers an investigation. By the time the situation gets to the point that an investigation is needed, it is too late for factory supervisors to effect the desired changes within production units.

When an incident discloses that a homemade tool had been used it is likely that many other homemade tools can be found. Chances are these have been used previously and for perhaps a considerable length of time. An investigation into an incident involving an improvised solution without a background of disciplinary action is a very strong sign of a failure-permissive organizational culture. For that reason, the investigation team would rightfully inquire about the factory's record of disciplinary

Table 3.4 Using the Process Safety Management Standard to Address the Causes for a Failure-Permissive Organizational Culture

	Employee Participation	Process Safety Information	Process Hazard Analysis	Operating Procedures	Training	Contractors	Pre-Startup Safety Review	Mechanical Integrity	Hot Work Permits	Management of Change	Incident Investigations	Emergency Planning and Response	Compliance Audits	Trade Secrets
Operating Limits		X						X						
Tribal Knowledge	X			X								X		
Casual Compliance											X		X	
Improvisation													X	
Transparency	X					X								X
Uniformity in Message							X							
Hazard Awareness		X	X	X	X				X	X				

action against employees who willfully violated policy with respect to the cultural causes under investigation. It is reasonable to anticipate receiving a long list of previous violations. If it determined that the culture cannot change, then it must be controlled. The PSM standard covers both interests.

It is appropriate to end our discussion about developing recommendations with a thought about preserving the knowledge you receive from or through the investigation procedure. A well-documented sequence of events together with a narrative report is essential. But now you must embed what you know in a sustainable and regularly referenced system that continues to benefit others upon your departure.

PRESERVING ACQUIRED KNOWLEDGE

People separate from companies for a variety of reasons. Retirements, life changes, and personal preferences are all expected and guarantee that at some point in time, the knowledge you possess will walk out the door. You owe a debt to the companies that paid the price for teaching you what you know. Pay back that debt by leaving behind a permanent legacy that continues to pay dividends (or benefits) upon your departure. One of the best ways to do this is by imprinting your experience in your employer's codes, specifications, standards, and policies. Policies, by the way, include operating limits. This leaves behind the legacy that you wish to be remembered by. At the same time, you must honor the legacy left behind by others.

To be clear on what we mean by this, consider the origin of industry codes. Industry offers a wide variety of services to sustain process stability and asset performance. Chief among these are codes, standards, and specifications issued through professional organizations such as the American Society of Mechanical Engineers (ASME), the American Petroleum Institute (API), the American Fuel & Petrochemical Manufacturers (AFPM), the Center for Chemical Process Safety (CCPS), and the Institution of Chemical Engineers (IChemE), just to name a few. These codes, standards, and specifications are developed by professionals who volunteer their time to serve on committees. These committees review lessons from incidents and determine the most effective specifications that can prevent their recurrence. Note that much of this activity culminates in free or reasonably priced information that saves literally millions of dollars when put to use as standards or related documents within local production sites. These writings came about as a result of understanding phenomena that were not previously understood. Prior to understanding the hazards involved, it was likely more "economical" to operate differently. Codes, specifications, standards, and policies are industry's way of documenting what is known so that others can benefit. In an ethical environment we wish to assist others before they experience disasters. Do they really deserve to be trapped into thinking that they are operating responsibly and economically when in fact they or their assets are on the verge of failure?

Perhaps chronic "uniformity in message" defects result from acting indifferently toward documented design codes, standards, and specification. For this reason, 29 CFR 1910.119 (i) specifies that the Pre-Startup Safety Review (PSSR) shall confirm that:

Construction and equipment is in accordance with design specifications.

Even when covered by an exception or waiver, any difference detected between the way the process is supposed to be designed and the way it is actually designed after commissioning creates a permanent discontinuity in "uniformity in message" that sets a precedent for all other process-oriented decisions that follow. Therefore, an organizational culture struggling with "uniformity in message" may have a defective PSSR program.

Many times it will seem more economical to act contrary to industry codes, specifications, standards, and policies. However, *never* issue an exception or waiver to a code, specification, standard, or policy unless following through as directed will cause a conflict with process safety. Do not do it because you believe the hazards do not apply under certain circumstances or can be adequately managed through a more "economical" option that does not meet codes or specifications. Remember that many in your shoes have suffered the consequences of not knowing what they did not know [47]. The hazard you must fear the most is the one that others encountered when they were in your position. The price you pay for discovering these lessons on your own might be unaffordable. It is easy to *say* that you will not issue an exception to a standard, code, specification, or policy when it seems justified to do so in the interest of economics. You will, however, find that it takes an extraordinary amount of personal strength and operational discipline to live up to your commitment.

STEP 18: INDEPENDENT INVESTIGATION REVIEW

Not all investigations require an independent review by an appropriate SME. In cases where the consequences either did involve, or likely could have involved, serious injuries, death, or significant property or environmental damage, it is usually appropriate to engage a reputable consultant who can validate your work. Their expectation would be to use their experience to disclose things that the investigation might have missed. Most of the time, however, the concurrence of an independent authority gives the investigation team the full confidence needed to defend a specific position about the most probable cause.

PROBABLE AND PRECISE CAUSE

An investigation into a catastrophic process release can touch off a firestorm of debates over an absolute, or precise, cause. It is important to note that even when all of the evidence is fresh, it is never possible to create an exact replica of the incident as if a video camera was capturing every moment in full detail. Depending on your sources and inputs, people will interpret and relate events and facts from different perspectives. What is important is that your timeline resolves all contradictions and is based on corroborating factual evidence from various independent sources. In doing so, there leaves no question about the triggering event that caused a regrettable sequence of events and what must be done to avoid future incidents. You have defined the most probable cause.

Anyone claiming to have identified the "precise cause" at the end of a reactionary investigation instantly loses credibility. There is no such thing as "precise cause" when an incident investigation must be performed to evaluate the causes of an actual disaster that happened in the past, no matter how soon the investigation is coordinated and how flawlessly the investigation is executed. Precise causes are only for those who conduct their investigations far enough in advance to *prevent* disasters. Once again, how we establish our investigation triggers will determine whether or not we are able to proclaim complete understanding over a precise cause, which of course will be proven after the corrective actions have been implemented.

Probable cause, however, is no less valuable than precise cause. When the investigation is complete and all contradictions have been resolved, you will have identified many parts of the process that did not function the way that they were expected to when the process was commissioned. This will create transparency over one or two direct causes that when addressed make everything function as expected. And when everything functions as expected, you have prevented a disaster.

LESSONS FOR US

In this chapter we demonstrated how basic investigation practices can be used to develop the knowledge needed to prevent catastrophic process releases. Following a stepwise approach without taking shortcuts helps direct the investigation to conclusions fully supported by facts and evidence. Digging deep enough to define the root cause for an incident prevents related occurrences on all types of assets and process technologies. Preserving lessons learned from incident investigations creates an umbrella of protection that can last forever. Other messages worth retaining from this discussion include:

- Basing your investigation triggers on asset performance expectations allows action to be taken far enough in advance to prevent a disaster.
- Strict adherence to a selected investigation procedure is more important than the procedure you decide to follow.
- The normalization of deviance is present in all industrial disasters. It allows undetected hazards to persist until an incident exposes it.
- Operating limits need to be defined to resist falling victim to the normalization of deviance after the process starts operating.
- Organizational culture is defined by seven common prevalent attributes.
- An investigation must always seek to describe the precedent for counterproductive decisions and actions.
- Applying recommendations to resolve the superficial causes for an incident results in multiple recommendations that make the process more complex and thus more dangerous.
- No solution is universally effective. When considering applying a solution based on lessons learned, validate that the intended action is appropriate for the specific process installation and design configuration.
- Discipline is effective only when applied far enough in advance to prevent a catastrophic process release.
- The investigation is complete only when there are no more inconsistencies, contradictions, or cavities remaining in the timeline.

REFERENCES

[1] H.W. Heinrich, Industrial Accident Prevention: A Scientific Approach, second ed., 1941, p. 27.
[2] D. Vaughan, The Challenger Launch Decision, 1996, p. 115. ISBN: 0-226-85176-1.
[3] T. D'Silva, The Black Box of Bhopal, 2006, p. 195. ISBN: 978-1-4120-8412-3.
[4] U.S. Chemical Safety and Hazard Investigation Board, Report No. 2005-04-I-TX: Refinery Explosion and Fire (15 Killed, 180 Injured), 2007, p. 143.
[5] Report of the Presidential Commission on the Space Shuttle Challenger Accident, Chapter V: The Contributing Cause of the Accident, 1986.
[6] Columbia Accident Investigation Board, Report Volume I, 2003, p. 199.

[7] Deepwater Horizon Study Group, Final Report on the Investigation of the Macondo Well Blowout, 2011, p. 10.
[8] Center for Chemical Process Safety (CCPS), Guidelines for Investigating Chemical Process Incidents, second ed., 2003, p. 47. ISBN: 0-8169-0897-4.
[9] M. Paradies, L. Unger, TapRooT® Changing the Way the World Solves Problems, 2008, ISBN: 978-1-893130-05-0.
[10] H. Bloch, F. Geitner, Machinery Failure Analysis and Troubleshooting, second ed., 1994, p. 506. ISBN: 0-87201-232-8.
[11] T. D'Silva, The Black Box of Bhopal, 2006, p. 43. ISBN: 978-1-4120-8412-3.
[12] A.S. Kalelkar, Investigation of large-magnitude incidents: Bhopal as a case study, in: IChemE: Symposium Series No. 110, Preventing Major Chemical and Related Process Accidents, May 10–12, 1988, p. 559.
[13] P. Shrivastava, Bhopal: Anatomy of a Crisis, 1987, p. 43. ISBN: 088730-084-7.
[14] W. Morehouse, M.A. Subramaniam, The Bhopal Tragedy: What Really Happened and What It Means for American Workers and Communities at Risk, 1986, p. 4. ISBN: 0-936876-47-6.
[15] Union Carbide Corporation, Bhopal MIC Incident Investigation Team Report, March 1985, p. 7.
[16] R. Willey, D. Hendershot, S. Berger, in: The Accident in Bhopal: Observations 20 Years Later, 40th Loss Prevention Symposium, Orlando, Florida, USA, April 23–27, 2006, p. 6.
[17] P. Shrivastava, Bhopal: Anatomy of a Crisis, 1987, p. 44. ISBN: 088730-084-7.
[18] W. Morehouse, M.A. Subramaniam, The Bhopal Tragedy: What Really Happened and What It Means for American Workers and Communities at Risk, 1986, p. 18. ISBN: 0-936876-47-6.
[19] T. D'Silva, The Black Box of Bhopal, 2006, p. 54. ISBN: 978-1-4120-8412-3.
[20] T. D'Silva, The Black Box of Bhopal, 2006, p. 55. ISBN: 978-1-4120-8412-3.
[21] W. Morehouse, M.A. Subramaniam, The Bhopal Tragedy: What Really Happened and What It Means for American Workers and Communities at Risk, 1986, p. 5. ISBN: 0-936876-47-6.
[22] W. Worthy, Methyl isocyanate: the chemistry of a hazard, Chem. Eng. News 63 (6) (1985) 29.
[23] H. Bloch, F. Geitner, Machinery Failure Analysis and Troubleshooting, second ed., 1994, p. 502. ISBN: 0-87201-232-8.
[24] T.R. Chouhan, Bhopal: The Inside Story, Carbide Workers Speak Out on the World's Worst Industrial Disaster, 1994, p. 34. ISBN: 0-945257-22-8.
[25] Supreme Court of India Criminal Appellate Jurisdiction, Application for Directions to Institute Charges U/S 302 (For Offence U/S 300(4)) Read With S. 35 of the Indian Penal Code, 1860 Against the Respondents Herein, Curative Petition (Criminal) No. 39-42 of 2010 in Criminal Appeal No. 1672-75 of 1996 in the Matter of Central Bureau of Investigation versus Keshub Mahindra & Ors., 2011, p. 16.
[26] C. Kepner, B. Tregoe, The New Rational Manager: An Updated Edition for a New World, 1997, ISBN: 0-9715627-1-7.
[27] Center for Chemical Process Safety (CCPS), Guidelines for Investigating Chemical Process Incidents, second ed., 2003, pp. 216–219.
[28] Center for Chemical Process Safety (CCPS), Guidelines for Investigating Chemical Process Incidents, second ed., 2003, p. 217.
[29] N. Taleb, The Black Swan, 2007, pp. 57–58. ISBN: 978-1-4000-6351-2.
[30] T.R. Chouhan, Bhopal: The Inside Story, Carbide Workers Speak Out on the World's Worst Industrial Disaster, 1994, p. 33. ISBN: 0-945257-22-8.

[31] R. Fullwood, Probabilistic Safety Assessment in the Chemical and Nuclear Industries, 1999, p. 253. ISBN: 0-750-6-7208-0.
[32] D. Lapierre, J. Moro, Five Past Midnight in Bhopal, 2002, p. 177. ISBN: 0-446-53088-3.
[33] District Court of Bhopal, India, State of Madhya Pradesh Through CBI vs. Warren Anderson & Others, Criminal Case No. 8460 of 1996, June 7, 2010, p. 29.
[34] A. Agarwal, S. Narain, The Bhopal Disaster, State of India's Environment 1984–85: The Second Citizens' Report, 1985, p. 219.
[35] W. Worthy, Methyl isocyanate: the chemistry of a hazard, Chem. Eng. News 63 (6) (1985) 32.
[36] S. Diamond, The Bhopal Disaster: How It Happened, The New York Times, January 28, 1985.
[37] D. McQuarrie, P. Rock, General Chemistry, 1984, p. 580.
[38] Union Carbide Corporation, Bhopal MIC Incident Investigation Team Report, March 1985, p. 4.
[39] United States District Court for the Southern District of New York, In re Union Carbide Corporation Gas Plant Disaster at Bhopal, India in December, 1984, 1986. MDL No. 626; Misc. No. 21-38, 634 F. Supp. 842, 12.
[40] T. D'Silva, The Black Box of Bhopal, 2006, p. 73. ISBN: 978-1-4120-8412-3.
[41] A. Agarwal, S. Narain, The Bhopal Disaster, State of India's Environment 1984–85: The Second Citizens' Report, 1985, p. 215.
[42] R. Reinhold, Disaster in Bhopal: Where Does the Blame Lie? The New York Times, January 31, 1985.
[43] T.R. Chouhan, Bhopal: The Inside Story, Carbide Workers Speak Out on the World's Worst Industrial Disaster, 1994, p. 35. ISBN: 0-945257-22-8.
[44] E. Foulke, Sweeping Workplace Safety Changes, TapRooT Summit, Horseshoe Bay, Texas (USA), April 2014.
[45] U.S. Chemical Safety and Hazard Investigation Board, Case Study 2004-08-I-NM: Oil Refinery Fire and Explosion, 2005, p. 10.
[46] ACS Joint Board/CCS Hazards Identification and Evaluation Task Force, Identifying and Evaluating Hazards in Research Laboratories, 2013, p. 13.
[47] T. Kletz, What Went Wrong: Case Histories of Process Plant Disasters, fourth ed., 1998, p. 57. ISBN: 0-88415-920-5.

CHAPTER

Role Statement and Fulfillment

4

Modern industrial disasters rarely involve unexplained phenomena. The cause-and-effect relationships that govern industrial disasters can be explained scientifically and, for the most part, have been experienced and documented before. Upon reflection, we find that a problem which seemed to take us by surprise was not too uncommon. We often note that we are dealing with a situation that had been developing over the course of many years. Frequently, the information we need to stay in control has been reported by others in the past. Others feel the obligation to compile information that prevents us from going to places where they have already been. Incorporating existing information from previous incident investigations might save us from having to deal with the same issues. At other times we may have access to the information we need, but lack full appreciation for either its value or relevance. It may take an incident to clear up confusion over how it applies to our situation. Additionally, critical experience-based information can be forgotten or allowed to fade as time passes by.

Many excuses can be made for dismissing the opportunity to learn from others. Concerned organizations are sensitive to this point. Accordingly, they define and assign specific responsibilities to collect, evaluate, preserve, and utilize practical knowledge to improve process design and operating procedures. Their goal is to prevent serious industrial incidents without having to relive painful experiences on systems of their own.

Understanding the basic principles of asset reliability (chapter: Equipment Reliability Principles) is the first step to preventing industrial disasters and most importantly the relatively minor incidents that lead up to them. The second step is using this knowledge to trigger a systematic and repeatable incident investigation procedure (chapter: Incident Investigation) in response to warning signals of an approaching catastrophic process release, while there is still time to prevent it. Demonstrating competence in these two areas prevents us from adapting to an unstable operating condition that could lead to more serious consequences. Success, however, depends on how well we can transfer this knowledge over to systems operating under our direct control. Specifically, we are referring industry's knowledge about responsible process design, operation, and maintenance that prevents industrial incidents. It is a topic that is rich with notable examples; delving into the far-reaching case studies preserved for us in thousands of pages of literature that is anything but boring.

INDUSTRIAL AMNESIA

Making responsible choices based on captured, preexisting, industry knowledge may sound like an easy task. However, accomplishing this kind of knowledge transfer has proven to be extraordinarily difficult for the manufacturing industry. Many disasters and numerous lesser incidents have emerged from hazards that were clearly identified in earlier investigations. In other words, certain hazards persisted well beyond the times when an investigation defined and should have addressed these hazards (see chapter: Incident Investigation). Looking at the public record, we find that most modern industrial disasters do not represent a failure to *understand* how an unstable operating condition might develop. Instead, most modern industrial disasters happened due to a failure to adequately apply the available knowledge from prior experience. The space shuttle Challenger (1986) and Deepwater Horizon (2010) are just two examples that readily come to mind. Interestingly, both of these disasters occurred in completely different industries. Obviously, the problem is not limited to one specific business sector. Similar difficulties are observed wherever humans are required to interface with industrial assets.

So, why is it exceedingly difficult to act upon available information about how to prevent failures for which there is already some precedent? For one thing, as time passes it becomes harder to remember why a process is designed in a certain way, and why specific operating procedures must be followed. Experiencing chronic process upsets only accelerates the rate at which we forget. While that might sound like an incredible paradox, it can be explained. The pressures to find remedies for chronic and costly events create ample incentive to apply changes. Such situations might make us doubt that adhering according to prescribed operating practices is really necessary. To consider how reconfiguring the process or deviating from policies and procedures could improve system performance, we must first forget how the process is *supposed* to work. This self-inflicted industrial amnesia helps avoid feelings of guilt if the path chosen creates unknown hazards that inflict more serious damage. If used constructively, this guilt should prevent us from moving in a direction contrary to policies and procedures. It should also create discomfort when thinking about implementing an improvised, workaround solution.

When a process upset occurs there is usually an instinctive and immediate impulse to take some sort of action, even if in hindsight our actions only equate to squirming around hoping to find a more comfortable position. Being caught in the middle of a process upset turns our thoughts to abandoning what *should* work to keep the process under control. Under these circumstances we are more inclined to trade cautious deliberation for something new, creative, and different. In the heat of the moment we probably would do just about anything to maintain process control where and when action is desperately needed. In such cases there is a tendency to do what comes naturally and according to "what makes sense," to compensate for unique situations not specifically covered in an emergency response plan. When trying to manage an unstable process, we might even be compelled to improvise using only what the process has to offer, since time is of the essence and a delayed response could increase the consequences. Nobody wants to be asked uncomfortable questions about why intuitive steps were not immediately taken to bring an unstable situation under control after the dust settles. Of course, the most acceptable option is to avoid any situation where our actions could be questioned about what made us decide to act in a certain way. This preferred option is only guaranteed when there are no process upsets that require a response. Putting it another way: avoiding unproductive discussions about what made us decide to operate the process contrary to approved practices is an eventuality that results only by maintaining continuous process control. Problems need to be prevented to avoid developing our own operating solutions that might prove ineffective or dangerous. That, then, brings us back to the question that has proven to be extraordinarily difficult to answer:

How can we effectively utilize existing knowledge to prevent industrial incidents?

On this point everyone can agree: it would not be acceptable to simply sit back and watch as an unstable process condition continues to deteriorate. Responsible process operation requires taking action to restore control as safely and quickly as possible. For that reason, many processes are designed with automatic interlocks and basic process control systems to readily detect and manage a process upset that has progressed to the point of an imminent failure. These types of advanced process control systems do not require human intervention to execute a defined course of action when a more drastic response is appropriate. Training programs, simulations, and drills are also developed to improve the accuracy and effectiveness of an emergency response plan in accordance with process safety management (PSM) standards. When faced with a process upset it is of utmost importance to respond in a way that does not make the situation worse or might only spread the problem to a different location. Becoming familiar with knowledge about maintaining process stability prepares us and others mentally to make responsible choices about how to effectively manage a process upset.

The best way to successfully manage a process upset, however, is to prevent any type of unstable operating condition that could require an extreme or possibly unconventional response. Since most industrial processes operate on a continuous basis, incident prevention is not a temporary assignment. Rather, incident prevention is a full-time commitment that involves literally everyone with manufacturing responsibilities. It is therefore appropriate to designate specific roles for the purpose of meeting process reliability performance expectations.

ESTABLISHING PROCESS RELIABILITY EXPECTATIONS

Organizations demonstrate a genuine interest in preventing incidents by assigning specific responsibilities to acquire, evaluate, preserve, and utilize practical knowledge about maintaining stable process operation. Resources can be dedicated to detect and control hazards that, if disregarded, could cause a process upset. Usually, these hazards are revealed by production constraints that appear long before a destructive sequence of events makes them impossible to ignore. Therefore, priority is first given to diagnosing and eliminating the causes for production constraints. In the context of incident prevention, production constraints are really warning signs of an approaching incident that could result in a catastrophic process release.

A typical process reliability professional's position summary (Table 4.1) divides the expectations needed to maintain continuous process control into three major responsibilities:

1. Reliability analyst: The purpose of this responsibility is to manage asset and process performance according to clearly defined life cycle expectations. At a minimum, asset life cycle expectations are specified for functional performance, maintenance plans, and total-expected life before replacement should be needed. Operating limits are at the core of this responsibility.
2. Executive investigator: This responsibility is designated to investigate any deviation of asset or process performance from clearly defined life cycle expectations. A systematic and repeatable problem-solving approach is used to develop and implement effective recommendations that can be reasonably expected to correct a problem. A verification step confirms that life cycle expectations are met after the solution is applied.
3. Knowledge transfer coordinator: This responsibility covers the acquisition and utilization of industrial knowledge on assets and processes under supervision's jurisdiction. It also defines expectations for transferring incident prevention knowledge to various groups that must act cohesively to be successful.

As with every specialized role in a manufacturing organization, successor planning must be given careful consideration. Ultimately, the incumbent who might vacate a position must groom a successor to take his or her place at the appropriate time. Successor planning in this case prevents personnel changes from interfering with the incident prevention program's continuity.

PRODUCTION CONSTRAINT REMOVAL

Production constraint elimination is one of the process reliability professional's foremost interests. Asset and process life cycle expectations are most consistently met by eliminating production constraints. As production constraints are removed, hazards that could undermine the successful PSM implementation are taken away with them. Although hazard elimination is always preferred, known hazards are normally covered adequately by process design. Such hazards are not likely a realistic production threat. Production constraints are perhaps the greatest proof that other hazards have not been adequately addressed through process design or operating methods. These are the kinds of hazards that deserve our immediate attention. Naturally, if these hazards were truly understood and adequately controlled, then there would be no reason to accept unnecessary production penalties.

To be effective, the process reliability professional must accurately diagnose the causes for production constraints by applying a systematic and repeatable problem-solving

Table 4.1 Generic Process Reliability Professional Position Summary

Role	Responsibility	Expectations	Measures
Reliability analyst	Maintain adequate control over inherent hazards that could result in premature asset functional failures or chronic repeat failures.	• Define and implement operating limits. • Review, update, approve, and enforce policies and procedures required for effectively controlling asset operation within prescribed limits. • Inspect assets protected by operating limits to validate protection consistent with expectations.	• Limit compliance (ratio of time in control over time in operation) • Chronic limit deviations (limit deviations reappearing between consecutive reporting periods) • Asset mean time between failure • Unplanned maintenance events
Executive investigator	Acquire, evaluate, preserve, and utilize knowledge needed to diagnose and address the causes for asset or process life cycle deviations.	• Define, implement, and enforce investigation criteria (triggers) that provide sufficient time to apply a solution prior to an incident. • Use a systematic and repeatable problem-solving approach to diagnose problems that meet investigation criteria. • Support the line organization by coordinating the removal of production constraints.	• Repeat failures • Reliability growth plots • Output loss resulting from production constraints
Knowledge transfer coordinator	Distribute and communicate the practical application of industrial technical knowledge at all levels in a manufacturing organization.	• Document technical information in an audience-appropriate language. • Justify cases for action to address actual process hazards detected.	• Ratio of hazards addressed by knowledge transfer, to total number of hazards addressed • New incident prevention lessons learned (from internal and external sources)

Job description: continually meets regulatory and economic commitments by managing asset and process performance according to life cycle expectations (mechanical integrity, maintenance requirements, total life expectancy, and productivity).

approach (chapter: Incident Investigation). Note that addressing production constraints requires developing a specific action plan to exercise continuous control over a process, according to its design intent. These plans might involve upgrades and projects to correct system performance issues resulting from process deterioration or neglect. Special attention must also be given to correcting inherent design defects that originate from miscalculations, invalid assumptions, or deviating from documented standards, specifications, and codes during the process design and construction phases.

The process reliability professional (engineer, specialist, or technician) position contributes to the profitability of a manufacturing operation by reducing the break-even quantity (BEQ). Controlling manufacturing costs simultaneously raises the production value and available capacity (chapter: Equipment Reliability Principles). Performance in these areas makes the role economically feasible. Although the principles of process reliability apply to everybody who is involved in a manufacturing operation, assigning the process reliability professional role to a single "owner" creates the most value. This prevents confusion over who is coordinating the industrial incident prevention program and what responsibilities are meant to govern that person's priorities.

Designating the process reliability professional as a separate job function apart from an operations-related role can be crucial. It helps to maintain an appropriate distance from daily production or output distractions that could interfere with the independent thinking needed to prevent more serious incidents. Along those lines, it may be appropriate for the process reliability professional to report directly to a level sufficient to control the routine allocation of often shifting priorities. But make no mistake about it: tactical and strategic positions both create value in a productive and lasting organization. Incident prevention is a full-time obligation that pays out tremendous dividends. The process reliability professional occupies a *strategic* position that operates with long-term process stability and regulatory compliance in mind. It is for this reason that the position must directly report to an upper-level supervisor—perhaps the reliability manager or equivalent, depending on how the organization is set up.

TRACKING PERFORMANCE MEASURES

One of the process reliability professional's primary job functions is to monitor and track a practical set of system performance statistics (measures). Collectively, these measures demonstrate the relationship and interaction between reliability, production, and process safety. They also quantify the contribution that the process reliability professional is making to the organization's overall success. To be productive, the process reliability professional must understand and analyze the factory's production constraints in a progressive order. Removing one production constraint simply moves another into its empty slot. The objective is to successively eliminate the most restrictive production constraint until none remain.

There always seems to be at least one constraint that dominates factory production. More realistically, there are probably several that are creating recognizable penalties at any given moment inside a typical manufacturing operation. By diagnosing the causes for persistent production constraints, the process reliability professional is again uniquely positioned to eliminate *actual* hazards embedded in a process. These hazards, of course, could potentially damage a factory's safety, environmental, asset protection, and economic performance. A company's brand, or reputation, is influenced by how well these priorities are actually being managed.

Perhaps the most effective way to determine what production constraint deserves the most attention is by regularly visiting the line organization. Ask the workers to explain the process problems that they are contending with and which is causing them the most distraction. Listen to their answer, devise an investigation plan, and then work together with them to solve the problem. This is really what employee participation (chapter: Employee Participation) is all about.

As actual process restrictions are lifted the BEQ goes down, productivity goes up, and the hazards responsible for process upsets and more serious process safety events are removed. The most impressive improvement, however, is the development of positive organizational culture attributes that almost certainly follows. Managing a recurring process problem day after day can create a calloused organizational culture that accepts process upsets as normal operation. Removing these distractions by demonstrating visible actions to eliminate production constraints can change an organizational culture to one that rejects thoughts about process upsets. The remarkable benefit that results from this change is that with time, everybody thinks like a process reliability professional even though they might not carry the title. In a relatively short amount of time, a healthy awareness of hazards evolves. Similar problems in other areas are addressed by layering knowledge from successful investigations upon unexplained process performance. This generates an enormous payback in terms of incident prevention, as PSM goes viral through repeated success stories.

DEVELOPING PROCESS SAFETY SKILLS

To be effective in the role, the process reliability professional must be able to efficiently acquire missing knowledge through a systematic and repeatable problem-solving approach (chapter: Incident Investigation). Every investigation should be viewed as a "stepping stone" to diagnosing a more complex production constraint or failure mechanism. This type of repetitive exposure to investigation principles shapes the process reliability professional into a true specialist in the craft of root cause failure analysis and production constraint removal. Ultimately, the role becomes critical in defining and addressing the causes for operating issues over a very wide range of process technologies and manufacturing operations. The key to success is retaining knowledge from previous investigations and following a prescribed problem-solving procedure that consistently produces actionable results.

Think also of the merits of a proper follow-up. One of the process reliability professional's primary functions is to verify that an unstable operating condition has been adequately resolved after a specific course of action has been implemented. There are several methods that can be used to perform this type of follow-up verification and analysis. Statistical methods are less subjective and are therefore preferred over their qualitative counterparts. Reliability growth plots are a useful approach to assessing the performance of manufacturing processes that might be operating under the influence of a failure mechanism.

ASSESSING PROCESS PERFORMANCE AND IMPROVEMENT

H. Paul Barringer, a professional engineer, has spent a considerable amount of time documenting and demonstrating the practical use of reliability growth plots to solve process reliability problems in the manufacturing industry. Barringer freely makes examples of his work available online (www.barringer1.com). Software can also be purchased to simplify the technical analysis of process information. The output from a reliability growth plot analysis can be used to determine if an unstable operating condition realistically represents a catastrophic process release hazard.

Top among the trending techniques extensively demonstrated by Barringer that apply directly to the process reliability professional's role are Crow-AMSAA (Army Materiel System Analysis Activity) reliability growth plots [1]. This analysis method applies mathematics to what can easily become an unscientific and unproductive assessment (what is commonly known as a "debate") over the direction in which a process reliability problem is headed. Is the problem getting worse, better, or staying the same? Many times the answer depends on whom we ask, and how much personal accountability that person senses in the matter. Regardless of where we might stand, our argument is much more convincing when it is supported by data. Confidence can be obtained by making a math problem out of a process reliability scenario. Otherwise, it becomes possible for assertions to be made purely on a subjective basis alone. Any mismatch of opinion about a problem's status would be of significant interest to the process reliability professional since it might delay taking decisive action to resolve a chronic performance issue. The prevention of industrial incidents requires demonstrating progress on eliminating the causes for persistent hazards.

A relatively straightforward fundamental principle is at the core of a Crow-AMSAA reliability growth plot. *Cumulative failures* according to *cumulative time to failure* are plotted on a log–log scale to create an almost linear relationship between the data points. The linear regression (best-fit) analysis including these plotted data points can then be used to estimate how many failures will occur at any given time, and when these failures are likely to occur, according to Eq. [4.1] [2]:

$$N_t = \omega t^{\beta} \tag{4.1}$$

where N_t is the number of failures expected at any point in time (t); t is the cumulative test time; ω is the "scale" parameter (in this case the failure rate, λ) [3]; β (beta) is the "shape" or "growth" parameter (in this case the slope of the "best-fit" line) [3].

Two important statistical relationships can be derived from this analysis [4]:

1. The slope of the line, β, which indicates whether or not the problem being evaluated is increasing ($\beta > 1$), decreasing ($\beta < 1$), or staying the same ($\beta = 1$).
2. The hypothetical Y-intercept (as a function of t and λ), which makes it possible to accurately predict when future failures will occur.

To illustrate, let us use a Crow-AMSAA reliability growth plot to analyze the maintenance history behind the chronic pump failures in the case study covered at the end of chapter "Equipment Reliability Principles." Let us say that the 12 repeat failures occurring in the 180-day period between April 17, 2003 and October 14, 2003 trigger an incident investigation according to the criteria defined in chapter "Incident Investigation." At this point there has been no "incident" other than using the installed spare to keep up with a chronic series of maintenance events resulting from repeat mechanical seal failures. Appreciating that repeat failures are a common pattern behind industrial disasters, our decision to activate an investigation at this time is a valuable defense.

To quantify the type of harm that might develop if this situation was allowed to continue, we first need to organize the maintenance record by cumulative days and cumulative failures (Table 4.2). This table contains the information that we will use to generate a log–log Crow-AMSAA reliability growth plot (Fig. 4.1). Next, we perform a linear regression analysis to generate the "best-fit" line

Table 4.2 Asset Performance Analysis Input, Based on Actual Seal Failure History

Seal Failure Event Date	Cumulative Time (Days)	Cumulative Failures (Count)	Mean Time Between Failure (Days/Failure)
April 17, 2003	**X-axis**	**Y-axis**	
May 9, 2003	**22**	**1**	**22**
May 23, 2003	**36**	**2**	**18**
June 9, 2003	**53**	**3**	**18**
June 9, 2003	**53**	**4**	**13**
June 18, 2003	**62**	**5**	**12**
June 20, 2003	**64**	**6**	**11**
July 31, 2003	**105**	**7**	**15**
August 22, 2003	**127**	**8**	**16**
August 25, 2003	**130**	**9**	**14**
September 26, 2003	**162**	**10**	**16**
September 26, 2003	**162**	**11**	**15**
October 14, 2003	**180**	**12**	**15**
December 6, 2003	233	13	18
December 9, 2003	236	14	17
December 9, 2003	236	15	16
December 15, 2003	242	16	15
December 15, 2003	242	17	14
January 28, 2004	286	18	16
March 22, 2004	340	19	18
April 1, 2004	350	20	18
April 3, 2004	352	21	17
April 7, 2004	356	22	16

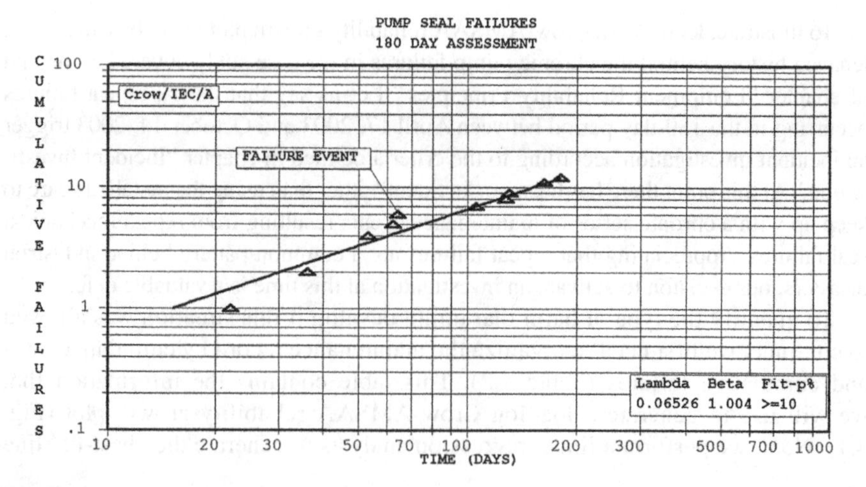

FIGURE 4.1

Reliability growth plot for seal failure history.

between the plotted points (note: this can all efficiently be done with the software mentioned earlier). Using only the first 6 months' worth of run data (180 days), we come up with:

- $\lambda = 0.065$
- $\beta = 1.004$

With this information, we can use Eq. [4.1] to calculate the number of failure events that would occur if the problem was allowed to persist for a full year:

$$N_t = 0.065 \times 365^{1.004}$$

$$N_t = 24 \text{ failures}$$

Just for the record, note how accurately this analysis corresponds to actual system performance. The calculated 24 failures in 365 days are not far off from the 22 repeat failures that did actually occur in 356 days. These data are especially compelling considering the fact that the last two actual failures occurred within 4 days of each other (day numbers 352 and 356). If this actual failure rate (one failure every 4 days) was to continue for the remainder of evaluation period there would have been two more mechanical seal failures (raising the total to exactly 24 failures) at day 365, had time not been cut short by the incident 9 days earlier on day 356.

Even if we exclude this mental exercise from the analysis, the example still demonstrates the accuracy that results when a scientific approach is used to quantify asset performance history. From this information, we can see that the actual failure rate at this pump location ($\lambda_1 = 6.53 \times 10^{-2}$) is much higher than what would be expected if the repairs (a major overhaul on both the primary and spare pump) took place at a frequency more consistent with industry statistics. In this instance we are referring to industry statistics matching a typical 4-year maintenance turnaround cycle ($\lambda_2 = 1.37 \times 10^{-3}$). In determining if this asset reliability problem represents a potential process-safety hazard, two criteria must be considered [5]:

1. The β (beta), which in this case is 1.004, is about as close to 1.0 as it can reasonably get. This slope is what would be expected when a chronic problem is not getting any better (or worse) with time. Therefore, this case likely involves an inherent failure mechanism that is being managed through recurring breakdown maintenance even though the system would not likely have been designed for this type of operation. Perhaps the normalization of deviance could explain the prolonged difference between actual and expected performance.
2. The failure rate ($\lambda = 0.065$) or mean time between failure (MTBF) ($\theta = 15$ days) falls seriously short of industry benchmarks [6]. In other words, the failure is creating a competitive *disadvantage* compared to other companies in the market who have figured out how to avoid excessive maintenance at this pump location. A more realistic failure rate might be based on a 4-year turnaround interval. The current situation represents a reliability difference that increases the potential for personnel exposure during maintenance by a factor of 4800% $\left(\frac{\lambda_1}{\lambda_2} \times 100\right)$.

Please note that, in all cases, good engineering judgment must be allowed to prevail over the statistical analysis. In this case, the coexistence of the two distinct conditions noted above provides compelling evidence that a process-safety hazard is involved. Not only is a chronic-failure mechanism dominating process performance, but major process-intrusive maintenance is occurring at an unsustainable rate. Therefore, the incentive to take remedial action is quite strong. Action is also strongly favored by the economic savings that would result from simply reducing the maintenance to a more reasonable rate.

JUSTIFYING CASES FOR ACTION

The statistical evaluation of process performance data demonstrated in this example provides all the ammunition that a process reliability professional would likely need to obtain permission to take action. In the assessment it was determined that the financial burden created by the chronic-failure mechanism responsible for frequent maintenance can be eliminated upon implementing an effective solution to the problem. For example, the following expression calculates the estimated economic savings that results from solving a process reliability problem (Eq. [4.2]):

$$\text{Savings} = \text{Current Cost} - \text{Future Cost} \qquad [4.2]$$

If we were to estimate the average repair or replacement cost for a typical centrifugal refinery process pump mechanical seal to be about $15,000, then increasing the MTBF to 4 years at both the primary and spare pump ($\lambda = 1.37 \times 10^{-3}$) would generate the following savings over the course of 1 year:

$$\text{Savings} = (24\ \text{failures})\ (\$15{,}000) - (0.50\ \text{failures})\ (\$15{,}000)$$
$$\text{Savings} = \$352{,}500$$

If we extend these savings over the targeted 4-year maintenance turnaround interval where an effective solution would dispense with the need for corrective maintenance between scheduled pump overhauls, then the total savings would reach $1.41 million. This economic basis could be used to justify implementing a specific plan to address the problem.

Notice that this justification covers only maintenance (parts and labor) costs. It does not include the safety or environmental benefits that would also result from addressing a repeat failure mechanism that is causing a very low MTBF. Neither does our justification quantify the benefits that are obtained as workers observe how we respond: we respond in a way consistent with the actions expected from a failure-restrictive organizational culture. Finally, we have not even considered the economic improvement that would result from lowering the BEQ, or eliminating production slowdowns or business interruptions if the maintenance burden is causing a production constraint. All of these arguments are valid points to include when developing the basis for eliminating a chronic process reliability problem.

VALIDATING PERFORMANCE RESULTS

We wish to restate that a process reliability professional provides a valuable service by verifying that corrective actions taken to solve a problem have been successful. It is easy to walk away from a problem after taking action, thinking that the job is complete, and things will get better as expected. Responsible operation, however, requires monitoring the solution's effectiveness or performance. We monitor the process to confirm that expectations are being met and that value is being generated according to the plan. This is especially important, considering the fact that implementing a successful solution involves three steps:

1. Investigate the problem according to a systematic, repeatable procedure.
2. Take action, and
3. Verify (by measuring) success.

After resources have been provided to correct a problem, there really is only one acceptable outcome: the problem has been solved; the hazard or constraint can no longer exist. By placing an even weighting on each of the three steps for success, our chances for success are only 33% if all we do is take action and do not actually follow up. Therefore, the odds for success would not be in our favor. Investigating a problem according to a systematic, repeatable procedure might increase our probability for success by another 33%. In this case our performance goes from an "F" to a "D," so although there is a higher chance for success, it probably is not sufficient for what most people would consider to be a good investment.

The final contribution to our success is obtained by validating that the selected action resulted in the anticipated performance improvement. The objective here is to verify and document that implementing the agreed-upon approach has fully solved the problem. Incorporating these three requisites for success into a cohesive problem-solving effort then raises our chances for success to 100% (A+). Once again, a reliability growth plot analysis is one way to objectively evaluate system performance in response to changes made. Because these changes were made to address a problem, a post implementation evaluation is of great importance. A diligent process reliability professional stays in the game, so to speak, and considers evaluation work as one of the most important follow-up activities.

To demonstrate, let us once again use our chronic pump mechanical seal failure data to simulate what would statistically happen upon implementing an effective solution (Fig. 4.2). In this example we have plotted the actual maintenance history through the last failure when the catastrophic process release occurred on April 7, 2004; our purpose is to show the difference in performance when an effective solution has been applied. Notice that adding more failure history to the analysis does not considerably change the β that was calculated when only the first 6 months were taken into consideration. This result favors the conclusions reached about a chronic repeat failure that might represent a process-safety hazard. After implementing a solution, we would expect to see an inflection point in the asset performance data. In this case, we have simulated what an improvement would look like by dropping the failure rate to 1.37×10^{-3} after the first 180 days. We see a cusp in the data, with the performance data thereafter following a new, more horizontal trend line. In real life,

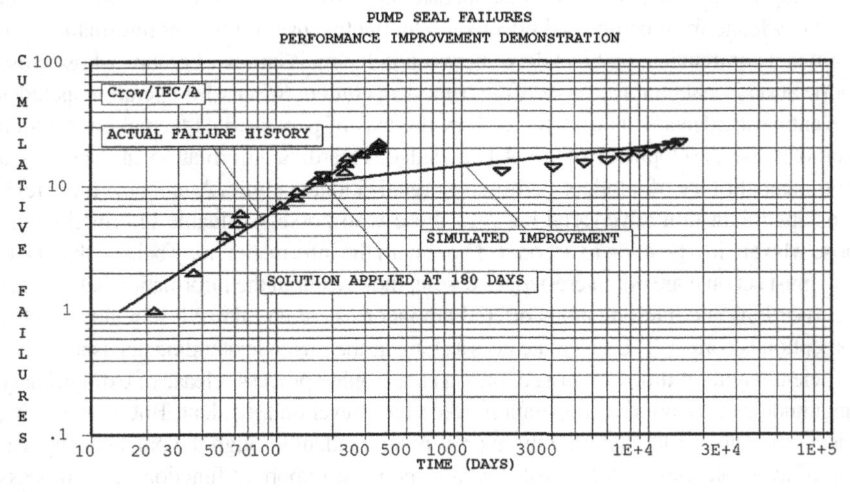

FIGURE 4.2

Simulated reliability growth plot after applying solution.

however, we observe the actual performance trend continuing on a nearly identical course as before. This is what we would expect if no change was made to address a chronic asset reliability issue.

After confirming that the asset reliability problem has been solved, the ongoing monitoring responsibilities should be transferred to the process engineer or technology specialist assigned to the operating area where the solution was implemented. If performance improvement is not demonstrated, then the problem must be reevaluated to determine what was missed so that appropriate corrections can be applied. Consistent with the incident investigation principles discussed in chapter "Incident Investigation," the effort to correct the problem should continue until performance improvement has been confirmed. Likewise, after the issue is resolved and becomes part of the normal process monitoring program, any unanticipated performance loss must be explained. If we wish to consistently maintain acceptable control, the explanation must be immediate and issues addressed without undue delay. Needless to say, reverting to complacency could undermine many otherwise effective solutions.

Once the information needed to remain in control has been obtained and a particular solution implemented, we have to adhere to it. *Lasting solutions* are possible only when everyone is dedicated to sustaining success. Conceivably checks and balances have to be devised and might have to become part of our permanent audit program to verify that processes are being operated according to policies and procedures.

PRESERVING THE VALUE OF ACQUIRED KNOWLEDGE

There are many cases, however, where the information that we need to stay out of trouble is already in our possession, but misunderstandings over its value or relevance prevent this knowledge from being used effectively. Ineffective use of relevant information can result in a communication breakdown between individuals and workgroups whose close cooperation is essential for successful business operation. Successful operation includes preventing incidents. It is easy to see how the loss of perceived or factual cooperation promotes the development of tribal knowledge. Hoarding information in "silos" is a prominent attribute of a failure-permissive organizational culture. After some time, relationships might turn adversarial between those who possess valuable information and those who are in a position to act on it. Those with the information may believe that those who must act on it are not interested. Likewise, those without the information believe that the assertions and interpretations offered are inconsistent and flawed. In reality, neither perception is correct. The stalemate lasts as long as the misunderstanding persists. After a sufficient length of time has passed, only a catastrophic process release of extraordinary magnitude can restore the cooperation needed to prevent an incident. But by that time, it is all too late and the opportunity to prevent an incident through the practical application of available knowledge is lost. Perhaps the most important function that a process reliability professional can offer is to prevent these types of *relationship-related* disasters from causing *physical* disasters directly involving a manufacturing process.

Not surprisingly, personal intervention is often needed to resolve any misunderstandings. Personal intervention to protect others from acting unsafely is a responsibility that applies to everybody in a manufacturing organization. However, finding the personal strength to intervene when necessary, and to do so without being invited, is exceedingly difficult. Yet, tactful intervention is an identifying marker in a failure-restrictive organizational culture. It is essential for anyone committed to consistently maintaining acceptable process control. While it is much easier to stay out of passionate turf battles over technical information, avoiding the necessary conversation may well lead to making a catastrophic error possible.

COOPERATIVE COMMUNICATION

To appreciate the importance of facilitating effective communication between different individuals, workgroups, and departments in an organization it is particularly useful to consider the events that preceded the space shuttle Challenger Disaster (1986). Diane Vaughan, through extensive research that she fully documented in *The Challenger Launch Decision*, creates a fact-based, dynamic, and compelling look at circumstances routinely encountered in the manufacturing industry [7]. In a production environment, it is quite common for one party's good intentions and another party's desire to stay out of disputes to lead to an incident of extraordinary magnitude. In many such cases, the information needed to avoid disaster is readily available. The hazards are fully identified along with acceptable ways to control them. Yet, the decision is made to push ahead without taking necessary precautions. Everyone seems to be in agreement over the selected course of action before the incident occurs. Then suddenly all is lost and everyone laments over why a more aggressive approach was neither advocated nor implemented in conscious efforts to control a foreseeable sequence of events. The question we might ask about this common occurrence is not so much, "How can situations like this be prevented," but, "What makes regrettable choices like this possible to begin with?"

The degree of performance achieved by a process reliability professional is reflected in these types of decisions. A process reliability professional is responsible for adequately conveying the meaning of critical information before a decision is made. It may be the decision-makers' prerogative to act or not act on the information conveyed by the process reliability professional. However, logic tells us that people cannot act responsibly with information that they do not understand.

A dose of realism allows us to conclude that those who have ultimate decision rights are not universal geniuses. They often include people higher up in the organization who may simply not understand the value and application of seemingly routine technical data. By the time an incident elevates the relevance of certain data to their personal attention it will be too late for its effective use.

The process reliability professional seeks ways to prevent this situation from developing. This is done by promoting an accurate understanding about the value of technical information so that responsible decisions can be made on the basis of preexisting knowledge. In administering this incident-preventing flow of information, the process reliability professional assists personnel at all levels within an organization to meet their personal and mutual performance expectations. The various job functions and organizational levels to whom this information may have to be explained include technical support, production, reliability, supervisors, and accounting. And much of the incident-preventing information originates not from within the factory's existing data files but from actively pursuing outside sources. The responsible gathering of applicable industrial knowledge is made possible through various systems (conferences, trade journals and magazines, Websites, books, standards, codes, specifications, formal and informal networking practices, etc.). Explaining how retrievable outside data and information relate to incident prevention at a local site is part of the process reliability professional's dedicated role.

The depth of the entire issue can be fully appreciated by carefully examining the way information was exchanged on the eve of the space shuttle Challenger launch (January 27, 1986). A meeting (teleconference) took place that night between the contractor responsible for the design and reliable operation of the solid rocket booster (SRB) field joints and supervisors with decision rights over the launch. On the agenda were concerns over the cold temperature forecast for the following morning when the launch was scheduled; this might require making the decision to postpone the launch. The contractors responsible for the mechanical integrity of the SRB field joint had the information they needed to convince supervision to cancel the launch. It made complete sense to the contractors. Why would it not? After all, they were the people in possession of the technical information about temperature's influence on field joint reliability and understood its significance to launch safety. Unfortunately, their argument did not make sense to the supervisors with decision rights, whom they needed to convince [8].

How can such a difference in opinion exist when a common set of technical information is involved? Everybody was looking at the same information to base their decision on. To understand how a conflict might develop in similar cases under our control, it is important to consider the way that technical information was described to supervisors as represented by the data displayed in Fig. 4.3 [9]. The mental image created by the contractors relates the number of observed field joint "incidents" (post launch defects) to the temperature of various launches. The impression that the information on the chart *intended* to leave is that a shuttle launch should not be attempted below a minimum temperature of 53°F. That was the argument made by the contractor on the eve of the launch. The supervisors, however, rejected that argument.

Without explaining the details in the chart any further, it is quite easy to understand why supervision did not understand the contractor's concern based on the information that was presented. Why? For several reasons, namely:

1. The information simply does not promote the concept of a relationship between the number of incidents and the launch temperature. For one thing, the data show that a launch at 75°F produced more failures than five others that took place at lower temperatures. Does this suggest that the incident rate increases above 75°F as well as below 53°F?

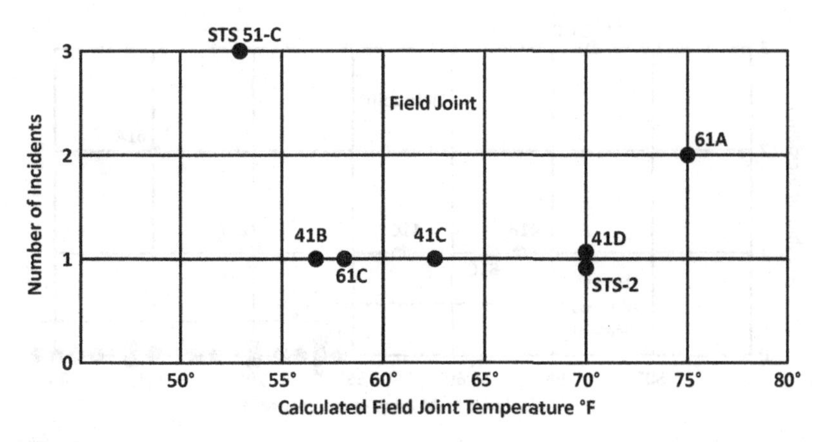

FIGURE 4.3

Technical basis for recommending a minimum temperature limit (expert knowledge required).

2. The hazard responsible for incidents does not appear to get any worse between 55 and 75°F. Every launch recorded in this range of 20°F resulted in exactly one defect.
3. There is only one data point that shows three incidents at 53°F. Is there any other evidence that the problem will get any worse below that temperature, or will the incident rate go down as it did when the temperature dropped below 75°F? Were the incidents at 53°F anomalies similar to those recorded at 75°F might have been? Are there any other factors that could account for more incidents on this specific launch or is temperature the only parameter that influences the integrity of the joint?

In the context of the concern expressed by the contractors, the information displayed in the chart seems to raise more questions than it answers. Delaying a mission, whether rescheduling a space shuttle launch or shutting down an industrial process due to safety concerns is a weighty decision that must be supported by a reasonable argument. Such an argument does not exist in the above example. Supervisors would not be inclined to take aggressive action on the basis of this kind of information.

Surprisingly and with great regret, the data do in reality support the argument made by the contractors. If we define a successful launch as one in which there are no incidents, then the contractors made an honest but critical mistake by excluding the history of successful launches from the data set. Perhaps the contractors considered only space flights that resulted in incidents to be relevant to the discussion. However, when the record of successful flights is added to the previous data, a completely different message is conveyed (Fig. 4.4). Notice now that there were no successful launches below 65°F, which is quite different from the consistent history of successful field joint performance above 65°F. Field joint failures are

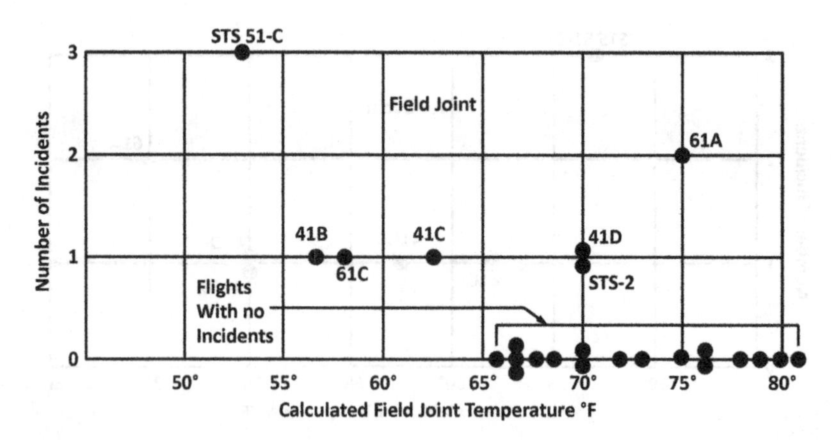

FIGURE 4.4

Technical basis for recommending a minimum temperature limit (novice knowledge required).

clearly a function of launch temperature in the new chart. The information has not changed—there is just more information. Based on this analysis it is reasonable to conclude that the chances for an unprecedented joint failure are much higher at launch temperatures below 53°F. The information that incorporates the data from all launches provides a compelling argument. Seeing the entire launch history eliminates the confusion that is generated when only some of the decision support data is examined.

Back to the topic of having the information needed to avoid an incident, but failing to act responsibly with the information, there is no reason to acquiesce when a credible technical basis for a certain decision is present but those with decision rights are not acting appropriately. If a situation like this develops, it should be taken as a signal that the message is not coming across clearly. Chances are that many problems involve situations as described above; the information can create value only when others understand what it really means.

Explaining to others what technical information means is sometimes very difficult. The manufacturing industry operates sophisticated processes that are not always the most intuitive. Many modern businesses are run by commercial analysts with Master of Business Administration (MBA) degrees, and not scientists. Getting to the point where the right people can act responsibly with the knowledge that industry has amassed over the decades takes much work. That work only pays off when individuals in possession of critical information are persistent with pointing to its meaning and value until the correct understanding is developed with those in authority.

With this insight into cooperative communication, be willing to modify the way you present technical information to others if at first they do not understand it. Changing your approach might seem uncomfortable and less logical to you, but the adjustment might be needed for your audience to comprehend what you are saying. It is acceptable to change the way you present information, but it is never acceptable to back down from it when others are not following you. Victims are created both literally and figuratively when yielding to authority allows a bad business decision to be made. Making responsible decisions requires acting according to what the technical record supports.

SUPERVISION'S RESPONSIBILITIES

A successful PSM program is contingent upon supervision's performance. Supervisors are ultimately held accountable for the success or failure of the programs and initiatives that operate under their jurisdiction. For this reason, they usually specify a practical set of system health and performance measures. Routinely monitoring these measures gives them the confidence they need to continue moving in a direction that works or make adjustments if needed to improve control.

The implications of supervision's influence over PSM's success are too significant to be measured by lagging indicators alone. Waiting for an incident to expose a performance issue cannot realistically be expected to acceptably prevent a catastrophic process release. A more conservative approach uses leading indicators to verify that hazards that could undermine acceptable PSM performance are under control.

It is appropriate to concentrate on four specific measures that are directly related to the supervisor's ability to promote a failure-restrictive organizational culture. These include the following [10]:

1. The management strategy, or the fraction of the system's failure rate resulting from failure modes that were not detected prior to commissioning a new, or modifying an existing, process.
2. The fix effectiveness factor, which is the incremental decrease in a failure rate after corrective actions have been applied.
3. The rate at which active, unmitigated failure modes are being detected.
4. The time it takes to develop and implement corrective actions.

These performance measures are dependent on another set of responsibilities reserved for supervisors. These responsibilities involve supervision's use of decision rights for the purpose of setting the manufacturing operation's priorities. Specific responsibilities related to supervisory functions include the following [11]:

1. Revising the production plan as needed, to provide safe operation in the interim period while corrective actions are being developed and implemented.
2. Increasing testing, research, and evaluation frequency of an unstable process, to create an acute awareness over a detected hazard.
3. Funding additional developmental efforts, to design the most effective approach for controlling hazards detected in a specific process.

4. Adding or reallocating manufacturing resources, to maintain an adequate amount of focus on managing a problem before the normalization of deviance obscures its recognition.
5. Interrupting production if needed until a later time when minimum asset reliability requirements can be demonstrated for safe operation.

Last but not least, supervisors must confer with appropriate human resources representatives to judiciously administer a progressive disciplinary program. This relates to instances where behavior consistent with failure-permissive organizational culture attributes is being demonstrated. Specifically, supervisors must support disciplinary action under the following circumstances:

1. Improvising with process hardware or procedures in any way, shape, or form.
2. Creating a workaround solution to overcome (bypass) a reliability or production constraint.
3. Using the management of change program to adapt to, incorporate, or accommodate process hazards.

To reinforce the basic concept of the *constructive* use of discipline as mentioned in chapter "Incident Investigation," disciplinary action is only effective when it is administered before an incident exposes the behavioral defects. For the PSM vision to succeed, discipline must take place independent from an incident investigation. Incident investigation programs that incorporate disciple will discourage employee participation while inhibiting the cooperative transfer of information. This ultimately damages transparency and creates tribal knowledge. These conditions undermine the development and continuity of a failure-restrictive organizational culture, which in turn makes the process more difficult to control.

Supervision's responsibility encompasses acting consistent with accrued knowledge to prevent incidents. This responsibility extends to preserving lessons for others to benefit on a consistent and perpetuating basis. However, as we discussed earlier, the process reliability professional's activities will ultimately determine how well supervisors can perform according to these responsibilities and expectations. Supervisors can only be expected to act responsibly with incident prevention information when they fully understand how it applies to systems under their control. It is not realistic to assume that supervisors can grasp the meaning of information that they can readily access without someone on the technical staff taking time to explain its significance. If an explanation or solution was inherently and intuitively obvious, then the incident would likely not have happened in the first place. With this thought in mind, the process reliability professional must be able to provide a compelling technical argument for supervisors to take action. Otherwise the opportunity to act responsibly with available knowledge will fail, and an incident will occur regardless.

When someone who understands the value of industry knowledge fails to defend its use for making an informed decision, where does the breakdown occur? Is it with the person in possession of the valuable information or those who require more help to fully understand its practical application? The responsible use of available industry knowledge requires taking an introspective look at how the transfer and communication of information is managed and retained for beneficial purposes. Perhaps this is why it has been extraordinarily difficult to apply lessons already learned about how to prevent industrial disasters. The power to act responsibly with information made available to protect us and others is placed squarely under our control.

LESSONS FOR US

Incident prevention is a full-time commitment that involves different people who collectively represent a manufacturing operation. Specific responsibilities can be assigned to a central owner, who coordinates activities directly related to maintaining effective communication and preserving knowledge that helps those in authority make responsible operating decisions. These responsibilities include explaining the value and practical use of technical information that applies to processes under their control. Other points relevant to responsibilities needed for a PSM program to operate effectively are:

- Eliminating production constraints offers a direct way to address process safety hazards.
- Most catastrophic process releases can be prevented by applying available industry knowledge.
- Industry knowledge must be communicated and understood for it to create value.
- The hardest part about taking advantage of lessons learned is speaking up when others are about to make a mistake.
- System performance improvement must be verified after a solution has been implemented.
- Never allow a misunderstanding to control asset or process performance.
- Consistent, uninterrupted incident prevention results when everyone thinks like a process reliability professional.
- Patience, persistence, and confidence are sometimes needed to teach others the meaning of important technical information.
- A progressive disciplinary program only supports the PSM effort when it is applied *before* an incident makes it necessary.
- Yielding to authority is appropriate in a manufacturing organization *unless* doing so allows others to make irresponsible decisions about how to design or operate a process.

REFERENCES

[1] H.P. Barringer, in: Use Crow-AMSAA Reliability Growth Plots to Forecast Future System Failures, International Maintenance Excellence Conference (IMEC), Toronto, ON, November 3, 2006, 2006.

[2] U.S. Army Materiel System Analysis Activity (AMSAA), Department of Defense: Reliability Growth Management, 75, 2011 (MIL-HDBK-189C).

[3] U.S. Army Materiel System Analysis Activity (AMSAA), Department of Defense: Reliability Growth Management, 79, 2011 (MIL-HDBK-189C).
[4] H.P. Barringer, in: Predict Failures: Crow-AMSAA 101 and Weibull 101, International Maintenance Excellence Conference (IMEC), Kuwait, December 5–8, 2004, 4, 2004.
[5] H.P. Barringer, in: Predict Failures: Crow-AMSAA 101 and Weibull 101, International Maintenance Excellence Conference (IMEC), Kuwait, December 5–8, 2004, 9, 2004.
[6] H. Bloch, Pump Wisdom: Problem Solving for Operators and Specialists, 184, 2011, ISBN:978-1-118-04123-9.
[7] D. Vaughan, The Challenger Launch Decision, 200, 1996.
[8] D. Vaughan, The Challenger Launch Decision, 355, 1996.
[9] D. Vaughan, The Challenger Launch Decision, 383, 1996.
[10] U.S. Army Materiel System Analysis Activity (AMSAA), Department of Defense: Reliability Growth Management, 6, 2011 (MIL-HDBK-189C).
[11] U.S. Army Materiel System Analysis Activity (AMSAA), Department of Defense: Reliability Growth Management, 6, 2011 (MIL-HDBK-189C).

CHAPTER 5

Spent Caustic Tank Explosion: A Case Study in Inherently Safer Design [1]

Two workers were implementing a normal operating procedure to transfer spent caustic into a storage tank when it exploded suddenly and without warning. The tank launched about 30 ft into the air before coming down close to where it originally stood. One of the workers was kneeling down behind a set of chemical storage totes when the tank exploded. These containers absorbed the impact of the explosion and shielded him from the hazardous, fiery process spray. However, the concussion of the blast knocked him to the ground, leaving him disoriented. Upon coming to his senses a few moments later, he realized what had happened and found a safe exit path from the fires that were burning around him. He then immediately radioed for help before joining the other workers who had already started activating local fire monitors to put out the fires.

Emergency response teams happened to be conducting training drills in the factory when the explosion occurred. They heard the blast and saw black smoke rising to the west of where they were located. Before receiving the distress call they began driving in the direction of the black smoke. In less than 5 min they arrived at the scene and were coordinating the effort to bring the situation under control. All fires were extinguished within 30 min. From that point the emergency response was systematically completed without further difficulty.

The incident ended without any injuries or fatalities. However, the property damage produced by the explosion was sufficient to interrupt spent caustic-disposal operations. Although the emergency response could not have been more effective, site supervision appreciated how randomness had made the difference between a near miss and a disaster. For example, any collateral damage that might normally be expected when large objects become airborne in an integrated, congested process unit was not experienced in this instance. Remarkably, the tank came back down upon its base without striking any other process-containing objects in the surrounding area. Also, neither of the two workers in the immediate area were harmed by the explosion. The worker closest to the tank was shielded by two portable storage containers stacked in front of him. These large plastic totes absorbed the full impact of the release, essentially removing the potential victim from the direct line of fire. Lastly, the emergency response team arrived within minutes because prescheduled drills and simulations were being conducted at the time of the incident.

Rethinking Bhopal.

Collectively, this series of fortunate events limited the consequences of the explosion to economic losses only.

The incident itself came as a complete surprise to factory personnel that took pride in their history of exemplary process-safety performance [2]. Prior to the explosion there was neither concern nor awareness over an explosion hazard inside the tank. After all, the process had operated safely since the factory was constructed about a half-century earlier. More recently the system had been fitted with a hydrocarbon removal process. This process upgrade instilled further confidence in safe process operation for the foreseeable future. If an explosion was possible, then the probability was higher before this upgrade was applied. These contradictions made the incident very difficult to accept. Immediately, a formal investigation was organized. The study was commissioned to resolve these inconsistencies and incorporate lessons that could be used to improve the factory's process safety management (PSM) performance moving forward.

WITNESS STATEMENTS

The incident occurred close to the end of a comfortable, late summer day. Many operations were taking place inside the factory during this time when the climate was more accommodating. For this reason, the quality of input from various observers was higher than what might normally be expected if a similar incident was to occur in the darkness of night or under inclement weather conditions.

According to the eyewitnesses, a sheer white cloud appeared above the tank just after the explosion caught their attention. Some witnesses also reported seeing the tank flip over once as it launched about 30 ft into the air, while spraying fire into the process unit below. When it hit the ground, a large yellow and orange fireball ascended into the sky. The tank landed on its side, almost exactly in the same spot where it had been standing.

PROCESS AND ASSET DESCRIPTION

Sulfur is poisonous to many catalytic industrial processes. This is one of the reasons why some form of pretreatment is usually needed prior to the catalytic phase of a reaction process where sulfur compounds might be present. A simple pretreatment approach (Fig. 5.1) involves "sweetening" the process by removing sulfur compounds from liquefied petroleum gas (LPG). After the sulfur is removed, the process can safely be introduced to a reactor that contains an appropriate catalyst.

To perform this purification step, propane containing a mixture of gases including H_2S is first cooled and pressurized before being introduced into a liquid–liquid contactor as "sour LPG." Once inside the contactor, the sulfur-containing H_2S reacts with sodium hydroxide (NaOH) according to the following reaction (Eq. [5.1]):

$$H_2S + NaOH \rightarrow NaHS + H_2O \qquad [5.1]$$

This conventional acid–base reaction produces a salt (NaHS) that remains in solution with the sodium hydroxide as it continues to dilute while the reaction

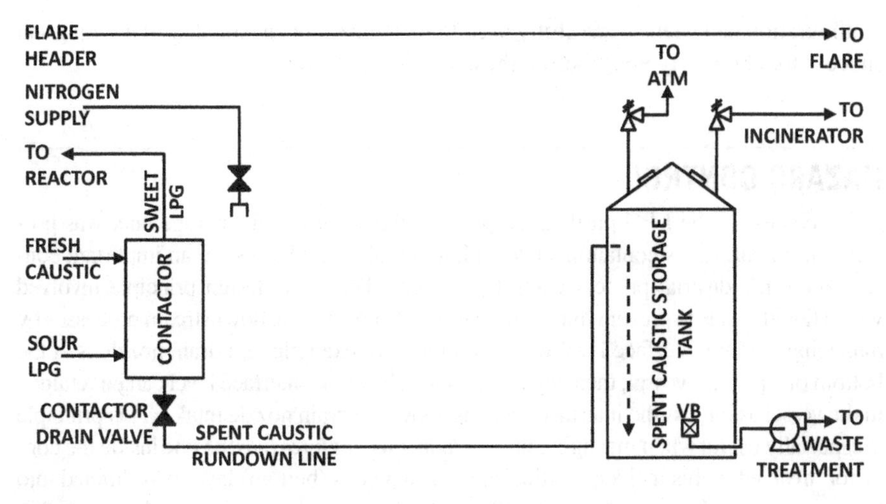

FIGURE 5.1

LPG sweetening process.

proceeds. The organic and aqueous substances inside the contactor separate to form an interface layer. The top organic layer (decontaminated propane) removed from the contactor is transferred to storage or to a catalytic reaction process. In the catalytic reaction process it combines with other hydrocarbons to manufacture higher molecular weight products of greater value.

The reaction process inside the contactor continuously decreases the strength of the circulating caustic solution. The efficiency of H_2S removal depends on maintaining a sufficient NaOH concentration inside the contactor. If the diluted NaOH solution concentration becomes too low, then the reaction performance might suffer, and contaminated LPG could exit the contactor. To prevent this from happening, a portion of the diluted NaOH ("spent caustic") is periodically drained from the contactor and replaced by an equal volume of "fresh caustic." Doing so maintains an adequate caustic strength inside the contactor for the LPG purification process to continue meeting its functional expectation.

The atmospheric storage tank that exploded was designed to receive spent caustic batches that were drained from the contactor and other assets in aqueous caustic service. The tank was a cone-roof asset designed to contain up to 35,000 gallons of spent caustic. The tank was located within a production unit far from the tank farm where larger finished product tanks were situated. A common rundown line routed spent caustic drained from other process locations into the tank. A pump at the base of the spent caustic storage tank slowly discharged the spent caustic into the wastewater treatment plant at a constant rate of about 3 gallons per minute. This process configuration absorbed large quantity liquid spent caustic slugs to protect a waste treatment and discharge system that was sensitive to sudden pH swings. Therefore, the spent caustic disposal system represented an environmental control system that any company might use to meet prevailing regulatory requirements. This process configuration had been successfully used to manage spent caustic disposal with great success for about fifty years before the explosion occurred.

HAZARD CONTROL

In the context of the LPG purification process, the spent caustic storage tank was protected from hazardous contaminants by a liquid seal. Liquid seals are an important consideration in industrial process control (Fig. 5.2). The basic design principle involved with a liquid seal is to prevent light hydrocarbons from entering downstream processes by managing a liquid interface level while draining. For example, if a drain nozzle is at the bottom of a process vessel, then any compounds above the interface level can be retained in the vessel as long as the interface remains above the drain nozzle intake. This principle is especially useful when managing organic/aqueous mixtures like the one inside the contactor involved in this incident. In this case, the aqueous (bottom) layer was drained into the storage tank while the organic (top) layer remained inside the contactor by preventing the interface level from dropping into the drain nozzle intake. Loss of level control while draining spent caustic into the storage tank could cause the interface level to enter the

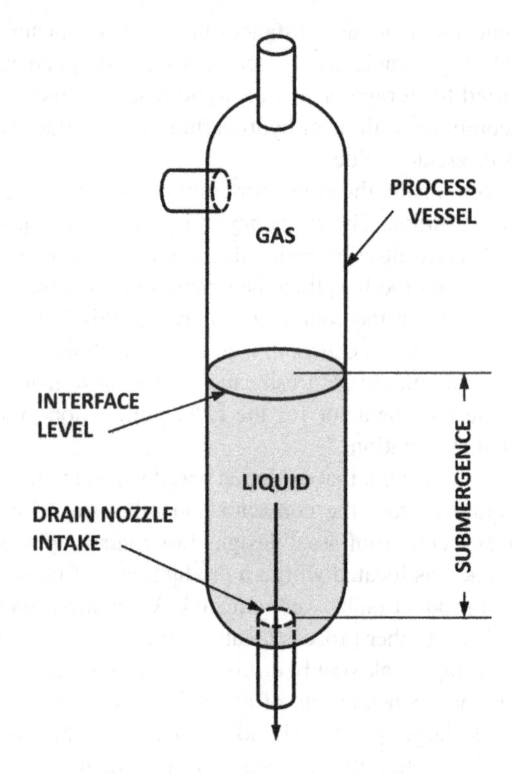

FIGURE 5.2

Basic function of process liquid seals.

drain nozzle intake. In this case, LPG would be routed into the spent caustic storage tank, which was not designed to contain volatile hydrocarbons. This unintended operating condition would increase the atmospheric tank's internal design pressure beyond what an atmospheric cone roof tank could withstand.

The storage tank's internal pressure was controlled by a conservation vent that allowed air to flow both in and out of the tank. Under normal operating conditions, a rising liquid level inside the tank would cause tank vapor to exit. This action would prevent internal tank pressure from exceeding its maximum limit. Likewise, as the tank level dropped, the conservation vent would allow air to enter the tank to prevent the sides from collapsing in. Under abnormal operating conditions, the tank's pressure relief system might automatically open. This scenario would happen if volatile hydrocarbon was to inadvertently enter the tank.

Failure to adequately control the tank's internal pressure could potentially lead to a wall rupture. These conditions could release the process into the atmosphere. This type of incident would represent an environmental hazard but could also create a safety hazard due to exposure concerns. The tank's caustic contents were not considered suitable for human contact. Therefore, special personal protective equipment was necessary in cases where intimate contact with the process was possible. The tank was also protected by an automatic pressure relief valve that would open if for any reason the conservation vent happened to fail. The tank was constructed with low-grade carbon steel, which was an appropriate choice for this particular service not expected to contain H_2S.

A vertical fill line extended from the top of the tank where spent caustic would enter and down to the bottom below the tank's normal liquid level. This design was consistent with industry practices to control static electric ignition hazards that could result by splashing liquid down into containment from an elevated location. This control method definitely applied to this particular service, where the spent caustic entering the tank might be expected to contain trace amounts of flammable propane and butane under normal operating conditions. For the extensive life of the process, this method of hazard control, combined with standard tank grounding and bonding practices, had effectively managed any potential ignition hazard inside the tank.

BACKGROUND

In compliance with industry regulations, the spent caustic disposal system was included in the factory's recurring process hazard analysis (PHA) program to detect and assess design and operating hazards that could potentially result in a catastrophic process release. Three years before the tank exploded, a normal PHA revalidation cycle occurred as required on a 5-year recurring basis. During that meeting the PHA team noted that spent caustic transfers into the storage tank were being performed by a manual procedure. A "low-level" scenario could be caused if the worker inadvertently left the contactor drain valve open while routing spent caustic into the atmospheric storage tank. This error would compromise the liquid seal inside the contactor by dropping the interface level below the contactor drain nozzle intake.

FIGURE 5.3

Refinery LPG fire.

Photo credit: U.S. Chemical Safety Board.

The resulting loss of the liquid seal would allow LPG to contaminate the tank. Any LPG entering the tank could potentially flash into the atmosphere through one of the tank's two roof hatches or its conservation vent. The ultimate outcome of this proposed scenario could be destructive to life and equipment. Fatalities could result from a vapor cloud explosion after a large-scale flammable vapor (propane) release from the spent caustic storage tank. After crediting all available safeguards to prevent this scenario, the PHA team determined that one additional independent protective layer was needed to acceptably control the hazard represented in this scenario.

The history of industrial propane releases into the atmosphere has not been comforting (Fig. 5.3). Similar hazards involving the loss of LPG containment continue to result in serious incidents that include injuries, extensive property damage, considerable economic losses, and even fatalities. It is therefore understandable why the PHA team paid careful attention to circumstances that could lead to similar incidents on systems under their direct control. Sensitivities to propane release incidents would responsibly be heightened on a process that for many years had operated without much concern. Taking defensive action against the proposed scenario was warranted based on the objective assessment provided by the PHA team.

PROCESS HAZARD ANALYSIS ACTION ITEM

In accordance with the company's risk calibration (see chapter 13: Process Hazard Awareness and Analysis), the PHA team assigned an action item to apply an additional layer of protection to prevent the proposed scenario. The action item elaborated on two possible solutions. One of the solutions was to install process instrumentation capable of detecting hydrocarbon in the spent caustic line reaching the tank. This sensing equipment would activate an alarm to notify workers if the valve was left open long enough to lose the protective liquid seal. The other recommendation involved inserting an intermediate flash drum in the rundown line to the storage tank (Fig. 5.3). The flash drum would vent any fugitive hydrocarbon into the flare header before draining the spent caustic into the storage tank. In this way the flash drum acted like a "degassing vessel" to divert any flammable hydrocarbon to a safe location rather than directly into the tank. The degassing vessel was an independent add-on system that would provide the safety function needed for the hazard to meet risk tolerance.

A project team was assigned the responsibility of designing a process that could address the PHA action item. The degassing vessel option was greatly favored by the project team. In comparison to the instrumentation option that would generate an alarm, the degassing vessel was a passive control. On the other hand, instrumentation tied to a process alarm would be considered an active control. Accordingly, a multimillion dollar capital project was funded to complete the PHA action item by upgrading the process with a degassing vessel. The upgrade's function would be to vent light hydrocarbon into the flare header; thereby containing flammable material that would otherwise be routed into the storage tank. In compliance with project management guidelines, two PHAs were completed on the conceptual system prior to its construction. Although only one PHA is customary after the initial process drawings are issued, a second PHA was conducted to evaluate a significant design change that was proposed before construction began.

DESIGN CHANGE

The project originally specified using a transfer pump to drain the degassing vessel contents into the tank as needed when the degassing vessel was full. However, the project was modified to incorporate an inherently safer design option prior to construction. The design modification eliminated the transfer pump by feeding nitrogen into the degassing vessel. The purpose for adding nitrogen was to create sufficient pressure differential for draining the vented spent caustic into the storage tank (Fig. 5.4). An inert gas (nitrogen) supply was to be used as the prime mover. Using nitrogen to transfer spent caustic out of the degassing vessel made it unnecessary to install a transfer pump.

This design change addressed the reliability, maintenance, and handling concerns that had been expressed about using a rotating asset in this application. A procedure was also written for the approved operation of the differential pressure transfer system. These procedures explained the steps and order needed to safely transfer spent caustic into the storage tank from the degassing vessel using nitrogen.

The modified spent caustic management system, including the new degassing vessel, was equipped with process instrumentation for performance monitoring and archiving purposes. The process operating parameters monitored were:

1. Spent caustic tank level
2. Degassing vessel level
3. Degassing vessel pressure

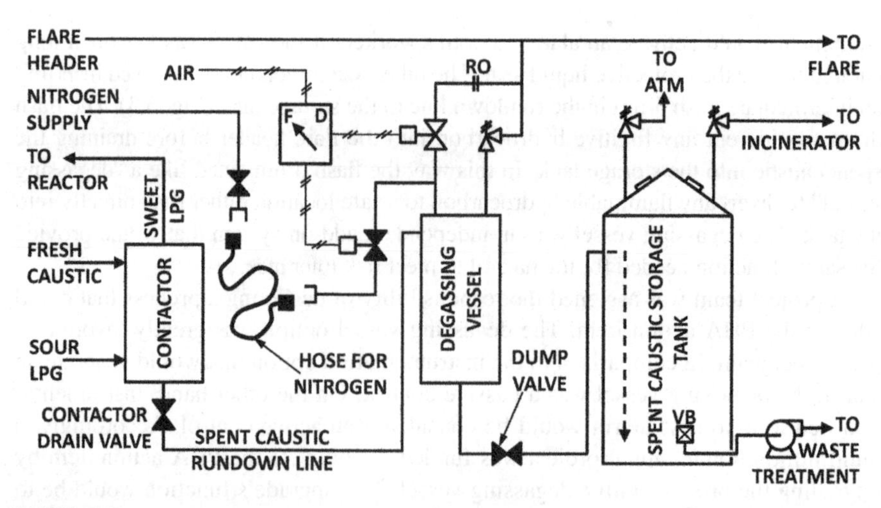

FIGURE 5.4

Final spent caustic disposal process configuration.

The finalized, amended design package was installed and commissioned according to plan. After the workers were adequately trained on its operation, the PHA action item was closed and the new process was integrated into normal operation. For about the next year and a half, the process continued operating without incident as it had before. Then suddenly and without warning the explosion occurred while two workers were performing a normal operating procedure. At the time of the explosion they were simply transferring spent caustic into the tank as they had done successfully many times before.

INCIDENT INVESTIGATION

Historical process information (PI) was immediately reviewed for any obvious explanation about what might have happened (Fig. 5.5). In the 10days leading up to the incident there were four instances where the spent caustic was observed draining into the storage tank. The information captured in PI showed that the explosion occurred moments after the exchange of inventory between the degassing vessel and storage tank began. The final transfer appeared very similar each of the four times that the degassing vessel was drained in the previous 10days before the explosion. In each case and as expected for any normal draining procedure, a pressure spike accompanied the exchange of level between the degassing vessel and storage tank. Before the final transfer procedure was complete, all three transmitters simultaneously fell below zero when the explosion occurred. In the context of the explosion, this type of transmitter behavior was both understandable and anticipated. With no PI directing the investigation down a specific path, the focus shifted to explaining how the three constituents of the "fire triangle" (oxygen, fuel, and heat) happened to be inside the tank at the exact same time.

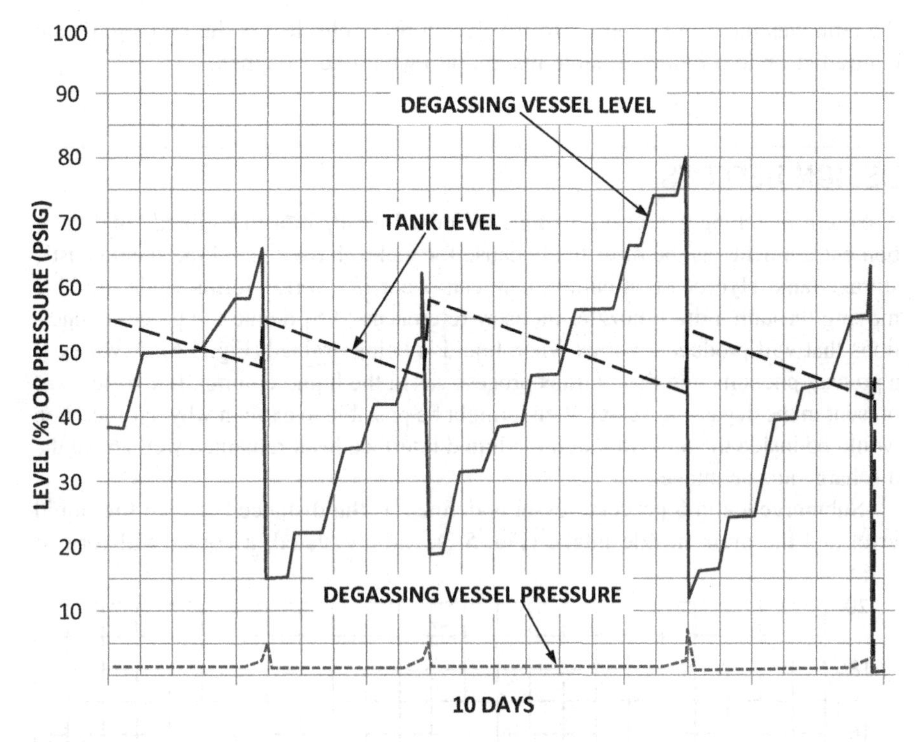

FIGURE 5.5

Postincident spent caustic disposal system process monitoring.

Oxygen could easily be explained. The tank was not isolated from the atmosphere. Air was therefore expected to enter the tank when its contents were being pumped out to waste treatment. Three vacuum breakers were set to open when the tank's internal pressure reached the small negative pressure of 1 in of water column (−0.03 psig) to prevent collapsing the walls inward. The tank was not equipped with an inert gas-blanketing system that would have suppressed a potential ignition by keeping oxygen out. At any time, sufficient oxygen to support combustion could be expected inside the tank due to the presence of air in the empty space above the liquid level. The tank had operated this way for over 50 years when the explosion occurred. This design did not change when the process was upgraded to mitigate the PHA scenario.

It was not quite as easy to explain how fuel had entered the tank. Process information recorded at the time of the incident supported the operators' insistence that the degassing vessel's liquid interface level did not have sufficient time to drop into the drain nozzle intake before the explosion occurred. This was especially concerning since the degassing vessel operated on the principle of a liquid seal to prevent any leakage of gas or liquid hydrocarbon that could represent a potential fuel source. The liquid seal appeared to have remained intact at all times. The presence of hydrocarbon in an amount that would create an ignition hazard inside the tank was therefore contrary to the design principle. It was the first significant contradiction to provide valuable insight into what likely happened.

DESIGN DEFECTS

A pattern matching exercise (chapter: Incident Investigation) identified "hydrocarbon entrainment" as the most likely cause for fuel to have entered the spent caustic storage tank. Hydrocarbon entrainment can occur if a vortex forms while liquid is moving through a drain nozzle. Industry references [4,5] define the process conditions that will produce a vortex when liquid is being drained (Fig. 5.6). A vortex is a process phenomenon that entraps process gas in the liquid column. If a vortex was present in the degassing vessel, then it might be possible to explain why hydrocarbon would be inside the tank although the liquid interface level remained well above the discharge nozzle intake.

Submergence in a process vessel is defined as the distance between the liquid level and the drain nozzle intake (Fig. 5.2). Simply stated, a vortex will form if

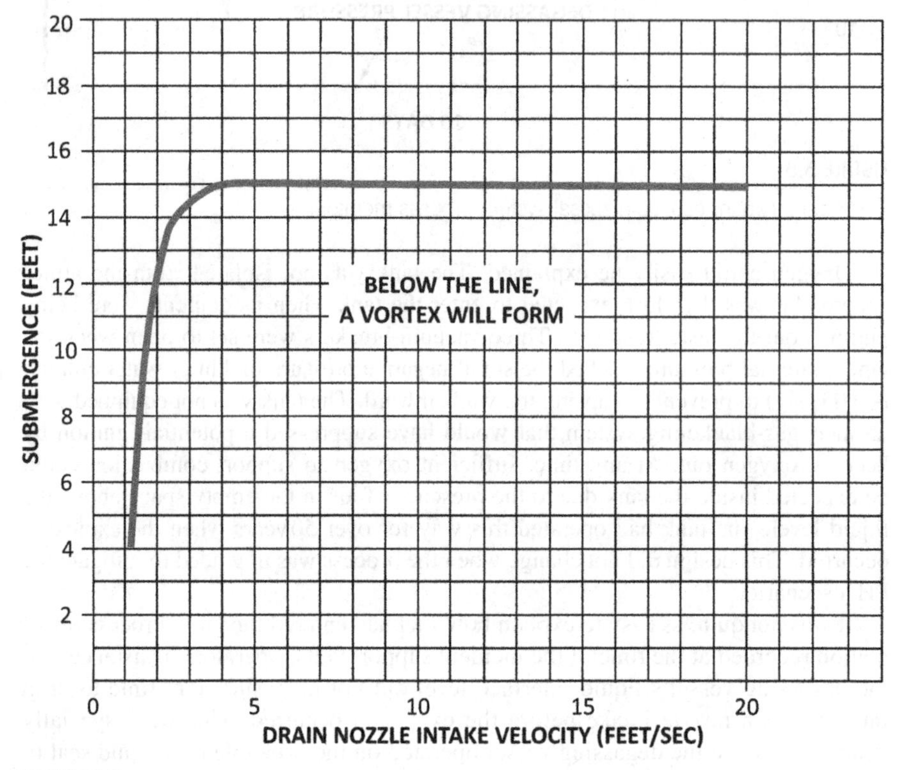

FIGURE 5.6

Submergence requirements for vortex prevention (drum-level dropping).

the submergence decreases below a minimum level determined by the discharge nozzle intake velocity. As the discharge nozzle velocity increase, a higher liquid level must be maintained to prevent forming a vortex. This relationship must be managed to prevent vapor from leaking into a downstream process where it could cause problems. Therefore, the inclusion of vortex breakers is usually specified for pumping applications. Vortex breakers avoid a design issue that could lead to premature and repetitive pump failures. However, the relationship applies to any process configuration that involves draining liquid from a vessel. If the actual submergence is less than the distance needed to prevent a vortex at a certain drain nozzle intake velocity, then a vortex breaker should be installed to avoid inadequate separation. Inadequate separation can create hazards, depending on the type of contaminant and its destination. For example, unknowingly carrying flammable hydrocarbon into a process that contains air would be a problem, whereas it might be acceptable if it is taken to a flare drum.

"As-built" photos of the degassing vessel internals prior to shipment confirmed that the installed asset was not equipped with a vortex breaker. Neither had a vortex breaker been specified as a design requirement. Therefore, depending on prevailing operating conditions, any hydrocarbon above the interface level could have been entrained in the liquid seal that was drained into the tank.

Following that line of reasoning, archived process data were examined to determine where the submergence actually operated relative to the conditions that would create a vortex. Every drain cycle experienced during the degassing vessel's abbreviated life cycle was plotted to determine if sufficient submergence existed to prevent a vortex (Fig. 5.7). It was particularly interesting to find that the calculated velocity at the 3-inch-diameter drain nozzle intake was as high as 20 ft per second. This excessive velocity could be explained by inspection records revealing that no provisions were made to prevent a siphon after establishing liquid flow through the tank's internal fill line. Under these circumstances, spent caustic would have traveled very fast, considerably faster than if the motive force was provided by nitrogen as the prime mover alone. While the high velocity would have reduced the time needed to transfer spent caustic into the storage tank, it would also have made it more difficult to isolate hydrocarbon from the tank unless a vortex breaker was used.

The submergence evaluation established that the storage tank was not adequately protected from hydrocarbon that might possibly leak in with spent caustic during a transfer procedure. However, the volume of hydrocarbon in the degassing vessel headspace during a normal spent caustic transfer was determined not to have been sufficient to generate an explosive air–hydrocarbon mixture inside the tank. This was especially true since operating procedures required essentially diluting the degassing vessel headspace with an inert gas (nitrogen) to transfer liquid into the storage tank without the use of a pump. The observed mismatch between actual and expected process behavior represented a second technical contradiction. Normal operating procedures limited the amount of hydrocarbon below what was needed to create an explosive air–hydrocarbon mixture inside the tank. Another pattern matching exercise was therefore organized to understand how excess hydrocarbon could have gotten into the tank.

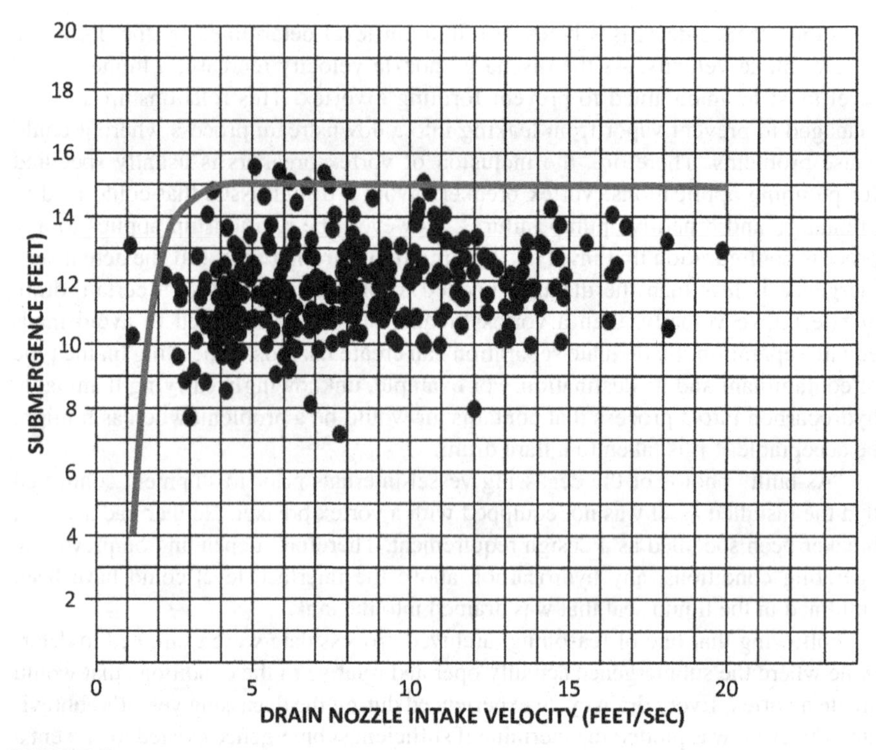

FIGURE 5.7

Degassing vessel drain liquid seal assessment.

PROCEDURE DEVIATION

The pattern matching exercise favored a procedure deviation to explain how extra hydrocarbon might have leaked into the tank during the last spent caustic transfer. Successfully transferring spent caustic into the storage tank involved an eight-step normal operating procedure:

1. Close the spent caustic inlet valve to the degassing vessel.
2. Connect a flexible hose between the factory nitrogen supply line and the degassing vessel.
3. Set the manual remote switch to the "Dump" setting to open the nitrogen supply valve and simultaneously close the flare vent valve.
4. Pressurize the degassing vessel to 30 psig with nitrogen.
5. Open the degassing vessel dump valve to expel the inert (degassed) spent caustic into the spent caustic storage tank.
6. Close the degassing vessel dump valve.
7. Move the manual remote switch back to "Fill" mode to open the flare vent valve and close the nitrogen supply valve.
8. Disconnect the nitrogen hose.

Looking back a second time at historical PI, two significant procedure deviations could now be identified:

1. The level inside the degassing vessel was increasing at the time that the liquid transfer into the storage tank was initiated. The first step required closing the spent caustic inlet valve to the degassing vessel. The level, therefore, should not have risen at the time that liquid was exchanged between the degassing vessel and tank.
2. The pressure inside the degassing vessel was well below 30 psig at the time that the liquid transfer into the storage tank was initiated. Engineering calculations showed that about 30 psig nitrogen pressure was needed to complete the normal exchange of inventory between the degassing vessel and tank. A pressure of 5–8 psig inside the degassing vessel would not be adequate to transfer spent caustic into the tank. Yet, PI showed consistently that the transfers were, in fact, complete.

The repetitive appearance of these procedure defects in the final 10 days of the life of the process suggested that an alternative operating procedure might have been in use at the time of the incident. At this point it became necessary to speak with the workers who were controlling the process in the final days leading up to the explosion. The twofold purpose for this conversation was to:

- Demonstrate transparency in accordance with investigation principles.
- Understand how the degassing vessel was actually being operated.

WORKER INTERVIEWS

At some point in time during an investigation it may become necessary to speak directly to individuals who can provide missing pieces of information. In this case, additional support was needed to explain what appeared to be a procedure deviation in the steps used to transfer spent caustic into the storage tank. The workers with experience operating the system involved in the explosion were therefore asked to explain exactly how the spent caustic was being drained into the storage tank.

The workers described a workaround solution that was implemented to transfer spent caustic into the storage tank. The workaround solution greatly improved the operability of the process after the degassing vessel was installed. Prior to turning the modified process over to operations, transferring spent caustic required opening and closing only the contactor drain valve. It was a simple and fast procedure. Things became much harder after the degassing vessel was placed in service. It then became necessary to connect a flexible nitrogen hose before waiting about 2 h to elevate the pressure inside the degassing vessel to 30 psig. The workaround solution could be implemented without using nitrogen while eliminating the 2 h wait period. Since the major incentive for selecting the workaround solution was to save time, it was a shortcut.

The shortcut (Fig. 5.8) could be performed by two operators. One of the workers would stand next to the degassing vessel to observe the sight glass liquid level and attend to the dump valve, while the other worker was stationed at the contactor drain

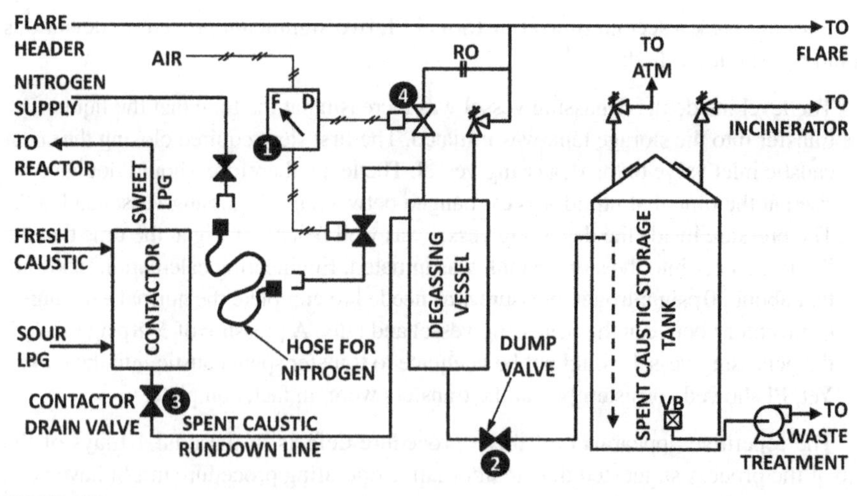

FIGURE 5.8

Workaround procedure (shortcut) used to drain the spent caustic vessel.

valve. The shortcut was executed by keeping the switch (1) in fill mode to continue venting the degassing vessel into the flare header. An open flare vent was needed to break the vacuum in the degassing vessel as the level dropped, in order not to inhibit the spent caustic transfer. The worker at the degassing vessel would then open the dump valve (2) leading to the spent caustic storage tank. After that the other worker would open the contactor drain valve (3) to begin draining spent caustic into the degassing vessel. In a few minutes, LPG entrained in the spent caustic would enter the degassing vessel, where a pressure spike would register on PI when the LPG vaporized. The slightly elevated pressure in the degassing vessel would satisfy the total head (elevation) requirement needed to initiate siphoning between the degassing vessel and the storage tank. When the operator at the degassing vessel heard a pounding noise and saw the liquid level start moving down in the sight glass, the worker at the contactor would be directed to close the contactor drain valve (3). With the degassing vessel flare valve (4) still open flare gas would fill the void created in the degassing vessel as its level dropped. When the degassing vessel level dropped to about 10%, the dump valve (2) would be shut and the degassing vessel would be returned to its normal idle condition, ready to receive more spent caustic batches before having to be drained again.

The workers cooperated fully during the interview. Upon telling them about the discovery that hydrocarbon would leak into the tank when draining the degassing vessel the workers insisted that it was impossible for hydrocarbon to exit through that route. They argued that the liquid seal was visually controlled every time that the degassing vessel was drained. Not once in the degassing vessel's history did the interface layer drop below the drain nozzle intake. It was always confirmed to be present inside the degassing vessel's sight glass.

The technical details behind forming vortices were then explained to the workers, together with the design omission of a vortex breaker. They were showed how the system operated relative to these conditions. This explanation compelled the workers to agree that the shortcut was one of the factors that contributed to the incident. Due to the shortcut, about five times the normal amount of hydrocarbon could be introduced into the process. Under these circumstances, sufficient fuel would have been available to create a flammable air–hydrocarbon mixture inside the tank.

The clearly defined hazard was unknown to anyone before the explosion. By resolving that hazard, a series of recurring unresolved tank odors and maintenance events also started making sense. These warning signals were explained upon validating the composition of the flare gas that would be transferred into the tank every time the degassing vessel was drained.

WARNING SIGNALS

The section of flare gas header piping connected to the degassing vessel contains about 4% H_2S at all times. Introducing flare gas into the tank would have caused the conservation vents to open either by raising the tank's empty-headspace pressure or by pushing vapor volume out of the tank as its level increased while receiving a spent caustic transfer. This venting of process vapor into the atmosphere would have created foul odors in the tank area.

For example, prior to the explosion two different contractors independently shut down work in the tank area at the same time due to foul odors. When workers arrived to investigate a few minutes later the odors had dissipated. Never was the mental connection made between hydrocarbon or H_2S in the tank's vapor space and odor reports near the tank. Many odor reports were dismissed on the assumption that they originated with work taking place in adjacent units. A more thorough investigation that might have traced the odors back to a definite source was never commissioned. Additionally, unresolved odor complaints were never reported. Only resolved odor complaints that could be traced back to a primary source were reported and tracked.

Perhaps the most distinct warning signal involved a series of maintenance events where hydrocarbon was detected leaking directly out of the tank. On one occasion work was scheduled to repair one of the tank's atmospheric pressure control devices after high hydrocarbon levels were detected during compulsory leak detection and repair (LDAR) monitoring. Accordingly, an initial repair attempt was made unsuccessfully to stop the leak. No improvement resulted from tightening down on the gasket bolts. Therefore, a second repair was attempted by replacing the leaking gaskets. Neither did this repair stop the hydrocarbon leak, which made it necessary to schedule a third repair attempt. On the day that the third repair was scheduled, the leak mysteriously stopped before another repair was attempted. The repair was canceled and the work order was closed with no further investigation.

Unfortunately the confirmed presence of hydrocarbon in the tank did not seem contradictory before the explosion. Neither did the unexplained disappearance of the hydrocarbon leak raise any concern before the explosion. Both these phenomena appeared to be normal inside the spent caustic storage tank, although by design it was not intended to contain hydrocarbon. Neither was the degassing vessel's liquid seal performance questioned despite the fact that hydrocarbon had randomly been detected exiting the tank.

IGNITION SOURCE

As explained previously, the project to install the degassing vessel added a flare line connection to the spent caustic disposal system. Flare gas entrained in the spent caustic draining out of the degassing vessel provided an unexpected route for H_2S to enter the tank. Under these conditions, pyrophoric iron sulfide deposits could reasonably be expected to form on bare mild steel tank walls (Eq. [5.2]):

$$Fe_2O_3 + 3H_2S \rightarrow 2FeS + 3H_2O + S \quad [5.2]$$

When exposed to air, these deposits likely decomposed exothermically in a reaction that generated enough heat to ignite the flammable air–hydrocarbon mixture inside the tank's vapor space (Eq. [5.3]):

$$4FeS + 7O_2 \rightarrow 2Fe_2O_3 + 4SO_2 + \text{heat} \quad [5.3]$$

The heat produced by this reaction is most likely responsible for igniting the flammable mixture that was inside the tank at the time of the explosion. This conclusion is indirectly supported by the rust that was found on the internal surface of the tank's 20-inch-diameter roof hatch stems and a sheer white cloud that suddenly appeared hanging above the tank immediately after the explosion occurred. The composition of the white cloud was likely SO_2 from sulfur combustion when the flammable mixture ignited.

EVALUATING CAUSES

The investigation was now complete. The timeline contained no inconsistencies or contradictions and resolved all the facts and evidence according to scientific principles. Nothing prevents the investigation from stopping here, to explain how the incident resulted from two credible causal factors:

1. An unresolved design defect that allowed hydrocarbon to penetrate the degassing vessel's liquid seal while embedding an ignition source inside the tank.
2. An unapproved procedure deviation that increased the concentration of fuel in the storage tank into an ignitable range.

The interaction between these two causal factors describes how the three constituents needed to ignite a flammable mixture were all present inside the tank and why they were present at the same time. However, neither of these two causal factors traces the incident back to its original source. There is a far more significant cause buried deep beneath the shallow surface created by these two superficial causal factors. The full protective value that this incident has to offer can only be extracted by tapping into its prevailing root cause. Based on the information we have already discussed, can you identify the root cause for the explosion?

Think for a moment about how the process might have worked if the transfer pump had not been replaced by a differential pressure transfer system midway through the project. The intent was to address the hazard involved with leaving the contactor drain valve open. Previous generations had preserved their knowledge in pump design specifications and standards. This collective process design knowledge was documented in the hope that others would act on it. Acting on it would shield future generations from calamity. The value of this preexisting knowledge was lost when the decision to replace the transfer pump with an inherently safer design option was made. More precisely, the vortex breaker specified in the original process design was eliminated when the process design no longer included a pump. The knowledge acquired from the incident investigation explained the relationship between submergence (liquid column height) and drain nozzle intake velocity that forms a vortex.

The lesson learned was that vortices are not exclusive to pumping applications. Their conditions apply equally to any situation where a liquid level is carried above a drain nozzle intake. The design team did not know this until the incident brought it to their attention.

In this context, the incident was a costly lesson relearned. In fact, it could possibly have been fatal. The incident could easily have been prevented by installing a vortex breaker inside the degassing vessel, which is why specific instructions to do so were included in pump specifications and standards. But prior to this incident, the knowledge needed to control this hazard was unavailable. Or perhaps it is more accurate to say that the knowledge needed to control this hazard *was* available but failed to offer any protection because it was only contained in pump specifications and standards.

Eliminating the pump removed specific design requirements that would have prevented the incident. Therefore, although the same process conditions applied to the inherently safer design choice, the documented knowledge that would have prevented the incident was hidden from those who needed it. No doubt, a more effective use of available knowledge could have prevented this incident. Despite the fact that acting upon this knowledge could have prevented the incident, we must still dig deeper to fully appreciate the incident's root cause.

It is appropriate at this point to examine more closely the decision to eliminate the transfer pump. Many skeptics might be misled into thinking that economics was what actually drove the decision to replace the transfer pump with a differential pressure transfer system. An investigation is not needed to recognize that the capital purchase price and ongoing maintenance and operating costs for the spent caustic disposal process could be reduced through this design change. However, economics never factored into the evaluation of inherently safer design alternatives that could meet the functional requirements demanded by this service. A process reliability concern provided a compelling argument for the project team to pause for a moment and examine how the issue might be effectively addressed by an inherently safer design option. The interaction demonstrated in this example is the perfect definition of "employee participation" as prescribed by the PSM standard. Tragically, in this case the decision to change the design to address a potential maintenance and reliability concern actually defeated the purpose for which employee participation is practiced. We have now finally penetrated deep enough to start seeing the incident's true root cause.

How can the selection of an inherently safer design solution possibly be considered the root cause for this incident? To understand you must not be distracted by the more superficial causes for the explosion and the multiple system failures that otherwise might have prevented it. The goal of process design in all sectors of the manufacturing industry has always been to *eliminate* hazards wherever practical [6]. In a sense, the decision to replace the transfer pump with a differential pressure transfer system satisfied this objective. It corresponded to recognized inherently safer design practices that are intended to simplify process design [7]. However, in the context of the service conditions under which the safer solution would operate, the decision proved flawed. In this instance, modifying the process design by *adding* an inherently safer design solution only gave *the appearance* of a process being made safer. In reality, the decision to install a differential pressure transfer system made the process significantly more hazardous than it ever had been in the past. Compared to the original design option that included a transfer pump it was also an inferior choice. Inadvertently, a project that was commissioned to make the process safer had almost caused a fatal process safety incident. Only after the incident was it recognized that any changes made to reduce one hazard will also assist with either gaining or losing control of other hazards [8]. It might even introduce other unknown hazards, as it did with an ignition source that had never existed before the add-on project was complete.

In assessing where the project to complete the PHA action item went wrong, insufficient attention was given to human factors related to equipment design that could lead to its improper operation [9]. Obtaining the highest level of protection available by eliminating a hazard does absolutely no good if the selected approach makes the process more difficult to operate. This is exactly what happened here. It was a painful lesson learned.

In the context of eliminating only the potential hazards associated with leaking pump seals, selecting an option to transfer the process using differential pressure was consistent with inherently safer design practices. However, the design change significantly increased the operating requirements for factory workers who would now be expected to manage the process differently. From that time forward they would have to manage a much more complicated and time-consuming procedure than before. In this way, the design change directly contradicted one of the foremost inherently safer design principles. Had the original transfer pump been installed, draining the tank would probably have added no more than two additional steps to the existing normal operating procedure. It also would have included a vortex breaker in order to comply with approved pump specifications and standards related to process design. Conversely, the decision to find an alternative to the transfer pump not only eliminated the vortex breaker, but also introduced a much more complicated normal operating procedure. The new and additional operating requirements involved with the modified system included:

- Connecting and disconnecting a flexible hose to keep possible contaminants out of the factory nitrogen supply;
- Opening and closing multiple block valves in a specific order and with precise timing;
- Manipulating a remote hand switch; and
- Waiting for about 2 h to prepare the system for transferring vented spent caustic from the degassing vessel into the tank.

Collectively, these operating requirements significantly increased the human factors that could potentially interfere with proper system operation. Had human factors been considered when evaluating potential ways to close the PHA action item as well as possible ways to address the concern over using a transfer pump, then the differential pressure transfer system would not have been installed. Again, the qualification for selecting an inherently safer design solution must always be to either reduce or maintain a certain level of anticipated human interaction with the process. If installing an inherently safer design option makes operating the process more complex, then the option will probably cause the opposite of its intended purpose. Pursuit of an option intended to control or remove a hazard will succeed only if an organizational culture adequately controls the temptation to create workaround solutions or shortcuts.

In this case, the workers developed a shortcut that eliminated all of the additional operating requirements. Unfortunately, the workers never communicated their local procedure modification to the technical staff, contrary to the approved operating procedure. Therefore, the hazards related to the new method of operating the system were concealed until an investigation was activated to explain why an explosion had occurred. Up to that point, the shortcut was implemented through tribal knowledge (chapter: Employee Participation). On the outside, it appeared that the project to resolve a PHA action item was a complete success. In reality, it inserted a defect that rather quickly resulted in a process safety failure.

The benefits of digging deep enough to tap into the incident's root cause are clearly evident. The corrective actions available by addressing the two preliminary causal factors are not nearly as comprehensive or effective as those that can be derived from a deeper analysis of their source. For instance, addressing the asset design and procedure implementation defects would impact a relatively narrow range of equipment and factory personnel. In contrast, addressing the root cause would place solid emphasis on practicing the principles that govern inherently safer design solutions. These principles apply to virtually every industrial process and everybody involved in its operation [6]:

> *Safety is built into the process or product, not added on. Hazards are eliminated or significantly reduced, not controlled, and the way they are eliminated or reduced is fundamental to the design that it cannot be changed or defeated without changing the process. In many cases this will result in simpler and cheaper plants.*

This guiding principle is far more valuable than any other lesson that might be obtained on an individual basis. The fact is that vortex breakers apply to a specific design function. The principle documented above, however, applies to all processes and all assets operating within those processes. It also applies to every manufacturing industry business sector, including others such as the transportation, medical, agricultural, and food and beverage industries.

This is not to say that knowledge about the conditions that apply to vortices in systems that require controlling a liquid interface layer is not important or does not belong in standard process training. Nor does it mean that workers should not be held accountable for following procedures, complying with policies, and communicating operating difficulties through appropriate or designated reporting channels. On the other hand, none of these human factors would have been involved if the design team had secured the PHA action item's success. Although the project team intended to build the safest system possible, the path they chose did not correspond with inherently safer design principles. As a result, their attempt to make the process safer instead made it more likely to fail.

The incident and investigation proved valuable to the effort to restore production in the safest manner possible. With these points in mind, a small maintenance project was organized to build safety into the process instead of adding it on. The vision therefore was clear about how to address the PHA concern and get the spent caustic disposal system back into a safe working condition.

FUNCTIONAL RECOVERY

Simplification in all aspects of process operation now finally and unswervingly became the goal for well laid-out actions to restore spent caustic disposal operations. Conveniently, a large pressure vessel that had been abandoned several years earlier was within close proximity of where the tank that exploded had been constructed. With little difficulty, a temporary piping configuration was assembled to route the spent caustic rundown line into the abandoned pressure vessel (Fig. 5.9).

The abandoned pressure vessel provided more storage capacity than the volume provided by the original spent caustic storage tank. It was already connected to the flare header and could be continuously vented to a safe location by simply opening a valve to keep the line open. An outlet pump complete with a suction nozzle intake vortex breaker was already connected to the abandoned pressure vessel. Not only were these assets readily available but they could easily be restored to full operation. Unlike the atmospheric tank, the pressure vessel could withstand higher pressures than it could reasonably be exposed to under normal operating conditions. In every sense of the word, reusing the preexisting pressure vessel fit the true definition of an inherently safer design. Most importantly, it was now a simple operation that reduced human factor concerns to the absolute minimum. From that time forward all that was needed was to open and close the LPG contactor drain valve to transfer spent caustic into waste treatment. The total cost for this design option was a fraction of

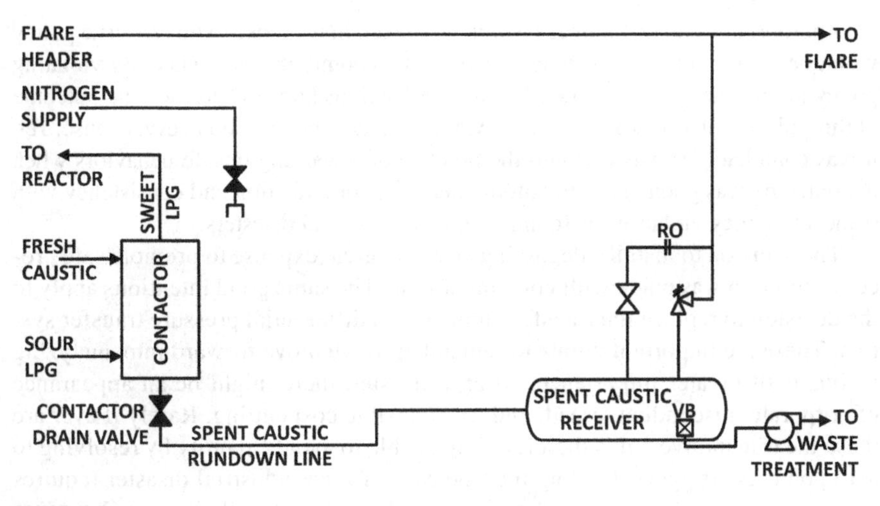

FIGURE 5.9

Final spent caustic disposal design solution.

the original project that had culminated in a very serious failure. Experience corroborated the expressions of a safety authority [6] who accurately suggested that installing inherently safer design technology in many cases will result in simpler and cheaper processes.

Full spent caustic processing functionality was restored within about 2 weeks after the explosion. In contrast, the earlier project to modify the process had taken over a year to finish. This is just one of many reasons why the rationale for installing an add-on system was questioned after the temporary system was in full operation. Note again that the new spent caustic management system built safety *into* the process. This is perhaps the main reason why the latest, albeit labeled "temporary" installation was soon upgraded to be permanent. The system now operates as reliably and uneventfully as it had for over five decades prior to the explosion. The only difference is that it now operates safer than it ever did before.

ORGANIZATIONAL CULTURE

In closing, based on our discussion how would you assess the prevailing organizational culture in this example? Consider the fact that a policy violation is probably what saved the worker from a painful injury and perhaps even death. When the explosion occurred, he was not standing where he should have been according to company policy. By policy he should have been attending to the valve that he opened right next to the tank that exploded. However, at the time of the explosion he was adjusting an asset behind the two temporary chemical totes conveniently separating him from the tank. These stacked totes shielded him from the fiery process that suddenly sprayed out of the tank in all directions with no time to escape.

Appropriately, no disciplinary action was taken in response to this specific policy violation. How could discipline be applied at this point? It was too late. By violating policy in this instance the worker had avoided serious harm. Therefore, the outcome of this policy violation was positive. It was a perfect contradiction in every sense. Yet, it was completely consistent with the practice of rewarding unsafe behaviors when the outcome was good. This dichotomy between contradiction and consistency with respect to policy violations is found within all industrial disasters.

The decision to install a degassing vessel at great expense to promote safe process operation was made with good intentions. The same good intentions apply to the decision to replace the transfer pump with a differential pressure transfer system. These are important points to remember as we move forward into analyzing incidents of greater importance. After a disaster, there might be an appearance of corporate misconduct, greed, and irresponsible cost cutting. Rarely if ever are these the true motives. It is therefore impossible to avoid calamity by resolving to act upon these triggers. Finding the true cause for an industrial disaster requires digging much deeper beneath the sensational exterior that shallow explanations usually create.

The good news is that disasters really follow patterns that anyone interacting with a manufacturing process can easily recognize. Every day we make decisions that we believe will serve the best interest of ourselves and the companies that we work for. Success or failure depends on organizational culture. Companies that investigate incidents to detect and then successfully address a defective organizational culture are those that will dominate and control the markets in which they compete.

LESSONS FOR US

The case history communicated in this chapter provides valuable insight into the complete life cycle of an industrial disaster. Of significant interest are the underlying inherently safer design principles that must be practiced to limit the complexity that makes a process more likely to fail. Design teams able to incorporate these principles in capital projects will be directed to the most effective, simple, and economical process solution. Other lessons worthy of our consideration are:

- Misinformed attempts to improve process safety can have counterproductive results.
- Industrial disasters originate with good intentions.
- Inherently safer design represents a balance between hardware selection and process operability.
- A problem not communicated cannot be solved.
- Effective hazard management involves adhering to operating procedures.
- Shortcuts and workaround solutions create unknown hazards.
- Warning signals provide an opportunity to take defensive action before an incident happens.

- Worker cooperation involves both knowing when to interview knowledgeable personnel and appropriate questions to ask them.
- The most protective knowledge resides deep beneath an incident's superficial exterior.
- Applying discipline based on the outcome of a policy violation is an inconsistent approach to process safety.

REFERENCES

[1] K. Bloch, D. Wurst, Process safety management lessons learned from a petroleum refinery spent caustic tank explosion, Process Saf. Prog. 32 (2) (2010) 332–339.

[2] D. Wurst, J. Cornelisen, Process safety leadership: becoming street smart on process safety, Process Saf. Prog. 29 (4) (2013) 148.

[3] U.S. Chemical Safety and Hazard Investigation Board, Report No. 2007-05-I-TX: LPG Fire at Valero – McKee Refinery, 2008, p. 1.

[4] N. Lieberman, Troubleshooting Refinery Processes, 1981, p. 272. ISBN: 0-87814-151-X.

[5] R.E. Syska, J.R. Birk, Pump Engineering Manual, third ed., 1976, p. 153.

[6] D.C. Hendershot, Inherently safer design: the fundamentals, Chem. Eng. Prog. 108 (1) (2012) 40.

[7] S.T. Maher, K.D. Norton, S. Surmeli, Design an inherently safer plant, Chem. Eng. Prog. 108 (1) (2012) 43–47.

[8] D.C. Hendershot, Inherently safer design: the fundamentals, Chem. Eng. Prog. 108 (1) (2012) 41.

[9] D.C. Hendershot, Inherently safer design: the fundamentals, Chem. Eng. Prog. 108 (1) (2012) 42.

PART

Research and Development

2

Process safety has always been a driving commitment within the manufacturing industry. Long before the Bhopal disaster, the importance of communicating lessons learned to prevent industrial incidents was both recognized and acted upon by at least one major oil refining company. Those in possession of experience-based knowledge considered it their obligation to impart what they knew to others who were willing to listen. This knowledge was, and still is, treasured by organizations who appreciate their responsibility to protect the lives of others.

Along these lines, the American Oil Company (AMOCO) set the example by publishing nine booklets to communicate life-preserving information about safe process design and operation. Twenty-nine years after the first booklet (*Hazard of Water in Refinery Process Systems*) was released in 1955, the Bhopal disaster reminded industry that incident prevention is a life-preserving science. Incident prevention represents a commitment that has no end.

In response to the Bhopal disaster, sweeping changes were introduced to make better use of existing programs that could manage the hazards inherent to manufacturing processes. Revolutionary new programs were also developed based on lessons learned that only a failure of catastrophic dimensions could provide. In many cases, these programs were more than just discretionary best operating practices.

The Bhopal disaster made these programs a legal obligation for those whose actions could potentially harm people or the environment. These programs sharpened industry's attention on hazard recognition. The Bhopal disaster taught industry that much more needed to be done to control process hazards that could result in a catastrophic loss of primary containment.

It is of compelling importance, therefore, to understand how the application of industry's current process safety management (PSM) practices before the Bhopal disaster could have changed history. Could industry's current application of the PSM standard have prevented the Bhopal disaster? The answer to this question is of significant interest to those who invest their confidence in PSM to avoid similar occurrences.

This assessment starts at the very beginning, which is when a new product was introduced to the market. Vital lessons can be learned from examining the research and development behind a product that solved a specific problem. Solving problems, as we shall see, initiates a sequence of events that can have unintended consequences. It is important to understand how the sequence of events we initiate by solving a problem can continue producing benefits.

CHAPTER

Inventing Solutions*

6

EVENT DESCRIPTION

The seed that ultimately grew into the Bhopal disaster was planted on September 8, 1959. This is when a new pesticide molecule was patented by a major multinational company headquartered in the United States of America [1]. Ironically, this new pesticide molecule was developed exclusively to replace a product that had won a Nobel Prize for a diligent researcher in 1948 [2]. This honor was certainly appropriate for the person credited with the discovery of the aging molecule's insecticidal qualities that made it a powerful and effective weapon in the world's battle against typhus and malaria [3].

But time has a way of exposing things that the eye cannot readily see. In this case, it took a while for the world to recognize that using the original product as an insecticide created an environmental hazard [4]. Still, the thought of living without it was perhaps even more objectionable. Pondering these complexities sent chemical manufacturers all over the world scrambling back into their laboratories. Their mission was to synthesize a new, more environmentally sensitive product to replace it. The first company to introduce such a solution would surely control the global pesticide market—a market devoted to killing insects so that people might live.

As to the company that actually discovered the viable chemical solution, the multitude of inventions credited to it was already incredibly impressive [5]. Its international presence, reputation, and name recognition (the "brand") made it likely that the company had very few, if any, serious competitors [6]. For decades, they had consistently stretched the limits of technology that constrained the rest of the chemical manufacturing industry. In essence, their introduction of a revolutionary new pesticide product was no real surprise. This development reaffirmed that the market rewards those willing to use their creative abilities to bring about effective solutions. While smaller chemical companies might come and go, this one was here to stay. The company in the middle of it all attracted and retained the professional talent that was needed to survive. On top of it all, the company had a vast network of factories able to put their creative solutions into the hands of consumers around the world.

Unlike the earlier product that was to be abandoned, the new product was a non-chlorinated water-soluble chemical compound that was easily broken down by metabolic processes in mammals. This made the new product non-persistent, whereas the

* For timeline events corresponding to this chapter see page 415 in Appendix.

former product collected in the fatty tissues of humans, animals, and insects; thereby magnifying its presence over time [7]. In 1961, news of how the new product had saved Egypt's economy by killing the cotton worm, a pest with a chemical resistance to chlorinated insecticides, spread throughout the world [8]. Had that infestation taken place only 2 years earlier, Egypt's cotton industry—the country's economic backbone—would have been completely decimated. It has been said that word of mouth is the best form of advertising. Along those lines, stories like this created a global interest in the new product. For that reason it became an overnight sensation and continues to be one of the most successful commercial pesticides ever invented.

After the new product hit the market, the former molecule was slowly phased out. The new product was warmly received for its ability to provide at least as good a performance without raising similar concerns about carcinogenic, bioaccumulation, and wildlife health hazards [9]. Time, however, was needed to determine how the hazards exchanged between the old and new products would be remembered.

TECHNICAL ANALYSIS

As we step into the sequence of events that led to the Bhopal disaster we already find several conditions that provide valuable insight into the development of an industrial disaster. The first of these conditions relates directly to the company that offered a technological breakthrough to address a specific demand created by a changing market. The company that solved a serious environmental problem was no ordinary company. Their track record was simply awe-inspiring, to put it mildly. For years they had capitalized on just about every opportunity that the chemical manufacturing industry could provide. They did this by:

1. Responding to specific market demands for products that the public knew was needed (as was demonstrated in this case), and
2. Creating markets for new technology that the public did not yet know was needed.

The results obtained by this approach to business growth and stability are reserved only for companies that are willing to push the envelope of commonly accepted industrial limits. Performing to the level demonstrated by the new pesticide product requires operating in uncharted territory where the competition is either unable or unwilling to travel. Companies capable of breaking through these technical barriers are the ones that consistently deliver performance that would otherwise be impossible. Developing new technology such as the one involved here requires ignoring predefined limits and being comfortable with operating outside of limits in the interest of serving the needs of the population. At the same time, this behavior relates to a specific failure-permissive organizational culture attribute (chapter: Incident Investigation).

Whether in a sports arena or a research laboratory, our free-market society rewards those who demonstrate a consistent ability to redefine the limits that restrain all others.

The success, reputation, and strength of the company described here—a company that invented the technology needed to replace a more hazardous pesticide molecule—defined it as one of these companies. Their formula for success was to encourage workers and professionals alike to break through the conventional walls that contained the chemical manufacturing industry. This allowed them to look where others could not; thereby discovering new ways to do things better than anyone else. It allowed them to develop a pesticide product that immediately became an international sensation.

The molecule they invented is the perfect incarnation of this aggressive business approach. This governing business model, however, makes it difficult to establish different expectations for the line organization that is responsible for safe and compliant factory operation. It is not impossible to delineate these expectations, but the distinction must be recognized and enforced among people with direct or indirect product manufacturing responsibilities. Awareness of this hazard is needed to control it. Promoting the same expectation throughout different departments within an enterprise may sound entrepreneurial, but it can be dangerous if not controlled properly as we shall observe in later chapters.

Moreover, in exchange for being rewarded by our free-market society by introducing valuable new technology that satisfies or creates a consumer demand, inventors are obligated to define specific manufacturing requirements for safe production. While all companies in the manufacturing industry are responsible for protecting the lives of others during factory operations, those who invent new products are in the best position to identify the hazards introduced by their technology. Their ability to share this knowledge with others promotes safety throughout the industry and the community.

Suffice it to say that some processes are less forgiving than others. Demonstrating a casual attitude toward process design, operation, and maintenance is not acceptable under any circumstances. An uncompromisingly serious attitude is especially important for a company willing to experiment with a potential technology breakthrough, where any minor incident could lead to discovering new, and possibly unforeseen, consequences. Responsible process control requires first making sure that the basic reliability, maintenance, and operating practices are adequately applied (chapter: Equipment Reliability Principles). If that is not enough, then additional protection can usually be justified. Additional protection, however, is not effective unless the basics are firmly established and working properly.

Another observation worth noting in the initiating sequence of events is the influence of process safety management (PSM), or more precisely inherently safer design, principles. Notice that a new pesticide molecule was needed to eliminate the hazards of the product it was to replace. These hazards were eliminated by substituting one chemical type for another. Consistent with the principles embodied in the Hierarchy of Controls (chapter: Incident Investigation), the preferred solution offered here was to eliminate these hazards through *substitution*. In this specific case a new chemical had to be invented from scratch. The new invention had effective insecticidal properties but disposed of the environmental issues that discredited the old product. This concept is quite similar to modern installations wherein bleach (sodium hypochlorite) is sometimes approved to replace chlorine gas.

However, it must be noted that any change of product, process, or technology provides an opportunity for hazard exchange (chapter: Equipment Reliability Principles). In such cases, the known hazard that is being eliminated might actually be more acceptable than a new hazard that could be introduced. Certainly, anyone involved in a manufacturing process must be aware of this relationship. Change of any sort is not to be taken lightly. In the rare cases where a change is absolutely necessary, due diligence beyond what is normal must be practiced to completely assess the hazards. It would not be acceptable to administer a management of change (MOC) program to simply "check the box." Neither would it be responsible to use the MOC program as an administrative tool designed to support a firmly established decision to make a change. Remember, hazard recognition applies at all phases of process operation. An MOC program must be authorized to send developers back to the drawing board if needed. The program should also be granted authority to reject change proposals that involve too many unknown hazards if it is determined that an adequate level of hazard control is not possible (chapter: Process Hazard Awareness and Analysis).

While on the subject of change, our entry into the Bhopal disaster demonstrates how a PSM program can become a hotbed for process changes. Although the developer's intentions for introducing the new product were good, a change was involved. Good intentions might end in a disaster, depending on how decisions are made throughout the life of a process (chapter: Spent Caustic Tank Explosion: A Case Study in Inherently Safer Design). By applying these principles we know that industry's PSM program is on solid footing. It represents a great roadmap for initiating changes intended to prevent incidents. On the other hand, if not executed properly and with full awareness of the circumstances underlying the Bhopal disaster, PSM merely scratches the surface. If that is the case, then PSM risks lulling its practitioners into a false sense of safety and security. Good intentions were definitely observed in this incipient event where a new molecule was introduced to solve an environmental problem. In this case, a solution was mandatory. Living with the environmental problem was as unacceptable as was living without the product that was causing the problem. A change was therefore necessary.

The company that introduced the new molecule had an important choice to make before introducing it to the market. How would they treat the intellectual property represented by this emerging technology? Would protection be attained through the laborious and generally expensive legal route needed to establish a patent, or would the company decide to market the product as a trade secret? There are advantages and disadvantages associated with either choice (chapter: Equipment Reliability Principles). For many decades, the developer of the new product had favored covering their many inventions with patents. Their choice was no different in this case. In exchange for publicly releasing the details needed to manufacture the product, the inventor was legally protected from competitors who might financially benefit from the inventor's intellectual property [10]. For a limited period of time, the parent company would be able to control the market without interference from any of its competitors.

But all business decisions have consequences; some more profound than others. The decision to market a product under a patent or as a trade secret may seem trivial to most factory workers and professionals with manufacturing experience. After all, what does the process care about how the owner decides to market and protect their inventions? The significance of this decision is of serious importance to those who wish to learn how to prevent industrial disasters from others' experience. We will explore this question more thoroughly as we examine other events leading up to the Bhopal disaster. A comprehensive analysis of the sequence of events provides new lessons for those who value the importance of building disaster resistance into product manufacturing processes.

LESSONS FOR US

The words in this chapter and in the chapters that follow directly relate to activities that routinely take place inside the manufacturing industry. The narrative starts by alluding to what made a certain company consistently successful in matters dealing with inventions, and how that success ultimately translated into problems. This particular company's success is the lamp that attracts our attention. However, the performance commitment behind their success story is familiar to perhaps every manufacturing company in the world. Anyone reading these words employed by a progressive manufacturing organization can easily relate to the story being told. Success is reserved only for those able to employ the creative talent needed to capitalize on changing market conditions. Although the finish line is reserved for only one winner, the starting block is the same for everybody. We should therefore have no difficulty finding multiple opportunities to apply what we can learn from a definitive analysis of the Bhopal disaster. With this thought in mind, pay close attention to the following lessons as the case study continues to develop:

- Time consistently reveals things hidden before an incident.
- Experimentation must remain in the laboratory. Strict adherence to operating limits and proceduralized operating practices is required from the line organization.
- A free-market society financially rewards those capable of introducing breakthrough technology.
- Introducing breakthrough technology requires breaking conventional rules to redefine acceptable limits.
- Inventors are best equipped to define specific production hazards and how to address them.
- Building systems in accordance with manufacturers' specifications is the most direct approach to controlling unknown hazards.
- The decisions we make from the moment that a new product is invented have a residual effect on manufacturing operations throughout the life of a process.
- PSM principles are a primary source of changes within industry.

- Substituting products and processes without careful study tends to create a hazard exchange whereby known problems are solved at the expense of introducing new, unknown, hazards.
- All business decisions, no matter how insignificant they may appear at first, can have serious consequences in the long-run.

But just in case you missed the build-up to one major message that serves as the connecting thread, the overarching lesson in all this narrative: This chapter attempts to express what separated a famous, productive, and innovative company from its competition. Even the most productive mindset will create issues if productive mindsets are not controlled properly. Again, that is a huge factor with ramifications slowly winding their way through the developing story. It is a story about both *what* happened and *why* it happened. The implication can be translated into the admonition to be careful with who you empower with certain responsibilities (chapter: Role Statement and Fulfillment). Ask yourself what mindset is allowed to prevail in the ones to whom you convey and entrust an entrepreneurial spirit. Their actions and inactions, their decisions and indecisions, could come back to haunt you. "The road to hell is paved with good intentions," says an old aphorism. To which we might add that even the best of intentions can inflict permanent damage from which there may be no satisfying recovery.

REFERENCES

[1] J.A. Lambrech, α-Naphthol Bicyclic Aryl Esters of N-substituted Carbamic Acids, Patent No. 2,903,478, 1959.
[2] S. Mukherjee, Bhopal Gas Tragedy: The Worst Industrial Disaster in Human History, 12, 2002, ISBN: 8186895841.
[3] National Pesticide Information Center (NPIC), DDT (Technical Fact Sheet), 1, 2000.
[4] D. Lapierre, J. Moro, Five Past Midnight in Bhopal, 21, 2002, ISBN: 0-446-53088-3.
[5] D. Lapierre, J. Moro, Five Past Midnight in Bhopal, 34, 2002, ISBN: 0-446-53088-3.
[6] D. Lapierre, J. Moro, Five Past Midnight in Bhopal, 35, 2002, ISBN: 0-446-53088-3.
[7] National Pesticide Information Center (NPIC), DDT (Technical Fact Sheet), 3, 2000.
[8] T. D'Silva, The Black Box of Bhopal, 31, 2006, ISBN: 978-1-4120-8412-3.
[9] National Pesticide Information Center (NPIC), DDT (Technical Fact Sheet), 2, 2000.
[10] T. D'Silva, The Black Box of Bhopal, 50, 2006, ISBN: 978-1-4120-8412-3.

CHAPTER

Product Manufacturing and Distribution* 7

EVENT DESCRIPTION

Not long after the patent was issued, the parent company began mass-producing the pesticide molecule (technical grade product) in its major domestic factory. By patenting the pesticide, the parent company was able to secure exclusive manufacturing, sales, and distribution rights until 1972 [1]. The domestic factory where the molecule was produced was appropriately sized for the world's anticipated and expanding needs for this essential product.

The technical grade product was originally manufactured through the "chloroformate" intermediate process. In this case, the intermediate product was a hazardous liquid lachrymator that was formed by reacting phosgene ($COCl_2$) with α-naphthol ($C_{10}H_7OH$) according to Eq. [7.1] [2]:

OH + $COCl_2$ ⟶ O-C(=O)-Cl + HCl

α-Naphthol | Phosgene | Naphthyl chloroformate | Acid [7.1]

Chloroformate intermediate synthesis route

The reaction produced the intermediate product (naphthyl chloroformate [3]) and hydrochloric acid (HCl). After separating the HCl, the naphthyl chloroformate was routed to intermediate storage. The final reaction involved combining the intermediate substance with monomethylamine (CH_3NH_2), also known as "MMA," to form the technical grade product and more HCl (Eq. [7.2]):

O-C(=O)-Cl + CH_3NH_2 ⟶ O-C(=O)NHCH$_3$ + HCl

Naphthyl chloroformate | Methylamine | Pesticide | Acid [7.2]

Original pesticide manufacturing route in the domestic factory

*For timeline events corresponding to this chapter see pages 416–417 in Appendix.

Rethinking Bhopal.

The primary hazards associated with manufacturing process related to the use of phosgene. Unlike *liquid* naphthyl chloroformate, phosgene is a *gas* at normal atmospheric pressure and temperature [4]. Therefore, phosgene represented a more immediate exposure hazard upon release compared to liquid naphthyl chloroformate, which has its own safety hazards [5]. Accordingly, minimization practices were incorporated in the phosgene manufacturing process, whereas fewer restrictions were imposed while manufacturing the intermediate phosgene derivative (chloroformate). This conservative phosgene manufacturing approach limited the amount of toxic gas that could potentially escape if a leak was to develop.

The domestic factory's size was such that it could manufacture the finished pesticide product in excess of the local demand. The parent company had an international subsidiary operating in India. Unlike the situation that put the parent company's domestic (USA) factory in an export position, India could not produce enough pesticide internally to satisfy its internal demand [6]. Moreover, poor harvests could only be avoided by counting on foreign supplies to make up the pesticide shortfall. The country was therefore dependent on outside suppliers to stabilize India's agricultural industry. The subsidiary in India recognized the opportunity to supplement India's agricultural production with surplus technical grade product (the concentrated active ingredient) produced by the domestic factory in the United States. Accordingly, in December 1960 the subsidiary applied for a license to import technical grade product from the United States [7].

The business arrangement between the parent company and its international subsidiary was a win–win situation for all involved. India was and perhaps still is the world's largest democracy. Capitalizing on the vast international market represented by the subsidiary was a very appealing prospect for the parent company [8]. In a way, the pesticide symbolized everything that the manufacturing industry stood for, which encompasses improving public safety, comfort, and security with environmentally safe products. In addition to these virtues, the product brought India one step closer to becoming a self-sufficient manufacturing nation able to satisfy the internal needs of its population without relying on outside support. Although the true realization of that goal would involve breaking India's dependence on supplemental imports, the parent company's international presence gave India direct access to the technology it needed to make such a goal technically feasible. In this context, transporting the pesticide into India was an important "first step" to establishing the industrial position that India desperately longed for [9].

The international subsidiary in India had little trouble obtaining an import license and immediately started receiving the technical grade product from surplus domestic factory production. The first shipment of technical grade product left the domestic factory's dock on January 5, 1961 [10]. Initially, India's demand for the technical grade product amounted to 1400 MT annually [6]. After demonstrating the product's effective use, India forecasted that the pesticide's annual usage rate would grow to 5900 MT by 1974 [11].

The international subsidiary in India set up basic operations to dilute the technical grade product into usable formulations ranging from 1% to 20% [11]. Dilutions of various ranges met various product grade specifications that could then be distributed throughout India [12]. Different product concentrations were formulated to protect various types of plants from different species of insects. Diluting the technical grade product in India served two important purposes:

1. It minimized international shipping costs.
2. It financially supported the local suppliers that could furnish the inert dilution ingredients.

TECHNICAL ASSESSMENT

HAZARD AWARENESS

A closer examination of the domestic factory's manufacturing process reveals a pattern of hazard awareness that permeates throughout the entire case study. The same concept extends deep into today's factories as well. Our study of this pattern will answer critical questions about process design and operating decisions which, viewed in hindsight, caused serious problems. No doubt, the technical principle observed here still governs hazard recognition and tolerance in modern industrial processes today.

It should be noted that the domestic pesticide factory operated directly by the parent company incorporated a specific approach to managing potential process hazards. We need to know and understand this approach which was perhaps considered quite appropriate at the time but later caused problems elsewhere. Once we know and understand this approach, it will become our future obligation to act responsibly with this knowledge. Simply put: It must be our obligation and purpose to prevent similar problems from developing in the processes that we operate today. Although a scientific basis may exist, the chosen implementation method for a transfer of science may lead to regrettable decisions. The seriousness of these unfortunate decisions is demonstrated later in the unforeseen sequence of events that brought the Bhopal factory closer to ruin.

Notice that the pesticide manufacturing process in the domestic factory took place in three distinct steps:

1. First, phosgene ($COCl_2$) was synthesized by reacting chlorine with carbon monoxide.
2. Next, naphthyl chloroformate was produced by reacting phosgene with α-naphthol.
3. Finally, the finished pesticide molecule was manufactured by reacting naphthyl chloroformate with MMA.

Special provisions were in place to manufacture gaseous phosgene (a hazardous substance) "in situ" [5]. Under these circumstances the phosgene was used and consumed immediately upon its production [13]. This approach to designing the process shows sensitivity for managing hazardous substances that could possibly cause harm upon their accidental release. However, the process was also designed with storage tanks to contain large volumes of the liquid naphthyl chloroformate, which was also recognized as a hazardous substance. Comparing these two handling methods creates a contradiction. Why was sensitivity shown toward the manufacture and immediate use of gaseous phosgene whereas similar concerns were not incorporated in the process design and storage of liquid naphthyl chloroformate?

In contemplating the answer to this puzzling contradiction, it is important to recognize that the relative hazard imposed by a compound is a function of its physical properties [14]. Generally speaking, gaseous compounds are considered more hazardous than their liquid or solid counterparts. The reasons for this distinction have a universal application in modern process design, operation, and control:

1. Gaseous hydrocarbons are typically easier to ignite.
2. Gaseous vapors can be inhaled while liquids and solids cannot be.
3. Gases released into the atmosphere are difficult to contain and control.

Due to these complications, industry tends to demonstrate more concern over the potential release of a gaseous substance while displaying a relative attitude of indifference for potentially hazardous processes that would be released as a liquid or solid. We observe the same attitude professionally represented in a recent industry report concerning modern refinery alkylation units. The scientific principles described in this study provide a basis for advocating substitution practices to manage specific hazards related to refinery alkylation processes (Table 7.1). Assessing the difference between managing HF (hydrofluoric acid) and H_2SO_4 (sulfuric acid) as a potential substitute, the report says as follows:

> *HF is much more dangerous when released because it readily forms dense, highly toxic vapor clouds that hover near land and can travel great distances. In contrast, sulfuric acid typically remains in a liquid state during upsets and releases. (Footnote: HF has a boiling point of 67°F and a vapor pressure of 783 mmHg. By comparison, sulfuric acid has a boiling point of 554°F and a vapor pressure of 0.01 mmHg.) [15]*

Like the spent caustic storage system involved in the incident discussed in chapter "Spent Caustic Tank Explosion: A Case Study in Inherently Safer Design," little concern was expressed over storing a hazardous *liquid* in a process unit. Any accidental leak could be easily contained without resorting to sophisticated control measures. However, the introduction of a gas (propane and flare gas) is what made the process unstable. In fact, the potential for a vapor release is what justified the adding of safeguards, which in turn introduced a design defect.

Table 7.1 Distinction Between HF and H_2SO_4 Chemical Hazards (See Also Table 8.1)

Product	Boiling Point (°C)	Vapor Pressure (psi at 20°C)
HF	19	15.1
H_2SO_4	290	0

HF, hydrofluoric acid; H_2SO_4, sulfuric acid.

With this distinction in mind, the domestic factory's practice of applying minimization techniques to phosgene (boiling point of 8°C) and not liquid naphthyl chloroformate makes sense. As a liquid, any spilled naphthyl chloroformate could easily and effectively be contained by workers wearing standard personal protective equipment. This basis for controlling hazards according to a compound's expected physical state is a ribbon of consistency that attaches the parent company to its international subsidiary, as we shall see in later chapters.

With this explanation that is somewhat easy to relate to, we find that the difference in the way that these two hazardous compounds were managed does not necessarily reflect a careless attitude toward release prevention. The fact remains that the domestic factory incorporated special design and operating practices to control the hazards inherent to the potential release of a *gaseous* compound. At the same time, less concern was demonstrated for the potential release of a hazardous compound in its *liquid* state. It was as if the prevailing concept for managing process hazards was to convert a hazardous gaseous substance into a liquid or solid derivative, similar to how we might manage process hazards through substitution practices—replacing *chlorine gas* with *liquid bleach*, for example. In this sense there is really little difference in the way that modern industry manages process hazards today. There is a scientific principle behind the practices that were used to manage hazardous compounds in the domestic factory. We would therefore want to be sure to avoid developing a false sense of security based on a compound's physical state by asking a very simple question (assuming no safeguards):

What is the worst thing that could happen?

The answer to this question may help to justify adding protection even if the compound's physical properties do not create an immediate concern. When contemplating how much protection is needed, note that gaseous process releases tend to dissipate more readily. Therefore, gaseous releases typically introduce an acute hazard whereas liquid or solid releases represent a more chronic threat. For that reason, liquid or solid compound releases tend to have a residual effect.

PROCESS OPTIMIZATION

From the day that the new pesticide was introduced, the parent company that owned the patent was in a good position to capitalize on their invention. The product was effective. It solved an environmental problem. People were willing to purchase it. The reputation of a former product that had once created a sizable demand was spoiled by the potential consequences associated with its continued usage. Holding the patent for an effective replacement product gave the parent company certain production privileges not available to other manufacturers. Among these was to set the selling price without interference from their competitors.

As time was limited before the patent expired and the market would be flooded with generic alternative products, there was a greater potential to profit from the new pesticide molecule immediately after its introduction. This accounts for establishing production inside a major domestic factory that had sufficient capacity to become the world's sole supplier of the pesticide's active ingredient. It also accounts for the interest shown by the international subsidiary that immediately applied for an import license. This license would allow the subsidiary to purchase any excess product made available to them. Finally, it guaranteed that the parent company would almost immediately begin looking for ways to optimize the process and to make production more efficient.

Although the invention solved a serious problem facing the global agricultural industry, substituting its use for the former product was not an even exchange. Manufacturing the new pesticide molecule was both labor-intensive [3] and generated large volumes of chemical waste. The production of excess waste promoted the resurgence of environmental concerns [16]. Additionally, the intermediate naphthyl chloroformate product was extremely corrosive and produced low yields (86%) upon reaction with MMA [17]. For these reasons, it was relatively costly to deal with [18].

As might be expected from any responsible owner, the parent company immediately began looking for a more cost-effective approach to manufacture the product. A change was needed to resolve the mandatory environmental concerns that were developing over the chloroformate intermediate route. A new method of synthesis might also help to address some of the more discretionary issues related to asset reliability (corrosion), labor requirements, and low yields.

As is true inside the manufacturing industry today, mandatory issues dominate the assignment and prioritizing of work. Mandatory issues are things that absolutely, positively have to get done. Safety and environmental commitments along with regulatory requirements fall into this category. Asset reliability and process productivity matters are among the presumed discretionary issues. Under that presumption, repeat failures and production constraints might receive less than adequate attention. A factory's resources are thus often allocated or redirected to resolve the more critical mandatory or deadline-driven issues. However, discretionary asset reliability issues can propagate into mandatory safety and environmental concerns if left alone long enough. Resolving the mandatory environmental issue created by excessive waste generation would require changing the manufacturing process somehow. As discussed earlier, even changes originating with good intentions can have counterproductive results (chapter: Spent Caustic Tank Explosion: A Case Study in Inherently Safer Design). It might even convert a relatively minor environmental or process performance issue into a more serious safety issue.

LESSONS FOR US

This chapter delves into the prioritization of manufacturing activities and hazard control as practiced within old and new factories alike. Factory operations are dominated by mandatory requirements. However, resource allocation must be effectively controlled to prevent relatively minor discretionary issues from developing into a serious, mandatory obligation. Resources must be properly allocated to keep the "small things small" to prevent bigger issues from developing. Additional lessons that we can learn from this discussion include the following.

- Gaseous compounds generally represent a more immediate hazard than their liquid or solid counterparts.
- While gas releases disperse after isolation, remediating a liquid or solid process release involves intentionally placing people in the direct line of fire.
- Liquid and solid releases can have a residual effect that lasts for an extended length of time.
- Assessing hazards on the basis of a compound's physical state can create a false sense of security.
- Consider the worst thing that could happen upon a process release to determine how much protection is needed.
- A careful study of an "as-built" process configuration discloses how the designers intended to control various process hazards.
- Process optimization might be necessary to address process inefficiencies that could hinder a product's long term success.
- A decision that seems to make perfect sense based on scientific principle can prove faulty in hindsight, after an incident occurs.
- Remaining focused on the most important process hazards requires minimizing the distractions created by industrial incidents.

REFERENCES

[1] T. D'Silva, The Black Box of Bhopal, 192, 2006, ISBN: 978-1-4120-8412-3.

[2] W. Worthy, Methyl isocyanate: the chemistry of a hazard, Chem. Eng. News 63 (6) (1985) 30.

[3] T. D'Silva, The Black Box of Bhopal, 34, 2006, ISBN: 978-1-4120-8412-3.

[4] A. Agarwal, S. Narain, The Bhopal Disaster, State of India's Environment 1984–85: The Second Citizens' Report, 219, 1985.

[5] C. Jiménez-González, D.J.C. Constable, Green Chemistry and Engineering: A Practical Design Approach, 398, 2011, ISBN: 978-0-470-17087-8.

[6] T. D'Silva, The Black Box of Bhopal, 31, 2006, ISBN: 978-1-4120-8412-3.

[7] Supreme Court of India, Criminal Appeal Nos. 1672, 1673, 1674 and 1675 of 1996, in the Matter of Keshub Mahindra vs. State of Madhya Pradesh, September 13, 1996. 9.

[8] T.R. Chouhan, Bhopal: The Inside Story, Carbide Workers Speak Out on the World's Worst Industrial Disaster, 19, 1994, ISBN: 0-945257-22-8.

[9] I. Eckerman, The Bhopal Saga: Causes and Consequences of the World's Largest Industrial Disaster, 22, 2005, ISBN: 81-7371-515-7.

[10] Committee on Inherently Safer Chemical Processes: The Use of Methyl Isocyanate (MIC) at Bayer CropScience, Board on Chemical Sciences and Technology, Division on Earth and Life Studies, National Research Council, The Use and Storage of Methyl Isocyanate (MIC) at Bayer CropScience, 157, 2012, ISBN: 978-0-309-25543-1.

[11] T. D'Silva, The Black Box of Bhopal, 32, 2006, ISBN: 978-1-4120-8412-3.

[12] District Court of Bhopal, India, State of Madhya Pradesh Through CBI vs. Warren Anderson & Others, Criminal Case No. 8460 of 1996, 8, June 7, 2010.

[13] S. Varadarajan, et al., Report on Scientific Studies on the Factors Related to Bhopal Toxic Gas Leakage, 9, 1985.
[14] B.J. Gallant, Hazardous Waste Operations and Emergency Response Manual, 41, 2006, ISBN: 978-0-471-68400-8.
[15] United Steelworkers (USW), A Risk Too Great: Hydrofluoric Acid in U.S. Refineries, 1, 2013.
[16] T. D'Silva, The Black Box of Bhopal, 49, 2006, ISBN: 978-1-4120-8412-3.
[17] Committee on Inherently Safer Chemical Processes: The Use of Methyl Isocyanate (MIC) at Bayer CropScience, Board on Chemical Sciences and Technology, Division on Earth and Life Studies, National Research Council, The Use and Storage of Methyl Isocyanate (MIC) at Bayer CropScience, 92, 2012, ISBN: 978-0-309-25543-1.
[18] T. D'Silva, The Black Box of Bhopal, 50, 2006, ISBN: 978-1-4120-8412-3.

CHAPTER

Process Selection*

8

EVENT DESCRIPTION

Aside from the new pesticide's undisputed commercial success, the process by which it was manufactured was neither cooperative nor efficient [1]. Shortly after initiating mass production in the domestic factory, a potential environmental problem developed. The concern was over the excessive amount of waste that was generated by the manufacturing process.

The major concern originated with the amount of hydrochloric acid (HCl) that was generated as a by-product of the manufacturing reaction (Eq. [7.2]). Recovering the technical grade product required separating the acid, which was then neutralized in a typical acid–base reaction. This waste management process produced a considerable volume of solids (salt). The disposal of this solid waste product became a major production issue. Because the process was so inefficient [2] (86% possible yield) more HCl was processed per unit of technical grade product manufactured than what would otherwise result from higher-yielding reactions. A solution was needed to address these concerns. Otherwise, the competition could easily swoop in and introduce a more efficient and environmentally friendly alternative. These complications prompted the parent company to consider different production routes [3].

Over the next 5 years, the parent company worked to develop an alternative method of producing the pesticide, to replace the inefficient "chloroformate" manufacturing process (chapter: Product Manufacturing and Distribution). In 1966, the domestic factory commissioned a new production unit to manufacture "methyl isocyanate" (MIC) [4]. Similar to naphthyl chloroformate, MIC was also a phosgene derivative. Ultimately, this intermediate product was selected to replace the uncooperative, inefficient, wasteful, and therefore unsustainable chloroformate process.

The alternative production method was perhaps the best solution that could possibly have been hoped for. The new process used the exact same ingredients that were incorporated in the original process. Converting to this process would raise the final reaction's yield to 92%—a 6% improvement over the existing reaction [2]. The change simply involved switching the order in which the intermediates were produced and reacted. This simple solution addressed the environmental concerns and other inefficiencies related to the original manufacturing process.

* For timeline events corresponding to this chapter see page 418 in Appendix.

Rethinking Bhopal.

The new process got started the same way that the original process did. It began by reacting carbon monoxide with chlorine gas to form the toxic, reactive intermediate phosgene ($COCl_2$) just the same as before. Next, however, the phosgene was reacted with monomethylamine to generate a highly reactive chemical liquid (39.1°C boiling point at 760 mm Hg [5]) intermediate called methyl isocyanate (CH_3NCO) or "MIC" for short (Eq. [8.1]) [6].

$$\underset{\text{Phosgene}}{COCl_2} + \underset{\text{MMA}}{CH_3N_2} \rightarrow \underset{\text{MIC}}{CH_3N{=}C{=}O} + \underset{\text{Acid}}{2HCl} \qquad [8.1]$$

The final manufacturing step involved a catalytic reaction between MIC and α-naphthol, which produced the technical grade pesticide product (Eq. [8.2]) [7,33].

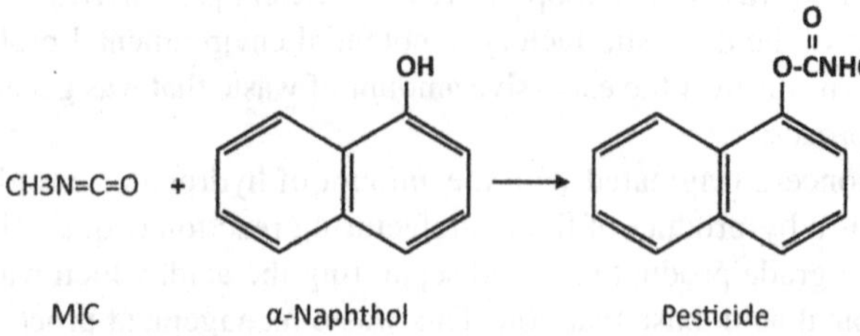

O
OH O-CNHCH$_3$
CH3N=C=O + →
MIC α-Naphthol Pesticide

Final pesticide manufacturing route (MIC process) [8.2]

Unlike the former process, reacting MIC with α-naphthol produced no by-products of concern together with the final pesticide product [8]. The new intermediate MIC process was not only less corrosive, but also generated less waste than the former chloroformate process. It was a practical way to minimize the waste produced by the manufacturing process. As a result of this development, the market remained firmly under the parent company's control.

But all changes have repercussions and even the best intentions can backfire unless careful thought is given to controlling process hazards. The problem, of course, is learning through trial and error that you missed something important—something that you needed to know earlier. In this case, however, the results of the process modification seemed relatively straightforward. The process was modified to account for design, maintenance, and operating differences that were needed to control the hazards that applied to the new manufacturing method.

It was somewhat reassuring that the naphthyl chloroformate was being replaced by a compound that was also a liquid at normal working temperatures. Had a gaseous substance replaced the chloroformate, then serious changes in process design and configuration would have been needed, similar to the way phosgene was being managed. However, the MIC intermediate was a highly volatile substance compared to its nonvolatile chloroformate counterpart [5]. Therefore, it would readily evaporate upon its generation [9]. Accordingly, controlling the product's vapor pressure was a serious concern [10]. A refrigeration unit (cooler) was needed to suppress evaporation while the intermediate was in storage. Without refrigeration, significant quantities of MIC would escape into the process vapor management system through the storage tank's normal vent line (Fig. 8.1).

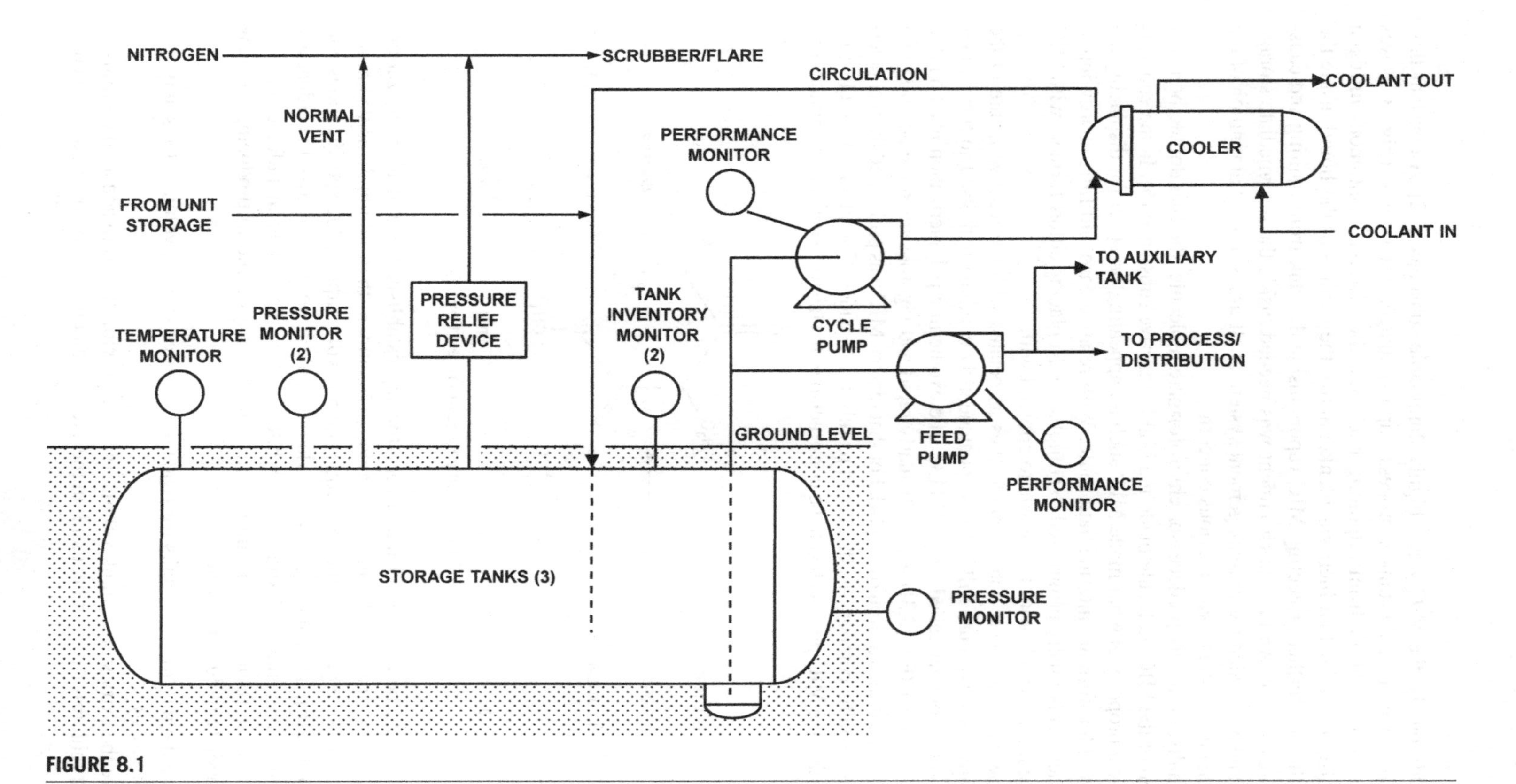

FIGURE 8.1
Domestic factory methyl isocyanate storage tank and venting system design [11].

Additionally, the MIC was a highly flammable substance [12]. A constant flow of nitrogen though the tank's process vapor management system (Fig. 8.1) was needed to prevent air from migrating back into the process. Continuous nitrogen flow also maintained an inert gas blanket above the surface of the liquid inside the tank; thereby further protecting MIC vapors inside the tank from igniting and causing an explosion. An inert environment was needed inside the intermediate storage tanks and associated vapor management system at all times. Potential ignition of the volatile tank contents was a serious concern.

Finally, unlike its predecessor chloroformate molecule, the two double bonds at the end of the MIC molecule made it a highly reactive substance [5]. In the end, this chemical property is what made MIC such an appealing prospect for displacing the former chloroformate intermediate route. With relative ease and little energy input, the reaction chemistry progressed to completion. Replacing chloroformate with MIC offered higher yields with less waste to contend with.

However, there were some complexities associated with the reactive nature of the alternative process. Pure MIC would spontaneously combine with itself to form a nuisance solid polymer called "trimer" [13]. The exothermic polymerization reaction that produced trimer (Eq. [8.3]) was particularly rapid and vigorous upon the addition of a catalytic agent such as iron (rust) [14]. For that reason, MIC was prohibited from contacting any metallic components that could readily form oxides. Compliance with this rule meant that the MIC process had to be constructed with stainless steel, at a minimum [6].

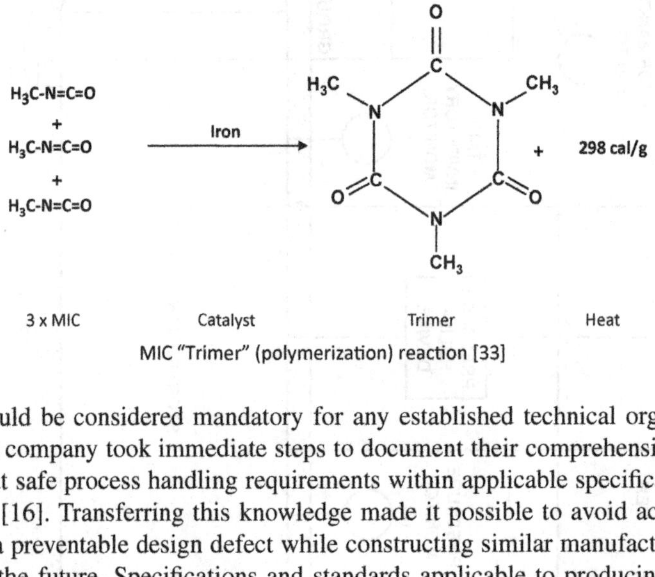

$$H_3C\text{-}N{=}C{=}O + H_3C\text{-}N{=}C{=}O + H_3C\text{-}N{=}C{=}O \xrightarrow{\text{Iron}} (CH_3NCO)_3 \text{ (trimer ring)} + 298\ \text{cal/g}$$

3 x MIC — Catalyst — Trimer — Heat

MIC "Trimer" (polymerization) reaction [33] [8.3]

As would be considered mandatory for any established technical organization, the parent company took immediate steps to document their comprehensive knowledge about safe process handling requirements within applicable specifications and standards [16]. Transferring this knowledge made it possible to avoid accidentally inserting a preventable design defect while constructing similar manufacturing processes in the future. Specifications and standards applicable to producing the new intermediate included instructions to:

- Size the intermediate tanks for twice the required storage volume (for safety purposes.)
- Use the extra storage tank volume (see the previous item) to add an inert diluent (chloroform) as a heat sink in the event of an emergency. It was also noted

that adding a diluent would not stop a reaction but would provide more time to control the problem.

- Keep an empty tank available at all times as an *alternative* to sizing the tanks twice their needed capacity. The empty standby tank could then be used for the same purposes as described previously.
- Maintain an atmosphere of dry nitrogen, under slight pressure, in the MIC tank's vapor space.
- Keep the circulation pumps switched "ON" to continuously cycle MIC through the refrigeration unit [17].
- Maintain the MIC tanks' temperature below 5°C (41°F) and preferably at about 0°C (32°F) [18].
- Equip MIC storage tanks with dual temperature indicators to sound an alarm and flash warning lights if an abnormal rise in tank temperature was detected.
- Inspect the vent valve and safety valve on a regular basis. Instructions were also provided to remove any deposits that could cause the vent system and emergency pressure relief system to malfunction.
- Use stainless steel for piping and valves.
- Use spiral-wound gaskets made of fluorocarbon resin and stainless steel.
- Use fluorocarbon resin-lined or flexible stainless steel hoses.
- Do not use iron or steel, aluminum, zinc, or galvanized iron, copper, tin, or their alloys in MIC service [19].
- Do not use polyethylene or any plastic or plastic-lined hoses, other than those made of fluorocarbon resin.
- Do not use quick-disconnect fittings.
- Seal threaded joints with fluorocarbon resin tape.
- Install relief valves on lines that could be isolated at both ends with block valves.
- Keep the storage system and transfer lines free of contaminants.
- Handle only MIC or dry nitrogen in the storage tanks and associated process lines.
- Install check valves or other devices to prevent the backflow of contaminants into the MIC process.

Although the MIC production unit had been brought online in 1966, the chloroformate manufacturing route remained operational through 1973 [8]. Until that time came, MIC was only used to manufacture newer pesticide inventions borrowing from the same technology. This would make sense, considering the design modifications that were likely needed to make the existing chloroformate process compatible with MIC. It is conceivable that at least one turnaround cycle would have been needed to bring a former chloroformate process into compliance with the specifications and standards governing the use of MIC. For example, a total system outage would reasonably be expected to replace lower grade piping and valve components (iron) with stainless steel. After the final conversion was complete, there was no reason to continue operating the inferior chloroformate process any longer and so it was abandoned [20].

TECHNICAL ASSESSMENT

At this point in the sequence of events, a consistent set of patterns begins to develop. Some of the patterns encountered in this event have appeared since the beginning and continue to do so as the timeline advances. Other patterns that carry forward originate strictly with this event. Collectively these patterns help to establish a basis for complicated actions that take place during later events, as we observe the manufacturing process become progressively less stable. It is of serious importance to recognize the relationship between these patterns and the decisions that are involved, in order to benefit fully from the lessons that this case study provides. The knowledge that is received through this study makes it possible to avoid making changes that might make a stable industrial process more difficult to control and more sensitive to error.

Referring back to chapter "Inventing Solutions," a new pesticide molecule was invented to replace an aging product that had been discredited over environmental concerns. A change in technology was used to solve the problem. A similar pattern is observed in this chapter. Likewise, concerns over the replacement product's potential environmental impact prompted the parent company to consider alternative production methods. The change in this case was strictly limited to the chloroformate intermediate process used to solve the original problem.

An element of competition spurred this change, as it did in the former instance as well. The first company that could introduce a practical solution to the chloroformate intermediate process would reap rich rewards for their invention. If another company could rise to the occasion, then the parent company might lose control of the market. The same competitive pressures existed between the first change where the alternative pesticide molecule was introduced, and the second change that affected how that specific molecule was being manufactured.

Undoubtedly with great relief to the parent company, an effective solution was found internally before the competition could introduce a more environmentally-sensitive substitute product. As a result, the parent company was able to hold onto its market position. In fact, the parent company could have even publicized this voluntary conversion as a demonstration of its environmental commitment. Actions speak louder than words. The conversion that took place here provides evidence that environmental issues were taken seriously by the parent company.

Notice, however, that the parent company used a different method to solve the internal problem that it had discovered. In contrast to the first solution, the introduction of breakthrough chemical technology was not involved here. In this case, the solution simply involved reordering the sequence of reactions originating with the same basic ingredients that were used before. In other words, the parent company was able to *improvise* a solution. This improvised approach solved an environmental problem by using only resources that were readily available. By doing so, there was no need to develop an entirely new method of manufacturing the same molecule. This approach was by far the most economical and expeditious way to address both the environmental and productivity concerns.

This is the first time that improvisation appears in the sequence of events leading up to the Bhopal disaster. An improvised solution prevented the parent company from

Table 8.1 Distinction Between MIC and Phosgene [26] Chemical Hazards (see Also Table 7.1)

Intermediate	Boiling Point (°C)	Vapor Pressure (psi at 20°C)
Phosgene	8	23.5
MIC	39	8

surrendering its market position to another competitor. Many actions that appear later in the timeline follow the same example; whereby serious production issues are addressed by simply modifying the existing process rather than introducing an new process.

With respect to managing specific process-related hazards, we also see a continuation of the same methodology before and after the development of the new intermediate (MIC) process. Unfortunately, the new manufacturing approach did not eliminate phosgene from the reaction. Phosgene was still considered to be the most hazardous product in storage. The phosgene-handling practices did not change. It was still being produced in situ and on an as-needed basis only. Neither did the storage practices of the new intermediate product change. Using the same principles described in chapter "Product Manufacturing and Distribution," phosgene's chemical properties would have raised more attention than its corresponding derivative (Table 8.1). In relative terms on the basis of its physical properties at room temperature, the MIC was safe compared to phosgene.

Just because a product is thought to be less hazardous than others does not explain why it would need to be stored in large quantities. There is another reason why the domestic factory considered storing large volumes of the intermediate product to be appropriate. The public record explains that reserve supplies of the intermediate were kept in storage to compensate for unplanned downtime due to MIC production issues [22]. Keeping large amounts of unreacted MIC on hand prevented an immediate supply chain interruption if for some reason the MIC unit tripped offline or had to be taken down for unplanned maintenance (chapter: Equipment Reliability Principles).

It is very important to note how expected asset reliability problems could justify storing large supplies of the hazardous intermediate products by design, as it did in this case. This pattern of manipulating the process to compensate for potential asset reliability problems is repeated throughout the sequence of events leading up to the Bhopal disaster. Special process considerations were integrated in the factory's design to continue production in the event of an unplanned maintenance interruption.

Reserving excess inventory to accommodate asset failures makes it clear that minimization practices are a function of asset reliability. The more reliable a process is, the better chances there are to maintain continuous production without stockpiling excess inventory. When excess inventory is not required to maintain production in the event of an unforeseen process shutdown, minimization practices can be implemented without fear of production shortages. However, this degree of manufacturing confidence is predicated on reliable process operation. It may involve investing more in asset reliability than would otherwise be considered necessary. Construction costs are likely to increase, but relative to the devastating penalties that chronic unplanned maintenance and production losses can impose, the economic trade-off represents an incredibly good and protective investment. Large storage tanks are just one of many "creative" workaround solutions that could be implemented when such process investments are not appreciated. Hoarding product inventory allows production to continue for a limited period of time while ongoing maintenance eventually restores required asset functions.

Examining the intermediate process conversion further provides vital information about the MIC storage tank's design, operation, and process complexities. While naphthyl chloroformate was regarded as a nonvolatile substance, MIC was quite the opposite (8 psi vapor pressure at 20°C.) Not only did this increase the process' flammability, but the conversion also made it more difficult to contain the intermediate product inside a standard atmospheric tank. An atmospheric floating roof tank is designed to accommodate high volatility liquid hydrocarbons (Table 8.2). However, the fact that phosgene vaporizes (boils) at 8°C is perhaps one of the complicating issues that made storing MIC in a conventional floating roof storage tank impractical.

The solution was to refrigerate the tank contents. In accordance with industry specifications (Table 8.2), refrigeration can be added as a solution for containing high vapor pressure and highly volatile hydrocarbons and specialty chemicals in low-pressure storage tanks. This action prevented phosgene in the tank from vaporizing. Safely containing the MIC process in atmospheric storage tanks involved continuous refrigeration. Otherwise the phosgene inhibitor would escape. Refrigerating the tank contents lowered the product's volatility sufficiently to contain it in a tank continuously venting into an atmospheric system.

The method selected to preserve the phosgene inhibitor in the MIC storage tank corresponds directly to instructions not to allow the temperature of the tank contents exceed 5°C at any time [24–26]. This practice inhibited dangerous side reactions if the MIC in the storage tanks was contaminated [27].

Here we find the purpose for an MIC refrigeration system, which was to reduce the intermediate's vapor pressure and thus fugitive vapor emissions. But the refrigeration system represents only one of the attenuation methods [28] that were adopted to normalize the process' extreme flammability, volatility, and reactivity (Table 8.3). Using these auxiliary systems to attenuate the process reminds us of the circumstances involved in the spent caustic storage tank explosion discussed in chapter "Spent Caustic Tank Explosion: A Case Study in Inherently Safer Design." Similar to that incident, we see how complexity was added to the intermediate manufacturing process while taking necessary steps to make it safer. Moving in this alternative direction defeated the principles of inherently safer design that made the process easier to operate and less dependent on mechanical availability [3]. There is nothing at this point to stop us from attenuating the MIC process to manage its properties, other than our concern based on knowledge about what eventually happens at the end of this case study and what continues happening in real life today (chapter: Spent Caustic Tank Explosion: A Case Study in Inherently Safer Design).

With that eventuality in mind, it is already appropriate to begin asking how we would be expected to act in a similar situation. Would we also attenuate our processes

Table 8.2 Hydrocarbon Storage Tank Selection Guidance [23]

Note:
This table serves as a guideline only based on typical product attributes under ambient and refrigerated operating conditions. Before selecting the final tank design for any storage project, test data should be collected for the fluid to be contained.

Tank Design	Diagram	Vapor Pressure psi[a]	Crude Oils	Condensate	Oils	Natural Gasoline	Butanes	Propane	Raw Natural Gas Liquid	Ethane	Petrochemicals	Natural Gas	Liquefied Natural Gas	Treating Agents	Dehydration Fluids	Specialty Chemicals	Solid Materials	Water
Cone Roof		0-2.5 psi	X	X	X	X	R	R	R	R	R		R	X	X	X	A	A
Floating Roof (IFR or EFR)		2.5-15 psi[b]	X	X		X	R	R	R	R	R		R			X		
Sphere		>5 psi		X			X	X	X	X	X	X	X					
Cylinder (Pressure Vessel)		15-1000 psi		X			X	X	X	X	X	X	X					

A = Atmospheric pressure
X = Acceptable storage at ambient temperature
R = Acceptable storage with refrigeration only
[a]*Vapor pressure represents pressure differential values. Thus, it is appropriate to express vapor pressure in units of psi.*
[b]*Floating Roof Tanks – never exceed applicable regulatory limits (usually less than 15 psi)*

Table 8.3 Storage System Provisions Made to Manage MIC Properties

MIC Property	Provided Asset(s)	Asset Function	Control Method
Volatility	Circulation pump, heat exchanger, and organic refrigerant	Refrigerate tank contents to 0°C, to suppress evaporation and flash point by reducing vapor pressure.	Attenuation
Flammability	Nitrogen supply, pressure control valves	Establish a nitrogen blanket inside the empty space in the storage tanks, to prevent igniting a flammable mixture inside the tank vapor space and continuously purge air (oxygen) out of the vapor management system.	Attenuation
Reactivity	Stainless steel tanks, pipes, and valves	Inhibit rust formation on MIC contact surfaces, to rid the process of a potential catalytic agent capable of triggering an exothermic polymerization reaction inside the storage tank and vapor management system.	Inherently safer design

with add-on design solutions; thereby increasing operational complexity with the good intention of making the process more controllable? Remember that safety is built into the process or product—not added on (chapter: Spent Caustic Tank Explosion: A Case Study in Inherently Safer Design). Unfortunately, building safety into the process is much easier said than done. We might be fooled into thinking that our process can become safer by adding more assets to manage specific chemical properties. However, this type of action subverts the principles of inherently safer design. We see this trend starting to develop here.

The provisions added to control the properties of MIC in storage (Table 8.3) had to remain operational at all times. If any of these assets were to fail, then the resulting control system functional loss might allow the product's flammability, volatility, or reactivity to cause problems. The more severe MIC process' properties would revert back to their natural state. The consequence for this involuntary conversion could range from anything between an economic or environmental incident (product evaporation) and a safety incident (an explosion or a thermal runaway reaction.)

This cause–consequence relationship is at the root of our ever-repeated emphasis on asset reliability. This example clearly demonstrates how the reliability requirement of a process system is dependent on the method chosen to manage its properties. In this specific case, the selected approach to managing the process in the storage tanks dramatically increased the amount of attention and care needed to support effective process control. This is not to say that the intermediate product could not be safely contained [29]. It did, however, mean that there was less tolerance for maintenance defects, unreliable assets, and maintenance practices contrary to best-in-class. Essentially, the valves, pipes, pumps, and utilities had to be available at all times. The manpower and staff needed to adequately support these functions would depend on the performance of the installed assets and their compliance with specifications and standards. In the end, we see how this process conversion magnified the potential consequence for human error. Everything established to attenuate the properties of the product in the storage tanks had to work perfectly.

Certainly these assets were operated to control the chemical properties of the product in the storage tanks. These assets needed to operate continuously for the new intermediate to behave more like the former intermediate in the storage tank. Stainless steel was the exception. Little effort was needed to control the intermediate product's reactivity in a process constructed with an appropriate corrosion-resistant alloy such as stainless steel. Stainless steel was, therefore, the inherently safer design choice. After it was installed, it would require far less care and attention than either of the attenuation methods would. It is for good reason, then, that stainless steel was specified as a mandatory design requirement for surfaces that might come in contact with MIC. Clearly specifying stainless steel as the design requirement in MIC service should have eliminated any argument over the decision not to install a less expensive design alternative.

Looking ahead, after the Bhopal disaster the parent company was sharply criticized for operating the refrigeration unit haphazardly [17]. The basis for this assertion implies that the refrigeration unit served as a safeguard to prevent a thermal runaway reaction. Hindsight is always 20/20 [30], and although there is no question that the refrigeration unit could, and perhaps should, have offered protection in the event of a contamination incident, that was really not its intended design function. The impression that developed in response to the Bhopal disaster closely follows the circumstances described in chapter "Spent Caustic Tank Explosion: A Case Study in Inherently Safer Design" where an adjunct system was provided to prevent a vapor cloud explosion. The support system described in that study should also have prevented the development of an explosive mixture *inside* the tank. That, however, was not its intended purpose. In fact, no consideration had been given to suppressing an explosive mixture inside the tank before the explosion occurred. Therefore, the process was not adequately protected from the failure mechanism that ultimately destroyed it. Had the focus been on preventing an explosion *inside* the tank, then it might have been possible to prevent a vapor cloud explosion *outside* the tank as well. The same is true in the case of the MIC refrigeration unit that attenuated the volatility of the tank contents. The intention here was to make it suitable for storage in a tank that vented into an atmospheric vapor management system. Had the focus been on preventing a thermal runaway reaction inside the storage tank, then the refrigeration unit might also have been designed to provide a specific safety function.

It is important to note that although explicit instructions were written into MIC process design specifications and standards to operate the refrigeration unit continuously "on," never was it stated *why* continuous operation was necessary. If the refrigeration system's purpose was to remove the heat of reaction upon a storage tank contamination incident, then it was not clearly documented as such. There is, however, a technical basis for lowering the product's vapor pressure according to information in the public domain. This scientific explanation corresponds with cited references indicating that the refrigeration system's purpose was to reduce the vapor pressure of the highly volatile product stored inside the tank to control the product's volatility and flammability (not reactivity) hazards. The impression that we are left with is that the parent company did not recognize the potential for a contamination incident to result in a thermal runaway reaction. This is really no different than the situation in chapter "Spent Caustic Tank Explosion: A Case Study in Inherently Safer Design" where *before* the tank explosion little concern was expressed over the potential ignition of a flammable mixture inside the tank.

This conclusion, however, is not supported by the fact that specific design instructions were documented regarding the use of extra tank volume for safety purposes if an inert diluent "heat sink" (excess chloroform [31]) was needed to dissipate the heat of reaction. This directive is consistent with comments written into the MIC unit design standards and specifications about sizing the tank for double its anticipated storage volume and *alternatively* supplying an empty standby tank. In these two examples, extra tank volume was reserved to provide a safety function in the event of a contamination incident. Moreover, the extra tank space was to be used for more time to remedy a situation, or bring it under control well before a full-scale thermal runaway reaction could take place. Obviously, the parent company was well aware of the hazard involved with a potential storage tank contamination incident. Special provisions were incorporated to control an exothermic reaction inside the tank should a contamination incident occur. Since these provisions were all passive controls, we might ask if any active controls existed to *prevent* a tank contamination incident. Interestingly, at this point in the case study we simply do not know. There appears to be no specific action taken to *prevent* a contamination incident. All we know at this point is that a combination of provisions was made to manage a contamination incident *after* it occurred:

1. Spiking the MIC with phosgene [32],
2. Doubling the necessary size of the storage tanks, and
3. Installing a standby tank.

In investigation terms, the inconsistency described here represents another contradiction in logic. This clearly signals that an important piece of information is missing. The unresolved contradiction causes us to seek a scientific clarification to reconcile the two opposing facts. At the end of the case study we must be able to explain why so much attention was paid to *responding* to a contamination incident whereas very little, if any, concern was expressed over *preventing* it.

Possessing this knowledge can help prevent us from acting upon faulty reasoning. In hindsight, we can readily see that a refrigeration system whose function was reserved for another purpose *should* have prevented a safety incident. Unfortunately, this type of "lesson" is neither transferrable nor actionable. We cannot prevent an incident on a process that is believed to be adequately protected, prior to an incident that proves us wrong. In a similar sense, the parent company appears to have been satisfied that there was no cause for concern over a potential incident that would require the refrigeration system's availability until it was proven wrong. Only by understanding why the parent company felt adequately protected without the refrigeration system can we derive a valuable lesson that will prevent us from making a similar mistake in the future. This question can be answered by examining specific process design restrictions detailed in chapter "Factory Construction."

LESSONS FOR US

This chapter describes patterns demonstrated by the parent company that set a precedent for future events. Seeing these patterns allows us to anticipate, view, and evaluate actions and decisions that occur later. With regard to competitive pressure we saw how, in chapter "Inventing Solutions," the parent company developed and introduced new technology. In doing so they were able to solve an environmental problem involving a competitive product. However, a different environmental problem involving its own brand was voluntarily solved internally through improvisation. Simply manipulating the sequence in which compounds were reacted shifted hazard control from favoring inherently safer design to human-dependent protection. As a result, the process became considerably less tolerant of what we might collectively describe as human factors. These interweaving patterns consistently apply throughout the sequence of events behind the Bhopal disaster. We see how the parent company regularly used improvised solutions to manage production constraints and other internal problems. Other lessons worthy of note include:

- Competition fuels the engine of change in the manufacturing industry.
- Cost reduction is rarely the primary cause for making process changes.
- Changes made to improve environmental performance tend to create unforeseen safety hazards.
- A product's physical properties define the appropriate storage system.
- Using attenuation methods to adjust a product's physical properties makes the process more complex.
- Attenuation practices conflict with inherently safer design principles.
- Modifying a process to address environmental concerns can make the process more dangerous.
- There is no way to prevent an incident on a process that you are convinced is already adequately protected.
- Minimization practices are only practical when the process reliability exists to support it.
- Overcoming expected or actual asset reliability issues to maintain production is a common excuse for manipulating the process.
- A chemical's physical properties do not make it hazardous. How we decide to manage those properties is what makes a chemical hazardous.

REFERENCES

[1] T. D'Silva, The Black Box of Bhopal, 49, 2006, ISBN: 978-1-4120-8412-3.
[2] Committee on Inherently Safer Chemical Processes: The Use of Methyl Isocyanate (MIC) at Bayer CropScience, Board on Chemical Sciences and Technology, Division on Earth and Life Studies, National Research Council, The Use and Storage of Methyl Isocyanate (MIC) at Bayer CropScience, 92, 2012, ISBN: 978-0-309-25543-1.
[3] Committee on Inherently Safer Chemical Processes: The Use of Methyl Isocyanate (MIC) at Bayer CropScience, Board on Chemical Sciences and Technology, Division on Earth and Life Studies, National Research Council, The Use and Storage of Methyl Isocyanate (MIC) at Bayer CropScience, 109, 2012, ISBN: 978-0-309-25543-1.
[4] Committee on Inherently Safer Chemical Processes: The Use of Methyl Isocyanate (MIC) at Bayer CropScience, Board on Chemical Sciences and Technology, Division on Earth and Life Studies, National Research Council, The Use and Storage of Methyl Isocyanate (MIC) at Bayer CropScience, 157, 2012, ISBN: 978-0-309-25543-1.
[5] W. Worthy, Methyl isocyanate: the chemistry of a hazard, Chem. Eng. News 63 (6) (1985) 27.
[6] Union Carbide Corporation, Review of MIC Production at the Union Carbide Corporation Facility Institute West Virginia, April 15, 1985. 1-2.
[7] District Court of Bhopal, India, State of Madhya Pradesh Through CBI vs. Warren Anderson & Others, Criminal Case No. 8460 of 1996, 30, June 7, 2010.
[8] W. Worthy, Methyl isocyanate: the chemistry of a hazard, Chem. Eng. News 63 (6) (1985) 32.
[9] T.R. Chouhan, Bhopal: The Inside Story, Carbide Workers Speak Out on the World's Worst Industrial Disaster, 33, 1994, ISBN: 0-945257-22-8.
[10] R.R. Fullwood, Probabilistic Safety Assessment in the Chemical and Nuclear Industries, 253, 1999.
[11] W. Worthy, Methyl isocyanate: the chemistry of a hazard, Chem. Eng. News 63 (6) (1985) 29.
[12] T.R. Chouhan, Bhopal: The Inside Story, Carbide Workers Speak Out on the World's Worst Industrial Disaster, 55, 1994, ISBN: 0-945257-22-8.
[13] T. D'Silva, The Black Box of Bhopal, 98, 2006, ISBN: 978-1-4120-8412-3.
[14] Union Carbide Corporation, Review of MIC Production at the Union Carbide Corporation Facility Institute West Virginia, April 15, 1985, 1-1.
[15] District Court of Bhopal, India, State of Madhya Pradesh Through CBI vs. Warren Anderson & Others, Criminal Case No. 8460 of 1996, 33, June 7, 2010.
[16] T. D'Silva, The Black Box of Bhopal, 54-55, 2006, ISBN: 978-1-4120-8412-3.
[17] S. Diamond, The Bhopal Disaster: How It Happened, The New York Times, January 28, 1985.
[18] T.R. Chouhan, The unfolding of Bhopal disaster, J. Loss Prev. Process Ind. 18 (2005) 207.
[19] T.R. Chouhan, Bhopal: The Inside Story, Carbide Workers Speak Out on the World's Worst Industrial Disaster, 44, 1994, ISBN: 0-945257-22-8.
[20] Committee on Inherently Safer Chemical Processes: The Use of Methyl Isocyanate (MIC) at Bayer CropScience, Board on Chemical Sciences and Technology, Division on Earth and Life Studies, National Research Council, The Use and Storage of Methyl Isocyanate (MIC) at Bayer CropScience, 158, 2012, ISBN: 978-0-309-25543-1.
[21] World Health Organization, Phosgene Health and Safety Guide, 8, 1998.

[22] W. Worthy, Methyl isocyanate: the chemistry of a hazard, Chem. Eng. News 63 (6) (1985) 32.

[23] Gas Processors Suppliers Association (GPSA), Engineering Data Book, FPS Version, vol. II, 1998, 6-2.

[24] J. Browning, Union Carbide Corporation Press Conference Transcript, March 20, 1985. 2.

[25] I. Eckerman, The Bhopal Saga: Causes and Consequences of the World's Largest Industrial Disaster, 26, 2005, ISBN: 81-7371-515-7.

[26] W. Morehouse, M.A. Subramaniam, The Bhopal Tragedy: What Really Happened and What It Means for American Workers and Communities at Risk, 19, 1986, ISBN: 0-936876-47-6.

[27] S. Varadarajan, et al., Report on Scientific Studies on the Factors Related to Bhopal Toxic Gas Leakage, 24, 1985.

[28] D.A. Crowl, J.F. Louvar, Chemical Process Safety: Fundamentals with Applications, second ed., 2002, ISBN: 0130181765. 22.

[29] Union Carbide Corporation, Review of MIC Production at the Union Carbide Corporation Facility Institute West Virginia, April 15, 1985. 1-3.

[30] Committee on Inherently Safer Chemical Processes: The Use of Methyl Isocyanate (MIC) at Bayer CropScience, Board on Chemical Sciences and Technology, Division on Earth and Life Studies, National Research Council, The Use and Storage of Methyl Isocyanate (MIC) at Bayer CropScience, 142, 2012, ISBN: 978-0-309-25543-1.

[31] Supreme Court of India, Criminal Appeal Nos. 1672, 1673, 1674 and 1675 of 1996, in the Matter of Keshub Mahindra vs. State of Madhya Pradesh, September 13, 1996. paragraph 24.

[32] A. Agarwal, S. Narain, The Bhopal Disaster, State of India's Environment 1984–85: The Second Citizens' Report, 219, 1985.

[33] District Court of Bhopal, India, State of Madhya Pradesh Through CBI vs. Warren Anderson & Others, Criminal Case No. 8460 of 1996, 31, June 7, 2010.

CHAPTER

Industry Compliance* 9

EVENT DESCRIPTION

For about 6 years the domestic factory in the United States was the only location both authorized and capable of producing the technical grade product that everyone wanted. The international subsidiary had been a loyal customer, importing technical grade product from the parent company from the very beginning. As time went on, the pressure to work within the framework of India's foreign trade policy increased [6]. In essence, these principles were directed at developing self-sufficiency and local manufacturing capabilities. It was thus only a matter of time before these circumstances would compel the international subsidiary to seriously consider manufacturing the technical grade product locally [1]. Establishing local manufacturing capabilities would enable the international subsidiary to not only formulate, but actually produce, pesticides that exclusively originated with indigenous raw materials.

However, continuing the existing customer–supplier relationship between the international subsidiary and its parent company created a problem. For one thing, India's foreign trade policy favored the consumption of products manufactured locally. These guidelines were established to minimize the economic losses that might result from a trade imbalance created by an uneven dependence on foreign imports [2]. Even though the subsidiary had direct access to one of the World's most popular and technologically advanced pesticide brands, it was only in the business of diluting the technical grade product obtained from abroad. While this put some money back into the local economy, it was hardly enough to stabilize or develop the economy as was intended. Much more money was leaving the country than could be recovered from local sales. This trade imbalance did not in any way support India's goal to become a competitive, economically stable, industrialized nation.

India had established a framework for governing foreign trade relationships for the specific purpose of protecting the economy. The relationship embodied in the pesticide formulations license (chapter: Product Manufacturing and Distribution) deviated from these principles. The international subsidiary honored the parent company's patent and, therefore, was not legally entitled to manufacture the molecule independently. As a result, maintaining a continuous supply of the pesticide to local farmers required demonstrating an indifferent attitude toward the principles designed to stimulate India's economy and chemical industry.

* For timeline events corresponding to this chapter see page 419 in Appendix.

This business approach was a sensitive matter. Restricting access to pesticide imports would hurt the population and thus defeat the intended purpose of India's foreign trade policy. On the other hand, allowing the unrestrained consumption of these imports created economic problems that made it even more difficult for India to achieve financial stability. The long-term solution was to legally obtain the necessary permits that would authorize the international subsidiary to manufacture the product locally. Obtaining these permits would eliminate the present outflow of hard currency and satisfy India's internal demand for agricultural products.

The savings that would result from this transition were significant for India. Recall that the parent company only accepted payment in the form of US currency. Purchasing products from the parent company therefore meant that the international subsidiary had to raise an appropriate amount of cash by selling local products to consumers in the United Sates. This put India at a serious disadvantage. India's industry was still in its primitive stages of development. It could not realistically compete for business by exporting locally manufactured products to the United States. It was likely that similar products could already be obtained from domestic producers at a discount. Manufacturing the technical grade product from start to finish within India would effectively solve this problem. It would make it unnecessary for India to raise the US currency needed to purchase an adequate supply of pesticide products. At least $11 million would be saved by successfully converting the formulations business into an outright manufacturing operation [3].

But the economic savings potential went further than that. Simply having the ability to control the supply and distribution of pesticide products locally would reduce the amount of supplemental imports needed to make up for any potential agricultural product shortfall. This scenario was even more consistent with the principles established to prevent trade deficits that could suppress India's economic stability and growth. Similar to the strategy used years ago to construct the pesticide manufacturing process in the domestic factory, the Bhopal factory was also planned to be oversized. This design would make it possible for the subsidiary to *export* products to international customers, just like the domestic factory could. Having the ability to place excess technical grade product on the international market would allow the international subsidiary to directly contribute to local economic growth after paying the parent company royalties as per the licensing agreement. This would advance the development of India's industrial sector. It would make the country similar to the United States by being able to capitalize on the global demand for products that served a moral humanitarian interest.

Collectively, these benefits made the prospect of manufacturing the technical grade product in India very attractive. Approval of the application would represent significant progress in transforming the subsidiary's formulations operation into a fully-functional and sustainable industrial process. More importantly, it would bring the international subsidiary into full compliance with regulatory framework designed to protect the economy. On April 7, 1966 the international subsidiary proceeded to file the paperwork to manufacture 5000 MT of the technical grade product annually [3]. This tonnage was based on market growth projected in 1974 [4].

TECHNICAL ASSESSMENT

This short description of the event that initiated the advancement of pesticide manufacturing technology in India provides critical insight into the complexities involved with managing industry policies. It is certainly of interest that economics did not seem to play a role in the international subsidiary's decision to apply for an industrial license to manufacture the technical grade product. At the very least, economics were not *directly* involved in that decision; regulatory or legislative compliance was the prevailing concern behind this event. More specifically, compliance with foreign trade policies developed to serve the "national interest" was involved [5]. In this particular case, the national interest was twofold:

1. Providing the steady supply of pesticide products needed to stabilize the agricultural industry, and
2. Preserving, maintaining, and growing the economy.

So, in the end economics were involved after all, albeit *indirectly*. This makes sense considering the complexities involved with serving the first national interest, which was to provide a steady supply of a product that could not easily be done away with.

Living without the product that was only available from a single factory operating in the United States had frightening implications. A decision prolonging the shipment of such imports could, potentially, cripple India's agricultural industry. Recall that India did not, at that time, have sufficient pesticide manufacturing capacity to meet its internal demand for agricultural products. Complying with India's industrial regulatory framework translated to withdrawing from the import business needed to maintain and develop the local agricultural industry. Lives would be lost, and only the higher-class citizens would be able to purchase the food needed for survival. Results like this were out of the question. Restricting product imports from the domestic factory would defeat the purpose for which regulations that served a national interest were created.

As industry professionals, we must judge exceptions to be appropriate in such cases where compliance with policy would be a threat to public safety. That appears to be the situation here. Either waivers were signed, authorizing the international subsidiary to continue its relationship with the foreign parent company in control of the supply chain, or a "blind eye" was turned that allowed the deviation to persist unchallenged.

Either situation might be considered unfortunate in the eye of the beholder who observes the authorized or unauthorized deviations that persist with impunity. As an example, assume that a premium watchmaker resides in a country with import regulations. The intent of the regulations is to protect the economic welfare of local businesses. But what if the watchmaker is forced to compete with a reseller of low-cost watches imported from a foreign supplier due to inconsistent application of the regulation? The situation breeds animosity as the premium watchmaker sees his business dwindle down to nothing while the reseller continues to prosper. Eventually, the local watchmaker closes down; he cuts his losses while lamenting that laws intended to protect him failed. It would be difficult to recover from a management system failure of this type and the loss of confidence and damaged reputation that ensues. "Uniformity in message" is needed to promote evenhandedness. Without "uniformity in message" there will be many unintended failures. Lack of uniformity tends to inflict serious damage to a culture that must operate according to plausible rules.

Ultimately, the economic backbone of an institution that fails to enforce applicable policies and regulations will collapse. There is no incentive to develop internal production capabilities when the enforcement of policies is unevenly applied or shows lopsided favoritism. When a policy, which includes regulations, is implemented, it must be enforced uniformly and without exception. Otherwise, serious damage can be done in a system where chaos rules. In such cases the most defiant competitor is (by default) rewarded.

This was definitely not the case here, where favoritism was not being shown. Although the business relationship between the parent company and its international subsidiary deviated from the intent of established foreign trade policies, lives were at stake. Restricting the availability of the pesticide obtained from the domestic factory would result in death. Here, and in all cases, good judgment must prevail. Deviations must be allowed sometimes for good judgment to prevail. Without question, limiting the supply of India's pesticide to comply with the regulatory framework described in this specific case would not have been exercising good (engineering) judgment. A policy deviation was, therefore, appropriate in this specific case—at least in the short term.

The business relationship, however, was extremely damaging to India's economy, which now places a demand on the second national interest involved in the foreign trade regulation. India consumed mass quantities of the pesticide that had to be imported from the domestic factory. The parent company would accept payment only in the form of hard US currency—cash equivalent for the purchase of the technical grade product that was diluted upon its arrival in India. In other words, India could only obtain this product if it could generate US dollars to pay for it. In a practical sense, this could only be done by selling products made in India to consumers in the United States. The competitive disadvantage represented here is analogous to expecting to watch David conquer Goliath in a battle for business. In this scenario, India was expected to outcompete the world's leader in technology. The leader had a quiver full of consumable products it could offer its homeland consumers at a considerable discount. As difficult as raising this necessary cash was, vast amounts of it were leaving India to continue this customer–supplier relationship. Ironically, purchasing the products that a nation needed to survive was financially killing the nation. Under examination, it could be said that the way the situation was being managed created another problem.

This basic rule applies practically everywhere you look inside the manufacturing industry. Getting what you want basically means giving up something that you might desperately need. What India needed was financial stability. Recall that it was now 6 years since the international subsidiary had been importing the technical grade product from the United States. After 6 years, this subsidiary would have been expected to show progress, or at least interest, in complying with their country's policy regarding foreign trade relationships.

On a smaller scale, similar complexities involved with the management of policies are routinely encountered inside the factories that we operate today. Likewise, today there are many policies that we must comply with to maintain safe, effective, and legal operation. Every year the list gets longer. However, no policy is perfect; policies cannot be expected to blanket every possible scenario. In cases where complying with a policy would defeat its purpose, adjustments must be considered so that the policy can be applied without creating predictable and preventable damage. In short term, it may be necessary to issue an exception to the policy that provides a reasonable temporary solution. Ultimately, however, a permanent and sustainable solution must be applied.

The importance of complying with policies is significant. Allowing a policy deviation to persist with or without an exception makes it impossible to retain a uniform and consistent message. Even if good judgment is preserved by deviating from a policy, continued deviation is not a good practice. It can lead to misalignment between departments—perhaps divisions between operations, engineering, procurement, and maintenance personnel. Infinitely more cultural damage is done when the acceptance of an unauthorized policy deviation persists. Under such conditions anything is possible; including the worst of all consequences. We shall see this in later chapters. As we proceed in our reading, keep in mind that the roots of this behavior can be linked to the event discussed in this chapter. Pay attention also to the complications that can arise when inconvenient policies are changed to make compliance more possible (chapter: Process Optimization).

Once again, however, a temporary exception to foreign trade policy was appropriate in this specific instance. Cutting off the supply of the product needed to preserve health, safety, and security to satisfy a compliance obligation would have been counterproductive. Permanent compliance with impending regulations soon to follow would require establishing product manufacturing capabilities on Indian soil. The application for a manufacturing license was the first step that was needed to achieve full compliance with the principles governing the activity of foreign businesses in India. It was the only practical solution capable of serving the public's needs without creating severe economic and cultural damage by continuing to demonstrate indifference toward national policy.

LESSONS FOR US

Complexities often develop when managing compliance matters. This can be expected, but must never be used as an excuse to disregard policies or regulations that we are obligated to comply with. In many cases, complying with a policy requires patience as we carefully assess options that will deliver the intended benefits. Temporary exceptions are appropriate only in cases where immediate compliance with a policy will create unacceptable physical harm to health, the environment, and/or physical property. However, policy exceptions should be rare and should expire upon the implementation of a permanent solution that makes full compliance possible. The attitude we display and express concerning compliance with internal and external policies determines how others will approach process reliability issues in the factories we operate. Other lessons we receive through this thought-provoking analysis include the following:

- In the manufacturing industry, getting something that you want might require giving up something that you desperately need.
- No policy is perfect, such that its broad application will produce desirable results under all imaginable situations.
- Good technical judgment must prevail in all instances involving compliance with policies.
- Policy exceptions must only be issued through an approved, documented deviation system.
- Prolonged policy deviations create cultural damage, even in cases where exceptions are approved.
- Good policies are easy to comply with.
- Process complexity is directly proportional to the amount of policies that apply to running the business.

REFERENCES

[1] United States District Court Southern District of New York, Janki Bai Sahu, et al., Against Union Carbide Corporation and Warren Anderson, No. 04 civ. 8825 (JFK), 15, 2012.
[2] T. D'Silva, The Black Box of Bhopal, 24, 2006, ISBN: 978-1-4120-8412-3.
[3] T. D'Silva, The Black Box of Bhopal, 33, 2006, ISBN: 978-1-4120-8412-3.
[4] T. D'Silva, The Black Box of Bhopal, 32, 2006, ISBN: 978-1-4120-8412-3.
[5] T. D'Silva, The Black Box of Bhopal, 25, 2006, ISBN: 978-1-4120-8412-3.
[6] R. Reinhold, Disaster in Bhopal: Where Does the Blame Lie? The New York Times, January 31, 1985.

CHAPTER

Production Commitments* 10

EVENT DESCRIPTION

It took only 8 months for the international subsidiary to receive a letter of intent in response to its petition for an industrial license. The letter of intent served as a formal notice that certain conditions had to be met before the requested license would be issued. The letter of intent also provided tentative approval to proceed with a project that would allow the subsidiary to manufacture the technical grade product in India. Pursuit of this project would force the partnership forming between the parent company and its international subsidiary to operate a business within the framework of India's foreign trade regulations. The objective was to build a chemical manufacturing factory capable of producing up to 5000 MT of pesticide products annually [1].

The letter of intent specified additional production details above and beyond the compulsory regulatory issues. For example, the basic terms of the agreement reserved 50% of the subsidiary's factory output for local formulators, who would then blend the technical grade product into different grades [2]. The international subsidiary introduced this blending practice 6 years earlier. The difference was that instead of importing the technical grade product from overseas like the subsidiary had been doing, local businesses would be supplied by the Bhopal factory. This arrangement would assist in the development of local businesses that served India's chemical and agricultural industries. The plan was to encourage others to follow the subsidiary's example of growth and development in these areas.

The letter of intent also required the development and approval of a foreign collaboration agreement. This agreement would define the ongoing relationship between the parent company and its international subsidiary after completing the factory construction project. The responsibilities included within this agreement would also become a matter of policy that the business unit would be expected to comply with. The foreign collaboration agreement would be expected to cover items such as the use of imported assets as well as technical support requirements [3].

The letter of intent would automatically be converted into an industrial license upon the subsidiary's fulfillment of the initial requirements. The letter of intent would expire, however, if the stipulated requirements were not met within a specified period of time [3]. In anticipation of receiving the industrial license, the international subsidiary began looking for an appropriate location to build a chemical manufacturing factory.

*For timeline events corresponding to this chapter see pages 420–425 in Appendix.

Rethinking Bhopal.

On June 28, 1968 the international subsidiary leased 5 acres of land in Bhopal, the capital city of Madhya Pradesh in central India [4]. By selecting Bhopal as the site to manufacture the technical grade product, the international subsidiary provided visible evidence of its intent to comply with foreign trade regulations. The acquisition of this factory site pushed the international subsidiary closer to obtaining an industrial license that would make it possible to open a chemical manufacturing facility.

The selected location was both an ideal and realistic choice for conducting factory operations. At that time in India's industrial development, Bhopal possessed the type of infrastructure one would hope to find in a major industrial hub [5]. For one thing, the land plot was capable of supporting local factory operation. It was large enough to contain an industrial process with at least enough capacity to meet the local demand. It was also situated within 2 miles of a major railway station [6], which made it readily accessible to those who might travel to and from it on a regular basis. Regular travelers might include workers and other visitors conducting official and unofficial business. Additionally, the site was located next to a railroad track. Provisions were thus available to accommodate the steady flow of products and supplies in and out of the factory.

However, selecting the factory site was by no means the only condition that needed to be satisfied prior to receiving permission to proceed with the construction project. The letter of intent specified other items that had to be addressed to obtain an industrial license. Since these additional obligations were not completely met by December 1967, the application expired [7]. If an industrial license was to be issued, then a separate request would have to be made at a later time. In the interim, the international subsidiary continued formulating pesticide blends by obtaining technical grade product imported from the domestic factory in the United States.

The parent company was comfortable with its level of accountability in this continuing relationship. Exporting product exclusively from its domestic site operating in the United States shielded the parent company from having to adjust its business plan. Instead it could focus its attention on how to maintain compliance with local state and federal regulations, while the international subsidiary had to deal with its own compliance matters. The technical grade product was still patented. If anyone wanted to buy it, then it was to be done on the parent company's terms only.

Failure to receive the industrial license did not stop the international subsidiary from making good use of the land it had leased for factory construction. Before the end of 1969 it had transferred its formulations division to the Bhopal site [8]. Afterward, the subsidiary continued to function exactly as it did before. Products imported from the domestic factory were blended into formulations that could be sold to local customers.

On January 1, 1970 the subsidiary reapplied for an industrial license that would authorize it to synthesize the parent company's trademarked technical grade product completely from indigenous raw materials [9]. Two years later the parent company's patent would expire. With the patent expiration date rapidly approaching, the parent company and its international subsidiary needed to cooperate to meet their mutual goals.

The international subsidiary was coming under constantly increasing pressure to comply with policies governing the use of foreign imports over local products. The parent company also had a problem brewing, with its specialty product soon to become a commodity. The parent company needed an international factory devoted to manufacturing the pesticide to realistically compete for the business it had created over a decade earlier [10]. The thought of being locked out of the international market upon the patent's expiration did not resonate well with the parent company. Local competitors would then be able to manufacture the same product using intellectual property that the parent company had disclosed in its patent application. The pesticide market would soon see competitive products offered at a discount by those who could avoid shipping charges, unlike the product exported from the domestic factory [10]. While customer loyalty might have a residual effect, it is perhaps more difficult to maintain when it exists in name only. A large factory would create the visible presence that the parent company needed for others to feel secure about keeping with the same brand name product, especially if it meant paying a bit more for it.

The license application proposed synthesizing the technical grade product according to the methyl isocyanate (MIC) process used in the domestic factory [11]. At the time when the application was submitted, the proposed technology had successfully been in use for 4 years. In 1973 MIC would replace the chloroformate process that the domestic factory was still using to synthesize the technical grade product [12]. After converting over to the MIC process in the domestic factory, the chloroformate process would be abandoned. Preparing to construct the MIC process in India was simply keeping up with advancing technology. Not only was the process more economical, but its real selling point was the fact that it generated less waste. This is why the parent company had looked for an alternative manufacturing process in the first place.

The second letter of intent was issued on March 13, 1972 [13]. Unlike the first letter of intent that was issued 8 months after the initial application was submitted, it took 2 years for the new letter of intent to be released. The patent expired that same year. When the patent expired the technical grade product instantly became a commodity that any independent chemical manufacturer owning the necessary assets could legally produce.

Similar to the first letter of intent, the second letter also specified certain conditions that had to be met before the industrial license would be issued. In harmony with principles governing the operation of the international business from its inception, these terms included clear directives to reduce the parent company's ownership, and therefore its influence, in their foreign investment. To reduce its import dependency that had prevailed for over a decade at great expense to the local economy, local services would be substituted where practical. "Practical" in this case would encompass any situation where a local service equivalent to an imported foreign service could be obtained [14]. Additionally, 50% of the factory's total output was still to be reserved for local formulators wishing to compete for agricultural industry business.

Unlike the first letter of agreement that left much of the terms of the foreign collaboration agreement up for negotiation, a more prescriptive approach was followed the second time around. This time, special language was incorporated into the agreement, stating that the contract would expire 5 years after factory production commenced [11]. A stipulation was appended to the agreement on August 25, 1973 [15]. This guidance further shifted the balance of power to the local business by specifying a plan to convert the subsidiary into an independent producer of technologically advanced chemical products. This would mean ultimately severing ties with the parent company at an appropriate time. In this capacity the subsidiary would finally achieve full compliance with foreign trade regulations. It would eliminate the need for imported technology—whether physical or intellectual. Most importantly, it would alleviate the constant and increasing pressure that the international subsidiary was under, to reduce imports [16] and the loss of foreign exchange [17].

TECHNICAL ASSESSMENT

The fact that the international subsidiary applied twice for a manufacturing license provides valuable insight into the customer–supplier relationship that existed between the parent company and its international subsidiary. The first application was allowed to expire, presumably due to concerns over market development that fell short of growth projections [18]. To understand why a second application was submitted, we must first consider the commitments that existed between the parent company and its international subsidiary. As explained earlier, the technical grade product was exclusively manufactured in the domestic factory, operated by the parent company in the United States. This factory's output was then exported to India, where the subsidiary was licensed to blend the product into different formulations.

Even though the international subsidiary carried the parent company's name, there was little professional involvement between them. Essentially, the international subsidiary was one of the parent company's loyal customers. On a *professional* level, the parent company had no vested interest or commitment in operating decisions made by the international subsidiary. Neither did it intend to enter into a partnership to construct a factory with the international subsidiary. This relationship protected the parent company from sharing accountability in matters of international policies and regulations governing the manufacture of its products. By operating the world's one and only technical grade product manufacturing process in the United States, the parent company could concentrate on managing its own business, which included complying with strict policies and regulations at home. Using this business model had historically been very successful for the parent company.

The same was not true for the international subsidiary that was obligated to comply with local policies and regulations it could not meet. These policies and regulations encompassed relatively strict guidelines designed to discourage the purchase of imported technology. The subsidiary's business operated exclusively on imported technology. Thus, it was under tremendous pressure to change the way it was operating.

Since the technical grade product was protected by a patent, the international subsidiary had no choice other than to accept the prevailing terms of the customer–supplier relationship. If the supplier chose not to cooperate by changing its business plan, then the customer would have to decide whether or not their relationship was worth continuing. In this case, it appears that the customer decided that the relationship *was* worth continuing. We say this because the international subsidiary continued to purchase the imported product from the parent company even in the midst of increasing pressure to reduce imports in accordance with applicable foreign trade policies.

The situation, however, was about to become different. The international subsidiary was well aware that the prevailing relationship had a limited life span. Within a short period of time, the patented product would automatically convert into a commodity. At that point the international subsidiary would be legally entitled to manufacture the molecule with or without the parent company's permission. The international subsidiary would convert from being a loyal customer to a competitor—selling identical products under differently named brands in compliance with trademark laws. This freedom would develop beginning in 1972 after the technical grade product's patent expired. The parent company had a decision to make. It could either lock itself out of the market or be involved with the construction of a new factory to continue benefitting as a licensed producer.

Even if the parent company chose *not* to cooperate after the patent expired, the subsidiary might still be able to purchase the product manufactured by a local wholesaler in India who will be willing to take on manufacturing responsibilities [19]. This would only require a bit more patience on their part—or maybe not, if a local independent producer could expedite manufacturing operations.

In the end, the international subsidiary could envision that within a short amount of time, the table would turn. They would be able to comply with the foreign exchange polices that applied to their business. This event would occur whether or not the parent company was willing to cooperate in a factory construction project. Based on how the subsidiary had observed the parent company deal with problems in the past, however, they might have bet that the parent company would reconsider their position as the patent's expiration date drew closer. This is a bet that they ultimately ended up winning.

LESSONS FOR US

Incentives play a tremendous role not only in *what* gets done, but also *when* it gets done. However, incentives provide a temptation for change. Drawing on lessons from previous chapters we know that changes can introduce unforeseen hazards. If capturing an incentive requires changing an approach in *how* work gets done, proceed with caution. The advice we have here is to resist implementing a change until a proper assessment can be made according to the example described in this chapter. You must quantify what you will be giving up in return for getting what you want (chapter: Industry Compliance). If the incentive is no longer there by the time you have completed this assessment, be content with the fact that you avoided making a potentially bad decision. The next incentive enabling you to act upon your assessment is likely around the corner. In other words, your success depends on choosing the right time to take action.

Notice that if our sole objective is to capture an incentive, no changes need to be made to performing fundamental asset reliability activities. When assets are responding to a standard regimen of process reliability (chapter: Equipment Reliability Principles), changing direction is the worst decision that can be made. This is one of the reasons that achieving expected asset reliability performance at all times is advocated; not only when the market is driving maximum process uptime. Incentives, however, can quickly vanish if proper attention is not given to operating production assets within limits. Other valuable lessons supported by these events include:

- Every business commitment you make lasts indefinitely.
- The decisions that you make from the instant you start your career affect your future performance and reputation.
- Competition entices you to do things that you would not otherwise consider.
- Visibility creates consumer confidence.

REFERENCES

[1] District Court of Bhopal, India, State of Madhya Pradesh Through CBI vs. Warren Anderson & Others, Criminal Case No. 8460 of 1996, 4, June 7, 2010.
[2] T. D'Silva, The Black Box of Bhopal, 34, 2006, ISBN: 978-1-4120-8412-3.
[3] T. D'Silva, The Black Box of Bhopal, 42, 2006, ISBN: 978-1-4120-8412-3.
[4] T. D'Silva, The Black Box of Bhopal, 40, 2006, ISBN: 978-1-4120-8412-3.
[5] P. Shrivastava, Bhopal: Anatomy of a Crisis, 39, 1987, ISBN: 088730-084-7.
[6] W. Morehouse, M.A. Subramaniam, The Bhopal Tragedy: What Really Happened and What It Means for American Workers and Communities at Risk, 3, 1986, ISBN: 0-936876-47-6.
[7] T. D'Silva, The Black Box of Bhopal, 182, 2006, ISBN: 978-1-4120-8412-3.
[8] United States District Court Southern District of New York, Janki Bai Sahu, et al., Against Union Carbide Corporation and Warren Anderson, No. 04 civ. 8825 (JFK), 7, 2012.
[9] District Court of Bhopal, India, State of Madhya Pradesh Through CBI vs. Warren Anderson & Others, Criminal Case No. 8460 of 1996, 4, June 7, 2010.
[10] P. Shrivastava, Bhopal: Anatomy of a Crisis, 41, 1987, ISBN: 088730-084-7.
[11] District Court of Bhopal, India, State of Madhya Pradesh Through CBI vs. Warren Anderson & Others, Criminal Case No. 8460 of 1996, 12, June 7, 2010.
[12] I. Eckerman, The Bhopal Saga: Causes and Consequences of the World's Largest Industrial Disaster, 25, 2005, ISBN: 81-7371-515-7.
[13] T. D'Silva, The Black Box of Bhopal, 197, 2006, ISBN: 978-1-4120-8412-3.
[14] T. D'Silva, The Black Box of Bhopal, 44, 2006, ISBN: 978-1-4120-8412-3.
[15] T. D'Silva, The Black Box of Bhopal, 45, 2006, ISBN: 978-1-4120-8412-3.
[16] United States District Court Southern District of New York, Janki Bai Sahu, et al., Against Union Carbide Corporation and Warren Anderson, No. 04 civ. 8825 (JFK), 17, 2012.
[17] R. Reinhold, Disaster in Bhopal: Where Does Blame Lie? The New York Times, January 31, 1985.
[18] T. D'Silva, The Black Box of Bhopal, 191, 2006, ISBN: 978-1-4120-8412-3.
[19] J.B. Law, A Review of U. C. India Ltd.'s Ag Products Business, 2, February 24, 1984.

PART

Process Configuration

3

Although creativity is primarily responsible for industry's continual success, its unrestricted pursuit can cause problems. Staying in control of a process requires knowing when deviating from basic design and operating requirements would be potentially dangerous. For example, it is important to exercise caution when dealing with a process design package. It would not be beneficial to construct a process contrary to prescribed design specifications, since this could introduce a defect. This situation could then be used as an excuse for other deviations to bring about a speedy recovery—perhaps adopting a trial-and-error approach to problem solving (brainstorming) instead of a systematic failure investigation method. This could lead to a mentally, physically, and economically exhausting effort to control the process for as long as the defect persists.

This section examines the impact that process configuration has not only on process containment, but also on its operability. The developing case history will show how deviating from specific design criteria can create problems. This analysis provides a compelling basis to reject design practices that conflict with applicable specifications and standards; constructing a process according to design specifications is especially important when a precedent for success already exists. In many cases, scientific principles might seem to support a more creative and efficient way to design a process. Under these conditions, it would be appropriate to consider how using creativity to design and configure a process might impact the lives of others. These concepts are clarified further by looking deeper into the sequence of events behind the Bhopal disaster.

CHAPTER

Licensing*

11

EVENT DESCRIPTION

Significant progress was again made to acquire the necessary industrial license when, on November 13, 1973, the parent company and its international subsidiary jointly signed a design transfer agreement [1]. Under the terms of this contract, the parent company agreed to furnish a technology package for a price of about $160,000 (USD) [2]. By purchasing the technology package, the subsidiary would be entitled to manufacture the technical grade product according to the parent company's intellectual property. In particular, the technology package would specify the minimum (basic) process design and construction requirements needed for safe, reliable, and stable factory operation [3].

The new process would be based on the design used to manufacture technical grade product in the domestic factory [4]. However, the process would be scaled to meet the international market's production demands [5]. For over a decade this manufacturing process had yielded successful results for the parent company. By the time that the application for an industrial license was filed, the process had proven to be both safe and economical [6]. Duplicating the design of this process was a realistic approach to expecting at least the same level of performance from new construction [7].

Indeed, the Bhopal factory would use the methyl isocyanate (MIC) process to manufacture the technical grade product [8]. Not only did chloroformate replacement correspond with the latest technology, but it was also the most economical and least wasteful manufacturing option available at the time. Besides that, the chloroformate process had been abandoned in the domestic factory during the same year. Under these circumstances, the partnership wisely resisted any inclination to act independently by using a process no longer supported by the parent company [7]. Designing and operating the same product manufacturing process that was used in the domestic factory was the most direct way to resolve any technical difficulties that might develop after commissioning new construction.

The partners also signed a foreign collaboration agreement that satisfied a requirement needed for an industrial license to be issued [7]. The terms of the foreign collaboration agreement specified exactly how the two parties involved in the manufacturing project would cooperate during factory construction and after commissioning the process. Under the terms of this agreement, the parent company negotiated a

* For timeline events corresponding to this chapter see pages 426–428 in Appendix.

2.5% royalty on all products sold in return for troubleshooting and technical support [9]. The parent company also agreed to provide training for local factory employees, by request, and for an additional fee [10].

According to the terms of the design transfer agreement, the parent company was obligated to provide process performance specifications and materials of construction for all major and minor assets [11]. These instructions would include specific asset design requirements to minimize the potential for leaks. For example, extended lengths of welded pipe sections were to be installed [12]. By designing the process in this manner, the potential for process leaks through flanges could be reduced. Similarly, the bleeder valves that could be used to flush debris out of process piping were also reduced in number. A conservative design approach was to limit installing unnecessary flanges and bleeder valves. This would avoid inserting potential discontinuities that could result in any accidental but completely avoidable loss of primary containment.

Building the process according to the design specifications documented in the technology package would significantly reduce the potential for unplanned maintenance. The MIC process was a clean service that could easily be managed by constructing the system with rust-resistant materials [13]. For this reason, stainless steel or higher alloy material was specified for the construction of assets that might come in contact with the intermediate product [14].

On January 1, 1974 revisions to India's foreign collaboration regulations went into effect [15]. These new rules made it more difficult for companies to resist complying with foreign trade regulations meant to develop India's economy and industrial independence. Specifically targeted were companies that hesitated to develop local manufacturing capabilities. Developing local capabilities was needed to reduce the nation's dependence on foreign imports [15]. The revised regulations made it mandatory for local businesses to become independent within 5 years [16]. During this time, foreign collaborators involved in a local business would be expected to transfer their technical and operating knowledge over to their local partners in preparation for transferring their ownership. The regulation also limited any foreign investor's ownership to no more than 40% of the business partnership [15].

Collectively, these policy revisions firmly established India's sincerity to become independent in matters of industrial manufacturing. The principles reflected in the stricter regulations had been in place for decades. The time had come to either contribute to India's economic and industrial growth or cease operating so as not to hinder progress toward industrial self-sufficiency.

India's foreign trade policy revisions were received with mixed emotions. The introduction of such laws caused a considerable dilemma for many affected businesses. Where trade secrets were involved, the revised regulations created uncertainty over shifting the balance of control over to local businesses. These concerns led to the exodus of at least two major companies that, over many years, had established lucrative operations in India [15]. For them, the uncertainty over having to change a business model that had worked extremely well up to this point forced them to make a difficult decision. In the end, they simply picked up their belongings and left. However, this was not the case for the parent company that, together with the international subsidiary, was attempting to qualify for an industrial license.

In February 1974 the parent company provided the technical drawings and design specifications in accordance with the design transfer agreement [17]. Satisfying the terms of the agreement allowed the industrial license to be issued on October 31, 1975 [18]. This event occurred about 3 years after the patent expired for the technical grade product. The industrial license permitted the subsidiary to construct and operate a factory that would produce up to 2000 MT of MIC and 5000 MT of MIC derivatives [19].

Whereas foreign partners could no longer own more that 40% of the local business, exceptions were allowed under certain conditions. The first condition was that sophisticated technology had to be involved in the manufacturing process [20]. Secondly, at least 60% of the factory's production volume had to be reserved for export sales [15]. The Bhopal factory's sophisticated process configuration would allow this export quota to be met [20]. Thus, an exception was granted and the parent company was allowed to retain the controlling interest with 50.9% ownership in the partnership [21].

However, there was no way to work around India's foreign trade regulations that limited the duration for the foreign collaboration agreement to no more than 5 years [22]. Therefore, specific wording was written into the contract; it stated that 5 years after the start of factory production, the subsidiary would break its affiliation with the parent company [16]. At that time the local partners would assume complete control over the factory.

TECHNICAL ASSESSMENT

A familiar pattern encountered in previous chapters continues to be fortified by the events related in this chapter. In these events the parent company consistently made changes to adapt to problems. The problem in this instance developed over new restrictions introduced in 1974 that would limit the operation of foreign businesses in India. Prior to this event, India's foreign exchange policies had very little to do with how the parent company managed its relationship with its international subsidiary. However, continuing to manage the business as it had in the past would severely limit its ability to benefit from the relationship. The turning point was in 1972, when the parent company's specialty product became a commodity.

With this event looming in the not-too-distant future, the parent company became seriously interested in the subsidiary's proposal to start manufacturing the technical grade product in Bhopal. In doing so, the parent company would assist in the subsidiary's continuing attempt to bring its activities into compliance with international law. In return, the parent company would be able to take advantage of a provision (a lawful exception) to retain the controlling interest in the partnership; not just the customary limit of 40%. The technical service agreement assured that the parent company would skim off a modest 2.5% royalty for every unit of product sold. However, the agreement would be set to expire 5 years after the start of production in accordance with the new regulations. At that time the partnership would end and the parent company would have no choice other than to transition ownership over to the international subsidiary.

Perhaps the thought was that an extension of the royalty arrangement could be negotiated in return for permission to continue marketing products under the recognized, popular brand name trademarked by the parent company [23]. In this way the independent manufacturer might be able to retain the competitive advantage over identical competitive products. Whatever the case might have been, the patent that covered the technical grade product expired before the design transfer and technical service agreements were signed. After the patent expired, it was only a matter of time before the market would be saturated by different brand names that were essentially the exact same molecule. If the parent company realistically expected to compete for this business it would have to support the project to construct a chemical manufacturing facility in Bhopal [24]. Otherwise it would have to be comfortable deciding to turn its back on the market that it had created and dominated for over a decade. Under the circumstances, divesting itself from the business would have been a painful decision for the parent company. Especially since the technical grade product was still the best pesticide that money could buy.

As mentioned earlier in this chapter, the parent company was not the only multinational technology developer in India faced with a tough choice when the foreign exchange regulations were revised in 1974. At least two other companies were deeply concerned over how these new restrictions might impact the stability of their businesses in the future. In the end, these two companies decided to cease operation in India rather than adjusting their business model to comply with the new policies. What differentiated how these companies responded to similar pressures?

The difference between the companies that stayed and those that left was how they had chosen to protect the products that they manufactured. In the case of the two companies that marketed their product as a trade secret, walking away was a painful decision that needed to be made in the interest of protecting their intellectual property. Complying with the new international trade regulations would have forced them to disclose secrets that others were never previously entitled to. For them, the thought of changing their business plan to accommodate the new regulations was out of the question. They would rather give up the profit represented by the international market which they were abandoning than experiment with a new business approach, and so they left.

For the parent company that stayed, however, notice that no interest had been demonstrated in building another factory before India revised its foreign policy (chapter: Production Commitments). Exporting product from the domestic factory with excess production volume was a business model that worked just fine for them. It was consistent with the parent company's business plan for product manufacturing and distribution. This is why the subsidiary's first attempt to obtain a manufacturing license failed. When the regulations became more restrictive, however, the parent company changed its business plan so that it would not have to divest itself from the overseas market. The parent company would later lament having made this choice.

This is a very important lesson that is often repeated today. When significant changes occur due to forces beyond our control, how willing are we to abandon a successful operating method to accommodate the disruptive influence? The way we answer this question can significantly impact our lives and the lives of others. With few exceptions, the best answer is to resist the natural impulse to change, hoping to accommodate what we cannot control.

Nevertheless, there are conditions that have to be met. Acceptable changes can usually be made by constructing a process according to design specifications. This fact illustrates the importance of the technology package that was delivered to the international subsidiary prior to constructing the Bhopal factory. The technology package contained very specific guidelines that were based on experience accrued at the domestic factory site [25]. Similar to the practices used to contain the process at that location, the Bhopal factory's process that might come in contact with MIC had to be constructed with stainless steel at a minimum. This included tanks, valves, and pipes in liquid or vapor MIC service (Fig. 11.1). Failure to comply with this design requirement might lead to the formation of trimer deposits (chapter: Process Selection). These deposits were recognized as a persistent cause of incidents involving choked process pipes [26]. At first glance and in cases where this known phenomenon occurred, water could be used to flush solid debris from the internal surface of affected assets [27].

However, flushing the pipes with water might introduce another problem. The MIC process incorporated inherently safer design principles. These principles involved a conservative approach to installing flanged pipe connections and bleeder valves. These optional components could provide a convenient source for untimely process leaks to develop. Successfully implementing this design practice was predicated on asset reliability that would support prolonged operation without performing invasive maintenance between scheduled turnaround intervals (chapter: Equipment Reliability Principles). These maintenance windows represented the only opportunity to execute a scheduled maintenance plan without having to manage the hazards related to potential process exposure. The exclusion of flanges and bleeders would not support the application of maintenance practices that might involve introducing foreign materials to flush debris out of process piping. Neither was the process designed to make utility connections that would facilitate this type of unforeseen maintenance to begin with. In every sense of the word, the MIC process was intentionally designed and operated in a manner that made any potentially unsafe maintenance practice more difficult [28].

In the context of process reliability, another dilemma could result from designing the process for an anticipated level of asset reliability that is not actually achieved. Bleeder valves are essential for safely isolating the process when invasive maintenance is needed. These circumstances normally involve replacing major assets to address aggressive or chronic failure mechanisms. A very basic isolation practice involves the use of a "double-block and bleed" procedure (Fig. 11.2). The purpose for this maintenance practice is not only to isolate, but also to *prove* the effectiveness of the applied isolation. Proving isolation is a very important industrial maintenance concept. Invasive maintenance can only be safely performed if effective isolation has been ascertained and verified *before* breaking containment.

The double-block and bleed practice simply involves first closing two block valves to isolate the process where the maintenance will occur. Then a bleeder valve

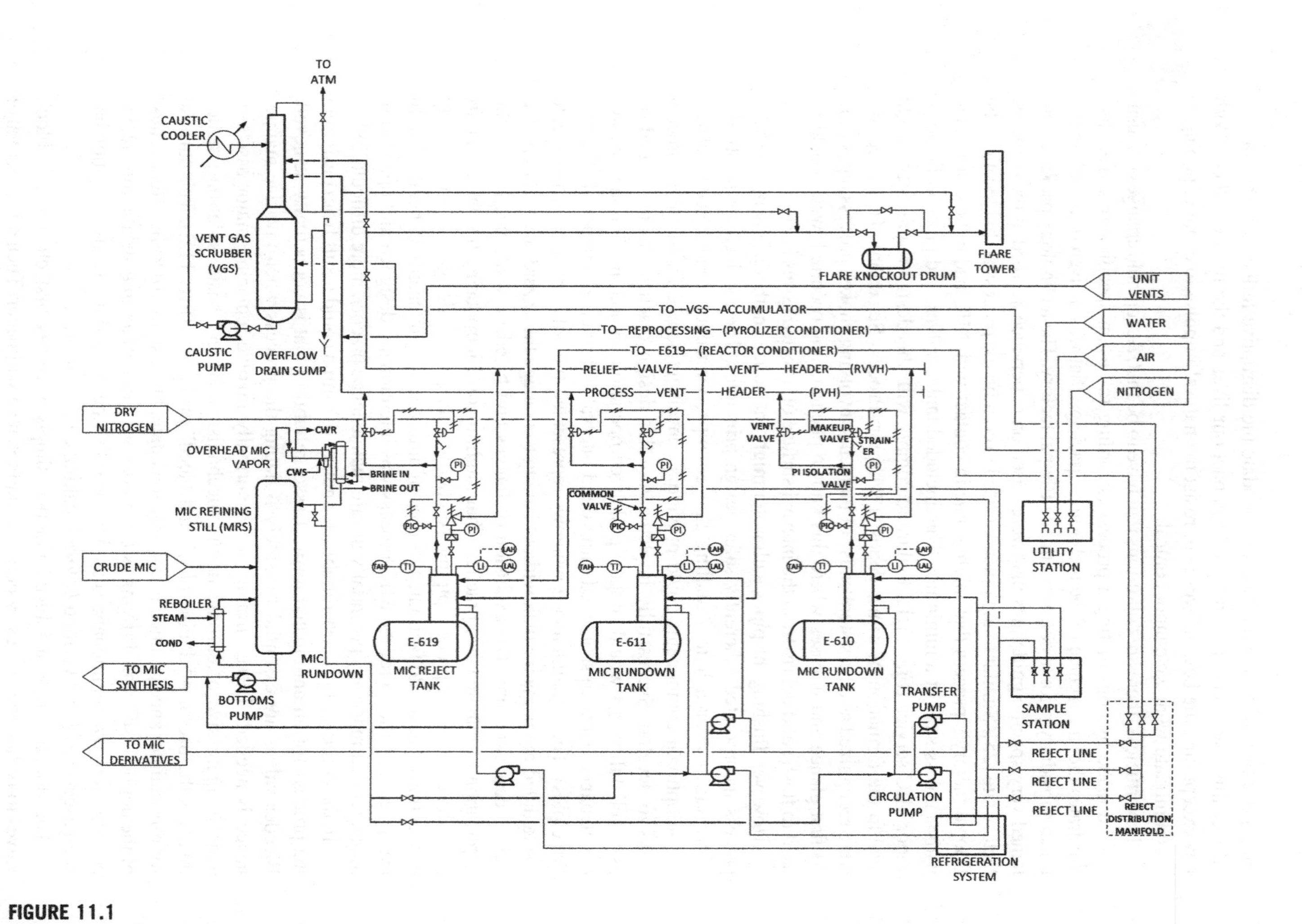

FIGURE 11.1

Process flow diagram of the Bhopal factory's methyl isocyanate (MIC) synthesis, storage, and distribution process. *CWS*, Cooling water supply; *CWR*, Cooling water return.

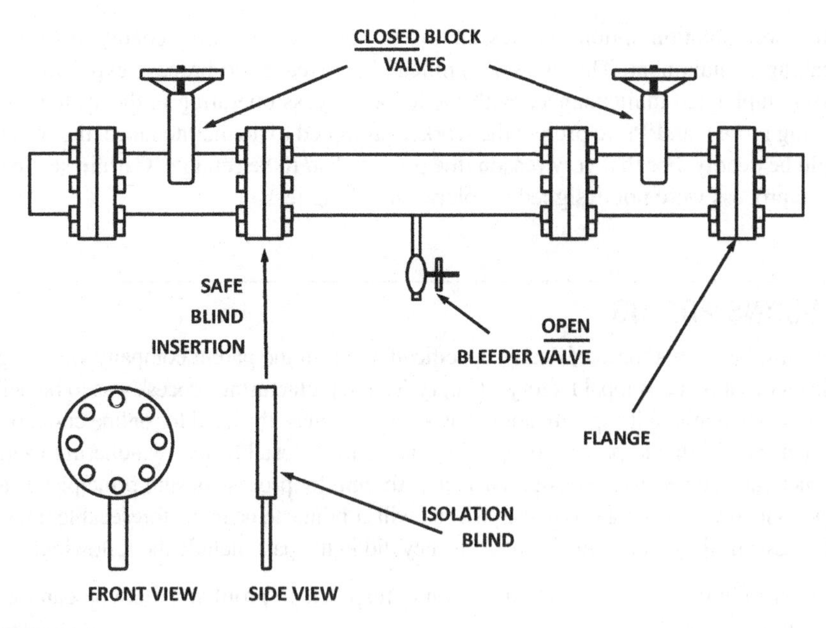

FIGURE 11.2

Double-block and bleed process configuration.

between the two isolation block valves is opened slowly to prove that the valves are not leaking process into the isolated section of pipe. If the process has effectively been isolated, then nothing other than a limited amount of residual process contained in the line when the block valves were closed should exit the bleeder valve.

If a bleeder valve is not present, then the only protection available is personal protective equipment (PPE). PPE, however, is really not intended to substitute for verifying process isolation. While verifying isolation with double-block and bleed does not eliminate the need for PPE, it definitely makes the use of PPE more effective. Bleeder valves are also used to drain and depressure process assets before inserting isolation blinds to perform online maintenance.

The conservative use of flanged connections, while intended to make the process inherently safer, could actually increase maintenance safety hazards. This would depend on whether or not blinding was required. Reducing the number of flanges at the Bhopal factory left fewer isolation options if blind insertion was needed to safely perform online maintenance between turnarounds. A system with fewer flanges presupposes a level of asset reliability that makes isolating the process for online maintenance unnecessary.

Based on operating experience at the domestic factory it was reasonable to expect that constructing the Bhopal factory in accordance with the technology package would provide a level of asset reliability that supported the inherently safer design selection. However, if invasive maintenance was, in fact, needed between major maintenance intervals, then designing the process for better containment would make the process significantly more dangerous. Assets would be taken down for repair with fewer isolation options and less ability to verify safe working conditions before breaking containment. This would significantly increase the process exposure risk during unplanned maintenance. With the lethal process operating in the system and nothing more than PPE to protect the workers assigned to do maintenance, the results could be deadly. For that very reason, the process *had* to be reliable. Certain sections of the process were not designed to tolerate anything less.

LESSONS FOR US

The subsidiary's purchase of design specifications from the parent company was a significant event in the Bhopal factory's history. The manufacturing process was to be built according to a reliability specification that would eliminate the need for online corrective maintenance. Reliable process operation could only be established by adhering to the instructions detailed in the design package. Abiding by process design principles is as important now as abiding was in the past and will continue to be in the foreseeable future. Other lessons that apply equally today as they did in the past include the following:

- Resist the impulse to change direction in response to problems that you cannot control.
- A good survival strategy requires having the courage to walk away from a deal that would require changing a successful business model.
- Inherently safer design practices are only effective when the asset reliability is there to support it.
- Decisions made during the life of a process have a permanent impact on asset performance.
- Processes can be designed to make unauthorized maintenance practices more difficult.
- Seemingly inherent forms of system protection can be intentionally defeated.
- Effective process isolation requires verifying that adequate protection has been applied before breaking containment.

REFERENCES

[1] District Court of Bhopal, India, State of Madhya Pradesh Through CBI vs. Warren Anderson & Others, Criminal Case No. 8460 of 1996, 17, June 7, 2010.
[2] T. D'Silva, The Black Box of Bhopal, 48, 2006, ISBN: 978-1-4120-8412-3.
[3] United States District Court for the Southern District of New York, In re Union Carbide Corporation Gas Plant Disaster at Bhopal, India in December, 1984, MDL No. 626; Misc. No. 21-38, 634 F. Supp. 842. 13, 1986.
[4] District Court of Bhopal, India, State of Madhya Pradesh Through CBI vs. Warren Anderson & Others, Criminal Case No. 8460 of 1996, 4, June 7, 2010.
[5] T. D'Silva, The Black Box of Bhopal, 82, 2006, ISBN: 978-1-4120-8412-3.
[6] United States District Court Southern District of New York, Janki Bai Sahu, et al., Against Union Carbide Corporation and Warren Anderson, No. 04 civ. 8825 (JFK), 27, 2012.

[7] Supreme Court of India, Criminal Appeal Nos. 1672, 1673, 1674 and 1675 of 1996, in the Matter of Keshub Mahindra vs. State of Madhya Pradesh, September 13, 1996. 9.
[8] Supreme Court of India Criminal Appellate Jurisdiction, Application for Directions to Institute Charges U/S 302 (For Offence U/S 300(4)) Read With S. 35 of the Indian Penal Code, 1860 Against the Respondents Herein, Curative Petition (Criminal) No. 39-42 of 2010 in Criminal Appeal No. 1672-75 of 1996 in the Matter of Central Bureau of Investigation Versus Keshub Mahindra & Ors., 8(g), 2011.
[9] T. D'Silva, The Black Box of Bhopal, 208, 2006, ISBN: 978-1-4120-8412-3.
[10] United States District Court Southern District of New York, Janki Bai Sahu, et al., Against Union Carbide Corporation and Warren Anderson, No. 04 civ. 8825 (JFK), 30, 2012.
[11] T. D'Silva, The Black Box of Bhopal, 206, 2006, ISBN: 978-1-4120-8412-3.
[12] A.S. Kalelkar, Investigation of large-magnitude incidents: Bhopal as a case study, in: IChemE: Symposium Series No. 110, Preventing Major Chemical and Related Process Accidents, May 10–12, 1988, 561, 1988.
[13] R. Van Mynen, Union Carbide Corporation Press Conference Transcript, March 20, 1985. 3.
[14] Supreme Court of India Criminal Appellate Jurisdiction, Application for Directions to Institute Charges U/S 302 (For Offence U/S 300(4)) Read With S. 35 of the Indian Penal Code, 1860 Against the Respondents Herein, Curative Petition (Criminal) No. 39-42 of 2010 in Criminal Appeal No. 1672-75 of 1996 in the Matter of Central Bureau of Investigation Versus Keshub Mahindra & Ors., 31, 2011.
[15] T. D'Silva, The Black Box of Bhopal, 35, 2006, ISBN: 978-1-4120-8412-3.
[16] T. D'Silva, The Black Box of Bhopal, 46, 2006, ISBN: 978-1-4120-8412-3.
[17] United States District Court Southern District of New York, Janki Bai Sahu, et al., Against Union Carbide Corporation and Warren Anderson, No. 04 civ. 8825 (JFK), 22, 2012.
[18] R. Reinhold, Disaster in Bhopal: Where Does Blame Lie? The New York Times, January 31, 1985.
[19] T. D'Silva, The Black Box of Bhopal, 196, 2006, ISBN: 978-1-4120-8412-3.
[20] W. Morehouse, M.A. Subramaniam, The Bhopal Tragedy: What Really Happened and What It Means for American Workers and Communities at Risk, 2, 1986, ISBN: 0-936876-47-6.
[21] T.R. Chouhan, Bhopal: The Inside Story, Carbide Workers Speak Out on the World's Worst Industrial Disaster, 19, 1994, ISBN: 0-945257-22-8.
[22] District Court of Bhopal, India, State of Madhya Pradesh Through CBI vs. Warren Anderson & Others, Criminal Case No. 8460 of 1996, 12, June 7, 2010.
[23] T. D'Silva, The Black Box of Bhopal, 195, 2006, ISBN: 978-1-4120-8412-3.
[24] Supreme Court of India Criminal Appellate Jurisdiction, Application for Directions to Institute Charges U/S 302 (For Offence U/S 300(4)) Read With S. 35 of the Indian Penal Code, 1860 Against the Respondents Herein, Curative Petition (Criminal) No. 39-42 of 2010 in Criminal Appeal No. 1672-75 of 1996 in the Matter of Central Bureau of Investigation Versus Keshub Mahindra & Ors., 10-11, 2011.
[25] District Court of Bhopal, India, State of Madhya Pradesh Through CBI vs. Warren Anderson & Others, Criminal Case No. 8460 of 1996, 6, June 7, 2010.
[26] T. D'Silva, The Black Box of Bhopal, 98, 2006, ISBN: 978-1-4120-8412-3.
[27] P. Shrivastava, Bhopal: Anatomy of a Crisis, 46, 1987, ISBN: 088730-084-7.
[28] Union Carbide Corporation, Cause of the Bhopal Tragedy, February 1, 2012. Retrieved from: http://www.bhopal.com/Cause-of-Bhopal-Tragedy.

CHAPTER

Factory Construction* 12

EVENT DESCRIPTION

The industrial license provided the permission that was needed to initiate a construction project to build the chemical manufacturing facility in Bhopal, according to the proposed design. The Bhopal factory's construction project was divided into three distinct phases [1] that took place from 1975 through 1979 [2]:

1. **The formulations section**. This location is where the active ingredient (technical grade product) generated by the manufacturing assets was diluted with inert ingredients. Various marketable product concentrations (grades) were then expected to be sold to international customers, to support the local agricultural industry. Any surplus product could be sold on the international market as per the terms of the industrial license. The formulations section was fed with technical grade product manufactured in the methyl isocyanate (MIC) derivatives section.
2. **The MIC derivatives section**. This process unit contained the assets needed to "derive" marketable pesticide products. Process operations in this area of the factory were conducted to manufacture the technical grade product using MIC and α-naphthol (chapter: Process Selection). The process was supplied with MIC from a charge pot located on the southern perimeter of the unit boundary. The charge pot (Fig. 12.1) was periodically replenished with 1-ton batches of MIC transferred in from the chemical storage area [3]. The chemical storage area and MIC synthesis section where commercial-grade MIC was produced were built at the same time.
3. **The MIC synthesis section**. The MIC synthesis section is where indigenous raw products were converted into commercial-grade (99.9% pure) MIC [4]. The reaction process also contained the assets that generated phosgene gas from carbon monoxide and chlorine. The phosgene was then rapidly consumed in the reaction with monomethylamine (MMA) to produce crude MIC liquid. Impurities in the crude MIC product were then separated in a distillation column to produce high-purity, commercial-grade MIC. The distilled MIC product leaving the unit was routed through a stainless steel rundown line, which led into the chemical storage area [5].

All of the various indigenous and intermediate chemical products that supported MIC production were contained in appropriate tanks located in the chemical storage

*For timeline events corresponding to this chapter see pages 429–431 in Appendix.

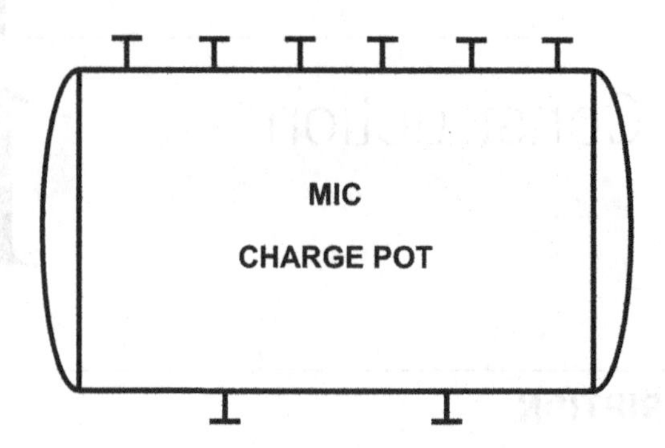

FIGURE 12.1

One-ton methyl isocyanate charge pot in the derivatives section [6].

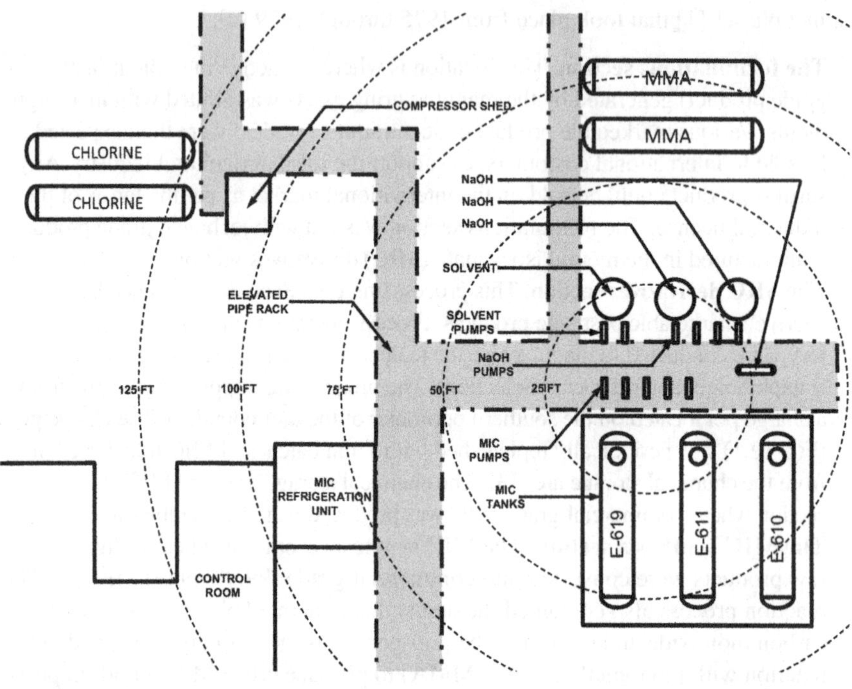

FIGURE 12.2

Bhopal factory chemical storage area [7]. *MIC*, methyl isocyanate; *MMA*, monomethylamine.

area (Fig. 12.2). Bulk product shipments arriving by rail were delivered to this area, as was the MIC product rundown from the MIC refining still (MRS). The various raw materials and intermediate products stored in this area consisted of:

1. MMA that was reacted with phosgene to produce MIC (chapter: Process Selection).
2. Chlorine that was combined with carbon monoxide to generate phosgene (chapter: Product Manufacturing and Distribution).

3. Caustic (NaOH) that was used to neutralize acidic waste products (chapter: Process Selection) and destroy any potentially fugitive MIC process vapor emissions that might create an environmental hazard while manufacturing or storing MIC.
4. An organic chloroform ($CHCl_3$) solvent that carried the product layer through the manufacturing process during MIC production. Neat organic solvent could also be used as a heat sink to quench hot MIC in the event of a process emergency. The manufacturing process was designed to recover as much solvent as the process would allow. This operating method was both the most economical and conservative manufacturing option.
5. Commercial-grade MIC that was to be transferred into the MIC derivatives section for production purposes.
6. Reject (contaminated) MIC that was then reconditioned, recycled, or destroyed upon determining if the contaminated product could be economically salvaged.

The factory construction project was completed in 1979 [8]. This development effectively eliminated India's dependence on technical grade product imports to sustain agricultural production. In reality, technical grade product imports ceased when Phase 2 was completed in 1977 [9]. At that time the derivatives section became fully functional. From that point forward, the technical grade product was manufactured by reacting imported MIC with α-naphthol in the MIC derivatives section [10].

Likewise, MIC imports were finally discontinued upon the completion of the third and final phase of factory construction, which brought MIC manufacturing capabilities into existence. This holistic "backward integration" approach [11] started with constructing the formulation section and ended with the MIC synthesis section. This plan was the most practical option to eliminate India's import dependency. In making this selection the construction project progressively weaned the factory off imports; valuable manufacturing experience was now generated by operating the MIC derivatives section while construction of the MIC synthesis section was in progress.

The α-naphthol manufacturing process was developed and implemented internally [12]. The scaled-up production unit went online in 1978 but quickly ran into performance problems that limited its sustainable use [13]. Regardless, MIC arriving in 55-gallon stainless steel drums was imported from the domestic factory while the MIC synthesis section was being constructed [10] from 1977 to 1979. In 1979, MIC imports stopped and the Bhopal factory was fully integrated. Afterward, the factory operated exclusively on internally produced MIC manufactured from indigenous raw materials. For the first time in almost 20 years, the local subsidiary was operating in complete compliance with local foreign trade regulations.

METHYL ISOCYANATE REFINING STILL

The MIC rundown system was equipped to capture the distilled product exiting the MRS. The MRS (Fig. 12.3) was a 45-tray fractionation column [14] measuring 5 ft in diameter and 65-ft tall [15]. More accurately, it was a splitter tower that

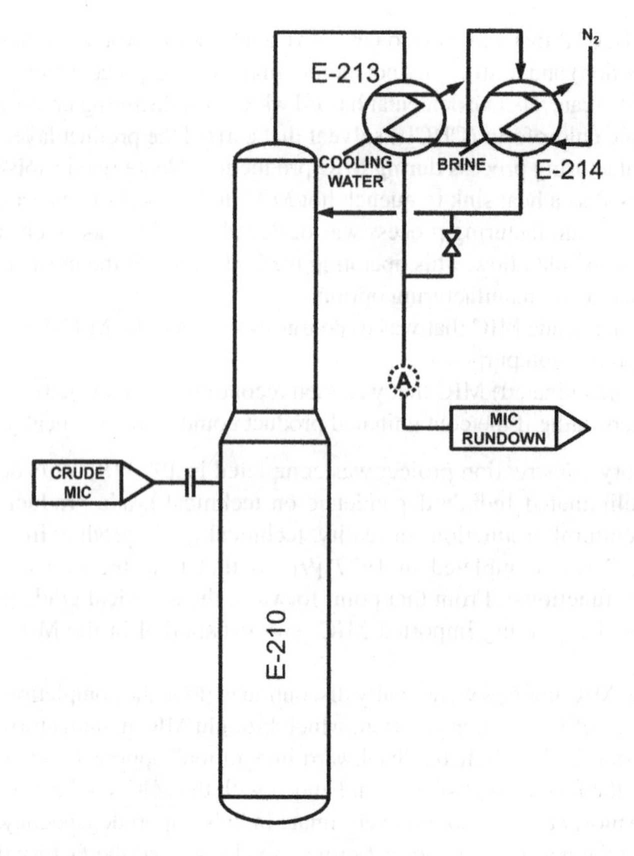

FIGURE 12.3

Methyl isocyanate refining still [6].

separated the top (reflux) product from impurities that exited at the bottom of the tower. The impure bottom stream was then recycled back into the manufacturing process.

The tower's performance was highly dependent on stable temperature and pressure control [14]. An upset in either of these two control parameters could cause flooding inside the tower. In such a case, impurities that would normally be removed from the bottom of the tower would be trapped in the overhead MIC product stream. This carry-over condition would result in the delivery of contaminated MIC in the rundown tanks. Depending on the impurity concentration, chloroform or other contaminants could spoil the MIC batch [14]. MIC samples could only be taken directly from the storage tanks [16]. Therefore, the quality of MIC samples in the rundown tank was frequently checked during an MIC production run [17]. Recalling the P-F interval discussion in chapter "Equipment Reliability Principles," this control method was implemented to detect a tower upset with sufficient time to remedy the problem. Otherwise, a complete batch of expensive MIC product might have to be destroyed at considerable economic loss.

Generally speaking, distillation columns perform best when operated at low pressure [18]. Fluctuating pressure inside the tower could make MRS temperature control extraordinarily difficult. Raising the overhead pressure correspondingly raises the boiling point of different chemical fractions contained inside the tower. The excess heat input resulting from a pressure backup anywhere in the rundown system could cause an overload or flooding condition within the tower [14]. This upset would allow impurities from the lower column trays to travel upward and into the reflux area. From there, these impurities would transfer into the rundown lines directed at the MIC rundown tanks. Any "off-spec" MIC product not meeting quality specifications would then have to be managed as reject, at additional cost. The most economical approach to MRS operation was to continuously control the process within specified limits in order to generate "on-spec" MIC product at all times.

The MRS overhead condensing system consisted of a cooling water heat exchanger (E-213) and a brine cooler (E-214), which converted distilled MIC product vapor back into a liquid stream [19]. A portion of this liquefied overhead product was then circulated back to the top of the column as reflux. Another portion was routed into the MIC rundown tanks to be consumed in the manufacturing reaction that produced the final product. In this particular application, a reflux ratio of 20:1 (reflux-to-rundown flow) was targeted as a critical process control parameter [14]. This high ratio was needed to maintain effective separation at the top of the MRS. High product flow or a low reflux ratio could entrain impurities in the MIC rundown stream. The MRS rundown product was routed to the MIC rundown tanks through a long, common stainless steel line running between the MRS and the chemical storage area.

METHYL ISOCYANATE STORAGE TANKS

The MIC storage system consisted of three 40-ft long by 8-ft wide diameter stainless steel storage tanks (Fig. 12.4) [20]. Two of the tanks (E-610 and E-611) were designated to receive rundown product from the MRS. These assets captured condensed MRS overhead product when the MIC synthesis section was operating. In this configuration the rundown tanks were really an extension of the MRS overhead process. Accordingly, the intermediate product storage tanks were designed to operate as process vessels, more than conventional finished product atmospheric storage tanks.

The other storage tank (E-619) was provided for the exclusive purpose of managing any contaminated MIC product that did not meet finished product quality specifications [23]. In most cases, a process upset inside the MRS would cause excessive chloroform levels in the overhead MIC product. A chloroform concentration in excess of 0.5% inside a rundown tank would cause the MIC batch to fail [24]. Under these circumstances the contaminated MIC product would likely be recycled back into the MIC synthesis section for reprocessing [5]. As long as the operating stability at the MRS had been restored, this procedure would effectively remove the excess chloroform that had originally carried over.

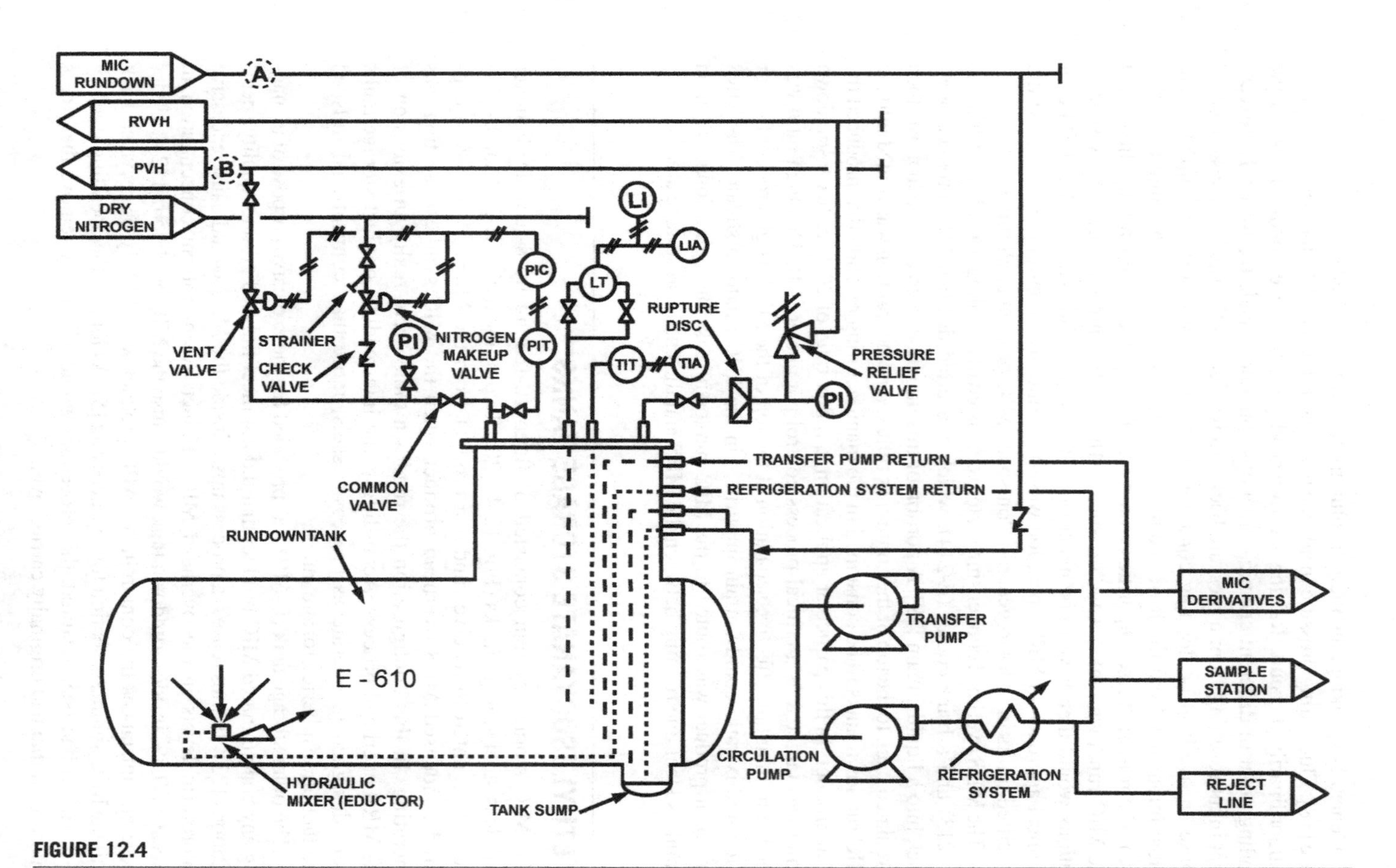

FIGURE 12.4

Methyl isocyanate (MIC) rundown tank [21,22]. *RVVH*, relief valve vent header; *PVH*, process vent header.

The MIC storage system included a versatile piping arrangement whereby tank contents could be interchanged at the reject distribution manifold (Fig. 11.1). MIC in a rundown tank not meeting finished product quality specifications could be transferred into the reject tank (E-619) for reconditioning. For this reason Tank E-619 was also called the "reactor conditioner" [25]. For example, depending on the type of problem, it might be possible to condition the product inside E-619 to meeting quality specifications by either:

1. Diluting the contaminated material with product that met quality specifications, or
2. Developing a specialized MIC reconditioning plan based on the actual composition of the tank contents.

If the reconditioning attempt was successful, then the product could be transferred back into the process for further use. In other cases, its disposal might be necessary. The MIC reject tank was also specified for use as an emergency receiver in the event that the product in either of the rundown tanks became unstable [26].

Several other options for managing a contaminated MIC batch existed at the reject distribution manifold. The manifold valve and piping configuration made it possible to transfer MIC between the rundown tanks. The preferred destination for any MIC routed into Tank E-619 was back into a rundown tank to be fed as normal product into the MIC derivatives section. Although a reconditioning activity would raise production (and breakeven) costs, it was a more economical and conservative option compared with wasting an entire batch of contaminated MIC. However, exercising uninterrupted process control was, again, by far the most economical approach to managing MIC production.

All three storage tanks could contain up to 15,000 gallons of MIC each [15]. However, a maximum level of 60% was initially specified [27]. The extra space inside the tank was to be reserved for safety purposes [28]. A low tank level limit of 20% [29] was also designated, presumably to avoid cavitating the MIC transfer and circulation pumps. Operating instructions further defined what was to be done in the event that the MIC storage tank level dropped below the minimum limit. Any storage tank entering a "low level" condition would be completely emptied so that its respective MIC pumps could be safely turned off. This approach to managing MIC storage tank assets was developed to prevent unstable hydraulics that could degrade mechanical seal reliability. A seal failure could produce a wide variety of consequences, ranging from minor leaks to a catastrophic process release.

METHYL ISOCYANATE CIRCULATION AND TRANSFER PUMPS

A total of five centrifugal MIC pumps were located in the "MIC pump area" (Fig. 12.5). These pumps were installed immediately in front of the MIC storage tanks (Fig. 12.2). Each MIC pump location was intended to operate continuously, returning MIC flow back into each tank at all times. In other words, at no time was it necessary to shut off the MIC

FIGURE 12.5

Methyl isocyanate (MIC) pump area, showing the five MIC pump pedestals (see bottom right corner).

Photo by Julian Nitzsche.

pumps. The only exception to this rule was if the product inventory level dropped below 20% and the MIC was removed; in which case the MIC pumps were to be shut off.

Two "transfer pumps" were located inside the MIC pump area. The purpose for these fluid movers was to deliver MIC into the derivatives section where it was used to manufacture MIC-based pesticide products. A single transfer pump was assigned to each individual rundown tank (E-610 and E-611). There was no transfer pump for the MIC reject tank (E-619) since its intended use was not for producing MIC derivatives. If through the conditioning process the contents in the reject tank could be recovered, then it could be directed back into an MIC rundown tank at the reject distribution manifold. At that point the MIC could be fed into the MIC derivatives section as any normal batch meeting finished product quality specifications would be.

The transfer pumps continuously kept discharge pressure on the line directed into the MIC derivatives section. During a pesticide production campaign, a worker simply had to open a valve on the discharge side of the transfer pump to fill the charge pot (Fig. 12.1). The MIC delivered to the charge pot could then be fed to the reaction vessels to produce the technical grade product.

The MIC pump area also contained three "circulation pumps." Unlike the transfer pumps that were only provided for the rundown tanks, each storage tank had its own individual circulation pump. Each circulation pump maintained continuous MIC flow through a fluorocarbon-based, 30-ton refrigeration system refrigeration system that served all three MIC storage tanks [30].

The presence of five (not ten) MIC pumps in the MIC pump area (Fig. 12.5) indicates that spare pumps were not installed in this particular application. The installation of two identical rundown tanks dedicated to produce MIC derivatives made furnishing identical spare pumps unnecessary. Perhaps the selected tank installation was the safest choice in the interest of accommodating any unplanned maintenance that might be needed between scheduled maintenance turnaround intervals. The selected rundown tank configuration made it possible to repair a damaged MIC pump "off-line" after completely draining the process from the tank and its associated piping. This course of action offered the greatest protection during maintenance without shutting down production. Conceivably, maintenance would never have to be simultaneously conducted on a live process. The contact hazards associated with MIC were well known [31]. Therefore, transferring MIC out of a storage tank was a prudent way to minimize the exposure hazard during unplanned pump maintenance or repair.

Each storage tank also contained a stationary hydraulic mixer, called an "eductor." The eductor's purpose was to keep the tank contents moving at all times. This function prevented the MIC from stagnating inside the tanks while awaiting transfer into the MIC derivatives section. The MIC circulation pump had to be operating for the eductor to work properly. If for some reason MIC circulation was interrupted, then the tank mixing would stop. The eductor was fastened to the discharge of the internal cold MIC pipe returning from the refrigeration system, at the bottom of the tank.

METHYL ISOCYANATE REFRIGERATION

The refrigeration system was designed to keep the temperature of the storage tank contents low. Under normal circumstances the temperature of MIC in the storage tanks was not to exceed 5°C (41°F). An alarm would sound inside the control room if the temperature of the tank contents increased above the maximum temperature set point [32]. Continuous refrigeration below 5°C was needed to contain the phosgene inhibitor (8°C boiling point) that was added to stabilize the MIC product inside the storage tank [33]. By maintaining a concentration of at least 200 ppm phosgene inside the storage tanks, there was essentially no chance of spontaneous MIC polymerization [34]. Refrigeration was also used to reduce the vapor pressure of the volatile storage material so that it could be contained inside the storage tanks. This approach was consistent with industrial practices applied to storing volatile process components at atmospheric pressure (chapter: Process Selection).

METHYL ISOCYANATE VAPOR MANAGEMENT

Due to the high volatility of MIC liquid, each MIC storage tank was equipped with its own variable-rate inert gas blanketing system (Fig. 12.6) [35]. This system's function was to absorb any flammable vapor inside the tank in a blanket of inert (nonflammable) gas. This inert gas mixture was then directed to a safe discharge location.

The inert gas blanket inside the MIC storage tanks was established by injecting nitrogen directly into each tank's empty head space above the liquid level. For safety purposes, the flow of nitrogen could not be interrupted. Failure to keep the nitrogen

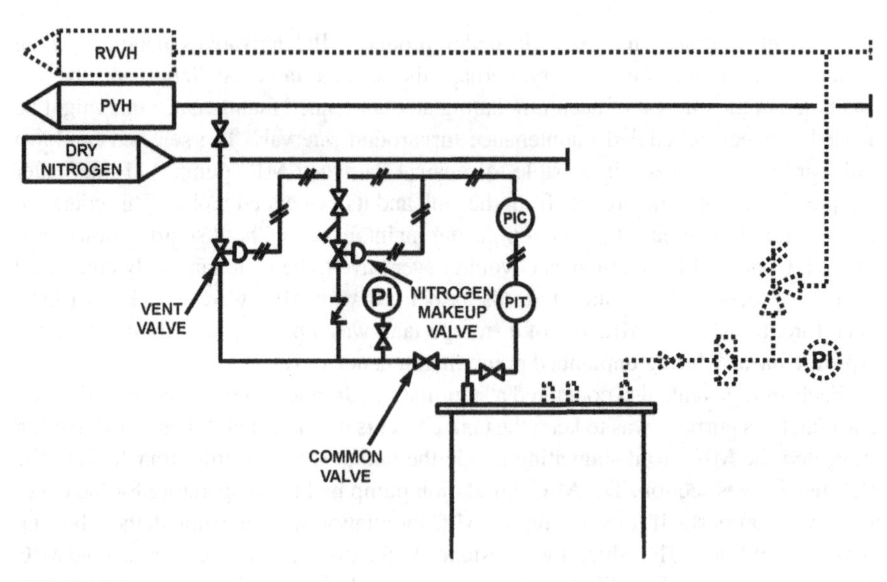

FIGURE 12.6

Methyl isocyanate storage tank variable-rate gas inert blanketing system (*solid lines*). *RVVH*, relief valve vent header; *PVH*, process vent header.

flowing would allow air (oxygen) to contact a process consisting of a volatile hydrocarbon (MIC). This system malfunction would create an ignition hazard inside the MIC storage system. For this reason, a backup supply of 450 cylinders of compressed nitrogen was provided [20]. This independent nitrogen reserve could be fed if the factory's normal source of nitrogen was lost. Under normal circumstances the factory's nitrogen was supplied by a third-party company operating an air separation business next door [36]. Keeping an independent supply of nitrogen bottles prevented problems beyond the factory's control from interfering with safe product manufacturing at all times.

Each of the MIC storage tanks was equipped with a pair of pressure control valves. The pressure control valves automatically regulated nitrogen flow into and out of the tanks, to maintain the tank pressure at about 2 psig [37]. The control valve configuration consisted of one nitrogen inlet diaphragm motor valve (DMV) and one discharge (tank vent) DMV [38]. The movement of these control valves was coordinated by a pressure indicator-controller that compensated for involuntary pressure changes expected under normal operating conditions. Normal operating conditions included fluctuations in tank level. The level inside the MIC storage tanks was expected to rise or drop due to controlled MIC movements into and out of the tank. A properly functioning vapor management system reduced the volatile MIC product's ignition hazard to an acceptable level.

The inert nitrogen blanket exiting the storage tanks continuously swept process vapor into the vent gas scrubber (VGS) or alternatively into a flare tower through the MIC vapor transfer headers (Fig. 12.7). The process vent header (PVH) pipes collected process vapors under normal operating conditions. The MIC storage tanks were

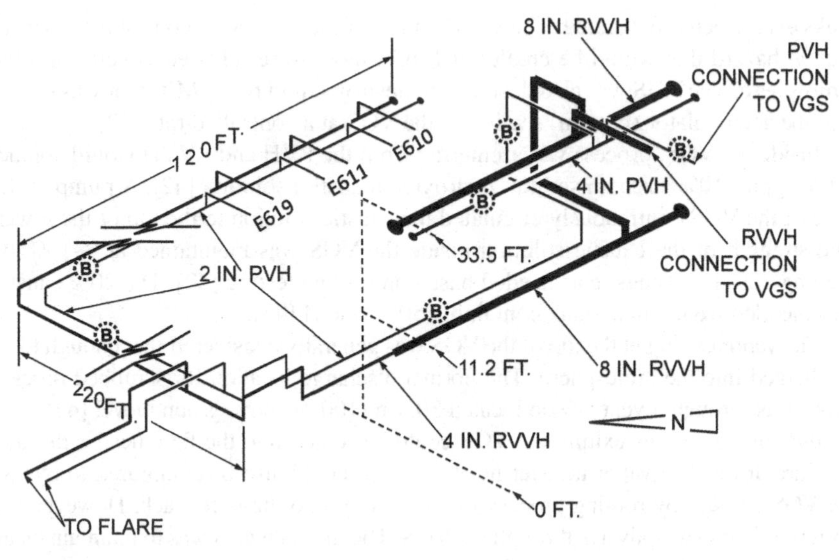

FIGURE 12.7

Bhopal factory vapor transfer header system (iron) [40]. *PVH*, process vent header; *RVVH*, relief valve vent header; *VGS*, vent gas scrubber.

protected by relief valves that would automatically dump the process into the relief valve vent header (RVVH) if for any reason the pressure inside the MIC storage tank exceeded 40psig [39]. The PVH and RVVH were designed to operate independently to prevent contaminating the MIC storage tanks with process relief vapors or liquid.

Unlike the stainless steel assets in MIC service from the MRS up to the storage tanks, the PVH and RVVH were made of iron [41]. Valves in these headers, including manual block valves and the DMVs that automatically controlled the pressure inside of the storage tanks, were also made of iron [39]. Selecting iron instead of stainless steel to operate in this service not only increased the potential for accelerated corrosion, but also conflicted with the applicable design specification that restricted the use of iron in MIC service. Neither did it correspond with the domestic factory's approach used to construct the MIC vapor transfer headers with stainless steel. In other words, the material chosen to construct the Bhopal factory's vapor transfer header system did not conform to applicable process design specifications [19].

VENT GAS SCRUBBER

The PVH and RVVH carried process vapors into a 60-ft tall, 5.5-ft wide in diameter VGS tower that was also made of iron (Fig. 12.8) [38]. Its purpose was to neutralize process vapor continuously exiting the MIC storage tanks as well as other assets that would generate vapor traffic only during MIC production [38]. The VGS was designed to operate continuously to neutralize any background MIC emitted from the tanks even when refrigerated. The VGS' primary function was to control the environmental hazard that would be created if MIC vapors were released directly into the atmosphere. The VGS was also designed to destroy liquid reject MIC that was routed into the accumulator (bottom) section of the VGS at a controlled rate [42].

Inside the VGS, process vapor entering from the PVH and RVVH would contact a 1200 gpm, 10% strength sodium hydroxide (caustic) solution [42]. A pump at the base of the VGS continuously circulated the caustic solution to the top of the tower. The strength of the caustic solution inside the VGS was maintained at 9–10% by feeding 20% fresh caustic as needed based on routine testing [42]. The 20% caustic was blended from a main tank containing 50% NaOH [43].

The vapor exiting at the top of the VGS was generally considered safe enough to be discharged into the atmosphere. The normal discharge location for scrubbed process vapor was through a vent nozzle located 30.5 m (100 ft) above ground level [42]. Any residual process gases exiting the VGS could be routed into the flare header through the flare drum. The vapor transfer header system could also be configured to bypass the VGS entirely by routing process vapor directly into the flare stack. However, the preferred lineup was always through the VGS. The ultimate goal was to maintain clean vapor flow into the atmosphere at all times, regardless of whether or not a process upset was in progress.

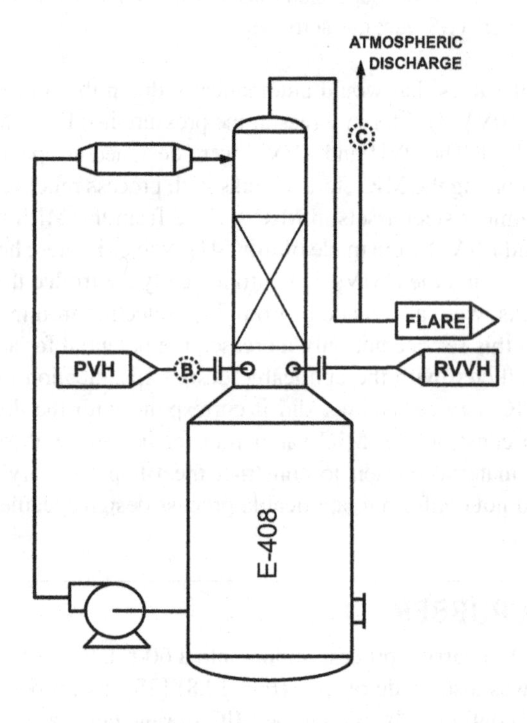

FIGURE 12.8

Bhopal factory vent gas scrubber tower [6]. *PVH*, process vent header; *RVVH*, relief valve vent header.

INSTRUMENTATION, GAUGES, AND ALARMS

The MIC process was equipped with instruments, transmitters, and gauges to continuously monitor and control the process. Continuous surveillance over critical process monitoring parameters was needed to maintain an acceptable level of process control. The more important control parameters that might signal a developing process upset were configured with alarm points. These provisions were made to call attention to an operating condition outside of predefined reliability or safety limits. In response to a process alarm, factory workers were expected to take immediate and continuous steps to bring the process back under control.

One of the most critical alarm points was the MIC storage tank high-temperature alarm. This alarm was configured to notify workers of an unstable condition inside the storage tanks any time the temperature reached above 11°C [30]. Upon receiving the alarm the workers were expected to troubleshoot the situation and be prepared to route the MIC into another part of the process through the reject manifold. This extreme measure might be needed to recover from a process upset.

Whereas the domestic factory was equipped with an MIC tank high-pressure alarm set at a 10 psig maximum [44], there was no corresponding pressure alarm on the Bhopal factory's MIC storage tanks (Fig. 12.9) [45]. Pressure inside the Bhopal factory's MIC storage tanks was only monitored at control room and field-mounted pressure gauges. The nitrogen supply header was, however, equipped with a low-pressure alarm [15].

Perhaps the high storage tank temperature alarm was considered to be an acceptable substitute proxy alarm for high tank pressure. If an unstable mixture inside the tank was

FIGURE 12.9

Bhopal factory methyl isocyanate storage tank pressure gauge.

Photo by Dennis Hendersot.

FIGURE 12.10

Bhopal factory methyl isocyanate storage tank temperature gauge.

Photo by Dennis Hendershot.

the concern, then tank pressure deviations would certainly occur with simultaneous temperature deviations. There would thus be a measure of redundancy here, although temperature would offer a more immediate form of detection. Regardless, the storage tank pressure gauges were scaled for operation between 0 and +45 psig. This pressure range was appropriate for the tanks that were rated to operate at a maximum allowable working pressure of 40 psig [39].

The MIC storage tanks were equipped with both local (field) and remote (control room) temperature and pressure gauges. Since the temperature of the refrigerated MIC inside the storage tanks was never expected to exceed 15°C, the temperature gauges were scaled for operation between −25 and +25°C (Fig. 12.10). However, the ambient temperature in the climate where the factory was constructed could very realistically exceed the maximum temperature range on the temperature gauges [2]. Therefore, any interruption in MIC refrigeration to repair a circulation pump mechanical seal could easily cause the associated storage tank's temperature gauges to fail above "high-range" [46].

TECHNICAL ASSESSMENT

In this chapter we are presented with evidence that process safety was incorporated into the design of the Bhopal factory. An example of this forethought can be observed in the way that the MIC rundown tanks were configured. Instead of installing identical spare MIC pumps for each asset location, an entirely independent spare rundown tank system was provided. Under these circumstances, the process exposure hazards that would apply during corrective maintenance could be effectively mitigated. The ability to transfer the tank contents away from where invasive maintenance might be needed displays a commendable degree of sensitivity toward protecting maintenance workers. Although the process was not intended to require corrective maintenance between turnaround cycles (chapter: Licensing), the availability of a spare tank demonstrates a definite concern over preventing potential process release incidents. The MIC storage tanks were designed to accommodate safe maintenance execution without interrupting production. The absence of spare MIC pumps also *forced* the affected personnel to perform maintenance in the safest manner possible. If an MIC pump needed to be repaired on an unavailable rundown tank, then there was no option other than to do it off-line after transferring production over to the available rundown tank.

Another commendable attribute of process design was the consideration given to protecting the storage tanks from liquid contamination. Notice that the MIC vapor transfer headers consisted of separated lines, with the PVH operating independently from the RVVH (Fig. 12.7). This process configuration protected the MIC storage tanks from any possible transient gas or liquid contamination resulting from a pressure relief scenario. The PVH section running between the VGS and the storage tanks was also designed and operated to keep out even the smallest amount of water [47]. Specific engineering controls were implemented as standard process design and configuration. This resolves the contradiction identified in chapter "Process Selection" where evidence indicates that the design reflected little concern over a contamination incident. Since a contamination incident was considered to be physically impossible [48], there was no need to extend the refrigeration unit's function to cover the hazard. There were perhaps other ways for a contamination incident to occur. But through the vapor headers, engineering and design controls made liquid contamination a remote possibility at best.

With these very notable process safety design examples in mind, it is somewhat disappointing to find areas where design decisions do not reflect the same degree of hazard awareness and protection. For example, one area that by today's standards would generate a discussion was the placement of the process control room relatively close to the MIC storage tanks (Fig. 12.2). A study of the factory's physical layout shows that the storage tanks were located within about 100 ft of the MIC control room. At this close proximity to the process, it was likely that an MIC leak in the pump area would create an immediate exposure hazard for anyone present in the normally occupied building.

Perhaps even more concerning was the deliberate use of iron pipes and valves to construct the Bhopal factory's MIC vapor transfer header system. Throughout this discussion we refer to "iron" as any ferrous material that readily forms rust upon contact with air. Selecting iron instead of stainless steel in this particular process was in direct conflict with documented design specifications that required using stainless steel for piping and valves in MIC service [49]. Stainless steel does not readily form metal oxides upon coming in contact with air. It was therefore safe for use in the MIC synthesis, storage, and vapor management systems.

At this stage of an incident investigation, it would be impossible to conclude what impact, if any, this defect had on the stability of the process. It is only appropriate to state the fact that iron was selected where stainless steel was specified. This fact raises a question over what made a lower cost, clearly inferior construction material acceptable for use in the MIC vapor transfer headers only. All of the other assets in the process circuit leading up to and including the MIC storage tanks were constructed with stainless steel in accordance with the design specification.

It is a scientific fact that iron offers *less* corrosion resistance than stainless steel. This fact explains why stainless steel was specified in MIC service. As discussed previously in chapter "Process Selection," rust acts as a catalyst that converts MIC into a solid polymer called trimer [31]. Trimer deposits were notorious for restricting vapor flow in process lines [50]. Construction materials that could form rust upon exposure to air were therefore prohibited in this specific application. Filling the vapor lines with trimer deposits could restrict vapor flow between the MIC storage tanks and the VGS (or flare). Therefore, stainless steel was identified as a compatible substance in MIC service whereas iron was not. The insistence on using stainless steel was intended to control the hazard of trimer formation in locations where MIC liquid or vapor could be expected.

In examining the Bhopal factory's vapor management system further, it can be seen why restricting vapor traffic out of the MIC rundown tanks would create a serious operational concern. Note that the condensed MIC flowed into the tanks through the rundown line (line "A" in Fig. 12.3). The rundown tanks' pressure was closely regulated at 2 psig with nitrogen flow. Since the rundown tanks vented into the vapor management system (line "B" in Fig. 12.4), sufficient blockage in the vapor lines (line "B" in Fig. 12.7) between the rundown tanks and the VGS (or flare tower) would create backpressure on the MRS while filling the tanks. This backpressure condition would potentially interfere with process control in the MRS. The result would be high pressure in the MRS, leading to higher heat input to boil the overhead product, which would contaminate the rundown product with excess chloroform. If the MIC in any rundown tank contained more than 0.5% chloroform, then the entire batch would have to be reprocessed or destroyed at considerable expense. Therefore, the vapor lines and valves had to remain clean all times [32]. In this application, rust could not be allowed to develop inside the pipes and valves where MIC was always expected. If rust was present, then the MIC in the vapor transfer headers would instantly deposit inside the pipe walls and valve bodies as solid trimer [51].

Recovering performance in a system congested with trimer deposits would require corrective maintenance to restore free passage of tank vapors through the MIC vapor transfer headers. Since the MIC vapor transfer headers were not designed for routine online maintenance (chapter: Licensing), performing this maintenance would represent a very complex, expensive, time-consuming, and dangerous activity. The vapor transfer pipes were common between all three MIC storage tanks. By design, a spare PVH or RVVH had not been provided for use if one or the other headers needed to be taken down for corrective maintenance [52]. Unless the formation of trimer deposits in the pipes could be reversed by some online cleaning process, then it might be necessary to cut into the vapor management headers for their physical extraction. Considering the potential complexity involved with this nonroutine operation, we would certainly expect to observe some consideration for inhibiting rust on the surface of the vapor lines where MIC vapor would travel in the iron headers.

Looking closer at the process configuration of the MIC storage tank head manifold, we can see that the inert gas blanketing system supplied an important secondary function. Its purpose was to inhibit rust inside the vapor transfer headers. Even though constructing the MIC vapor transfer headers with iron deviated from the process design specification, an exception could be justified on the basis of keeping the storage tank contents under a nitrogen blanket at all times. The inert gas blanket's *designated* function was to eliminate oxygen from the flammable process. However, industry often uses nitrogen to inhibit iron corrosion. In the continuous presence of nitrogen, the internal surface of the iron pipes would not corrode. Thus, rust would not form and the hazard related to forming trimer deposits in the MIC vapor transfer headers did not apply. Although the selected design alternative conflicted with the process design specification, the decision to construct the MIC vapor transfer headers with inferior metallurgy was consistent with scientific principles that most observant engineers would likely consider today under the same circumstances. At a minimum, an industry professional would likely inquire why it would not be possible to issue an exception before settling for the more expensive option directed by the applicable design specification. Upon learning the complexities associated with managing trimer deposits, a choice would have to be made. Either follow-through with the inherently safer design specification, or introduce multiple complex human-dependent processes to reduce the probability for interrupting nitrogen flow. Considering the time, energy, and uncertainties involved with developing that level of confidence, the extra expense involved with constructing a process to design specifications would be money well-spent. The purchase protects an employer's long-term investment. In this case, it would have resulted in selecting the inherently safer design choice.

Upon further analysis we find that the decision to construct the MIC vapor transfer header system with iron was consistent with earlier decisions regarding the storage of MIC in an atmospheric tank system. By the same token, attenuation was selected over the inherently safer design option (chapter: Process Selection). It could be said that a layer of protection was stripped away by constructing the process with iron instead of stainless steel. Stainless steel did not require the continuous presence of a corrosion inhibitor to avoid rust formation as iron did. Similar to the other attenuation methods used to control the hazards of storing a volatile liquid in an atmospheric storage tank, any interruption in nitrogen flow would cause the process (in this case, iron) to revert back to its natural form (in this case, rust). In this particular instance, loss of the nitrogen flow at the storage tanks would corrode the MIC vapor transfer header system all the way to the VGS/flare. This nitrogen "low-flow" condition would create the potential for many undesirable side-effects, including the following:

1. **Production constraints** resulting from accumulating trimer in the vapor lines. Congested vapor traffic in clogged vapor transfer headers would result in backpressure on the MRS overhead process. In turn, the overhead boiling point would rise causing tower flooding. This would lead to excessive chloroform concentrations in the overhead product.
2. **Nonroutine maintenance** services to remove trimer deposits from the inside of the vapor transfer headers.

3. **Property damage** and associated replacement costs from accelerated internal pipe corrosion.
4. **Personal safety issues** resulting from vapor leaks from corroded vapor transfer headers.
5. **Process safety events** due to the loss of the inert gas blanket used to suppress MIC ignition.
6. **Personal exposure incidents** due to an elevated potential for isolation defects in process locations not designed to accommodate invasive maintenance (chapter: Licensing).

Considering the totality of the problems that could result upon losing nitrogen flow through the MIC vapor transfer headers, it was somewhat reassuring and understandable that a backup supply of nitrogen bottles was made available. These assets were kept on-site to sustain nitrogen flow if for some reason the third-party producer could not deliver. On the other hand, all but perhaps one of the hazards resulting from the loss of nitrogen flow could have been eliminated by simply choosing to construct the process as directed by the inherently safer design specification. Constructing the vapor transfer headers with stainless steel components would have tremendously simplified process design, reliability, operation, and control. Simple processes are safer processes (chapter: Industrial Learning Processes).

Digging even deeper, the decision to install iron pipes protected by a nitrogen corrosion inhibitor was consistent with a pattern observed in previous chapters. Once again, improvisation was directly involved in the decision to install iron pipes instead of stainless steel as directed by the design specification. In doing so, the nitrogen supply's function was extended from controlling the MIC's ignition hazard to also acting as a corrosion inhibitor. Improvising with factory construction in this manner followed the precedent set forth when the chloroformate manufacturing process was converted to the MIC production route (chapter: Process Selection). Indeed, the specified stainless steel design was intended to prevent adverse reactions involving the use of MIC. This design intent was inadvertently undermined by improvisation that led to the downgrading of construction materials in the MIC vapor transfer headers.

Before improvising with process design, remember that your decision will not be free of consequences. For example, converting from the chloroformate to the MIC manufacturing process demanded more expensive construction materials. On the other hand, with nitrogen present, there appeared to be no cost associated with selecting iron instead of the specified material. In the context of what can be expected when a process change is made, the thought of saving money by downgrading the construction materials was unrealistic. Always remember that no changes in the manufacturing industry are free of cost or free of consequences. When a change is made, there will always be a corresponding penalty that must usually be offset financially. If this correction is not made, then something that is desperately needed to maintain process control can be sacrificed. The problem is that when you finally appreciate the importance of what you have given up, it is usually too late to get it back.

The same concept applies to the decision to forego the installation of high-pressure alarms at the MIC storage tanks. Although this decision also distinguished the Bhopal factory from the domestic factory, an argument could be made that installing a high-pressure alarm would be unnecessary in this particular application. Since a product's vapor pressure is a function of its temperature, the high-temperature alarm could substitute as a proxy alarm for high tank pressure. If there was a concern about high pressure caused by an upset inside an MIC storage tank, then the high temperature alarm alone would suffice. Since the storage tank pressure was automatically regulated, temperature was definitely a more direct and immediate way to detect a problem inside the tanks.

From a process safety perspective, however, the loss of an independent alarm point could make the difference in a situation involving double jeopardy. The probability for an incident caused by double jeopardy is extraordinarily low. It is therefore appropriate to disregard these scenarios when evaluating risk. However, the decision to implement a proxy alarm to optimize the process could convert a double jeopardy scenario into a distinct possibility. Money was saved in the decision not to install a high-pressure alarm although a corresponding high-pressure alarm was used in the domestic factory. Again, when you find out how much you really need what you have lost, it might be too late to get it back.

The process safety management standard implemented in response to the Bhopal disaster protects industry from making decisions that could deprive a process of adequate protection, or adequate warning (redundant or not.) Of course, it would neither be safe nor practical to outfit a process with every form of protection available. To avoid unnecessary complexity, combining layer of protection analysis elements in a standard hazard and operability study defines the protection one needs to effectively mitigate specific hazards. Chapter "Process Hazard Awareness and Analysis" uses standard industrial hazard analysis methods to predict how the MIC rundown tanks would be expected to perform, according to how the process was configured.

LESSONS FOR US

Starting with the process design and throughout its progression through the various construction phases, special attention must always be given to defect elimination. While that sounds like an obvious requirement, it often competes with budget constraints. Budget constraints favor studies whereby industry professionals probe into the apparent adequacy of more economical design alternatives, causing an often imperceptible drift toward presumably "just as safe, but less expensive" options. But the implementation of such options may actually force the design into the often overlooked requirement that instruments be far more reliable, operators be expected to act with precision at all times, and that decisions under pressure can be made with zero hesitation and zero defects. In short, special attention will be needed so we do not end up stripping back one or more of the many layers of protection. Using specified construction materials that are fully compatible with the process service including corrosion-resistant piping and valves can be equated to adding layers of protection. Chances are one could come up with a wide assortment of examples in either the direction of weakening or strengthening process safety protection.

Reliability principles and practices must dominate process operation from factory design onward. Safe operation is a mandatory requirement. As such it must be achieved instead of simply being spoken about. Because we recognize that design defects can, and often do, originate with a decision to modify some aspect of process design or a construction project, we must scrutinize these decisions very carefully. The point is made that modification-induced design defects are potentially dangerous. Since they originate with an intentional decision, they can become very persistent. The fact that they can usually be managed "for a period of time" makes them no less dangerous. It could even be argued that "can be managed" makes compromised thinking and the "normalization of deviance" possible (chapter: Commissioning Phase). Compromised thinking often increases the potential for process upsets and, ultimately, events that make it possible to lose primary containment.

It is reasonable to expect that at some point in a factory's life cycle, a project to optimize the process will be initiated. Process optimization is a continuous, never-ending effort to improve production efficiency by simplifying unit design and operation. It is during this time that harmful and persistent design defects are more likely to be introduced. A successful process optimization program would never abandon an inherently safer design approach in exchange for a more human-dependent or maintenance-intensive construction alternative. Acting contrary to this advice can establish a causal factor for a catastrophic system malfunction at a later time. Other lessons highlighted by this analysis are the following:

- Process design specifications authorize construction expenses and performance standards that might otherwise be considered cost-prohibitive after startup.
- Constructing a process according to a prescribed design specification promotes consistent hazard awareness throughout the life of a process.
- In an integrated process, an upset in one part of the system can create instabilities in another, seemingly unrelated, part of the system.
- The safest approach to constructing a new process is to duplicate the design of an existing reliable process. This includes materials of construction, transmitters, gauges, controllers, and alarms.
- When you finally appreciate the importance of what you have lost by disregarding process knowledge, it is usually too late to get it back.
- Deviating from process specifications and standards sets a permanent example that others will follow.

REFERENCES

[1] T. D'Silva, The Black Box of Bhopal, 42, 2006, ISBN: 978-1-4120-8412-3.

[2] District Court of Bhopal, India, State of Madhya Pradesh Through CBI vs. Warren Anderson & Others, Criminal Case No. 8460 of 1996, 17, June 7, 2010.

[3] A.S. Kalelkar, Investigation of large-magnitude incidents: Bhopal as a case study, in: IChemE: Symposium Series No. 110, Preventing Major Chemical and Related Process Accidents, May 10–12, 1988, 559, 1988.
[4] Committee on Inherently Safer Chemical Processes: The Use of Methyl Isocyanate (MIC) at Bayer CropScience, Board on Chemical Sciences and Technology, Division on Earth and Life Studies, National Research Council, The Use and Storage of Methyl Isocyanate (MIC) at Bayer CropScience, 86, 2012, ISBN: 978-0-309-25543-1.
[5] S. Varadarajan, et al., Report on Scientific Studies on the Factors Related to Bhopal Toxic Gas Leakage, 11, 1985.
[6] Indian Institute of Chemical Technology (CSIR), Technical and Tender Document for Detoxification, Decommissioning, and Dismantling of Union Carbide Plant, vol. I, 2010.
[7] Indian Institute of Chemical Technology (CSIR), Technical and Tender Document for Detoxification, Decommissioning, and Dismantling of Union Carbide Plant, vol. II, 2010.
[8] T.R. Chouhan, Bhopal: The Inside Story, Carbide Workers Speak Out on the World's Worst Industrial Disaster, 19, 1994, ISBN: 0-945257-22-8.
[9] T.R. Chouhan, Bhopal: The Inside Story, Carbide Workers Speak Out on the World's Worst Industrial Disaster, 25, 1994, ISBN: 0-945257-22-8.
[10] S. Varadarajan, et al., Report on Scientific Studies on the Factors Related to Bhopal Toxic Gas Leakage, 8, 1985.
[11] P. Shrivastava, Bhopal: Anatomy of a Crisis, 41, 1987, ISBN: 088730-084-7.
[12] T. D'Silva, The Black Box of Bhopal, 67, 2006, ISBN: 978-1-4120-8412-3.
[13] T.R. Chouhan, Bhopal: The Inside Story, Carbide Workers Speak Out on the World's Worst Industrial Disaster, 29, 1994, ISBN: 0-945257-22-8.
[14] Union Carbide Corporation, Bhopal Methyl Isocyanate Incident Investigation Team Report, March 1985. 5.
[15] R. Van Mynen, Union Carbide Corporation Press Conference Transcript, March 20, 1985, 3.
[16] T.R. Chouhan, Bhopal: The Inside Story, Carbide Workers Speak Out on the World's Worst Industrial Disaster, 56, 1994, ISBN: 0-945257-22-8.
[17] R. Van Mynen, Union Carbide Corporation Press Conference Transcript, March 20, 1985, 4.
[18] N.P. Lieberman, E.T. Lieberman, Working Guide to Process Equipment, second ed., 2003, 28.
[19] T.R. Chouhan, The unfolding of Bhopal disaster, J. Loss Prev. Proc. Ind. 18 (2005) 206.
[20] Union Carbide Corporation, Bhopal Methyl Isocyanate Incident Investigation Team Report, March 1985, 6.
[21] S. Varadarajan, et al., Report on Scientific Studies on the Factors Related to Bhopal Toxic Gas Leakage, 15, 1985.
[22] A.S. Kalelkar, Investigation of large-magnitude incidents: Bhopal as a case study, in: IChemE: Symposium Series No. 110, Preventing Major Chemical and Related Process Accidents, May 10–12, 1988, 575, 1988.
[23] P. Shrivastava, Bhopal: Anatomy of a Crisis, 42, 1987, ISBN: 088730-084-7.
[24] R. Van Mynen, Union Carbide Corporation Press Conference Transcript, March 20, 1985, 7.

[25] W. Morehouse, M.A. Subramaniam, The Bhopal Tragedy: What Really Happened and What It Means for American Workers and Communities at Risk, 6, 1986, ISBN: 0-936876-47-6.
[26] T. D'Silva, The Black Box of Bhopal, 136, 2006, ISBN: 978-1-4120-8412-3.
[27] Supreme Court of India Criminal Appellate Jurisdiction, Application for Directions to Institute Charges U/S 302 (For Offence U/S 300 (4)) Read With S. 35 of the Indian Penal Code, 1860 Against the Respondents Herein, Curative Petition (Criminal) No. 39-42 of 2010 in Criminal Appeal No. 1672-75 of 1996 in the Matter of Central Bureau of Investigation Versus Keshub Mahindra & Ors., 30, 2011.
[28] T. D'Silva, The Black Box of Bhopal, 54, 2006, ISBN: 978-1-4120-8412-3.
[29] S. Diamond, The Bhopal Disaster: How It Happened, The New York Times, January 28, 1985.
[30] S. Varadarajan, et al., Report on Scientific Studies on the Factors Related to Bhopal Toxic Gas Leakage, 77, 1985.
[31] W. Worthy, Methyl isocyanate: the chemistry of a hazard, Chem. Eng. News 63 (6) (1985) 28.
[32] T. D'Silva, The Black Box of Bhopal, 55, 2006, ISBN: 978-1-4120-8412-3.
[33] S. Varadarajan, et al., Report on Scientific Studies on the Factors Related to Bhopal Toxic Gas Leakage, 5, 1985.
[34] District Court of Bhopal, India, State of Madhya Pradesh Through CBI vs. Warren Anderson & Others, Criminal Case No. 8460 of 1996, 36, June 7, 2010.
[35] G.R. Kinsley, Properly purge and inert storage vessels, Chem. Eng. Prog. 97 (2) (2001) 60.
[36] S. Varadarajan, et al., Report on Scientific Studies on the Factors Related to Bhopal Toxic Gas Leakage, 6, 1985.
[37] W. Morehouse, M.A. Subramaniam, The Bhopal Tragedy: What Really Happened and What It Means for American Workers and Communities at Risk, 5, 1986, ISBN: 0-936876-47-6.
[38] S. Varadarajan, et al., Report on Scientific Studies on the Factors Related to Bhopal Toxic Gas Leakage, 16, 1985.
[39] District Court of Bhopal, India, State of Madhya Pradesh Through CBI vs. Warren Anderson & Others, Criminal Case No. 8460 of 1996, 14, June 7, 2010.
[40] A.S. Kalelkar, Investigation of large-magnitude incidents: Bhopal as a case study, in: IChemE: Symposium Series No. 110, Preventing Major Chemical and Related Process Accidents, 10–12 May 1988, 571, 1988.
[41] Supreme Court of India Criminal Appellate Jurisdiction, Application for Directions to Institute Charges U/S 302 (For Offence U/S 300 (4)) Read With S. 35 of the Indian Penal Code, 1860 Against the Respondents Herein, Curative Petition (Criminal) No. 39–42 of 2010 in Criminal Appeal No. 1672-75 of 1996 in the Matter of Central Bureau of Investigation Versus Keshub Mahindra & Ors., 32, 2011.
[42] S. Varadarajan, et al., Report on Scientific Studies on the Factors Related to Bhopal Toxic Gas Leakage, 18, 1985.
[43] Union Carbide Corporation, Bhopal Methyl Isocyanate Incident Investigation Team Report, March 1985, 10.
[44] Union Carbide Corporation, Review of MIC Production at the Union Carbide Corporation Facility Institute West Virginia, April 15, 1985, 4–2.
[45] Supreme Court of India, Criminal Appeal Nos. 1672, 1673, 1674 and 1675 of 1996, in the Matter of Keshub Mahindra vs. State of Madhya Pradesh, September 13, 1996, 15.

[46] Supreme Court of India, Criminal Appeal Nos. 1672, 1673, 1674 and 1675 of 1996, in the Matter of Keshub Mahindra vs. State of Madhya Pradesh, September 13, 1996, 16.
[47] Union Carbide Corporation, Cause of the Bhopal Tragedy, February 1, 2012. Retrieved from: http://www.bhopal.com/Cause-of-Bhopal-Tragedy.
[48] A.S. Kalelkar, Investigation of large-magnitude incidents: Bhopal as a case study, in: IChemE: Symposium Series No. 110, Preventing Major Chemical and Related Process Accidents, May 10–12, 1988, 553, 1988.
[49] Supreme Court of India, Criminal Appeal Nos. 1672, 1673, 1674 and 1675 of 1996, in the Matter of Keshub Mahindra vs. State of Madhya Pradesh, September 13, 1996, 9.
[50] T. D'Silva, The Black Box of Bhopal, 98, 2006, ISBN: 978-1-4120-8412-3.
[51] W. Worthy, Methyl isocyanate: the chemistry of a hazard, Chem. Eng. News 63 (6) (1985) 29.
[52] W. Morehouse, M.A. Subramaniam, The Bhopal Tragedy: What Really Happened and What It Means for American Workers and Communities at Risk, 7, 1986, ISBN: 0-936876-47-6.

CHAPTER

Process Hazard Awareness and Analysis

13

SUMMARY

The Bhopal factory was constructed to provide safe and reliable operation at maximum production rates. History demonstrates that the process failed to meet this functional expectation. To what extent did process-design complexities contribute to the catastrophic system malfunction that permanently shut down factory production? Could the hazard mitigation practices implemented after the Bhopal disaster have prevented it? Can lessons from a purposeful analysis of the Bhopal factory's process design be used to protect similar processes operating in the manufacturing industry today? These are just some of the questions that can be answered by applying fundamental process hazard analysis (PHA) concepts on the process directly involved with the Bhopal disaster.

Before evaluating the more obvious and superficial human causes for industrial incidents we must first examine how the Bhopal factory was designed (not intended) to operate. The purpose of this exercise is to determine how the process could be expected to perform under specific operating conditions. Human determination can, and often does, compensate for processing difficulties encountered in a manufacturing operation. Therefore human factors can only make sense in the context of the specific process where they were encountered. Clearly defining the complexities that workers can expect to confront during normal and abnormal operation puts mysterious actions and decisions into perspective. A process failure investigation can only be successful by taking the operability of the process into consideration.

Common cause failures [26] can defeat safeguard protection during a process upset. The ability to access critical system functions when it is most important requires identifying unintended dependencies before an incident discloses them. Retaining system functions is crucial for recovering from a process upset. A systematic evaluation of process design to identify unresolved dependencies establishes process control by:

1. Developing a process design package that makes faulty operation more difficult.
2. Adding sufficient forms of protection to prevent a system malfunction due to human error.
3. Explaining possible causes for human performance difficulties when conducting an incident investigation.

Rethinking Bhopal.

In accordance with the process safety management (PSM) standard, industry uses basic hazard assessment practices to detect and control process safety hazards during the design, construction, and operational phases of factory development. These same basic principles apply to diagnosing system deficiencies that can complicate process operation upon start-up. This approach will be used to create awareness over the expected performance of the industrial process constructed to manufacture the technical grade product at the Bhopal factory.

TECHNICAL ASSESSMENT

The Bhopal disaster unified industry's commitment to implement specific programs designed to identify, assess, and resolve unmitigated hazards prior to commissioning a new process. This sense of responsibility led to the widespread implementation of industry's PHA program to adequately manage potential hazards that could undermine the success of a manufacturing process. The intent of PHA is to address hazards that could otherwise interfere with maintaining an acceptable level of process safety, asset reliability, environmental protection, or factory economics. A significant amount of confidence is invested in industry's PHA program. It is essential for avoiding process upsets that could result in a catastrophic process release.

Industry's PSM standard does not specify a particular method that must be used to conduct a PHA. It simply states that the PHA method shall be appropriate to the complexity of the process and shall identify, evaluate, and control the hazards involved in the process. In the manufacturing industry, the hazard and operability (HAZOP) study is the most commonly used method to satisfy this PHA requirement.

Progressive companies take advantage of integrating layer of protection analysis (LOPA) elements into their respective PHA programs. Some of the benefits provided by this modern approach to hazard analysis include:

- Better alignment over what constitutes a "tolerable risk."
- Assignment of specific recommendations needed to achieve an acceptable level of risk tolerance.
- Defining a basis for immediate action if needed to address a specific hazard that represents an intolerable risk.
- Improved manufacturing process productivity, and therefore a competitive advantage.

The "Technical Assessment" section of this chapter demonstrates the application of PHA principles on the methyl isocyanate (MIC) storage system. This analysis can be used to determine if the system involved in the catastrophic process release would represent an acceptable risk tolerance by today's standards. It then becomes possible to determine if the policies and practices introduced in response to the Bhopal disaster could realistically have prevented it. Since the safety and environmental control assets encountered in the Bhopal factory (process scrubbers, flares, tanks, an inert gas blanketing system, etc.) are similar to those in widespread use throughout the manufacturing industry today, this analysis can help us to better develop reliable safeguards to prevent industrial incidents. At a minimum it will demonstrate the importance of incorporating elements of the PSM standard into modern incident investigations. This analysis can be used to determine how asset design may influence the potential source and final outcome of specific mechanical, and human, failures.

RUNDOWN TANK NODE

A PHA typically starts by dividing the production unit into process sections called "nodes." For example, the Bhopal factory's MIC rundown tank piping and instrumentation diagram (P&ID) (Fig. 13.1) provides sufficient detail to assess potential hazards contained within the MIC rundown tank design. For PHA purposes, the P&ID represents the MIC rundown tank node. Some experienced PHA facilitators might prefer to evaluate the refrigeration system as its own, separate node. For the purpose of this discussion, however, it is perhaps least confusing to incorporate the MIC rundown tank together with the refrigeration system as a single node. A PHA can be done either way, but in the interest of those who might be less familiar with the PHA program, this specific example is kept as simple as possible by visualizing the tank and refrigeration system as a single, cohesive system.

HYPOTHETICAL SCENARIO

Using a HAZOP study to perform a PHA involves developing a hypothetical scenario based on some deviation from normal operation. If that scenario results in a consequence of interest—usually loss of primary containment (LOPC) or an environmental event—then the hazard is evaluated further to determine what conditions would have to occur for that scenario to mature to its endpoint. For example, in the case of our tank node, the HAZOP team would be expected to identify and evaluate the MIC reactivity hazard that would result upon water, iron, or some other contaminant entering a rundown tank. The sequence of events involved in this scenario might realistically be documented something like this:

> *(Scenario 1) Exothermic reaction inside the MIC storage tank resulting in high temperature, leading to high pressure inside the MIC storage tank. MIC storage tank overpressure resulting in tank rupture and the catastrophic loss of MIC containment. Potential exposure of workers and community personnel to toxic process vapor and liquid, resulting in multiple inhalation injuries and fatalities.*

Since this scenario clearly involves a consequence of interest (catastrophic LOPC), the HAZOP team would then be expected to evaluate the hazard further to determine if the system is adequately equipped with safeguards to interrupt the sequence of events leading up to this scenario. In a conventional, *qualitative* HAZOP study the team would consider the multiple safeguards provided to avoid

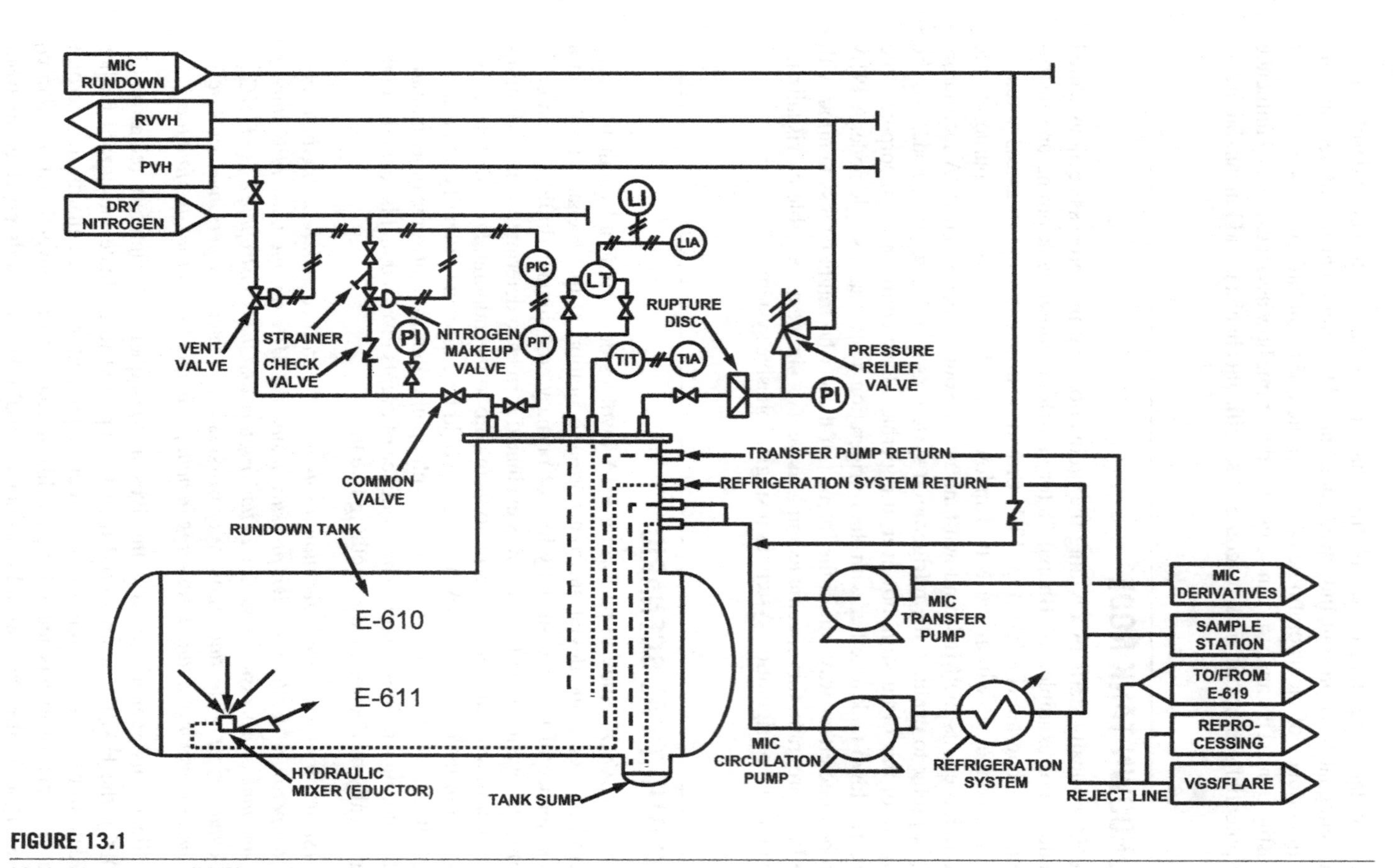

FIGURE 13.1
Methyl isocyanate (MIC) rundown tank node.

unacceptable consequences, which in this case might realistically involve inhalation injuries and possibly death. These safeguards include:

- Storing the MIC with a small quantity of phosgene to push the reaction equilibrium over to the products side of the chemical equation [1].
- A refrigeration system that constantly cools the contaminated storage tank contents below 5°C [2].
- A high-temperature MIC storage tank alarm set at 11°C, which would indicate if for some reason the refrigeration system has unexpectedly failed [3].
- An empty MIC reserve tank (E-619) to receive unstable hot material from either of the two MIC rundown tanks (E-610 or E-611) for additional cooling [4].
- The ability to dilute a contaminated mixture with an inert compound acting as a heat sink to slow down the reaction rate [5].
- The ability to transfer unstable MIC storage tank contents into the vent gas scrubber (VGS) or flare for final destruction [4].
- Rupture discs and pressure relief valves set at 40 psig to avoid an MIC storage tank overpressure incident that could lead to anything from a minor MIC release to a catastrophic, full-scale LOPC incident [6].

Taking into consideration the availability of the multiple different safeguards listed previously, the HAZOP team would have a basis to conclude that the system is adequately protected, and LOPC would likely not occur if for some reason a contaminant should enter the tank. However, the HAZOP team would also be expected to recognize that much of their confidence was credited to the tank's automatic pressure safety relief system. The pressure relief system's purpose was to dump the unstable process into the MIC vapor transfer headers (specifically the relief valve vent header (RVVH)). This action would be taken only if all the other safeguards failed and the tank's internal pressure elevated to a level high enough to penetrate the rupture disc and relief valve (40 psig).

The involuntary exit of process gas from the tank through the pressure relief system creates a potentially new hazard that the team would also be required to consider. The sequence of events behind this new scenario might be recorded something like this:

> *(Scenario 2) Exothermic reaction inside the MIC storage tank resulting in high temperature, leading to high pressure inside the MIC storage tank. Pressure relief valve automatically opens at 40 psig to avoid exceeding the tank's MAWP, resulting in hot process vapor flow into the RVVH. Process vapor traffic in the RVVH enters VGS and/or the flare tower for safe destruction and containment. Potential process upset resulting in economic losses, but no LOPC is anticipated. No safety or environmental consequences of interest.*

In this scenario, notice that the pressure relief device is no longer accepted as a safeguard. In fact, its designated function is what creates the hazard described in Scenario 2 (process getting out of the tank). However, the HAZOP team would likely conclude that the combination of remaining safeguards is still adequate protection from an uncontrolled process release. This is especially true because the process entering the process vent header (PVH) and RVVH would be directed to the VGS, and maybe the flare tower if needed for complete destruction. With this dual control system in place, there appears to be little reason for concern over a catastrophic MIC release into the atmosphere. On a *qualitative* basis, this combination of multiple safeguards still provides an adequate degree of protection over any MIC storage tank contamination scenario. However, adding LOPA elements to the assessment produces a significantly different result.

RISK CALIBRATION

LOPA methods offer a statistical approach for HAZOP teams to compare a predefined risk calibration standard to the level of protection achieved by adding safeguards. According to LOPA methodology, a safeguard cannot be credited unless it conforms to certain rules that distinguish it as an independent protective layer (IPL) [7]. If a safeguard meets the requirements of an IPL, then it can be credited against a certain sequence of events that could lead to a process release.

Corporate guidance is normally communicated to HAZOP teams in the form of a risk calibration matrix (Table 13.1) that defines their tolerance for specific consequences that can occur when a system malfunctions. HAZOP teams use this specific information to determine whether or not a hazard identified during a PHA meeting represents a tolerable risk. The basic premise is that higher consequence scenarios require a greater number of IPLs to reduce the likelihood of an actual

Table 13.1 Generic Risk Calibration Matrix [8]

		Consequence Category				
		Category 1	Category 2	Category 3	Category 4	Category 5
Frequency of Consequence (per year)	1×10^{0}	Optional (evaluate alternatives)	Optional (evaluate alternatives)	Action at next opportunity (notify corporate management)	Immediate action (notify corporate management)	Immediate action (notify corporate management)
	1×10^{-1}	Optional (evaluate alternatives)	Optional (evaluate alternatives)	Optional (evaluate alternatives)	Action at next opportunity (notify corporate management)	Immediate action (notify corporate management)
	1×10^{-2}	No further action	Optional (evaluate alternatives)	Optional (evaluate alternatives)	Action at next opportunity (notify corporate management)	Action at next opportunity (notify corporate management)
	1×10^{-3}	No further action	No further action	Optional (evaluate alternatives)	Optional (evaluate alternatives)	Action at next opportunity (notify corporate management)
	1×10^{-4}	No further action	No further action	No further action	Optional (evaluate alternatives)	Optional (evaluate alternatives)
	1×10^{-5}	No further action	No further action	No further action	No further action	Optional (evaluate alternatives)
	1×10^{-6}	No further action	No further action	No further action	No further action	No further action

event down to a tolerable level. Realistically, the likelihood of an incident can never be driven to zero. The objective, therefore, is to achieve a tolerable risk by applying an adequate number of IPLs. These IPLs must be available when they are actually called upon in order to interrupt the sequence of events that could end with consequences of interest.

A risk calibration matrix is based on the simple relationship that defines risk $\left(R_k^C\right)$ as a function of the frequency $\left(f_k^C\right)$ and consequence (C_k) for a possible incident (Eq. [13.1]) [9]. Note that this sample matrix is for illustration purposes only and does not represent the risk tolerance of any specific company, past or present. It does, however, provide a standard basis to determine how well the MIC storage system design conforms to a set of criteria that establishes its ability to contain the process. Determining if the MIC storage system includes enough IPLs to achieve an adequate level of protection involves establishing both the consequence category and the frequency of consequence for the specific hazard.

$$R_k^C = f_k^C \times C_k \qquad [13.1]$$

where, R_k^C is the risk of an incident (k) capable of producing a consequence of interest (C); f_k^C is the frequency of an incident (k) capable of producing a consequence of interest (C); C_k is the magnitude of a consequence of interest, produced by an incident (k).

CONSEQUENCE CATEGORY

The example risk matrix consequence categories (Table 13.1) range from one to five, with five being the highest possible consequence (a fatality). Some companies may assign more than five consequence categories, depending on how they define their consequence thresholds. However, there should be no argument over the consequence level represented by any scenario involving the release of hot MIC gas into the atmosphere. Simply stated, gaseous MIC could not be safely released into the atmosphere. Doing so might produce toxic inhalation effects that could realistically produce the highest consequence possible [10]. For this reason, the HAZOP team would likely assign a category 5 consequence to any scenario involving the release of MIC gas into the atmosphere. In fact, any scenario ending with the escape of MIC gas to the atmosphere would probably qualify for the most severe category on any risk matrix, regardless of how many consequence categories are defined.

In order for process design to meet risk tolerance criteria, it must include sufficient IPLs to reduce the frequency of consequence down to a degree that requires no further action. According to the sample risk matrix selected for this analysis (Table 13.1), a category 5 incident requires a frequency of consequence of 1×10^{-6} years or less to achieve an acceptable level of risk tolerance (no further action required). Any higher frequency would require the HAZOP team to formally recommend the installation of additional IPLs.

FREQUENCY OF CONSEQUENCE

The frequency of consequence for a specific scenario is the calculated probability for an incident to occur, in years. It is the product of the initiating event frequency (IEF) [11] and the probability of failure on demand (PFD) [12] for all the credited IPLs (Eq. [13.2]) [13]. Attention is first given to assigning the IEF, which is the likelihood for a deviation or a failure to occur. For any scenario that might be involved with an MIC storage tank contamination incident, the conservative approach might be to assign an IEF of 1/10 (or 1×10^{-1}). This seems reasonable based on the wide range of common compounds that could catalyze an exothermic reaction upon entering a storage tank containing MIC. Knowing that at least one contamination incident actually did occur in the course of 6 years at the Bhopal factory, this estimate may not be too far off.

$$f_i^C = f_i^I \times \mathrm{PFD}_{i1} \times \mathrm{PFD}_{i2} \times \cdots \times \mathrm{PFD}_{ij} \qquad [13.2]$$

where, f_i^C is the frequency for a release scenario with a specific consequence endpoint; f_i^I is the IEF; PFD_{ij} is the probability for failure on demand of an IPL credited against a specific scenario.

Next, we would assign a PFD for each of the IPLs. The PFD is the probability that an IPL will *not* be available when it is needed. The PFD for most basic process control systems is 1/10 (or 1×10^{-1}) [14], which means that the IPL will not be available once out of every 10 times that it is called upon for use. It might be acceptable to assign a 1/100 PFD (or 1×10^{-2}) to a properly-sized pressure safety relief valve [14], operating in clean MIC service. Regardless, it is important to remember that no IPL is perfect, and there is always a chance that it will not work for one reason or another [15]. That is why multiple IPLs are needed in systems where a release can be expected to generate extreme consequences. This would be the case where the release of toxic MIC vapor might be possible. If for some reason one of our IPLs should fail, there should still be more backup IPLs available to protect us.

In order for an IPL to be credited it must be properly sized, maintained, and tested for the scenario under evaluation [16]. Unless the sizing is confirmed and documented to be adequate for the scenario being considered, it is only a safeguard that cannot be credited as an IPL to reduce the frequency of consequence. However, let us say that the HAZOP team in our example has found the information they were looking for to become satisfied that the pressure relief valve is sized correctly and can handle the highest load possible following a contamination incident. Table 13.2 summarizes the IEF for a contamination event while at the same time taking credit for all the safeguards we identified earlier (as IPLs). This provides sufficient information to calculate a frequency of consequence.

For Scenario 1 (tank overpressure), notice that the most reliable safeguard, the automatic pressure relief system, would be expected to dump the storage tank contents into the RVVH if all of the other safeguards somehow failed to prevent a high-pressure event inside the MIC storage tank. To determine how this matches with acceptable risk tolerance, simply compare the frequency of consequence (1×10^{-10})

Table 13.2 Frequency of Consequence Determination (Scenario 1)

Summary Sheet for LOPA Method

Scenario Number: 13.2	Equipment Number: E-610	Scenario Title: MIC storage tank overpressure upon process contamination incident.	
Date:	Description	Probability	Frequency (per year)
Consequence Description/Category	Category-5 incident (MIC gas release): Potential for inhalation injuries and fatalities.		
Risk Tolerance Criteria (category or frequency)	Maximum tolerable risk for a fatal injury		1 x 10^{-6}
Initiating Event (typically a frequency)	Entry of iron, water, or any other compound incompatible with MIC process		1 x 10^{-1}
Enabling Event or Condition		N/A	
Conditional Modifiers (if applicable)			
	Probability of ignition	N/A	
	Probability of personnel in affected area	1	
	Probability of fatal injury	1	
	Others	N/A	
Frequency of unmitigated Consequence			1 x 10^{-1}
Independent Protection Layers			
	Phosgene spike	1 x 10^{-1}	
	Refrigeration system	1 x 10^{-1}	
	High temperature storage tank alarm	1 x 10^{-1}	
	Transfer into Reserve Tank (E-619)	1 x 10^{-1}	
	Transfer into parallel MIC Rundown Tank	1 x 10^{-1}	
	Dilution with inert liquid heat sink	1 x 10^{-1}	
	Route MIC into VGS Accumulator section	1 x 10^{-1}	
	Automatic tank pressure relief system	1 x 10^{-2}	
Safeguards (non-IPLs)			
Total PFD for all IPLs		1 x 10^{-9}	
Frequency of Mitigated Consequence			1 x 10^{-10}
Risk Tolerance Criteria Met? Yes (1 x 10^{-10} is less than 1 x 10^{-6})			
Actions Required to Meet Risk Tolerance Criteria: None required. System meets specified risk tolerance as-is.			
Notes: Analysis performed without applying LOPA rules governing the credit of IPLs.			
References (links to originating hazard review, PFD, P&ID, etc.):			
LOPA analyst (and team members, if applicable):			

MIC, *methyl isocyanate;* IPL, *independent protective layer;* VGS, *vent gas scrubber;* LOPA, *layer of protection analysis;* PFD, *probability of failure on demand;* P&ID, *piping and instrumentation diagram.*

with the specified action required for a category 5 incident. In this example we find that 1×10^{-10} is less than 1×10^{-6} so no further action is required. Mission accomplished. The team's original qualitative assessment appears to correspond with a more statistical approach to assessing the scenario defined in Table 13.2. In the extremely rare case that none of the preventive safeguards worked to avoid building sufficient pressure inside the tank for the rupture disk to burst, the unstable process would, instead, escape into the PVH and RVVH to prevent the tank from exploding. These circumstances reflect an incident that actually happened on the system under review.

However, based on the way that our HAZOP team credited various safeguards, the odds for an MIC contamination scenario to create a demand on the pressure relief system were extremely low (1-in-10,000,000,000 years). Although an incident, no matter how remote it may seem, can happen at any time, nobody could reasonably have expected a 1-in-10,000,000,000 year event to occur within only 6 years of commissioning the MIC unit. It would be a bet that few would be willing to take. A more reasonable expectation would be for an incident with a 1-in-10 year frequency to occur within the first 6 years of operation. The disparity between actual and estimated system performance raises a legitimate question over the credibility of the analysis. Are adjustments needed to reconcile the difference between actual and expected process behavior?

INDEPENDENCE

A modern approach to analyzing this scenario would involve applying LOPA rules that credit only safeguards that can be validated as IPLs. A very important LOPA principle involves taking credit only for safeguards that are truly independent [17]. In order to be *truly* independent, the functions of a safeguard cannot be associated with either the initiating event or the action of another protective layer credited in the scenario. Otherwise the safeguard may help to prevent an incident, but there will never be enough confidence to apply IPL credit for it. If a dependency exists, then only one failure is capable of rendering one or more other protective systems useless. On the other hand, if multiple protective systems are independent of each other, then backup protection will exist if one of them was to fail.

Up to this point in the analysis, IPL credit has been taken for each safeguard without verifying its independence. In this example the node drawing must be examined to verify that proper credit has been applied only to independent safeguards. Looking closely at the design configuration of the MIC-circulation loop that passes through the refrigeration system, a hidden dependency can be found. This specific dependency disables most of the safeguards that earlier created the perception of adequate system protection. This dependency significantly changes the system's expected design performance.

Notice that the MIC circulation pump discharge feeds both the refrigeration system and the reject line (Fig. 13.1). The reject line is the pathway through which contaminated MIC would exit the storage area to a designated alternative processing location in the event of a contamination incident. Note, however, that a failure at this specific pump location would interrupt MIC flow through the recirculation line.

Under these conditions, the refrigeration system becomes useless although it technically still works. In asset reliability terms, although the refrigeration system is available, it cannot be utilized (chapter: Equipment Reliability Principles). Therefore, the MIC in the storage tank cannot be cooled. Process cooling only takes place if the MIC circulation pump is working properly.

This dependency is the perfect example of a common cause failure. Losing the circulation pump correspondingly causes a loss of MIC cooling, which in turn causes the temperature of the MIC inside the storage tank to rise. Upon reaching 11°C the high-temperature alarm would activate. The alarm would signal the workers' attention to a problem inside the tank (elevated temperature) that needs to be addressed. Under the circumstances, they might want to transfer the MIC into an alternative storage tank with an available circulation pump. Unfortunately, the circulation pump failure withdraws this option from the workers. Since the MIC circulation pump also transports contaminated material out of the tank through the reject line, any alternative processing location including the other storage tanks, the VGS, and the flare tower become inaccessible (Fig. 13.2). At this point the workers are left with no option other than to leave the warm MIC stagnating inside the storage tank until the circulation pump can be repaired or replaced. This specific asset failure could represent a very dangerous hazard. If a contamination incident should simultaneously occur before the pump can be repaired, then the

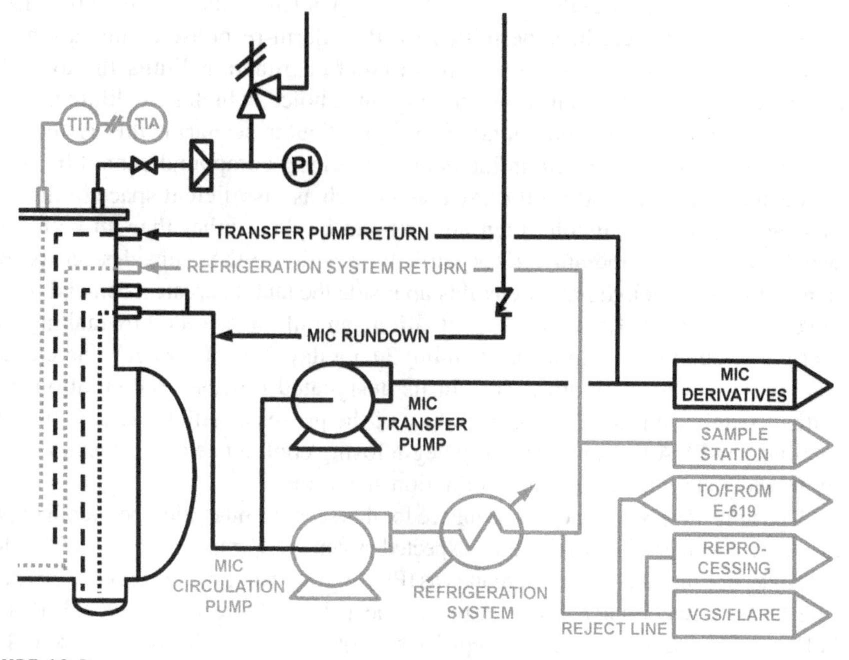

FIGURE 13.2

Common cause failures resulting from a methyl isocyanate (MIC) circulation pump failure.

ensuing reaction might cause enough pressure to build inside the tank for the pressure relief system to automatically dump the hot MIC gas into the RVVH.

To summarize, an MIC circulation pump failure in this specific process configuration would cause the corresponding loss of multiple storage tank system functions including:

1. Safeguard: Phosgene spike ($COCl_2$ vaporizes at >8°C).
2. Safeguard: MIC refrigeration system.
3. Safeguard: Disposal of unstable MIC into the VGS or flare.
4. Safeguard: Transfer unstable MIC into E-619 for cooling, expansion, reprocessing, or disposal.
5. Process monitoring due to temperature gauge failure (automatic reading above +25°C high scale).
6. Reprocessing impure MIC back into the MIC synthesis section.
7. MIC storage tank sample station flow.
8. MIC return flow through the eductor, resulting in the loss of internal tank mixing.

In this scenario, an MIC circulation pump failure instantly disables at least four other safeguards that were previously credited (albeit improperly) as IPLs. The only remaining safeguards aside from the automatic pressure relief system are the storage tank's high-temperature alarm and the ability to use the extra unused storage tank volume to dilute an unstable MIC mixture.

Unfortunately, since adding the inert diluent to the tank would occur in response to the high-temperature alarm, a dependency also exists here. Therefore, only one IPL credit is permitted for this alarm-response combination; not two. In other words, losing the MIC circulation function limits the available alarm-response mechanisms down to only one choice, which is to dilute the tank contents with an inert compound. In this particular scenario, this is the only option that exists to *prevent* an incident. If the inert compound cannot be added to the unstable tank mixture for any reason such as insufficient space in the tank or if no diluent is available, then nothing can be done other than sit helplessly watching as the temperature rises until the reaction either subsides on its own (a near miss) or enough pressure builds up inside the tank to create a demand for the tank's pressure relief system. This situation can only be avoided by making sure that the circulation pump keeps running 24 h a day, 7 days a week. The circulation pump can literally never stop in its designated process configuration. It is critically important for the overall safety of the process. With these complexities in mind, the PHA team would likely begin losing confidence in the system's ability to adequately manage a contamination scenario.

Updating the frequency of consequence for this scenario quantifies the overall impact that these dependencies will have on expected system performance (Table 13.3). When only independent safeguards are credited as IPLs, the frequency of consequence increases 1×10^{-5} years. According to this adjustment, one additional safeguard (with at least 1/10 PFD) would be needed to satisfy acceptable risk tolerance for a category 5 incident. This result is quite different from that obtained by doing the analysis without consideration toward independent layers of protection. In order to make each of the available safeguards

Table 13.3 Updated Frequency of Consequence Determination (Scenario 1)

Summary Sheet for LOPA Method

Scenario Number: 13.3	Equipment Number: E-610	Scenario Title: MIC storage tank overpressure upon process contamination incident.	
Date:	**Description**	**Probability**	**Frequency (per year)**
Consequence Description/Category	Category-5 incident (MIC gas release): Potential for inhalation injuries and fatalities.		
Risk Tolerance Criteria (category or frequency)	Maximum tolerable risk for a fatal injury		1×10^{-6}
Initiating Event (typically a frequency)	Entry of iron, water, or any other compound incompatible with MIC process		1×10^{-1}
Enabling Event or Condition		N/A	
Conditional Modifiers (if applicable)			
	Probability of ignition	N/A	
	Probability of personnel in affected area	1	
	Probability of fatal injury	1	
	Others	N/A	
Frequency of unmitigated Consequence			1×10^{-1}
Independent Protection Layers			
	Refrigeration system	1×10^{-1}	
	High temperature storage tank alarm, leading to dilution with inert liquid heat sink	1×10^{-1}	
	Automatic tank pressure relief system	1×10^{-2}	
Safeguards (non-IPLs)			
	Phosgene spike		
	Transfer into Reserve Tank (E-619)		
	Transfer into parallel MIC Rundown Tank		
	Route MIC into VGS Accumulator section		
Total PFD for all IPLs		1×10^{-4}	
Frequency of Mitigated Consequence			1×10^{-5}
Risk Tolerance Criteria Met? No (1×10^{-5} is greater than 1×10^{-6})			
Actions Required to Meet Risk Tolerance Criteria: Evaluate options to provide one additional layer of protection and initiate a project to address remaining risk.			
Notes: Items listed as Safeguards are not independent from a refrigeration system failure (due to loss of MIC Circulation Pump).			
References (links to originating hazard review, PFD, P&ID, etc.):			
LOPA analyst (and team members, if applicable):			

MIC, *methyl isocyanate;* IPL, *independent protective layer;* VGS, *vent gas scrubber;* LOPA, *layer of protection analysis;* PFD, *probability of failure on demand;* P&ID, *piping and instrumentation diagram.*

independent, the design team would be tasked with finding a way to separate at least one additional safeguard from the influence of an MIC circulation pump failure, so that it could be credited as an IPL. Alternatively, the design team could install more IPLs to adequately raise the level of protection to an acceptable level. Either way, there is more work to be done before the system design reflects an adequate level of risk tolerance.

PASSIVE PROTECTION

The foregoing adjustment also changes the outcome of the second scenario that involves sending boiling MIC into the RVVH in response to a contamination incident that creates a high-pressure excursion inside the MIC storage tank (Table 13.4). LOPA rules covering capacity [18] apply to this scenario. The same principle used to credit the pressure-safety relief device applies here as well. In order to take credit for passive protection offered by a continuously-operating flare or scrubber [19], the asset must be appropriately sized, designed, and maintained. In the case of passive protection, testing is usually not required since the safeguard always runs in its protective state. However, no (zero) IPL credit can be taken unless the sizing has been confirmed and documented to be sufficient [20]. Whereas credit could be taken for the MIC storage tank pressure relief valve because it was sized properly [21], the downstream systems that would receive its discharge must also be adequately sized, designed, and maintained. For now, let us assume that the HAZOP team reviewing the MIC storage tanks' design has not located the required documentation. Until this information is confirmed and documented, an atmospheric release is this scenario's endpoint. It is as if the VGS and flare tower simply do not exist. Although the overpressure hazard may be tolerable, the frequency of a consequence of interest may be more than allowable [22].

After adjusting the analysis to account for invalidated relief system capacity, the scenario that involves managing MIC vapor traffic in the RVVH in response to a contamination incident is seriously short of IPLs. Essentially, the only available credits are two initiating event frequencies (a contamination event and a circulation pump failure) and one IPL (the high-temperature alarm combined with an available response). No other IPLs can be credited to prevent the atmospheric MIC release once contaminated tank vapors start flowing into the PVH and RVVH. Under these circumstances, the HAZOP team's attention would no doubt immediately turn to performing sizing calculations. Their purpose would be to confirm and document the flare tower and perhaps the VGS (depending on the VGS' functional expectation) were both adequately sized to manage the scenario. According to the number of credits that the HAZOP team can take for a properly sized vapor-disposal system, this may be all that is required to achieve acceptable safety and environmental risk criteria. Operating without an adequately sized flare tower would require adding at least three additional IPLs (each with a 1/10 PFD) for this scenario to meet acceptable risk tolerance criteria. Without a flare there is essentially zero tolerance for a high-pressure upset leading to MIC vapor traffic inside the RVVH. Therefore, the combination of any available IPLs would have to adequately prevent a high-pressure upset in the MIC storage tank that would place a demand on the pressure relief system.

Table 13.4 Frequency of Consequence Determination (Scenario 2)

Summary Sheet for LOPA Method

Scenario Number: 13.4	Equipment Number: E-610	Scenario Title: Boiling MIC release to atmosphere due to use of automatic pressure relief system upon process contamination incident following a Circulation Pump failure.	
Date:	Description	Probability	Frequency (per year)
Consequence Description/Category	Category-5 incident (MIC gas release): Potential for inhalation injuries and fatalities.		
Risk Tolerance Criteria (category or frequency)	Maximum tolerable risk for a fatal injury		1×10^{-6}
Initiating Event (typically a frequency)	(1) Entry of iron, water, or any other compound incompatible with MIC process		1×10^{-1}
	(2) Circulation pump failure		1×10^{-1}
Enabling Event or Condition		N/A	
Conditional Modifiers (if applicable)			
	Probability of ignition	N/A	
	Probability of personnel in affected area	1	
	Probability of fatal injury	1	
	Others	N/A	
Frequency of unmitigated Consequence			1×10^{-2}
Independent Protection Layers			
	High temperature storage tank alarm, leading to dilution with inert liquid heat sink	1×10^{-1}	
	Automatic tank pressure relief system	1	
Safeguards (non-IPLs)			
	Phosgene spike		
	Refrigeration system		
	Transfer into Reserve Tank (E-619)		
	Transfer into parallel MIC Rundown Tank		
	Route MIC into VGS Accumulator section		
Total PFD for all IPLs		1×10^{-1}	
Frequency of Mitigated Consequence			1×10^{-3}
Risk Tolerance Criteria Met? No (1×10^{-3} is greater than 1×10^{-6})			
Actions Required to Meet Risk Tolerance Criteria: Add three additional layers of protection at next opportunity.			
Notes: Process design falls short of specified Category-5 consequence risk tolerance. Corporate management notified before ending the PHA meeting.			
References (links to originating hazard review, PFD, P&ID, etc.):			
LOPA analyst (and team members, if applicable):			

MIC, *methyl isocyanate;* IPL, *independent protective layer;* VGS, *vent gas scrubber;* LOPA, *layer of protection analysis;* PHA, *process hazard analysis;* PFD, *probability of failure on demand;* P&ID, *piping and instrumentation diagram.*

EVALUATING DOUBLE JEOPARDY

At this point it becomes obvious that two independent failures must occur in order for this scenario to develop. The majority of HAZOP teams would likely stop here with a deep breath of relief; claiming that the entire scenario can be dismissed on the basis of double jeopardy. The general concept behind double jeopardy is that the possibility for two totally unrelated failures to overlap one another, in a way that would create a hazard, is so statistically improbable that considering it is a waste of time. This can be proven for actual situations that represent *true* cases of double jeopardy. For example, a 1/10 year failure event that can be "revealed" (detected) and addressed in 3 days would create an annual time-at-risk probability of 0.00082 (72 h divided by 87,600 h—the total number of hours in 10 years). The likelihood for another *revealed* 1/10 year failure event to overlap the 72-hour repair period created by the first failure is so remote (less than 0.00082^2, or 6.8×10^{-7}) that the scenario automatically achieves an acceptable risk tolerance on its own *without* any IPLs. For that reason, HAZOP teams are typically given permission to ignore these scenarios and move on to more realistic hazards.

In the contamination scenarios discussed previously, an MIC circulation pump failure must correspond with a contamination event (another independent failure) to create a hazard. There is no question that the pump failure and contamination incident are two independent failures that must at some point occur simultaneously at some period of time for the scenario to develop. Therefore it would seem that a PHA team would be correct in claiming that this situation involves double jeopardy and thus can be disregarded.

However, discounting this scenario would not be acceptable. The truth is that this scenario is, in fact, *not* double jeopardy. This example illustrates one of the most common errors that HAZOP teams routinely make during actual PHA meetings. Before dismissing any scenario on the basis of double jeopardy, the HAZOP team would have to act responsibly with the process safety information provided to be sure that the rules of double jeopardy are not misapplied. One of these rules is that *both* independent failures must be revealed [23]. Looking more closely at the system under consideration, there is at least one more dependency that creates a problem. Unfortunately, the system is designed so that only one independent failure is revealed, not both. Therefore the scenario is valid and must be analyzed to verify that it achieves acceptable risk criteria.

Walking through the sequence of events explains why this is not a case of double jeopardy. Let us say that the MIC circulation pump fails (Failure 1). This ceases MIC refrigeration, which soon generates a high MIC storage tank–temperature alarm. As discussed earlier, the pump failure makes it impossible to push the tank contents into a safe area.

Now the second independent failure occurs when a contaminant accidentally enters the MIC storage tank (Failure 2). This catalyzes an exothermic chemical reaction inside the contaminated tank, which starts heating the tank contents further. The rising temperature inside the unrefrigerated tank creates an unstable process condition that initiates a thermal runaway reaction. The crisis developing inside the tank is not revealed because the high-temperature alarm is already active due to the loss of refrigeration. The temperature gauge does not offer any protection since the needle was already pegged against the upper range (+25°C). To the naked eye, everything appears normal, in the sense of what the condition looks like when a circulation pump failure has already been revealed.

Workers may have written a work order to repair the broken pump 24 h earlier, but the alarm remains active for an extended period of time while other priorities critical for production are attended to. Perhaps the alarm has been temporarily disabled, or acknowledged but not reset, to avoid creating a constant distraction inside the control room. While the system continues operating with a known problem, pending the necessary repair, what benefit does a constantly sounding alarm serve? In this case there might even be a management system to allow the alarm to be disabled and reactivated upon the return of the circulation pump. Regardless, there is no new information being transmitted to indicate that a contamination event has taken place.

None of the systems provided are therefore capable of revealing the concurrent contamination event and circulation pump failure. Therefore, the scenario will either resolve itself without anyone's knowledge or will be revealed through abnormal system operating conditions. Unfortunately, that time will arrive only when the incident has matured—perhaps past the point of no return.

With no secondary high-temperature or independent high-pressure alarm, the situation could persist undetected until it becomes too late to reverse the spontaneous generation of heat inside the tank. In this context, the circulation pump failure creates a high-temperature alarm scenario that hides the contamination event. Similar to the "check engine" warning light in most cars today, the high-temperature alarm in this specific process configuration is designed to cover a wide range of different causes that might indicate a nonspecific problem. Since the single high-temperature alarm is the basis to detect either an MIC circulation pump failure or an uncontrolled contamination event in this particular system, the two failures are not really independent. With this dependency in mind, the high-temperature alarm is really just a common trouble alarm for both failures. In this HAZOP node, only one failure can be revealed at a time. Therefore, this is not a case of double jeopardy. Failure to reveal both simultaneous failures makes this scenario valid. Therefore, the scenario must be evaluated.

Although a high MIC storage tank–pressure alarm could at first have seemed redundant with an existing high-temperature alarm (chapter: Factory Construction), adding a high-pressure alarm might make this scenario a true case of double jeopardy. Under these conditions, it would be permissible to ignore the proposed scenario. Remember that for double jeopardy, all concurrent failures required to trigger the scenario must be detectable. In our scenario, the MIC circulation pump failure would have activated the high-temperature alarm upon interrupting the MIC flow through the refrigeration system. At that point, the tank vapor management system would automatically compensate to maintain the target 2 psig pressure inside the tank. Should the tank contents then also become contaminated, this would generate more heat and the vapor management system would continue to make automatic adjustments until reaching maximum output. At that time the pressure would exceed 2 psig and a high-pressure alarm would call the workers' attention to a different condition influencing process stability, perhaps with sufficient time to apply appropriate corrective measures.

At the end of this discussion the HAZOP team would formally recommend evaluating options for independent warning devices that could reduce the probability of two concurrent incidents overlapping each other. The recommendation would trigger action to consider whether or not a high-pressure alarm would satisfy the response time requirements to receive credit as a valid IPL [24]. Regardless, the HAZOP team would be expected to fully evaluate this scenario to verify that it is properly addressed by adding appropriate IPLs.

HUMAN INTERVENTION

Discussing the possibility of double jeopardy helps a HAZOP team get a better sense of how independent systems operate together, and the way that potential hazardous scenarios are likely to unfold. Based on this understanding, the HAZOP team would now realize that there are still too many credits being taken for safeguards previously thought to be IPLs. There is still one final dependency that disables an additional safeguard previously credited as an IPL. This final dependency removes the high-temperature alarm credit that was taken earlier. For an alarm to be credited, the alarm must be capable of triggering a response. In the case of a circulation pump failure, there is no response option. Once again, the alarm will activate when refrigeration ceases. The only remedy is to perform the repair needed to get the circulation pump running again. Until that time, the alarm will remain active or will be disarmed through an approved management system.

In the scenario under evaluation, the circulation pump fails first, independent of a contamination event. This interrupts MIC flow through the refrigeration system, which causes the MIC temperature to rise above 11°C. Next, the high-temperature alarm activates, calling the workers' attention to a failure in the MIC storage system. Workers would investigate the alarm and find that the pump needs to be repaired. A work order would be written, but since this is considered a discretionary reliability incident, there is no reason to dilute the material inside the tank with an inert ingredient. Doing this would not serve any purpose, other than to spoil what otherwise might be a good MIC batch in storage. Therefore, it would not be done. The problem is with the circulation pump, not the material inside the tank. Under these circumstances, the decision could be made to temporarily operate this way, with the alarm sounding or disabled until the pump can be repaired or replaced. Any IPL credit for the alarm-response combination could not be taken under these circumstances. The alarm would not result in adding an inert ingredient into the storage tank since the material in the tank is stable and is simply following ambient temperature.

However, if a contamination incident severe enough to defeat the system's cooling capability occurs first, this would also trigger the high MIC storage tank–temperature alarm. In this situation the pump continues operating, at least up to the point that the contamination event occurs. Workers would investigate the high-temperature alarm and discover that the tank contents have been contaminated. As long as the MIC circulation pump continues working, there are three available options for regaining control over the situation:

- Transfer the contaminated tank contents into the emergency reserve tank (E-619).
- Dilute the contaminated tank contents with an inert compound.
- Transfer the contaminated tanks contents into VGS or flare tower for destruction.

If the pump fails before action can be taken, then it is still possible to dilute the unstable tank contents with an inert compound. The alarm can still be credited in this situation although overall the scenario still does not achieve an acceptable risk tolerance according to the guidance communicated to our HAZOP team in the risk calibration matrix.

In reality, all of these alarm-response options are linked to the high-temperature alarm. In LOPA studies, additional rules must be satisfied in order to take credit for an alarm [25]. On paper, alarms may create a false sense of security. Alarms require appropriate human intervention. Human response to a process alarm is usually considered less reliable than an engineering control. When an alarm is credited, we are really taking credit for both the persons responding to the alarm and also the proper functioning of that alarm. Only one IPL can exist for all of the alarm-response combinations available for a particular alarm. The same rule also applies to worker surveillance rounds where someone must be able to both recognize and respond appropriately to an abnormal operating condition. Increasing the number of response options to a single alarm function does not normally add more IPL credit. For that reason, all three of the available response options are dependent on a single high-temperature alarm point. Therefore, only one IPL credit is available.

In the end, the HAZOP team evaluating this scenario updates the analysis with the results of both evaluations (Tables 13.5 and 13.6). Scenario 1 (MIC tank overpressure) requires adding three IPLs to meet acceptable risk tolerance. Scenario 2 (MIC release to atmosphere) requires four additional IPLs to meet acceptable risk tolerance, pending vapor management system calculations to confirm that the flare tower and VGS are properly sized to mitigate this specific scenario.

If the process was still in its design phase, it would be appropriate to communicate the HAZOP study results back to the project team. The project team would then be expected to address the possible design inadequacies detected during the PHA. In the case of industry's ongoing PHA program applied to operating processes involving a category 5 incident with such a high frequency of consequence (Scenario 2), immediate risk mitigation would be necessary. If an immediate solution was not available, then the process would be intentionally constrained or even shut down until the PHA team's concern could be validated and addressed. Taking decisive action in this manner is a responsible approach in preventing industrial incidents that could later be classified as disasters.

Table 13.5 Final Analysis for Scenario 1 With LOPA Rules Applied

Summary Sheet for LOPA Method

Scenario Number: 13.5	Equipment Number: E-610	Scenario Title: MIC storage tank overpressure upon process contamination incident following a Circulation Pump failure.	
Date:	Description	Probability	Frequency (per year)
Consequence Description/Category	Category-5 incident (MIC gas release): Potential for inhalation injuries and fatalities.		
Risk Tolerance Criteria (category or frequency)	Maximum tolerable risk for a fatal injury		1×10^{-6}
Initiating Event (typically a frequency)	(1) Entry of iron, water, or any other compound incompatible with MIC process		1×10^{-1}
	(2) Circulation pump failure		1×10^{-1}
Enabling Event or Condition		N/A	
Conditional Modifiers (if applicable)			
	Probability of ignition	N/A	
	Probability of personnel in affected area	1	
	Probability of fatal injury	1	
	Others	N/A	
Frequency of unmitigated Consequence			1×10^{-1}
Independent Protection Layers			
	Automatic tank pressure relief system	1×10^{-2}	
Safeguards (non-IPLs)			
	Refrigeration system		
	High temperature storage tank alarm, leading to dilution with inert liquid heat sink		
	Phosgene spike		
	Transfer into Reserve Tank (E-619)		
	Transfer into parallel MIC Rundown Tank		
	Route MIC into VGS Accumulator section		
Total PFD for all IPLs		1×10^{-2}	
Frequency of Mitigated Consequence			1×10^{-3}
Risk Tolerance Criteria Met? No (1×10^{-3} is greater than 1×10^{-6})			
Actions Required to Meet Risk Tolerance Criteria: Add three additional layers of protection at next opportunity.			
Notes: Process design falls short of specified Category-5 consequence risk tolerance. Corporate management notified before ending the PHA meeting. IEF credit taken for loss of MIC Circulation Pump creates dependency with listed Safeguards, so credit is denied.			
References (links to originating hazard review, PFD, P&ID, etc.):			
LOPA analyst (and team members, if applicable):			

MIC, *methyl isocyanate;* IPL, *independent protective layer;* VGS, *vent gas scrubber;* LOPA, *layer of protection analysis;* PHA, *process hazard analysis;* PFD, *probability of failure on demand;* P&ID, *piping and instrumentation diagram;* IEF, *initiating event frequency.*

Table 13.6 Final Analysis for Scenario 2 With LOPA Rules Applied

Summary Sheet for LOPA Method

<table>
<tr><td>Scenario Number:
13.6</td><td>Equipment Number:
E-610</td><td colspan="2">Scenario Title: MIC storage tank overpressure upon process contamination incident following a Circulation Pump failure.</td></tr>
<tr><td>Date:</td><td>Description</td><td>Probability</td><td>Frequency (per year)</td></tr>
<tr><td>Consequence Description/Category</td><td>Category-5 incident (MIC gas release): Potential for inhalation injuries and fatalities.</td><td></td><td></td></tr>
<tr><td>Risk Tolerance Criteria (category or frequency)</td><td>Maximum tolerable risk for a fatal injury</td><td></td><td>1×10^{-6}</td></tr>
<tr><td rowspan="2">Initiating Event (typically a frequency)</td><td>(1) Entry of iron, water, or any other compound incompatible with MIC process</td><td></td><td>1×10^{-1}</td></tr>
<tr><td>(2) Circulation pump failure</td><td></td><td>1×10^{-1}</td></tr>
<tr><td>Enabling Event or Condition</td><td></td><td>N/A</td><td></td></tr>
<tr><td colspan="4">Conditional Modifiers (if applicable)</td></tr>
<tr><td></td><td>Probability of ignition</td><td>N/A</td><td></td></tr>
<tr><td></td><td>Probability of personnel in affected area</td><td>1</td><td></td></tr>
<tr><td></td><td>Probability of fatal injury</td><td>1</td><td></td></tr>
<tr><td></td><td>Others</td><td>N/A</td><td></td></tr>
<tr><td colspan="2">Frequency of unmitigated Consequence</td><td></td><td>1×10^{-2}</td></tr>
<tr><td colspan="4">Independent Protection Layers</td></tr>
<tr><td></td><td>Automatic tank pressure relief system</td><td>1</td><td></td></tr>
<tr><td></td><td></td><td></td><td></td></tr>
<tr><td></td><td></td><td></td><td></td></tr>
<tr><td></td><td></td><td></td><td></td></tr>
<tr><td></td><td></td><td></td><td></td></tr>
<tr><td colspan="4">Safeguards (non-IPLs)</td></tr>
<tr><td></td><td>Refrigeration system</td><td></td><td></td></tr>
<tr><td></td><td>High temperature storage tank alarm, leading to dilution with inert liquid heat sink</td><td></td><td></td></tr>
<tr><td></td><td>Phosgene spike</td><td></td><td></td></tr>
<tr><td></td><td>Transfer into Reserve Tank (E-619)</td><td></td><td></td></tr>
<tr><td></td><td>Transfer into parallel MIC Rundown Tank</td><td></td><td></td></tr>
<tr><td></td><td>Route MIC into VGS Accumulator section</td><td></td><td></td></tr>
<tr><td></td><td></td><td></td><td></td></tr>
<tr><td colspan="2">Total PFD for all IPLs</td><td>1</td><td></td></tr>
<tr><td colspan="2">Frequency of Mitigated Consequence</td><td></td><td>1×10^{-2}</td></tr>
<tr><td colspan="4">Risk Tolerance Criteria Met? No (1×10^{-2} is greater than 1×10^{-6})</td></tr>
<tr><td colspan="4">Actions Required to Meet Risk Tolerance Criteria: Safely shut down process unit until sufficient protection can be added to meet minimum risk tolerance, under corporate management's direction.</td></tr>
<tr><td colspan="4">Notes: Process design is four layers of protection short of specified Category-5 consequence risk tolerance. Corporate management notified before ending the PHA meeting. IEF credit taken for loss of MIC Circulation Pump creates dependency with listed Safeguards, so IPL credit for all preventive Safeguards was denied. Initial calculations do not allow credit to be taken for VGS/ flare system.</td></tr>
<tr><td colspan="4">References (links to originating hazard review, PFD, P&ID, etc.):</td></tr>
<tr><td colspan="4">LOPA analyst (and team members, if applicable):</td></tr>
</table>

MIC, *methyl isocyanate;* IPL, *independent protective layer;* VGS, *vent gas scrubber;* LOPA, *layer of protection analysis;* PHA, *process hazard analysis;* PFD, *probability of failure on demand;* P&ID, *piping and instrumentation diagram;* IEF, *initiating event frequency.*

Applying LOPA rules to the analysis gets us much closer to statistical reality. Instead of estimating that a category 5 incident involving the catastrophic release of boiling MIC could occur once in every ten billion years, taking credit only for confirmed IPLs gets us down to a more accurate probability of once in every 100 years. Still, the outcome proposed in Scenario 2 did, in fact, occur within 10 years after the process was constructed. Are still too many credits being taken? Perhaps—but any final adjustment could only take place after observing factory's *actual* performance.

LESSONS FOR US

Multiple safeguards were included in the Bhopal factory's process design to prevent a scenario that could result in a catastrophic process release. All but one of these safeguards could be defeated if a single asset not typically associated with a safety purpose was to fail. Likewise, today there are many modern examples of industrial processes that have failed despite the inclusion of many safeguards designed to prevent a catastrophic release. A systematic approach to analyzing specific design hazards demonstrates the importance of integrating LOPA concepts into modern hazard analyses. Applying the same principles on modern industrial processes can prevent complete system failure in the event of a process upset. Breaking dependencies embedded within process design prevents asset malfunctions that can defeat adequate system protection. Other lessons to take away from this analysis include:

- A single asset failure can disable multiple safeguards.
- Asset reliability directly influences process safety, productivity, profitability, and human performance.
- When evaluating process performance, a credible explanation can usually be discovered for disabling safeguards prior to an incident.
- True cases of double jeopardy are extremely rare. All instances must be fully examined to prevent disregarding a scenario that could be both entirely possible and highly probable.
- Incorporating LOPA elements into the HAZOP program justifies adding redundancy where needed to achieve a satisfactory level of safety performance.

REFERENCES

[1] District Court of Bhopal, India, State of Madhya Pradesh Through CBI vs. Warren Anderson & Others, Criminal Case No. 8460 of 1996, 36, June 7, 2010.
[2] T.R. Chouhan, The unfolding of Bhopal disaster, J. Loss Prev. Proc. Ind. 18 (2005) 207.
[3] S. Varadarajan, et al., Report on Scientific Studies on the Factors Related to Bhopal Toxic Gas Leakage, 77, 1985.
[4] R. Van Mynen, Union Carbide Corporation Press Conference Transcript, March 20, 1985, 4.

[5] T.R. Chouhan, Bhopal: The Inside Story, Carbide Workers Speak Out on the World's Worst Industrial Disaster, 45, 1994, ISBN: 0-945257-22-8.
[6] District Court of Bhopal, India, State of Madhya Pradesh Through CBI vs. Warren Anderson & Others, Criminal Case No. 8460 of 1996, 14, June 7, 2010.
[7] Center for Chemical Process Safety (CCPS), Layer of Protection Analysis: Simplified Process Risk Assessment, 80, 2001, ISBN: 0-8169-0811-7.
[8] Center for Chemical Process Safety (CCPS), Layer of Protection Analysis: Simplified Process Risk Assessment, 135, 2001, ISBN: 0-8169-0811-7.
[9] Center for Chemical Process Safety (CCPS), Layer of Protection Analysis: Simplified Process Risk Assessment, 118, 2001, ISBN: 0-8169-0811-7.
[10] Union Carbide Corporation, Review of MIC Production at the Union Carbide Corporation Facility Institute West Virginia, April 15, 1985, 1-3.
[11] Center for Chemical Process Safety (CCPS), Layer of Protection Analysis: Simplified Process Risk Assessment, 2001, ISBN: 0-8169-0811-7, pp. 63–74.
[12] Center for Chemical Process Safety (CCPS), Layer of Protection Analysis: Simplified Process Risk Assessment, 2001, ISBN: 0-8169-0811-7, pp. 75–113.
[13] Center for Chemical Process Safety (CCPS), Layer of Protection Analysis: Simplified Process Risk Assessment, 115, 2001, ISBN: 0-8169-0811-7.
[14] Center for Chemical Process Safety (CCPS), Layer of Protection Analysis: Simplified Process Risk Assessment, 96, 2001, ISBN: 0-8169-0811-7.
[15] Center for Chemical Process Safety (CCPS), Layer of Protection Analysis: Simplified Process Risk Assessment, 106, 2001, ISBN: 0-8169-0811-7.
[16] Center for Chemical Process Safety (CCPS), Layer of Protection Analysis: Simplified Process Risk Assessment, 88, 2001, ISBN: 0-8169-0811-7.
[17] Center for Chemical Process Safety (CCPS), Layer of Protection Analysis: Simplified Process Risk Assessment, 81, 2001, ISBN: 0-8169-0811-7.
[18] Center for Chemical Process Safety (CCPS), Layer of Protection Analysis: Simplified Process Risk Assessment, 83, 2001, ISBN: 0-8169-0811-7.
[19] Center for Chemical Process Safety (CCPS), Layer of Protection Analysis: Simplified Process Risk Assessment, 78, 2001, ISBN: 0-8169-0811-7.
[20] Center for Chemical Process Safety (CCPS), Layer of Protection Analysis: Simplified Process Risk Assessment, 89, 2001, ISBN: 0-8169-0811-7.
[21] S. Berger, Status of AIChE initiatives to promote effective management of chemical reactivity hazards, in: 3rd International Symposium on Runaway Reactions, Pressure Relief Design, and Effluent Handling, Cincinnati, OH, November 1–3, 2005, 2.
[22] Center for Chemical Process Safety (CCPS), Layer of Protection Analysis: Simplified Process Risk Assessment, 102, 2001, ISBN: 0-8169-0811-7.
[23] J.W. Chastain, S.A. Urbanik, R.W. Johnson, J.F. Murphy, New CCPS guideline book: guidelines for enabling conditions and conditional modifizers in LOPA, in: 9th Global Congress on Process Safety, San Antonio, TX, April 28–May 1, 2013, 11.
[24] Center for Chemical Process Safety (CCPS), Layer of Protection Analysis: Simplified Process Risk Assessment, 103, 2001, ISBN: 0-8169-0811-7.
[25] Center for Chemical Process Safety (CCPS), Layer of Protection Analysis: Simplified Process Risk Assessment, 2001, ISBN: 0-8169-0811-7, pp. 103–104.
[26] Center for Chemical Process Safety (CCPS), Layer of Protection Analysis: Simplified Process Risk Assessment, 82, 2001, ISBN: 0-8169-0811-7.

CHAPTER

Commissioning Phase* 14

EVENT DESCRIPTION

The factory construction project was completed in 1979 [1] and the commissioning phase officially started. Within days a problem surfaced that would undermine factory operation until production was permanently shut down on December 3, 1984. The problem involved a chronic series of methyl isocyanate (MIC) pump mechanical seal failures [2]; these pumps were located about 30 m from the factory's occupied control room .

A mechanical seal lasting 45 months would be considered "average" (not great) performance for a typical pump in the chemical manufacturing industry [3]. In comparison, an MIC pump seal would last no more than 24 days at the Bhopal factory. With a total of five MIC pumps running simultaneously, MIC leaks were occurring at an alarming rate [4]. The workers were responding to MIC pump leaks about once every 5 days.

Parts and labor costs to manage repeat MIC pump failures at the Bhopal factory proved to be extraordinarily penalizing. Factory workers toiled away at the process to keep up with the additional maintenance. But despite the effort to stay in front of the problem, the high failure rate and its corresponding mean time to repair (MTTR) took a devastating toll on factory output. Replacing an MIC pump seal was a costly undertaking that required temporarily switching to the alternate rundown tank (chapter: Factory Construction). If corrective maintenance was overlapping on both tanks at the same time, then factory production would stop. MIC stagnating inside a rundown tank was worthless. For it to be of any value whatsoever it had to be transferred into the derivatives section. For production to continue, at least one of the two transfer pumps had to be available.

Interrupting production for unplanned maintenance was bad for business. With the expiration of the technical grade product's patent 7 years earlier, loyal customers would definitely begin looking for alternative suppliers before doing without the product they desperately needed for agricultural production.

Although the economic penalty resulting from MIC pump seal failures was hard to ignore, a more serious situation was developing unnoticed by those affected by it. Before long, workers learned that they could respond to MIC leaks in the pump area without experiencing any immediate complications. As a result, MIC leaks really did not bother them. In their minds, MIC was nothing like its toxic counterpart,

* For timeline events corresponding to this chapter see pages 432–433 in Appendix.

Rethinking Bhopal.

phosgene. It was simply a lung and eye irritant [6], which made it easy to detect when it leaked [5]. Over a relatively short period of time, MIC pump seal leaks had become nothing more than a component of routine factory operation. If the factory was in the business of manufacturing technical grade product, then MIC seal leaks were an aspect of production that needed to be managed. In these terms, MIC pump seal failures were no different from any other operating difficulty that might be encountered in the Bhopal factory.

However, the problem was not only affecting the staff working inside the factory. The community living outside the factory was equally affected by the MIC leaks. Chest and eye irritation soon became a normal part of life for those living in the outlying area around the factory [7]. Shortly after these familiar physical symptoms appeared, they would vanish without explanation. Hidden behind the walls of the factory, workers were isolating production assets to control the problem. Due to the predictable frequency of the MIC pump seal failures, the community living near the factory soon learned what to expect. They knew both when to expect a release and that it would soon go away. Time after time they experienced the same routine. Much like the workers inside the factory, those dwelling outside the gate were conditioned by the same problem.

The asset reliability problem proved to be extraordinarily difficult to solve. Indeed, there was no immediate solution. Unfortunately, nobody could explain why the MIC pump seals kept failing. But the failures and leaks did, in fact, continue happening [8]. Initial complaints about the frequency of MIC pump failures were not taken seriously. After all, the purpose for a commissioning phase is to address these types of problems. The troubleshooting effort was far from complete. Eventually the problem would have to be solved. Complaining about it prematurely would certainly not hasten the implementation of an effective solution. On this point everyone could agree.

According to the fundamental principles of pump hydraulics, the MIC pumps should have been stable. At least there was no scientific basis to expect unstable hydraulics. Perhaps the problem involved an unidentified design defect that affected all five pumps equally? This did not make much sense either since the seal in place was of a conventional design that should have worked fine [9]. Additionally, the pumps consisted of materials compatible with the MIC process in accordance with design specifications. This did not, however, change the fact that MIC pump seals failures were occurring frequently and for no apparent reason.

But thankfully, when a mechanical seal *did* leak the release was relatively minor [10]. Perhaps it was even just a drip that provided indication that something had to be done to prevent an almost guaranteed catastrophic seal failure. Even if the failure was significant, however, the consequences were limited to a liquid spill inside a containment area—a “near miss” at the most. The fugitive process would simply evaporate in a relatively short period of time after shutting off and isolating the leaking MIC pump. Factory economics and potential loss of market share were by far the most immediate concerns related to this failure mechanism. Nobody liked the personal effects of recurring MIC leaks, but the chemical exposure issue was minor compared with the high cost of unplanned maintenance.

Although MIC pump seal failures were quite regular, the integrated factory’s inaugural year passed with no major injuries or fatalities. The potential safety hazard appeared to be completely under control. Apart from the enormous costs related to production losses and maintenance (parts and labor) for managing MIC pump mechanical seal failures, the hazard was tolerable. However, the circulation pump failures were creating an entirely different set of secondary issues that too went unnoticed. Or perhaps it is more accurate to say that the source of these secondary issues was hard not to notice. This took the mystery out of the observed cause-and-effect when an MIC circulation pump failed.

Circulation pump failures would create irregular readings on otherwise-functional process instruments. For example, when a circulation pump failed on any of the three storage tanks, the temperature would closely track outside (ambient) conditions. Many times, the outside temperature would climb above the maximum range on the temperature gauges. As a result, the gauge would fail "high-range" and would not provide accurate temperature readings above +25°C [11]. Without the circulation pumps, the temperature gauge readings were useless and irrelevant. There was really no reason to monitor the MIC temperature gauges when a circulation pump failed.

Within a short period of time, meaningless information became normal for MIC tank temperature gauges inside the Bhopal Factory. Why? Because the circulation pump frequently failed. The circulation pump fed MIC through the refrigeration system. With no MIC circulation, the tank contents could not be refrigerated. The only way to remedy the situation was to restore MIC refrigeration by repairing the circulation pump. At the same time, preparations would have to be made to replace the seal and watch the temperature gauges fail again within the next 24 days. The same cycle repeated itself over and over.

Similarly, the storage tanks were equipped with high-temperature alarms that would make an annoying noise above +11°C. During most of the year the ambient temperature in the geographic area where the factory was located was above the alarm set point [11]. Whenever a circulation pump failed, the temperature alarm would stay on. This situation would create a distraction inside the control room. Therefore, the temperature alarms were disabled early in the factory's life [12]—likely before the commissioning phase ended after about one year had passed without being able to correct the situation.

To reemphasize: With knowledge about the source of the problem and no way to control it unless the pump was repaired, the alarm served no useful purpose. Therefore, it was disabled. This extreme action at least returned peace to the control room. Under the circumstances, there was no need to continue monitoring the storage tank temperatures either. With refrigeration disabled for extended periods of time the temperature would remain high. There really was no questioning why the MIC storage tank temperature was elevated when an MIC circulation pump failed. High temperatures were expected under those conditions. Therefore, the workers stopped recording temperature readings [12].

When the commissioning period ended a year later on February 5, 1980 [13], the Bhopal factory had only manufactured 1534 MT of technical grade product [14]. This total production volume represented only about one-third of the factory's designated output capacity. The factory's performance was limited by the production constraint that resulted from persistent mechanical seal failures. Unfortunately, the problem was not corrected prior to handing production responsibilities over to the operations department. Any solutions therefore would be forthcoming only after the commissioning phase ended.

TECHNICAL ASSESSMENT

Chances are that few—if any—factories ever built in the history of the manufacturing industry have performed exactly as expected upon initial feed introduction. The Bhopal factory was no different in this regard. For this specific reason it is not only customary, but also expected for a "break-in" period to immediately follow the final phase of factory construction. This "grace period" allows appropriate adjustments to be made outside of the typical production pressures that dominate factory performance upon handing the process over to operations. Depending on the complexity of the issues encountered during the commissioning phase, it might only take a few days to verify that the process is ready for production. In extreme cases where serious deficiencies exist, the commissioning phase can last considerably longer—perhaps even up to a full year. After that, any patience exercised up to that point typically runs out. If production is at all possible when the commissioning phase ends, then any remaining operating issues are usually addressed online. The time invested while commissioning the process is intended to create the confidence needed to achieve an acceptable level of safety, environmental, reliability, and production performance. The length of time it took to commission the Bhopal factory, coupled with its low-production volume, indicates that the process was crippled by a chronic failure mechanism upon start-up. Indeed, from the very beginning factory throughput was limited by a persistent asset reliability issue with serious, widespread implications.

Some of these implications included the interruption of both MIC sampling and tank mixing. The MIC sample point was on the circulation pump loop. Therefore, the quality of the MIC inside a storage tank could not be checked when a circulation pump was removed from service for maintenance. No other provisions were available for safely obtaining an MIC sample from another location [15].

Notice how the misalignment of priorities observed in this portion of the Bhopal disaster's timeline corresponds to a pattern encountered earlier (chapter: Process Selection). A production constraint developed in this section of time soon after the commissioning phase started. The constraint resulted from a chronic mechanical seal failure mechanism that affected all five MIC pumps equally. The production constraint demanded the full attention of factory workers from the moment that feed was introduced to the process. Through direct experience, the workers figured out that liquid MIC leaks represented little threat to their personal health and safety. Wearing basic personal protective equipment (PPE) and applying simple hazard management practices was adequate protection for preventing an incident that would have to be reported internally. This shortcut allowed the workers to immediately attend to the process without fearing harm or discipline [16]. This helped them expedite the repair process and minimize leakage-related MIC losses.

Management would have found it difficult to administer discipline in this case, since the workers' actions were directed at enhancing production. Disregarding the PPE policy would limit the loss of MIC prior to isolating the process, and would also reduce the MTTR. Under the circumstances, the workers would more likely have been rewarded than reprimanded for their self-sacrificing actions. On a system so prone to failure, nothing but brute force could sustain production. Still, this is exactly the situation where discipline (in a punitive sense) is needed the most—before an incident creates a double-standard with respect to the constructive use of discipline. Well-directed punitive discipline is an essential component of operational discipline. The impact of the attitude demonstrated here, however, produces a negative effect as we shall see in Chapter "Process Isolation and Containment."

Compare the prioritization of hazards demonstrated here with earlier events involving the handling of phosgene and MIC at the factory. In that case we saw the same inconsistency in the way that two possibly dangerous products were being managed. In the case discussed in this chapter, we find an inconsistency in how production and safety commitments are prioritized. On the hierarchy of industrial priorities, safety is always "number one"—at the top on the list [5]. Yet, in practice, it is not at all uncommon to uncover situations like the one described inside the Bhopal factory involving MIC pump leaks, where safety took the back seat to production.

The relative degree of pain is what normally directs the setting of priorities under any circumstances. This relationship is not limited only to the manufacturing industry. If a situation is more painful than another, then the more painful scenario is the one that floats to the top. Anyone trying to remind someone else about the "spoken" priority message runs the risk of being immediately shut down. Where the pain is, attention usually follows. Upon recognizing this tendency, resourceful supervisors make a special effort to remain consistent on matters between spoken words and demonstrated actions.

The production constraint resulting from repeat MIC pump seal failures was an immediate source of economic pain. Frequent exposure to the leaking process may have produced temporary discomfort, but it was not even close to the acute pain that the production constraint created due to low MIC pump availability. Serious factory production losses were the major concern after start-up. It was clear to everyone that if production could not be recovered, then the business might not be able to survive. Indeed, the workers proved multiple times over that intimate contact with liquid MIC was possible without suffering the type of consequences that would be expected upon intimate contact with phosgene gas. Therefore, workers and maintenance crews could focus on keeping at least one rundown tank in service at all times. They became used to servicing the MIC process without taking the same precautions that were necessary with phosgene system maintenance.

If maintenance was overlapping on both rundown tanks, then factory production would stop until a transfer pump could be made available. With each MIC pump experiencing an MTBF of about 24 days, repairs were occurring at a rate of about 75 times more than what was needed if the process could operate at a more acceptable 5-year turnaround cycle. At the actual repair rate, the cost for additional maintenance (parts and labor) was neither realistic nor sustainable. More people were needed to interact with the uncooperative process. More staff only added to the budget imbalance. To make things worse, the repeat failures were causing a production constraint. Not only did breakeven costs escalate in response to excessive MIC pump repairs, but breakeven production volume was impossible due to the excessive downtime [13]. As a result, the factory began losing money from the very start, before the commissioning phase ended.

Because the immediate economic penalty incurred by an MIC pump seal failure was more damaging than the chronic process safety issue, another problem was rapidly developing. The leaks were becoming normal to the workers. But the workers were not the only ones growing used to the factory's performance in this regard. Any MIC released to the atmosphere was an environmental nuisance that created an off-site impact as well. The community surrounding the factory also came to expect leaks and unexplained smells on a regular basis. These odors would soon subside without explanation, and life would go on as normal [17]. Therefore, random odors from the factory were a common occurrence to the outlying community. After a short period of time, familiar MIC odors were an expected part of living near the factory. Again, there were no dire consequences involved with MIC pump leaks either inside or outside the factory.

It is important to recognize the impact that routine factory performance was having on the outside community. In later chapters we are confronted with information that does not make sense regarding how the community responded to workers' concerns and more direct warning signs of an impending disaster. In the context of the repeat MIC pump seal failures, these mysterious community actions and decisions make sense.

The conditioning process described here represents the normalization of deviance. This dangerous pattern has developed into a topic of serious process safety concern numerous times in the history of the manufacturing industry. We can observe it starting here, in the initial phase of factory production. Similar to the space shuttle explosions, the normalization of deviance presented here relates to an asset reliability issue. In the context of the normalization of deviance, mechanical seal failures, O-ring blow-by incidents, and foam strikes on the leading edge of a wing are all equal. They all represent chronic failure mechanisms that became a normal part of asset operation through a conditioning process that resulted in the normalization of deviance.

Along those lines, the storage tank instrument and gauge failures we referenced earlier were also accepted as normal operation in a relatively short period of time at the Bhopal factory. This is understandable in consideration of the type of failure mechanism that limited production during the first year of factory operation. Any circulation pump failure would cause the corresponding MIC storage tank's temperature gauges and alarms to fail. If the ambient temperature increased above +11°C, then the constantly-sounding alarm would be of no value. If the ambient temperature crept further above +25°C, then not only would the high temperature alarm already be active (and doing no good), but the temperature gauges would also fail. This relationship seriously limited process monitoring sensitivity. No longer could minor temperature deviations provide advance warning of a potentially dangerous situation developing inside the storage tanks. With the gauges and transmitters no longer functioning as intended by design, only a more direct warning signal could evoke a human response. This direct signal could only involve something happening outside the tanks, perhaps long after recovery was a practical option.

The circulation pump was critical in this aspect. Accurate temperature sensing required that the circulation pump had to remain operational at all times. However, the MIC storage tank high-temperature alarms were disabled very early into the Bhopal factory's life in order to eliminate the nuisance alarm [18]. This is understandable since a continuously failed (active) alarm could provide absolutely no benefit if an upset was to occur inside the tank. After all, if a circulation pump failed then the reason for the high temperature alarm was well-known:

1. A circulation pump had failed.
2. The circulation pump that failed had to be shut down and isolated to stop the MIC leak.
3. The stagnant storage tank contents were not being refrigerated.
4. The only way to get the alarm to clear was to cool the tank contents.
5. Cooling the tank contents required repairing the circulation pump.
6. Repairing the pump was of temporary benefit since the new seal would fail again within 24 days of being replaced.
7. The cycle would repeat every time that a circulation pump failed.

This sequence of events explains why so much attention was devoted to repairing the pumps. Here we see a potentially very dangerous situation evolving. If a circulation pump was to fail, then not only would the MIC temperature rise, but there would be no way to detect an unsafe operating condition inside the storage tanks. High temperature inside the tank was normal—to be expected. If an unsafe situation was detected in the tank, then there would be no way to access the MIC reject line (chapter: Process Hazard Awareness and Analysis).

In hindsight, the decision to disable the MIC storage tank high-temperature alarms might rightfully be criticized. But again, in the context of the recurring MIC circulation pump failures the high temperature alarms served absolutely no purpose other than to annoy anyone assigned to control room duties. Disabling the alarms was inevitable. The only solution was to address the chronic failure mechanism active at all five MIC pumps so that silence, according to design, was the norm. Allowing the pumps to operate continuously without shutting down for unplanned maintenance was the only way for the process to function in accordance with the process design basis. If this was not possible, then listening to the constant alarms offered absolutely no protection. Listening to a constant alarm did not make it possible to detect an unsafe condition inside the storage tanks. It was the same as if there was no alarm at all; thus, it was purposefully disabled and not simply increased to a higher temperature to make it stop [19]. The action taken, which involved disabling a safeguard, was a vivid demonstration of the normalization of deviance. High temperature inside the MIC tanks was so frequent and uneventful that it had become acceptable.

With regard to chronic MIC pump seal failures, the public record does not contain specific details about the failure mechanism. Based on the persistence of the problem and actions that will be discussed in greater detail later, it is likely that the problem was very difficult to solve. Chapter "Repeat Mechanical Seal Failures: A Case Study in Pump Reliability Improvement" documents lessons learned from a case history involving a chronic pump mechanical seal failure that defied all reasonable explanation for a prolonged period of time. A considerable amount of time and patience were needed to solve the problem. This information can be used to diagnose a specific cause for unexplained, repeat seal failures. In that case, similar to this case, the normalization of deviance resulted in a series of repeat failures. Once established, the normalization of deviance makes it impossible to distinguish right from wrong. Responsible process operation requires insisting that repeat failures be eliminated. Information made available about improving mechanical seal reliability assists in eliminating persistent repeat failures that can cause considerable harm—both physically and mentally.

In modern history, we observe how chronic failures involving the normalization of deviance ultimately end. Based on this assessment, we understand why the same fate would eventually apply to the manufacturing process constructed in Bhopal, India. In all cases where the normalization of deviance evolves, bad things happen. It must therefore be avoided at all costs. If the way you are managing a process is hazardous, do not believe that you can get away with mismanaging it without consequence permanently. Your decision will eventually impact both you and those surrounding you.

While on the topic of protecting those around you, the Bhopal factory provides the ultimate example of involuntary community participation. While the normalization of deviance had taken root inside the factory, it also branched out into the area surrounding the factory. We are gifted with an ability to sense and respond to harm. It is only normal to expect your neighbors to look out for your best interest and protect you, but personal preservation is always more reliable. The fact remains that the process releases inside the Bhopal factory were detectable by people living outside the factory.

In the refining industry, it is much the same with hydrogen sulfide (H_2S). H_2S is a toxic gas that claims the lives of multiple victims in the global manufacturing industry each year. The compound is a substance with a distinctive rotten egg smell. For that reason, it is very hard to ignore when it is present at low concentrations; this would be the situation as someone approaches a leak. When the odor is detected, immediate action is required. Protect yourself by moving in the upwind or crosswind direction or use a self-contained breathing apparatus until the leak source is identified and isolated.

Repeated exposure to any harmful substance without consequence creates a false sense of security. People in the community who are subjected to regular release incidents from a nearby factory may lose their fear of the process, similar to how these events also affect those working inside a factory. The normalization of deviance defeats our sensory perception of potential harm. When we detect a familiar occurrence, instead of taking immediate action to protect ourselves and others, we might wait to see what happens. In the case of an H_2S or MIC leak, waiting to see what happens is waiting too long. The point is that there is never an acceptable reason to allow a problem to persist without resolving it. Our ability to detect danger and take immediate action requires not growing accustomed to a potentially unsafe, but recurring, operating condition.

Since the community outside the Bhopal factory had lost their fear of the process, only alarms and sirens could offer protection in the event of a more serious process release.

Alarms and sirens, however, are controlled by individuals whose personal judgment might also be influenced by the normalization of deviance. These public warning devices are only available if they are independent; that is, they are not disabled by someone conditioned to accept the hazards related to a specific process. Unfortunately, those in control of the public warning signals that could be activated from inside the Bhopal factory's control room were those who felt comfortable disabling the MIC storage tank high-temperature alarms. This connection needs to be made in order to explain the context of similar actions that would take place later.

It is again important to note that many problems in the Bhopal factory were passed along to operations at the end of the commissioning period. MIC leaks, loss of storage tank mixing, disabled alarms and gauges, excessive maintenance costs, and tank sampling interruptions were just a few of the difficulties that were recognized before the commissioning phase ended. However, the most immediate problem was the devastating production constraint created by repeating MIC transfer pump mechanical seal failures. Due to this specific problem, the factory was not profitable and would have to find a way to increase production to at least break even financially. This would require eliminating the production constraint.

In the events discussed earlier in the timeline, we observed how problems were solved by introducing changes. The only change that could bring this case to a satisfactory conclusion would involve adjustments to increase the MIC pump MTBF. However, looking at the precedent set in the way previous problems were managed, we anticipate specific actions in agreement with a pattern. In other words it would be reasonable to predict that the problem with the MIC pumps would likely be addressed with an improvised solution. As we have already observed, improvised solutions introduce new hazards.

LESSONS FOR US

Asset reliability problems encountered during the commissioning phase result from unintentional process configuration, asset design, construction, or operating defects. Unless these defects are properly diagnosed and corrected, system performance will continue to suffer. If production constraints are the immediate impact, then the normalization of deviance creates insensitivities to the less consequential health and safety hazards. However, these issues cannot be ignored indefinitely. Even greater hardships can be expected when a production constraint creates a mismatch in business priorities. Other lessons resulting from this analysis include:

- Solving a process reliability problem requires first figuring out what is wrong, and then taking decisive action to correct it.
- Reject repeat failures. Repeat failures are urgent warning signs. Even minor incidents can create serious complexities both inside and outside the factory.
- Communities can be conditioned to accept factory performance. They must be conditioned to expect the best, which requires continuous process containment.

- Reliability problems draw more people into a potentially hazardous working environment.
- Well-directed punitive discipline is an essential component of operational discipline.
- Overstaffing can be used to temporarily manage a process reliability issue, but will prove uneconomical in the long run.
- Hiring more people to cope with an asset reliability problem magnifies the exposure hazard when the process escapes.
- The commissioning phase must proceed according to a predefined schedule to avoid tolerating an unacceptable situation. Tolerance leads to the normalization of deviance.
- There is always a credible basis for voluntarily disabling a safeguard, yet in hindsight there are often complications.
- Manual warning devices are dependent on the judgment of those who control the devices.
- Seek to define the scientific explanation for why an asset is not performing according to its designated life cycle expectation instead of resorting to improvised solutions.
- Dedicated investigation resources demonstrate their value by constantly removing the most limiting production constraint. The practice of removing production constraints is inextricably linked to fully examining and understanding all possible consequences.

REFERENCES

[1] District Court of Bhopal, State of Madhya Pradesh Through CBI vs. Warren Anderson & Others, Criminal Case No. 8460 of 1996, 18, June 7, 2010.
[2] Supreme Court of India, Criminal Appeal Nos. 1672, 1673, 1674 and 1675 of 1996, in the Matter of Keshub Mahindra vs. State of Madhya Pradesh, September 13, 1996. 14.
[3] H.P. Bloch, Pump Wisdom: Problem Solving for Operators and Specialists, 185, 2011.
[4] Supreme Court of India Criminal Appellate Jurisdiction, Application for Directions to Institute Charges U/S 302 (For Offence U/S 300 (4)) Read With S. 35 of the Indian Penal Code, 1860 Against the Respondents Herein, Curative Petition (Criminal) No. 39-42 of 2010 in Criminal Appeal No. 1672-75 of 1996 in the Matter of Central Bureau of Investigation Versus Keshub Mahindra & Ors., 8, 2011.
[5] S. Diamond, The Bhopal Disaster: How It Happened, The New York Times, January 28, 1985.
[6] R. Reinhold, Disaster in Bhopal: Where Does the Blame Lie? The New York Times, January 31, 1985.
[7] A. Agarwal, S. Narain, The Bhopal Disaster, State of India's Environment 1984–85: The Second Citizens' Report, 219, 1985.
[8] Supreme Court of India Criminal Appellate Jurisdiction, Application for Directions to Institute Charges U/S 302 (For Offence U/S 300 (4)) Read With S. 35 of the Indian Penal Code, 1860 Against the Respondents Herein, Curative Petition (Criminal) No. 39-42 of 2010 in Criminal Appeal No. 1672-75 of 1996 in the Matter of Central Bureau of Investigation Versus Keshub Mahindra & Ors., 50, 2011.

[9] T. D'Silva, The Black Box of Bhopal, 73, 2006, ISBN: 978-1-4120-8412-3.
[10] D. Lapierre, J. Moro, Five Past Midnight in Bhopal, 177, 2002, ISBN: 0-446-53088-3.
[11] Supreme Court of India, Criminal Appeal Nos. 1672, 1673, 1674 and 1675 of 1996, in the Matter of Keshub Mahindra vs. State of Madhya Pradesh, September 13, 1996. 16.
[12] T.R. Chouhan, Bhopal: The Inside Story, Carbide Workers Speak Out on the World's Worst Industrial Disaster, 33, 1994, ISBN: 0-945257-22-8.
[13] T. D'Silva, The Black Box of Bhopal, 58, 2006, ISBN: 978-1-4120-8412-3.
[14] National Environmental Engineering Research Institute (NEERI), Assessment and Remediation of Hazardous Waste Contaminated Areas in and around M/s Union Carbide India Ltd., 4, 2010.
[15] T.R. Chouhan, The unfolding of Bhopal disaster, J. Loss Prev. Proc. Ind. 18 (2005) 206.
[16] District Court of Bhopal, India, State of Madhya Pradesh Through CBI vs. Warren Anderson & Others, Criminal Case No. 8460 of 1996, 65, June 7, 2010.
[17] A. Agarwal, S. Narain, The Bhopal Disaster, State of India's Environment 1984–85: The Second Citizens' Report, 206, 1985.
[18] T.R. Chouhan, The unfolding of Bhopal disaster, J. Loss Prev. Proc. Ind. 18 (2005) 207.
[19] Union Carbide Corporation, Bhopal Methyl Isocyanate Incident Investigation Team Report, March 1985. 23.

CHAPTER

Repeat Mechanical Seal Failures: A Case Study in Pump Reliability Improvement

15

ACCELERATION BASICS

"Vibration" is a catchall term used to describe incipient failure of high-speed rotating machinery. Indeed, vibration increases the load acting on bearings and thereby causes many asset malfunctions and process releases. However, simply diagnosing vibration as the root cause for machinery failure does nothing to prevent the same failure from happening again. The simple fact is that all vibration is caused by something. Unless the specific cause for vibration is identified, a vibration problem can wreak havoc on all types of industrial assets. Misdiagnosed vibration problems are responsible for many repeat failures within the modern manufacturing industry.

But defining and controlling the specific cause for excessive vibration can be difficult, although a host of predictive monitoring tools is available to the user industry. Suitable tools can readily be obtained to both monitor and diagnose vibration excursions. Software-based continuous or intermittently scanning programs are feasible; they are especially useful for evaluating large amounts of process data. However, persistent vibration problems are still common even in cases where monitoring programs are regularly applied. After years of fighting a chronic vibration problem, we might get comfortable with it and decide to simply give up. We might become receptive to the idea of admitting defeat and accepting the problem because it may appear as nothing more than an economic inconvenience. However, in choosing to live with a chronic asset reliability problem (such as vibration), we must also accept that we now incur greater risk. This modified approach to managing the business (by merely coping with an unsolved problem) requires becoming insensitive to additional process hazards.

Coping with a chronic asset reliability problem is financially and emotionally destructive. Those who give up on finding answers might even find it gratifying to observe an uncooperative process responding to sheer determination, muscle, and sweat. In many cases, it is possible to coexist with an unruly and unmaintainable process—perhaps only for a short amount of time. This is especially true if the perceived downside is limited to discretionary maintenance costs.

Therein lie the seeds of disaster; it is now only a matter of time before the dragon we have struggled so hard to contain loses its fear of the whip. A number of adverse circumstances converge, and deviations combine and overpower our ability to control

Rethinking Bhopal.

the failure mechanism that we accepted earlier. At that point the reality of a serious process safety incident becomes abundantly obvious. Maybe circumstances give us one more chance to solve the problem—or maybe not. Either way, we contend that it is essential to remember that all failures have a cause. For that very reason it is unacceptable to give up under any circumstances. Interrupting the search for credible answers to a chronic reliability problem is a guaranteed way to lose control of the process.

Ignoring an unexplained machinery vibration phenomenon serves as one of many dangerous examples. Controlling a vibration problem requires determining what is causing it. A formal investigation might be required. Depending on the actual or potential safety, environmental, and economic harm produced by tolerating the problem, an investigation can begin paying dividends from the moment it starts.

DETECTING AND INVESTIGATING A VIBRATION PROBLEM

Diagnosing the specific cause for a chronic vibration problem requires a basic understanding of relevant parameters. Basic vibration monitoring often looks at the displacement of a machine component with respect to time. A relatively large machine operating at 1000 rpm and exhibiting a bearing housing displacement of 0.0002 in (=0.0098 in/s peak velocity) would be considered smooth, but the same bearing housing displacement at 30,000 rpm (0.314 in/s peak velocity) would be labeled rough. The displacement and vibratory movement of the rotor is in two or sometimes three directions (horizontal, vertical, axial), and its amplitude can range from inconsequential to severely-damaging. Speed thus plays a role in the matter, as do discontinuities in, say, a rotating mass. If a rotating impeller has seven vanes and rotates at 60 cycles per second (cps), the seven "discontinuities" will contribute an amplitude excursion at a frequency of 420 cps. If an unbalanced mass is fastened to a location near the impeller tip, it will create an amplitude at once and (likely) twice-per-revolution frequencies. Vibration velocity is a useful monitoring parameter, which captures such mass reversals. But velocity transducers are not very useful at extremely high frequencies. They will not respond fast enough to tell us that we have, say, a million vapor bubbles collapsing within an impeller during a single revolution of the pump shaft.

Elementary physics explains the basic relationship between force (F), mass (m), and acceleration (a). $F = ma$ would be an expression of interest because as an object vibrates, it is ultimately the resulting force that causes component distress. In any particular machine the moving mechanical mass is a fixed quantity; discontinuities or mass unbalances will cause certain displacements, vibration amplitudes, and velocities. Their significance has to do with available bearing clearances and the vibratory frequencies found in the vibration amplitude versus frequency spectrum. But recall that while mechanical issues can often be spotted by looking at the two primary parameters amplitude and frequency, an accelerometer is needed to sense or display the very high frequencies caused by millions of bubbles collapsing in the time it takes for a rotor to make a single revolution.

An accelerometer measures acceleration (in meters per seconds squared), which is the derivative of velocity with respect to time. Therefore, velocity (in meters per second) and position (in meters) can both be determined by simply measuring acceleration. That is why industry widely uses accelerometers to monitor machinery vibration. Together, these vibration monitoring inputs help us to diagnose the specific causes for asset malfunctions. We relate m/s^2 acceleration to the g-force (in units of gravity or "g") produced by acceleration due to the Earth's gravitational pull [1] at $9.8\,m/s^2$. Therefore, acceleration measured at $19.6\,m/s^2$ produces 2 g's.

ACCELERATION LIMIT

It can be inferred from this discussion that vibration displacement and velocity limits are usually administered to prevent a sudden failure that could wreck a fluid machine. For example, if the vibration-induced displacement of a machine component takes up the gap between shaft surface and bearing bore, there might no longer be an oil film. This could then lead to a compromised machine and its progressive degradation would likely cause an unstable operating condition. Although there is no industry consensus on what constitutes "acceptable" acceleration, industry agrees that large acceleration increases indicate nonspecific changes in mechanical or process operating conditions [2]. This was the situation observed at a refinery pump installation where a catastrophic failure came on the heels of a significant increase in vibration acceleration. The incident occurred before the maximum vibration velocity limit was exceeded. Had the maximum vibration velocity limit been exceeded, the asset would have been immediately shut down to avoid a catastrophic failure.

In response to this incident, a discretionary (voluntary) vibration acceleration limit of 20 g's was established to prevent future incidents. Site supervision arbitrarily selected the maximum 20 g limit based on the sequence of events leading up to the incident. Prior to the failure that resulted in a catastrophic process release, acceleration was recorded rapidly transitioning above 20 g's. The purpose for establishing an acceleration limit was to:

1. Increase the P–F interval (chapter: Equipment Reliability Principles) to coordinate an effective repair plan prior to experiencing a catastrophic failure.
2. Implement a more conservative approach to containing the process.
3. Incorporate lessons learned from a previous incident involving a catastrophic mechanical seal failure in a service that was operating above autoignition temperature.

After the acceleration limit was imposed, 20 pumps (roughly 1% of the affected population) received immediate attention. None of these pumps were found operating in accordance with the 20 g maximum acceleration limit at the time that the limit was applied. All but one of these assets were brought into compliance either by making a simple process adjustment or through a logical mechanical upgrade (new rolling elements, for example.) The exception proved to be particularly frustrating.

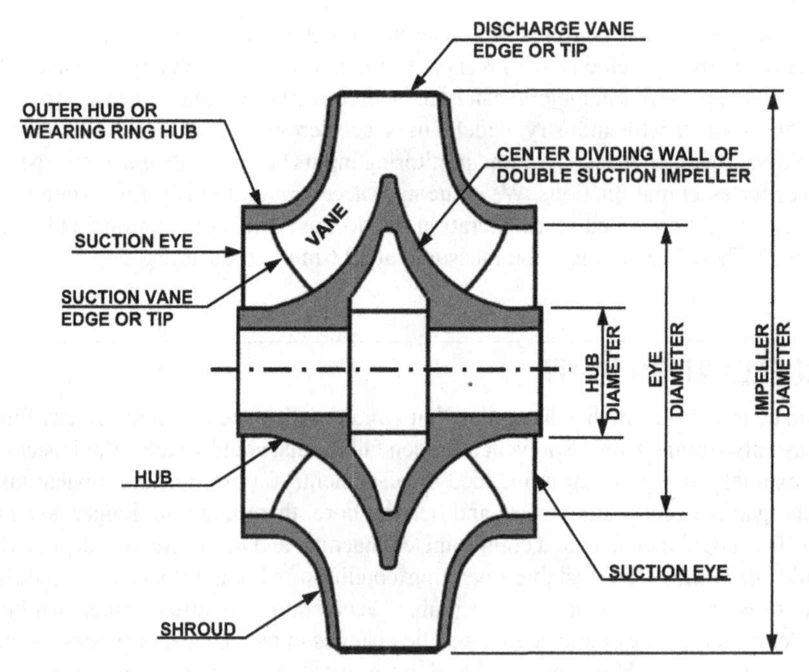

FIGURE 15.1

Double-suction pump impeller cross-section.

Acceleration at one high-speed (3600 rpm) double-suction pump (Fig. 15.1) location remained above 20 g after initial attempts to comply with the new limit failed. To supervision's credit, the technical staff's initial request to issue an exception that would allow operating the uncooperative pump above the 20 g limit was rejected. Indeed, the volume of nonspecific process and mechanical scenarios capable of causing high vibration acceleration was virtually unlimited. Furthermore, in terms of conventional vibration monitoring criteria, the problematic pump was "stable." Embarking on a study to resolve why a pump was not performing according to self-imposed expectations not formally recognized by professional industrial organizations or industrial peers was not to be confused with appealing work. However, site supervision was insistent upon understanding why this single asset location had not responded to practical adjustments to drop its vibration acceleration below 20 g's. They demanded a credible explanation that would either solve the problem or provide the confidence needed to issue an exception and tolerate the problem. By holding firm, supervisors empowered the investigation team to determine why the pump's acceleration could not be reduced below 20 g's.

The particular pump location where improvement was not immediately observed had a history of premature, repeat mechanical seal failures (Table 15.1). When the acceleration limit was implemented, the asset location was experiencing a mean time between failure (MTBF) of less than 6 months. A properly maintained centrifugal pump

Table 15.1 Double-Suction High-Speed Pump Reliability History

Cumulative Time (Days)	Cumulative Failures (Count)	MTBF (Months)
X-axis	Y-axis	
245	1	8.1
432	2	7.1
923	3	10.1
1040	4	8.5
1323	5	8.7
1397	6	7.7
1612	7	7.6
1707	8	7.0
1708	9	6.2
1769	10	5.8

built to API Standard 610 should provide closer to 5 years uninterrupted service in this specific process application. Of even greater concern was the fact that this particular pump location contained a process operating above its autoignition temperature. Due to these complexities, a catastrophic seal failure would surely result in a fire. In fact, setting the acceleration limit had been prompted by problems experienced in an identical process. A catastrophic seal failure that resulted in a fire had been involved in that incident.

REVISITING RELIABILITY GROWTH PLOTS

Recall our discussion about using reliability growth plot analyses to evaluate asset reliability in chapter "Role Statement and Fulfillment." Applying this analysis approach on the pump location's failure history (Fig. 15.2) provides critical information that explains why more superficial attempts to reduce acceleration had failed. The analysis shows consistent performance deterioration during the 5-year period leading up to when the acceleration limit was implemented ($\beta > 1$). The pump's performance at the time of the analysis indicates that a chronic failure mechanism was responsible for the observed series of repeat mechanical seal failures. These circumstances strongly favored an internal defect related to the asset's design, operating conditions, or a combination of both. Indications of a chronic issue inside the pump suggested that the acceleration problem might somehow be related to the failure mechanism responsible for low pump MTBF. Under these circumstances, diagnosing and addressing the cause for repeat mechanical seal failures that had become acceptable at this particular pump location would also likely solve the persistent vibration acceleration problem. In other words, the analysis indicated that a dependency existed between two seemingly unrelated problems at this particular pump location.

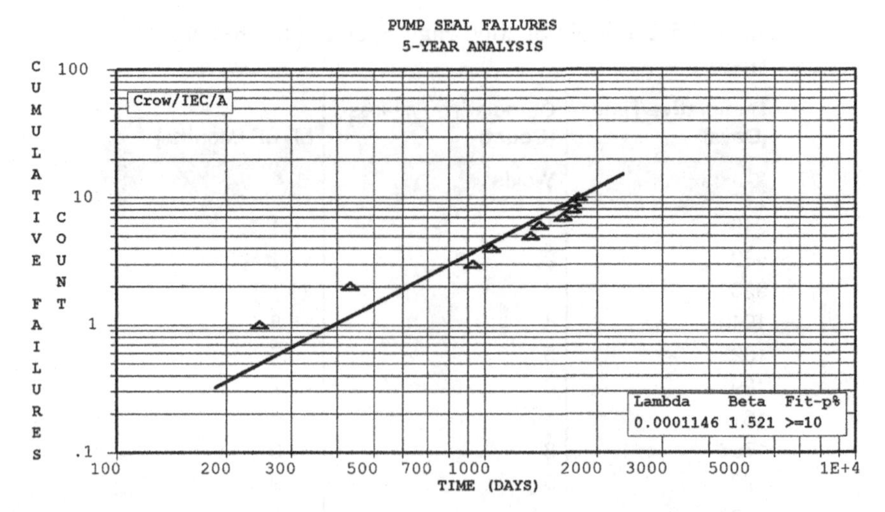

FIGURE 15.2

Reliability growth plot analysis of high-acceleration pump location reliability history.

UNSTABLE HYDRAULICS

Whereas vibration acceleration could be caused by just about any source of a high-frequency waveform, the causes for repeat mechanical component failures are limited to (chapter: Equipment Reliability Principles):

- Force,
- Reactive environment,
- Temperature, or
- Time.

After reviewing an initial list of potential causes of trouble for centrifugal pumps [3], unstable hydraulic force caused by insufficient "available Net Positive Suction Head" (NPSHa) was removed by a process of elimination. Conventional wisdom dictates that hydraulic stability results whenever a pump's NPSHa is at least one foot (0.3 m) greater than its *required* Net Positive Suction Head (NPSHr) [4]. A quick analysis confirmed a total of 80 ft of NPSHa. This NPSHa was 45 ft more than the 35 ft of NPSHr. Therefore, attention immediately turned to identifying other possible causes.

Although cavitation was initially ruled out as a possible cause for repeat seal failures as well as the unexplained acceleration phenomenon, familiar cavitation noises could be heard while standing near the pump. The physical damage observed in seals that had failed was also indicative of cavitation. For these reasons, the focus shifted on other explanations for unstable hydraulics that might promote a false the impression of cavitation.

CASTING DEFECTS SUSPECTED

Casting defects are a known potential source of persistent hydraulic instability [5]. Double-suction pumps like the one that had not responded well to corrective actions to reduce acceleration are even more susceptible to such problems because of their higher potential for casting defects. Examining the possibility of casting defects as the cause for high acceleration revealed that the pump casing metallurgy was incompatible with the process that it was contacting. Therefore, the entire pump was overhauled. The overhaul included replacing the pump's original case with a new one that met design specifications.

Upon opening the old case, widespread damage was found throughout the internal casting. On the other hand, the new case (designed according to process specifications) was completely intact with well-defined contours and edges in all areas. The old pump's internal condition caused a sense of relief over what appeared to be the discovery of a hidden explanation. The inspection results created optimism that the pump's vibration acceleration reading with a brand-new case would drop to an acceptable level below 20 g. It also corresponded with an impression left by the reliability growth plot analysis (Fig. 15.2) that the problem was getting worse over time.

Unfortunately, the hopes of completing the investigation based on this finding were instantly dashed upon energizing the pump. In reality replacing the pump case with a new one that met process design specifications did nothing to correct the problem. The acceleration did not change upon returning the overhauled pump to service.

At least one investigation team member was insistent that cavitation was the root cause for the problem, regardless that the pump was operating with excess available NPSHa. Other team members were skeptical. Certainly, the frequent mechanical seal damage events combined with the characteristic and persistent noise pattern at that pump location seemed to indicate that cavitation was involved. Therefore, a more comprehensive hydraulic study was commissioned. This study attempted to use the principles of falsifiability (chapter: Incident Investigation) to discredit the team's impression that sufficient NPSH was available.

This analysis confirmed the traditional notion that cavitation was not the problem. After all, the pump's suction takeoff was from an elevated section on the side of a process distillation tower. The pump on the ground had at least 80 ft of NPSHa compared to its 35 ft NPSHr. In other words, more than double the specified NPSHr was available at any given moment.

Desperate for answers, the investigation team decided to examine the pump location's warm-up lines. In a primary-spare configuration, warm-up lines allow a portion of the on-line pump's discharge to bypass the off-line (spare) pump's discharge check valve (Fig. 15.3). This provision prevents heavy process fluids from setting-up in the suction and discharge lines when a spare pump is in idle mode. Installing warm-up lines is one of two practical ways to keep a spare pump available between service runs. A suggestion was made that perhaps the ¾-in diameter bypass line configuration around the pump discharge check valve

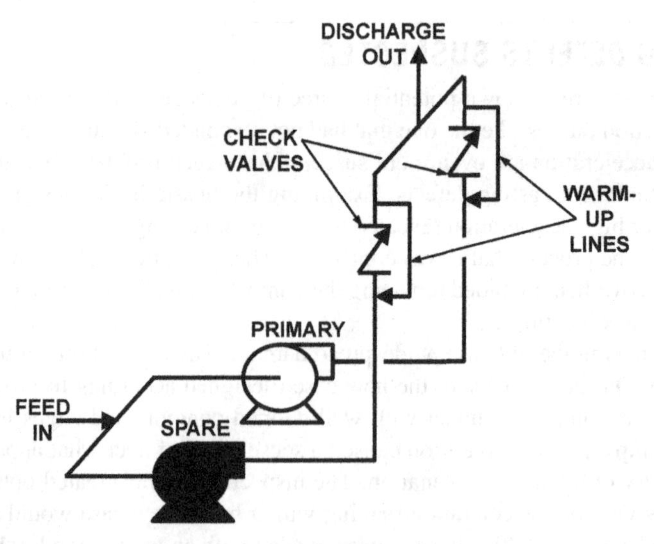

FIGURE 15.3

Standard warm-up line configuration in heavy process primary-spare centrifugal pump application.

was somehow related to high acceleration at the pump location. This explanation seemed to have merit because:

1. Prior to establishing the acceleration limit, the manual block valves installed on the warm-up line were removed due to high-frequency vibration that had caused them to fail repeatedly.
2. The pump had proven to be completely resistant to any form of maintenance correction directed at reducing the vibration acceleration level. All attempts to solve the problem thus far assumed that a pump-internal issue was involved. Perhaps the problem had something to do with an external problem instead. If so, then the investigation team may have been looking in the wrong place.

Similar to each of the previous attempts to solve the problem, running vibration tests at the warm-up lines failed to produce evidence of vibration originating outside of the pump. Surveying different points along the suction and discharge piping indicated that acceleration was actually lower at the warm-up lines compared to measurements taken directly at the pump. All efforts to solve the problem in the context of conventional knowledge about hydraulic stability were hitting dead-ends.

At that point the investigation had been in progress for about 8 months. A large volume of information and test data had been collected, but no real progress had been made. The cause for repeat seal failures or high acceleration at this single pump location was a mystery. In a concession to our own inadequacies, one of the team members proposed that we did not understand cavitation as well as we thought. From that moment onward, progress was steadily made. Recognizing that something was being missed, the team paused to conduct more research on causes for cavitation.

APPLYING MODERN HYDRAULIC SELECTION CRITERIA

Modern reliability authorities have noted that conventional wisdom regarding the required head for stable hydraulics may need to be replaced [4]. There are many instances where the stated minimum-required suction head pressure is simply insufficient to avoid severely damaging cavitation. For example, carbamate pumps are a specific application in the chemical manufacturing industry where even 25 ft more surplus head than required might not be sufficient to control cavitation [6]. Depending on specific operating conditions and design factors, an NPSH margin ratio ($NPSH_{ratio}$) up to 20 might be needed to suppress cavitation [4]. NPSH margin ratio is simply the pump's NPSHa divided by the pump's NPSHr (Eq. [15.1]):

$$NPSH_{ratio} = \frac{NPSHa}{NPSHr} \qquad [15.1]$$

where, NPSHa is the *available* suction head; NPSHr is the suction head *required* to avoid cavitation that will reduce the pump head by 3%.

Unstable hydraulics is a common cause for premature, repetitive pump mechanical seal failures. Avoiding excessive internal forces that can seriously reduce a pump's MTBF involves carefully selecting the right pump design for a specific installation. Making this determination requires going beyond simply verifying that NPSHa exceeds NPSHr by at least 1 ft (~30 cm). Let us say this in different words and rephrase the issue in no uncertain terms: There are services, fluids, and process applications where the 1-ft NPSH rule is inadequate. The question, then, becomes, "How do I determine the minimum NPSHa my pump needs to meet its performance expectations?"

Pump hydraulic selection criteria have been published in several popular industry references. These references are next used for the purpose of demonstrating how pump hydraulic selection criteria must be applied to specify a pump that can be expected to meet reliability expectations. These principles can also be used to diagnose potential design complexities that might shorten a mechanical seal's life or the life of rolling elements inside an existing process pump. Incorporating this information into a reliability program can make the difference between effectively solving and dragging out a persistent reliability problem. The driving principle behind communicating this information is to prevent similar incidents that could result in excessive maintenance costs, production constraints, and catastrophic process releases.

SUCTION RECIRCULATION

The investigation strongly pointed to suction recirculation (Fig. 15.4) as the most probable cause for repeat seal failures and high vibration acceleration observed at the subject pump. Suction recirculation (or simply "recirculation") is a hydraulic phenomenon that occurs inside the impeller of a centrifugal pump, particularly at some reduced flow rate to the left of the pump's best efficiency point (BEP). It represents a phenomenon whereby a portion of flow entering a pump impeller reverses direction and moves back in the

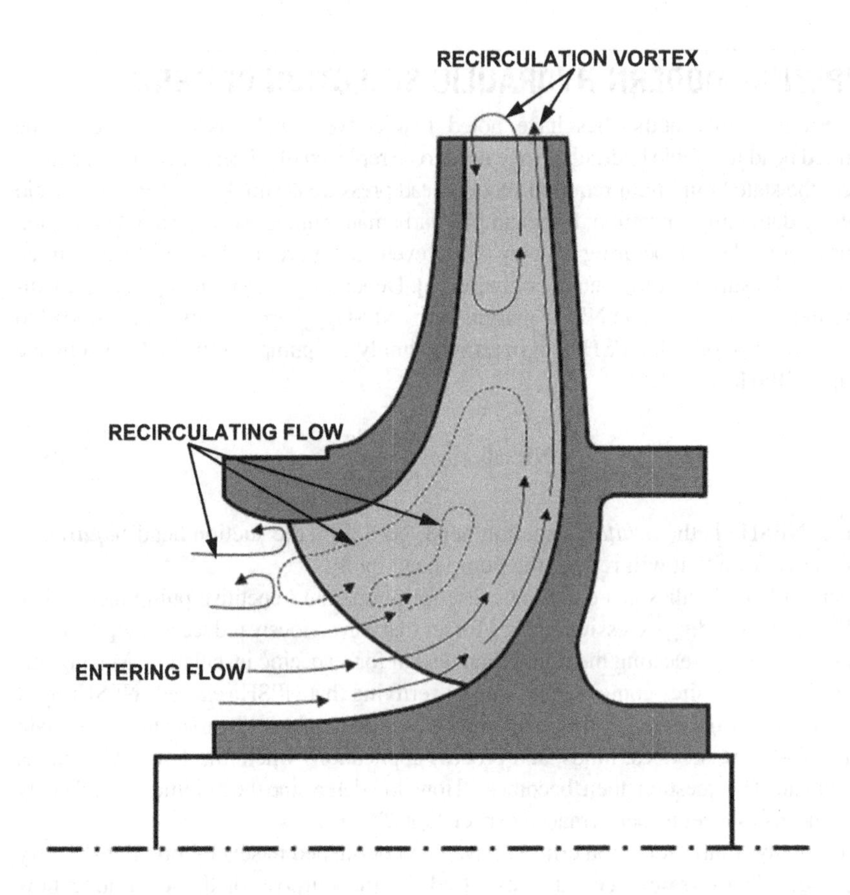

FIGURE 15.4

Suction recirculation can create excessive hydraulic forces inside pump impellers.

upstream or suction direction. This potentially destructive condition results in unstable flow and violent hydraulic changes akin to stall, backflow, eddy currents, turbulence, vortices, and cavitation. Mechanical seal and bearing life can be reduced by these often violent fluctuating forces, similar to what might result from conventional cavitation. References note that undesirable internal recirculation (hydraulic) forces are present in every pump [5]. However, the intensity of these hydraulic forces differs in various pumps and can often be controlled through effective pump design (selection criteria).

The four hydraulic selection factors that govern hydraulic forces inside a pump are the following:

1. Speed,
2. Flow (as a percentage of best efficiency flow),
3. Suction energy, and
4. NPSH margin ratio.

Table 15.2 Operating and Design Parameters for a Generic Double-Suction Refinery Pump

Parameter	Symbol	Value
Speed	N	3600 rpm
Flow at BEP	Q	4800 gpm
Available NPSH	NPSHa	80 ft
Required NPSH	NPSHr	35 ft
Impeller eye diameter	D_e	6.5 in
Specific gravity	S.G.	0.78

Suction energy and NPSH margin ratio are combined to provide a reliability factor [7]. This factor is provided to reasonably predict centrifugal pump performance before a purchase is made. However, determining a pump's reliability factor is equally useful for troubleshooting possible causes for unstable hydraulics in situations where vibration does not follow a typical "wear out" pattern and therefore does not respond to replacing components, corrective maintenance, and process adjustments. Screening design factors that could explain a persistent pump reliability issue involve first quantifying the pump's *actual* suction energy (ASE) (Eq. [15.2]) [7].

$$\text{ASE} = (D_e)\,(N)\,(N_{ss})\,(\text{S.G.}) \qquad [15.2]$$

where, D_e is the impeller eye diameter in inches; N is the pump speed; N_{ss} is the suction specific speed; S.G. is the specific gravity of the liquid in service.

Suction energy is a constant for a given pump operating at a constant speed. The impeller eye diameter (D_e), pump speed (N), and specific gravity of the liquid in service (S.G.) are all constants. Most of this information can be obtained from original equipment manufacturer (OEM) literature. Suction specific speed (N_{ss}), however, is a calculated value dependent on the following inputs (Eq. [15.3]):

$$N_{ss} = \frac{(N)\,(Q)^{0.5}}{\text{NPSHr}^{0.75}} \qquad [15.3]$$

where, N is the pump speed; Q is the flow at the pump's BEP *note: use ½Q for double-suction pumps*; NPSHr is the feet of suction head *required* at the pump's BEP.

Ascertaining the suction energy for the hypothetical pump with the specifications provided in Table 15.2 begins by calculating the suction specific speed as follows:

$$N_{ss} = \frac{(3600\text{ rpm})\,(2400\text{ gpm})^{0.5}}{35\text{ ft}^{0.75}}$$

$$N_{ss} = 12{,}250$$

The N_{ss} tells us something quite important about the pump we are screening. In general, industry has become acutely aware that pumps operating with greater than 9000 N_{ss} are deserving our attention if operated at flows away from BEP [8]. In such cases operating anywhere other than the BEP can considerably reduce the pump's life expectancy and increase the frequency of repetitive seal and bearing failures (and replacements.) Therefore, cautious pump users often consider 9000 N_{ss} to be the upper design limit during the initial selection period. While there are exceptions where pumps are designed to provide reliable service at N_{ss} values above 9000, recognizing this value as a triggering mechanism will help the user to perform due diligence with the aid of the OEM, so that misunderstandings do not develop after the pump is installed. Most importantly, discussions like this need to take place to achieve continuous process containment and to control maintenance costs. Note also that even though many petrochemical companies use an N_{ss} value of 9000 as an upper design limit, it really only applies for high-suction energy pumps.

But determining the N_{ss} alone is only part of the exercise. Continuing on, we calculate the pump's ASE:

$$\text{ASE} = (6.5\ \text{in})\ (3600\ \text{rpm})\ (12{,}250)\ (0.78)$$
$$\text{ASE} = 224 \times 10^6$$

To truly appreciate the impact that the pump design has in the service it is operating under, we must relate the pump's suction energy ratio (SE_{ratio}) to statistics provided in Table 15.3 that apply to specific pump types [4]. The SE_{ratio} must be determined to define whether or not conventional cavitation and/or low-flow suction recirculation might realistically be a chronic issue in a pump's anticipated or observed performance. The SE_{ratio} can be assessed by calculating (Eq. [15.4]):

$$SE_{ratio} = \frac{\text{ASE}}{\text{HSE}} \qquad [15.4]$$

where, ASE is the pump's actual suction energy; HSE is the start of "high-suction energy" for the particular pump being screened.

Since the generic pump used in this representative example was specified as a double-suction pump, the corresponding SE_{ratio} would amount to:

$$SE_{ratio} = \frac{224 \times 10^6}{120 \times 10^6}$$
$$SE_{ratio} = 1.9$$

Table 15.3 Suction Energy Classifications for Various Pump Types

Pump Type	Start of "High-Suction Energy" (HSE)	Start of "Very High-Suction Energy" (VHSE)
2-Vane sewage pumps	100×10^6	150×10^6
Double-suction pumps	120×10^6	180×10^6
End-suction pumps	160×10^6	240×10^6
Vertical turbine pumps	200×10^6	300×10^6
Pumps with inducers	320×10^6	480×10^6

RELIABILITY FACTOR

The influence of suction recirculation at the generic pump used in this example can now be estimated by applying the NPSH margin reliability factors provided in Fig. 15.5 [7]. A 1.0 reliability factor corresponds to no failures in 48 months or a mean time between repair of 72 months [9]. This information was developed from field experience and laboratory studies [7]. The relationship is based on the fact that the greater the suction energy, the more important it becomes to avoid damage by suppressing the residual cavitation that exists above the NPSHr.

The SE_{ratio} calculated in this example (2.5) is at the upper range of available data. The NPSH margin ratio is determined by dividing NPSHa by NPSHr. Looking at the information in Table 15.2, we determine that the NPSH margin ratio at this pump location is 2.3 (80/35). This amounts to a reliability factor of only about 0.23 (red line (gray in print version) on Fig. 15.5). Since a 1.0 reliability factor corresponds to a life expectancy of 48 months, damage due to recirculation in this pump would need to be repaired approximately every 11 months. This would, therefore, not be an appropriate choice for a pump in this particular service. Those who appreciate the relationship between asset reliability and process safety would accept nothing less than a 4-year MTBF at an absolute minimum. However, designing a pump for higher reliability may offer even less frequent turnaround intervals - if not simply fewer overhauls.

In this example, achieving a minimum reliability of 48 months at this pump location would require increasing the NPSHa to at least 130 ft, to achieve an NPSH margin ratio of 3.5. A 3.5 $NPSH_{ratio}$ at an SE_{ratio} of 1.9 would provide a reliability factor of about 1.0 as per the guidance provided in Fig. 15.5. A higher $NPSH_{ratio}$ would be needed to confidently operate the pump between 5-year turnaround cycles

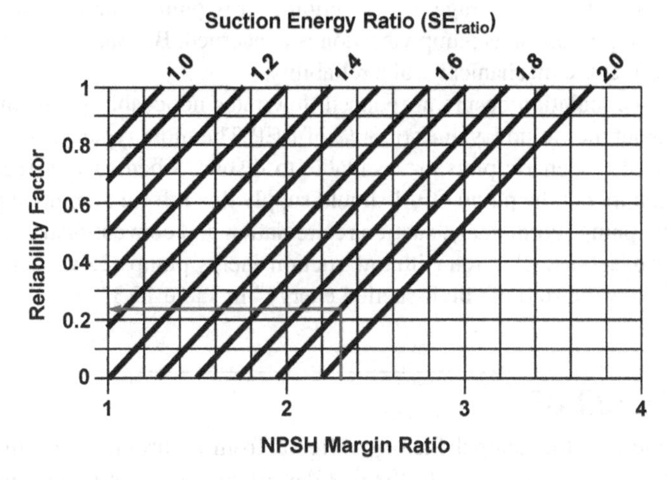

FIGURE 15.5

NPSH margin reliability factor determination.

without potentially interrupting production for unplanned maintenance. On average, the NPSHa must be at least 4–5 times the pump NPSHr to suppress all cavitation.

BUSTING CAVITATION MYTHS

Cavitation in a pump does not start at the pump NPSHr. At the NPSHr, there is enough cavitation in the pump to lower the developed head by 3%. It can take 2–20 times the NPSHr value ($NPSH_{ratio}$) to completely eliminate all cavitation inside a centrifugal pump. Suction recirculation also generates cavitation in the pump, which can be even more damaging. This problem often occurs in pumps with high-suction energy and insufficient NPSH margin ratios. Reliability experts refer to cavitation that occurs in a pump at flow rates above the start of suction recirculation as "conventional cavitation." Also, due to factory test tolerances and site differences, the actual installed NPSHr could be up to 20% higher than that published by pump manufacturers.

When cavitation bubbles form in a pump, any entrained air is released. This action adds to the vapors in the pump. Air and water vapors, being lighter than the liquid being pumped, will be forced to the center of the pump where the mechanical seal resides, when the pump impeller is rotating. Therefore, if the mechanical seals do not have a sufficient external flush of liquid (such as with dual seals), excessive cavitation can cause the mechanical seal faces to run dry resulting in a rapid failure. Mechanical seals need a thin film of liquid between the faces to avoid this dry operation.

Suction energy, which is a constant for a pump at a constant speed, cannot be looked at alone in regards to a pump failure. It has to be considered in the context of the NPSH margin ratio—as was demonstrated by the example provided in this chapter. The higher the suction energy, the higher the NPSH margin needed, until the point of zero cavitation in the pump has been reached. This occurs at NPSH margin ratios of 4–5, on average. Low-suction energy pumps can operate with very low NPSH margins as far as pump vibration is concerned. But vapor and released air might still affect the mechanical seal's reliability.

Suction recirculation occurs at reduced flow rates noticeably below the pump's BEP, yet sometimes at flow values of 80% of BEP. The more typical start values are 40–60% of BEP. Generic plots are available to display when suction recirculation starts in a pump, but the pump OEM should supply this information upon procuring a centrifugal pump. Fortunately, suction recirculation and conventional cavitation do not produce excessive vibration with low-suction energy pumps. Low-suction energy is a value below the start of "high-suction energy" in Table 15.3.

LESSONS FOR US

It is possible that the Bhopal factory suffered from a chronic pump mechanical seal failure mechanism similar to the one described here. The failure mechanism responsible for unstable pump hydraulics can remain a mystery over a long period of time. Solving the problem requires becoming familiar with specific principles governing sufficient NPSHa on centrifugal pumps. Thankfully, a screening method exists to diagnose problems that can remain unsolved for a long period of time. Using this information can significantly increase the life expectancy of fluid machines. Applying the lessons communicated in this case study reduces the potential for seal leaks and bearing failures that could result in a catastrophic process release.

Patience and persistence are needed to identify a chronic asset reliability failure mechanism. In most cases, determining the root cause for an unexplained process phenomenon requires resetting our knowledge on matters we think are under control. Progress is only made after admitting that we do not know as much about conditions that cause unsatisfactory process behavior than we originally thought. Under no circumstances is it appropriate to give up the search for definitive answers. In all cases, scientific principles govern industrial asset performance. Machines function exactly as designed under the operating conditions that they are subjected to. Stopping the search for definitive answers and settling for an alternative solution might introduce new hazards that can be even more destructive in the long run. Other lessons from this case history to remember as we move forward are as follows:

- A trial and error approach to solving a chronic asset or process reliability problem is both unrewarding and physically exhausting.
- Everything makes complete sense after you figure out what you have been missing.
- Mightiness and determination can only overcome persistent mechanical difficulties for a limited period of time.
- Some problems cannot be solved unless they are formally investigated.
- The purpose for investigating a repeat failure is to learn something about the process that you do not already know.
- Failure mechanisms often produce multiple effects that might seem unrelated at first.
- The high price paid for tolerating a problem today may become unaffordable tomorrow.
- Operating with conventional wisdom alone can significantly reduce an asset's life expectancy.
- Up to 20 NPSH margin ratio might be needed to completely eliminate all cavitation inside a centrifugal pump.
- Mechanical seals need a thin film of liquid between the faces to avoid running dry, which would result in a rapid seal failure.
- The higher the pump's suction energy, the more NPSH margin ratio is needed to achieve stable hydraulics that prevents repeat mechanical seal and bearing failures.
- Machines and processes do exactly what they were designed to do, under specific operating conditions.

REFERENCES

[1] C.D. Johnson, Process Control Instrumentation Technology, sixth ed., 2000. 236.

[2] H.P. Bloch, F.K. Geitner, Major Process Equipment Maintenance and Repair, 269, 1994.

[3] H.P. Bloch, F.K. Geitner, Machinery Failure Analysis and Troubleshooting, second ed., 1994, ISBN: 0-87201-232-8. 631.

[4] H.P. Bloch, A.R. Budris, Pump User's Handbook: Life Extension, fourth ed., 2014, ISBN: 0-88173-720-8. 122.

[5] W.E. Nelson, J.W. Dufour, Pump vibrations, in: Proceedings of the 9th International Pump Users Symposium, March 3–5, 1992, 1992, pp. 137–147.

[6] H.P. Bloch, A.R. Budris, Pump User's Handbook: Life Extension, fourth ed., 2014, ISBN: 0-88173-720-8. 485.

[7] H.P. Bloch, A.R. Budris, Pump User's Handbook: Life Extension, fourth ed., 2014, ISBN: 0-88173-720-8. 123.

[8] H.P. Bloch, Breaking the cycle of pump repairs, in: 43rd Turbomachinery & 30th Pump Users Symposia (Pump & Turbo 2014), September 23–25, 2014, 2, 2014.

[9] H.P. Bloch, A.R. Budris, Pump User's Handbook: Life Extension, fourth ed., 2014, ISBN: 0-88173-720-8. 121.

PART 4

Operations and Production

Relatively soon after factory construction ends, attention turns to production. A factory must be able to demonstrate its ability to compete, survive, and thrive on its own. Sustaining adequate performance translates into continuously meeting specified health and safety, environmental, asset protection, and economic business commitments. Any operating difficulties not addressed during process commissioning must be remedied. Any unintended operating condition disturbing the balance between these four performance commitment areas must be managed appropriately. Nothing can be allowed to undermine these performance commitments.

Situations where performance commitments are not being satisfied are more likely to experience drastic and sudden changes at some unexpected point in time. It will then take considerable effort to bring system performance back under control. It follows that when problems develop, solutions must address their root causes. Merely treating the symptoms is wasteful and unproductive. While it may be difficult to resist the impulse to inject creativity and personal strength to overcome mechanical adversity, ask yourself if these traits truly address the root causes. Unless your efforts are directed at managing the root causes for unacceptable process performance, you might simply be creating a false sense of security. If the action you take is misguided, then the problem you are trying to solve can get worse.

This part of the Bhopal factory's case study covers its brief operating period from 1980 to 1984. Although the process survived for only a comparatively short period of time, transferrable and widely applicable lessons were learned that have rich meaning throughout the global manufacturing industry. These lessons demonstrate how short-term solutions to process reliability problems can prolong, worsen, and even create other long-term performance penalties. Students, workers, and supervisors alike must pay serious attention to these patterns. Describing the sequence of events behind the Bhopal disaster provides an important reference point. Taking the time to fully examine the causes of industrial disasters can continually condition our personal behaviors, decisions, and actions in a direction away from risk and calamity. We will impart and build tangible failure resistance into a process by applying the knowledge that we have acquired. This resistance can become an impenetrable shield against a potential process safety incident.

The lessons communicated in this section can be put to immediate use by those with direct manufacturing responsibilities. Interested parties outside of the manufacturing industry also benefit from this discussion. As the timeline to an industrial disaster shortens and conceivably moves through escalating levels of concern, consider how you can make the difference when you encounter these situations in your assigned role.

CHAPTER

Management of Change* 16

EVENT DESCRIPTION

The commissioning phase ended on February 5, 1980 and the operations department assumed control over factory production. During the year leading up to the turnover date [1], a chronic production constraint had dominated factory performance. Repeat methyl isocyanate (MIC) pump mechanical seal failures seriously curtailed factory output. But despite these complexities, the factory was able to satisfy its export quota for the production year that ended on December 31, 1979 [2]. Thus, the parent company retained its slight majority ownership (50.9%) in the partnership. These terms were stipulated in the foreign collaboration agreement that would expire on January 1, 1985—5 years after production officially started [3]. At that point in the not-too-distant future the parent company would walk away from the business. The international subsidiary would then assume exclusive control over factory operations.

Due to the high frequency of MIC pump mechanical seal failures, personal exposure incidents were common. At the Bhopal factory, AVO (audio/visual/olfactory) monitoring was the standard practice for detecting MIC leaks [4]. Tears welling up in the workers' eyes indicated the presence of leaking MIC [5]. Accordingly, an MIC pump seal leak was reported as a "material" loss instead of a "near injury" on the incident report form [6]. This determination accurately reflected the foremost, economic-related concerns involved with recurring MIC pump leaks. Possible safety issues were dwarfed by the more immediate economic and production penalties resulting from the persistent asset reliability problem.

Unplanned maintenance costs were extraordinarily high for a factory operating in a remote part of industrialized society. Obtaining local parts for asset repairs was difficult in an industrial area still in its early stages of development. This explains why the Bhopal factory was not only *expected*, but *needed*, to be reliable. Factories that are expected to provide reliable production are not necessarily built to accommodate corrective maintenance. Corrective maintenance in a factory like the one constructed in Bhopal could be the kiss of death. The process simply was not designed to support a chronic, unresolved asset failure mechanism (chapter: Licensing). In many ways, the economic burden created by the low MIC pump mean time between failure (MTBF) made it impossible for the factory to even think about breaking even financially.

*For timeline events corresponding to this chapter see pages 434–437 in Appendix.

For example, it was particularly frustrating to lose valuable commercial grade MIC product when a circulation pump failed. Refrigeration would also stop upon the interruption of MIC circulation. The resulting failure to suppress the vapor pressure of highly volatile MIC would send excessive amounts of MIC vapor into the process vent header (PVH); only to be destroyed at the vent gas scrubber (VGS). Wasting valuable intermediate product in this manner was not part of the business plan. Nor did it help to manufacture enough product for the Bhopal factory to break even financially.

Of course, the greatest concern was over periodic, extended production delays that would occur when multiple MIC pumps were shut down for overlapping unplanned maintenance. During those times, factory production would come to a complete stop. Unless one of the two rundown tanks was available, MIC stagnating in the storage area could not be transferred into the derivatives section. This situation could give local competitors an opportunity to snatch up some of the coveted business belonging to the industry leader and technology developer. Losing loyal customers was a hard pill to swallow in addition to not generating production income. It was especially demoralizing for the incumbent that had developed and dominated the market for over 20 years.

Continued poor financial performance through the end of 1980 led to the appointment of a task force to implement a strategic plan for the Bhopal factory [7]. Despite the fact that the commissioning phase had ended a full year earlier, factory production for the second year could not be raised above 30% [8]. Two years had now gone by with unimpressive financial performance. The factory needed help living up to its economic business commitment if survival was to be expected. It was still not breaking even economically. Operating expenses were much higher than expected. Considering the financial implications involving the acceptance of repeat MIC pump failures, the problem could no longer be ignored. Therefore, a study team consisting of employees from both the parent company and international subsidiary was assembled to develop a solution. If the problem could not be solved then an alternative solution would have to be introduced.

It was not long before a potential solution was proposed. The solution essentially involved modifying nitrogen's original function. According to standard design, the purpose for providing a dry nitrogen blanket was to protect the highly flammable, reactive product in the MIC storage tanks from contacting air [9]. Alternatively, the nitrogen supply's function could be expanded to act as the prime mover for MIC in the storage tank. Pushing the MIC out of the storage tanks under nitrogen pressure would allow the MIC Transfer Pumps to be abandoned (Fig. 16.1).

The alternative operating method involved simply closing the MIC storage tank vent valves to isolate the tanks from the PVH. An MIC storage tank isolated from the PVH could be pressurized by feeding excess nitrogen in through the supply header. Upon reaching sufficient pressure, the MIC contained in the tank could be made to flow at a reasonable rate into the one-ton MIC charge pot in the derivatives section [10]. This workaround solution was an attractive option for eliminating the production constraint caused by the high-maintenance and production-limiting transfer pumps.

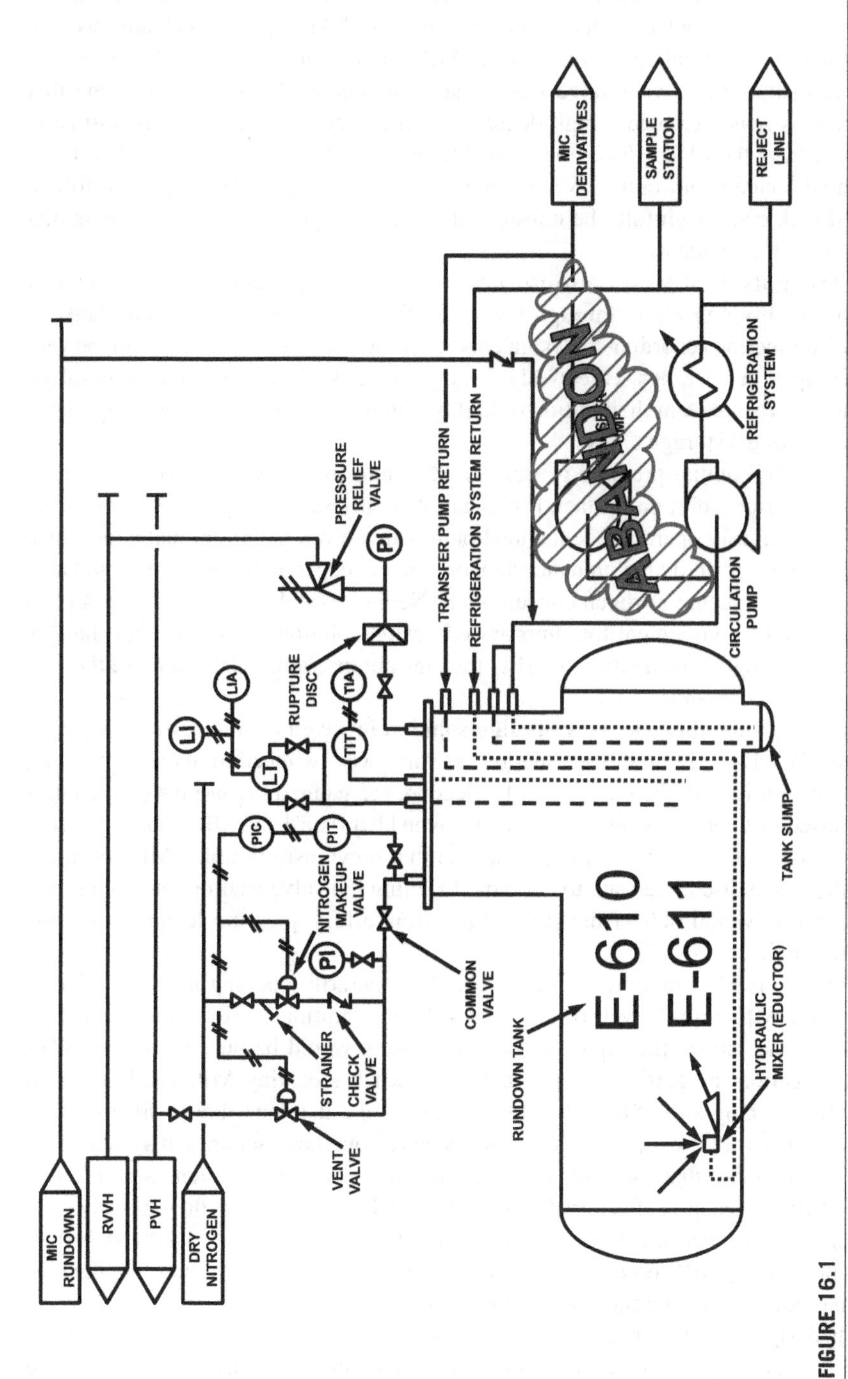

FIGURE 16.1
The methyl isocyanate rundown tanks could be reconfigured to operate without the transfer pumps.

Isolating the MIC rundown tanks from the PVH required closing the vent valves. The vent valves (chapter: Factory Construction) were DMVs [11], which are specially designed for leak-tight operation [12]. An MIC storage tank capable of holding pressure against a closed vent valve was considered adequately "sealed." Proving that a tank was sealed created a high degree of confidence that the tank was positively isolated from the PVH. Under these conditions it was physically impossible for any potential contaminant in the PVH to enter the MIC storage tanks [13]. Therefore, a sealed tank could rightfully be considered adequately protected against a potential contamination incident.

Closing the vent valves also isolated the MIC storage tanks from any air that could possibly migrate in through the VGS. This preserved the nitrogen blanket's intended function according to design conditions involved with running with the vent valves open at all times. Effectively sealing the tanks by closing the vent valves would likewise prevent the ignition of highly flammable liquid or vapor components contained in the storage tanks.

Using differential pressure to transfer MIC into the derivatives section seemed to be the perfect solution. In fact, it was an inherently safer design choice compared with continuously operating centrifugal pumps that, even under optimal conditions, would require periodic maintenance support and upkeep. Operating costs would also decrease by reducing nitrogen consumption. Neither would a capital project have to be approved to implement this improvised design solution. As if these advantages were not enough, abandoning the MIC transfer pumps would also decrease the factory's energy demand.

Implementing this alternative design solution involved a simple, three-step procedure (Fig. 16.2). First, the automatic vent valve was switched to manual operating mode so that it could be closed tightly. Second, the gauge pressure inside the tanks was raised to at least 14 psig by feeding nitrogen [10]. Third, the MIC was transferred into the derivatives section through the line that previously returned MIC from the transfer pump discharge back to the tank. Opening the valve leading into the derivatives section would deliver the stored MIC from a high pressure location to a low pressure point.

Filling the MIC rundown tanks involved temporarily opening the vent valve to permit the MIC liquid level to rise. Under these conditions, MIC vapor occupying the headspace above the liquid surface in the tank would be pushed into the VGS for destruction. Forgetting to open the valve when receiving MIC rundown from the MIC refining still (MRS) would not only make it impossible to fill the tank, but would also create a process upset in the MRS overhead separation system (Fig. 16.3). Remember that the discharge pressure in the rundown system was never to exceed 2 psig (chapter: Factory Construction). Obstructed vapor flow anywhere in the process between the MRS and VGS could elevate the MRS overhead system pressure. During MIC production, high levels of chloroform would boil into the rundown product if the pressure in the MRS was to rise [14].

By design, the MIC storage tanks were configured to operate slightly above atmospheric pressure [15]. According to design specifications, the transfer pumps were

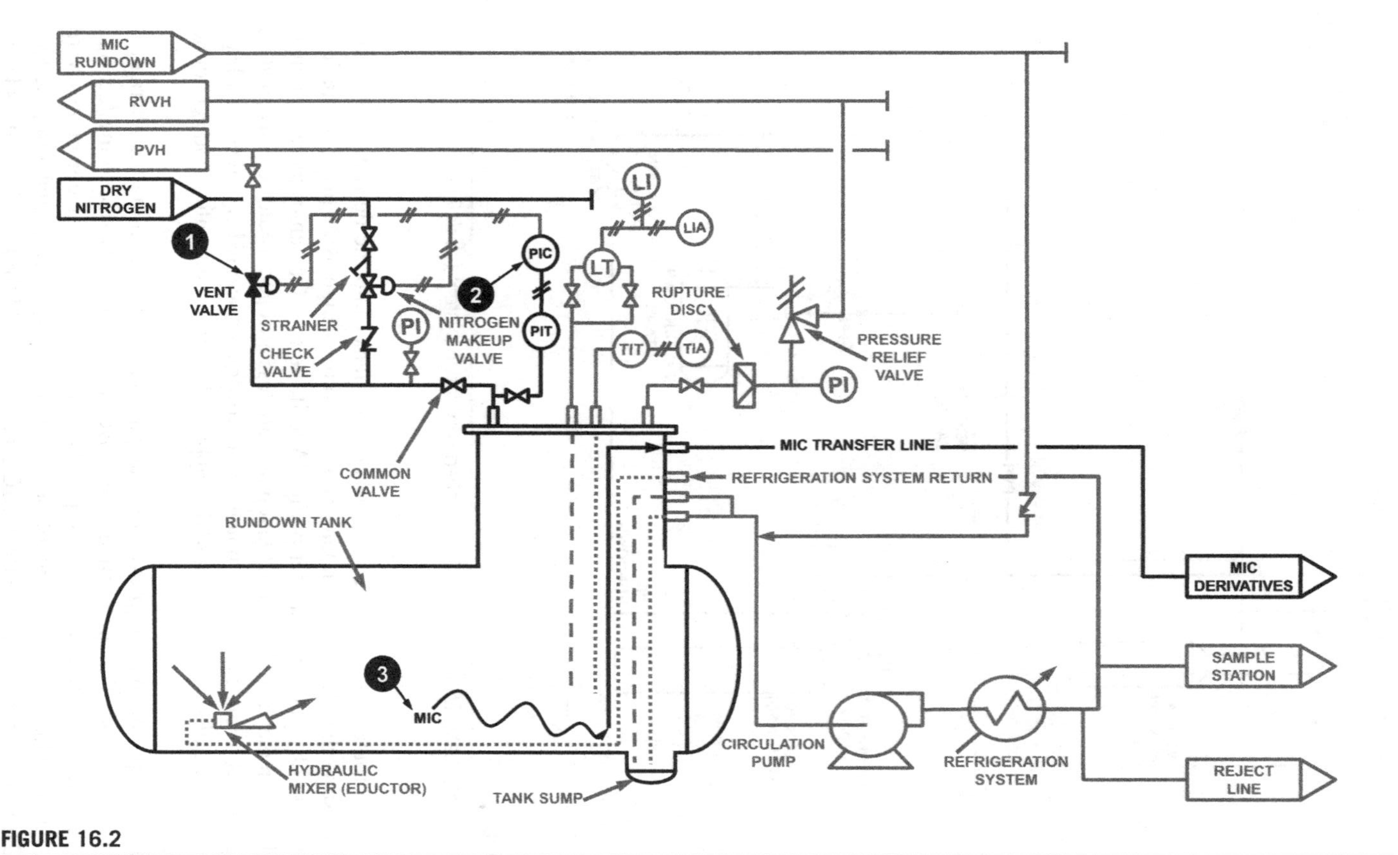

FIGURE 16.2

Procedure to transfer methyl isocyanate into the derivatives section using differential pressure.

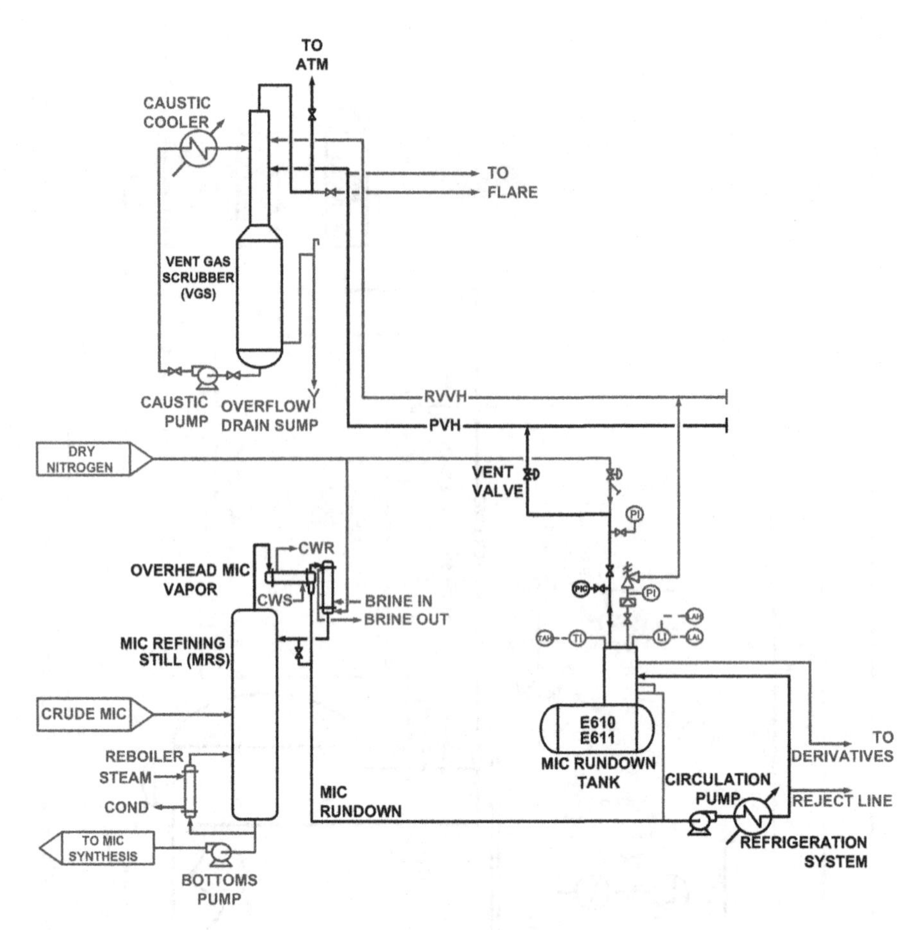

FIGURE 16.3

Methyl isocyanate (MIC) refining still overhead separation system (*dark lines*).

needed to route MIC contained in low pressure tanks into the derivatives section. But boosting the MIC rundown tank pressure up to as high as 25 psig [16] made it possible to eliminate the transfer pumps. And if the transfer pumps were no longer needed, then they could not fail. On top of it all, less nitrogen would be consumed on a regular basis. Once again, the potential improvements seemed significant. At least when superficially viewed, the solution appeared to offer both safety and economic benefits.

Factory productivity improved dramatically upon implementing the alternative operating method. This performance improvement was mainly due to the fact that MIC transfer pump failures were completely eliminated. The circulation pumps were still failing within every 24 days of being repaired. However, abandoning the two MIC transfer pumps raised the total MIC pump group MTBF from 5 to 8 days [17].

After abandoning the MIC transfer pumps, the MIC storage tank pressure gauges no longer provided much useful information. Prior to changing the MIC transfer method, storage tank pressure above 2 psig was considered abnormally high. Previously, MIC storage tank pressures above 2 psig might also have meant that the nitrogen control valves were malfunctioning, or perhaps even a contamination incident had occurred. After changing the way that MIC was transferred into the derivatives section, high tank pressure well above 2 psig became normal operation. Therefore, there really was no longer a basis for concern if up to 15 times the previous maximum tank pressure (or more) was observed. Elevated tank pressure in the range of 2–25 psig was completely normal. It was to be expected during routine process operation [18]. Ascertaining why the pressure inside the storage tank should read high at any given moment was from then on irrelevant. The new MIC transfer procedure required operating the tanks under high pressure.

Shortly after the MIC transfer pumps were abandoned, another problem developed. During normal MIC production campaigns, backpressure in the PVH would transmit back to the MRS overhead system whenever the vent valve was opened to fill the MIC rundown tanks. This backpressure raised the boiling temperature inside the MRS, causing loss of MIC separation efficiency. The problem was traced back to trimer buildup inside the iron PVH between the storage tanks and the VGS. Any blockage capable of creating more than 2 psig pressure inside the storage tanks could elevate the chloroform level in the MIC rundown stream. At about the same time, block valves in the MIC vapor transfer headers started failing miserably [19]. Many of these valves had worked properly up until the time that the new MIC transfer procedure was introduced. Now, however, process isolation could not be trusted.

Operating costs increased again, as trimer began choking vapor flow through the 2-inch diameter PVH. Reprocessing or destroying contaminated product was becoming a major inconvenience. Neither of these options was particularly economical. The best way to operate the process was at low pressure without disabling the transfer pumps. At least when the transfer pumps were operating, trimer choking was not an issue and the MRS would continuously produce pure MIC.

But going back to the old operating method was now out of the question. The third year of production was looking much better financially. The factory's production output increased significantly after eliminating the MIC transfer pumps. Dealing with the rapid buildup of trimer debris on the inner wall of the PVH was now the top priority.

The trimer fouling problem was addressed by introducing a new procedure to clean process pipes and valves in MIC vapor service. The procedure involved attaching water hoses to pressure gauge taps that could substitute as unofficial hose connections. Screwing off a pressure gauge would provide a threaded connection to which a hose adapter could be fastened. Connecting a water source from a nearby utility station allowed the trimer deposits to be flushed out with relative ease.

Water was a readily-available trimer solvent. MRS process control could be restored by using water to flush trimer debris out of MIC process lines [20]. Flushing out the vapor lines with water allowed process control at the MRS to return to normal for a brief period of time. But shortly thereafter, the trimer deposits would return in sufficient quantity to choke off vapor traffic again. Therefore, the PVH needed to be washed out regularly.

The flash rust that catalyzed the formation of trimer was insignificant compared to the severe corrosion that immediately followed. Deep gouging pits inside the PVH suddenly began perforating the vapor line. But MIC vapor could not be safely released into the atmosphere. Therefore, PVH leaks would have to be repaired before refilling a tank. If the repair could not be completed before the rundown tank inventory was depleted, then production would again stop until the repair was finished. Honoring the factory's overarching safety and environmental commitments required shutting down production if a PVH leak existed at the time that the rundown tanks needed to be refilled. Filling both rundown tanks provided about 30 days of production inventory [21]. Therefore, production could last no more than 15 days if the PVH leak and an MIC tank were undergoing maintenance at the same time.

Production was on target to exceed 50% of the factory's designated capacity as the third manufacturing year came to an end. The combination between hard work, considerable determination, and a new MIC transfer method had finally made it possible for the factory to break even [22]. However, a serious new problem developed only a week before 1981 was officially over. What would be recorded as the best year of manufacturing that the Bhopal factory would ever see ended with a fatality.

TECHNICAL ASSESSMENT

Closing the MIC storage tank vent valves was the solution to recurring production interruptions caused by MIC transfer pump failures. Earlier, we assigned constructing the PVH and relief valve vent header (RVVH) with iron as the first of two causal factors (chapter: Factory Construction). The fact that the vent valve was closed *independently* from any issue concerning the use of a restricted material to construct the PVH and RVVH makes closing the vent valve the second causal factor involved in the Bhopal disaster. From this point on, numerous defects propagate out from the sequence of events. All of these remaining defects, however, are dependent on the two causal factors that have now been defined.

An important reliability principle that was first covered in chapter "Licensing" appears again in the events described in this chapter. We see this principle weaving its way through the case study in later chapters as well. Notice that the Bhopal factory was constructed to a reliability design specification. The design specification was predicated on the factory's ability to contain the process. Based on this performance standard, specialized leak detection devices were unnecessary. The exclusion of these devices at the Bhopal factory translated into an asset reliability commitment. If the process leaked, then the workers would have no way to detect its release other than by personal exposure. Here we find the source of casual compliance (chapter: Incident Investigation). The fact that this became the accepted way to detect MIC leaks made the constructive use of disciple even more difficult and likely to cause bigger problems later. Unless leak detection assets were installed as the first line of defense, the problem could only be solved by improving asset reliability to match the corresponding process design specification. This process design specification did not match actual system performance involving repeat MIC pump mechanical seal failures.

At least two other familiar patterns are hidden within the sequence of events described in this chapter. First, we see *improvisation* again being used to address a problem (chapter: Process Selection). Second, we see a *hazard exchange* taking place, where one hazard that is considered unacceptable is being replaced by other hazards that are, on a relative scale, more acceptable and manageable (chapter: Inventing Solutions).

The immediate issue that received the most attention was the factory's financial performance. For two consecutive years after construction was completed the Bhopal factory had been unable to manufacture enough technical grade product to break even. A task force was appointed to develop a performance improvement plan. The joint team consisted of professionals from both sides of the partnership [7].

For the two consecutive years prior to the task force's assignment (1979–80), factory production had stagnated at about 30% of total output capacity. In the year that followed (1981), the factory output was raised to 53% of total production capacity. One of the foremost changes made during this period of improvement initiatives was the elimination of the MIC transfer pumps. Removing the transfer pumps had a number of economic benefits that made it possible for the factory to break even on production for the first (and only) time in its history.

There were, however, complications involved with this process configuration change. Notice that establishing pressure inside of a storage tank required closing the vent valve (Fig. 16.2). The original design intent for this valve was to remain open. Prior to making this change, tank vapors would freely pass into the PVH for destruction at the VGS. Maintaining low pressure (2 psig maximum) on the downstream side of the MRS overhead (rundown) system was needed for process control. The vent valves were closed to create about 25 psig pressure inside the storage tanks between MIC production runs; recall that material inside the storage tanks could then be fed into the derivatives section without the transfer pumps. The vent valve was opened only when MIC product was running down into the tanks, so that low pressure would again exist. Operating the tanks at low pressure was needed to manufacture pure MIC that met commercial grade product quality specifications of less than 0.5% chloroform.

While this alternative approach to supplying MIC into the derivatives section seemed advantageous, there were some complexities involved with isolating nitrogen inside the storage tanks. Implementing this change eliminated not only the unreliable transfer pumps, but also the secondary function of nitrogen inside the PVH and RVVH that was to inhibit rust. Afterward, opening the vent valves to fill the rundown tanks would once again allow MIC vapor to travel into the PVH. With the corrosion inhibitor's withdrawal, the rust that formed in the pipes catalyzed the trimer reaction, which rapidly choked flow through the narrow 2-inch diameter PVH (Fig. 16.4). This

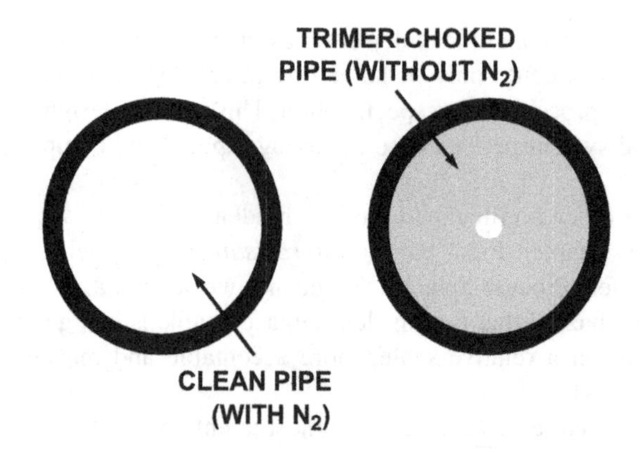

FIGURE 16.4

Cross-sectional view of process vent header, clean (left) and choked with trimer (right).

condition would restrict vapor traffic through the vent pipes while MIC was running down into the tanks from the MRS. Due to the backpressure created by trimer accumulating in the PVH, higher levels of chloroform would appear in the rundown stream. Thus, the PVH needed to be cleaned in order to keep the downstream pressure low and avoid excess production costs affiliated with reprocessing or destroying chloroform-contaminated MIC.

MIC trimer fouling in the vapor vent header system created a new problem that ultimately made the process less safe and more difficult to maintain. Trimer was a very stable polymer, with hard qualities similar to plastic [20]. Trimer debris trapped inside iron wedge gate valve bodies made it impossible to establish a reliable seal. Malfunctioning valves therefore became a common nuisance within the MIC process [23]. Gate valves (Fig. 16.5) were used because with proper mating between the disk and the seat ring, little or no leakage was expected to move across the disk [24]. However, deposits can collect in the space where the disk (or gate) contacts the seat ring. When that happens the valve often malfunctions and leaks, even though the valve's stem position gives the appearance that the valve is fully closed.

A major report describing the sequence of events behind a series of maintenance-related incidents responsible for multiple fatalities noted [25]:

> *Wedge gate valves can be closed through minor accumulations of scale or sludge, but eventually build up will prevent them from closing fully. They then no longer seal effectively and hence pipeline isolation may be prevented even though the valves are apparently closed.*

This process-related valve failure mechanism documented in the cited report is common throughout the manufacturing industry, especially in processes involving heavy hydrocarbons and fouling services. Therefore, it is not uncommon for leaking valves to be purged with water or steam prior to performing corrective maintenance. After implementing the alternative MIC transfer method that required closing the vent

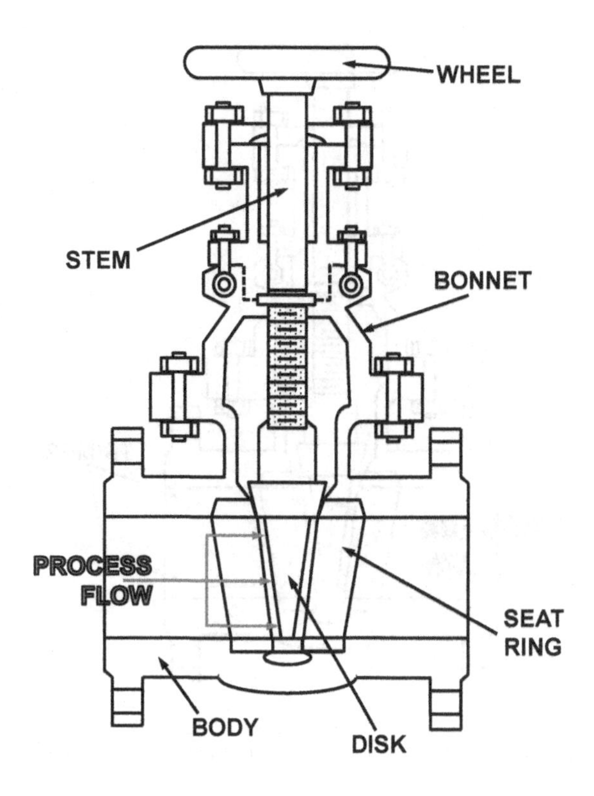

FIGURE 16.5

Gate valves require properly mating the wedge (disk) into the seat rings to provide a reliable seal.

valves, trimer deposits accumulated inside the iron valves in MIC vapor service. Afterward, the valves began exhibiting sealing problems (Fig. 16.6). They leaked frequently. To the naked eye and to the touch, the valves appeared to be closed. Inside, however, tenacious trimer deposits prevented them from closing properly. Under these circumstances, passing water or steam through an internally leaking valve without interrupting production satisfied two objectives:

1. An initial attempt to remove the temporary substance (trimer) that prevented the leaking valve from fully closing.
2. Clearing the valve of residual process in preparation for invasive maintenance if the valve's sealing function did not return after purging was complete.

Steaming out the leaking valves was a standard practice at the Bhopal factory [26]. Likewise, this common maintenance procedure is firmly implanted within the manufacturing industry today. Managing the trimer in this way obviously represented a potentially dangerous activity. The online cleaning method required introducing a substance that could potentially cause a contamination issue. Failure to properly

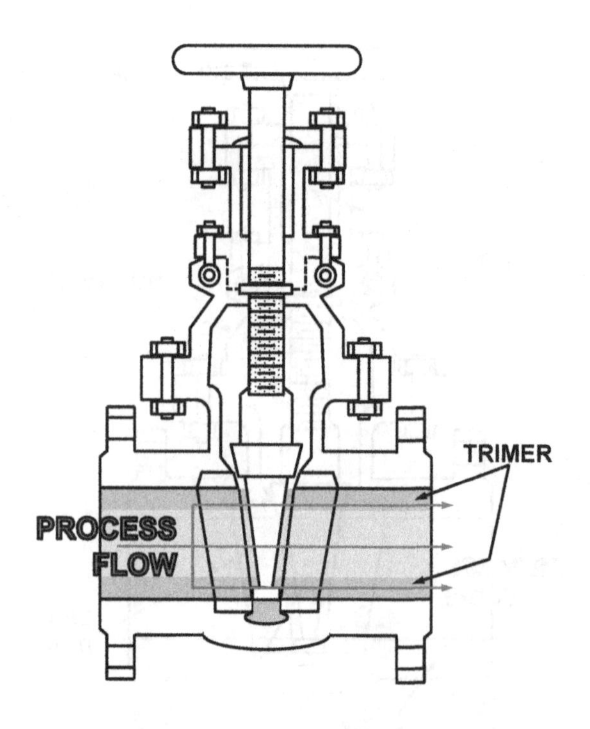

FIGURE 16.6

Valve malfunction caused by trimer accumulation in iron valves in methyl isocyanate service.

isolate the process could result in a runaway reaction inside the MIC storage tanks. Although hardly preferred, the practice of washing out the pipes and valves with water online seemed unavoidable to achieve the primary goal: Eliminating the transfer pumps so that the factory could realistically compete for, and retain, business. Therefore, it became the workers' responsibility to clean the pipes on a routine basis and to perform the procedure safely.

As if these inconveniences were not enough, there was yet one more complexity that developed in response to the water washing procedure. Initially, flash rust formed on the surface of the PVH when nitrogen flow was interrupted. This rust led to the accumulation of trimer deposits that would choke off vapor flow into the VGS and caused process control problems at the MRS. Flushing out the PVH with water to resume normal unrestricted MIC vapor flow accelerated the corrosion problem that was causing considerable and rapid PVH damage. Residual water in the PVH lingering in the presence of air aggressively dissolved the PVH [23]. Therefore, it became necessary to replace corroded sections of leaking vent header piping on a rather routine basis [27].

Notice that earlier the decision to use differential pressure to transfer MIC into the derivatives section seemed to correspond with inherently safer design principles [28]. However, under closer analysis we see how eliminating the transfer pumps instead made the process much more human dependent (chapter: Spent Caustic Tank Explosion: A Case Study in Inherently Safer Design). Implementing the new MIC transfer procedure required introducing a repetitive manual online process cleaning task. Avoiding a potentially serious process safety incident now required more worker diligence while increasing the demand for human intervention. In other words, the human probability of failure on demand (chapter: Process Hazard Awareness and Analysis) had to be reduced to maintain an equal amount of safety protection. Thus, these complexities really made the process *less* inherently safe than the former operation with unreliable machinery, where the human demand was strictly limited to process isolation after detecting an MIC leak.

Reconfiguring the MIC transfer method introduced the hazard (water) that would ultimately result in the disaster that would forever follow and become indelibly associated with the factory. Prior to eliminating the transfer pumps, there was no incentive for introducing water to the process. If necessary, the MIC vapor transfer header pipes could be cleaned during planned shutdowns and only after completely removing and isolating the process. By changing the MIC transfer method, the simultaneous use of water or steam on the live process became normal and customary. It became a part of routine maintenance, executed to maintain MIC purity and enable the level in the tanks to rise while receiving rundown from the MRS. Eliminating the transfer pumps shifted the accountability for effectively controlling online process cleaning hazards to the workers.

Ironically, the failure mechanism that required isolating the process effectively now made it more difficult to isolate the process. In other words, the valves that needed to seal off water from the process were more likely to leak due to the problem that necessitated online cleaning in the first place. With these complications in mind, valve leaks and the maintenance needed to restore their sealing function represented a significant hazard. Even if blinds were installed, the valves would have to seal firmly in order to prevent a release upon breaking containment. Thus, the only way to mitigate the hazard was through personal protective equipment (PPE), under the acceptance that a process release would be likely upon breaking containment. Mitigating this hazard depended on the performance of PPE and the staff who wore it. Referring back to our discussion in chapter "Incident Investigation," PPE is recognized as the least effective form of protection. Therefore, with repetitive maintenance events an exposure defect could be anticipated. Any permutation of this scenario could be deadly. An inevitable process exposure incident could be predicted on any maintenance-dependent process where PPE is the only form of reliable protection. Referring back to our discussion on safeguards, independent protective layers and safety credits (chapter: Process Hazard Awareness and Analysis), PPE, and human performance alone could never statistically adjust the risk of a consequence category 5 incident to an acceptable level by modern industrial standards.

Pipe cleaning at the Bhopal factory was a completely human-dependent activity. Thus, the net result for eliminating the transfer pumps was more routine corrective maintenance. Since this maintenance procedure was carried out by workers already on the hourly payroll, they were expected to provide this function as part of their normal and customary duties while on the clock. Chances are that prior to being assigned this additional responsibility, the workers already had no trouble staying occupied attending to other process needs during their normal shift. For example, the workers still had to contend with repeat circulation pump failures, even after the transfer pumps were no longer used. Depending on the circumstances on any given day, shortcuts were more probable after implementing the alternative MIC transfer method. Adjustments might be needed to schedule the additional work into a normal shift. Unless the workers' routine responsibilities were distributed more evenly, the limits of human performance could be exceeded (chapter: Equipment Reliability Principles). Human error capable of irreversible acts of commission and omission will occur upon exceeding these limits. This explains why the Bhopal factory's work force swelled to over 1000 employees during its peak year of production [29]. Increasing the work staff numbers to accommodate more maintenance and operating responsibilities might have helped to avoid making irrecoverable mistakes.

However, employing more people to implement excessive, unbudgeted, and unforeseen corrective maintenance is only temporarily effective. Increasing the worker count might be acceptable for a limited amount of time; it might get a facility through a certain difficult operating period needed to restore normal operation. However, the alternative MIC transfer method was intended to be *permanent*. Adding more workers to the factory payroll to manage the increase in normal, routine maintenance may have seemed favorable in the eyes of those who were hired. But for the Bhopal factory, it was an economic death sentence. The factory did not break even until 1981 when the change in MIC transfer methods was implemented. Although the factory did in fact break even after making the change, an extraordinarily large number of workers were needed to support basic maintenance and operation functions inside the factory. Examining the repeating patterns demonstrated by the partnership, it was only a matter of time before the next major problem would get the attention of investors, who expected their investment to generate a profit. Breaking even financially is required for a business to operate. But in a free market society, unsubsidized commercial businesses must generate a profit if they are to survive. Therefore, at some point in the near future, something would have to change. The Bhopal factory would have to generate a profit and not just break even. Layoffs could certainly be expected if a profit was not generated. The unprofitable factory might even have to be sold. Avoiding either of these outcomes would require a capital project to reconfigure the factory with stainless steel PVH and RVVH pipes and components according to original design specification. Alternatively, the problem responsible for low MIC pump MTBF would have to be diagnosed and corrected to return design operation to the factory. Those were really the only two options. Abandoning the MIC transfer pumps solved one problem in exchange for multiple other problems. As we will see the selected short-term, improvised, "economical" workaround solution would seriously impact the factory's operability in the long run.

LESSONS FOR US

Management of change (MOC) programs are widely practiced throughout modern industry. However, MOC programs are human-dependent processes that are just as likely to fail as any other industrial process under human control. Properly administering a change avoids numerous other problems; some potentially worse than the purpose for the original change. MOC programs that are implemented for the mere purpose of complying with a regulatory edict often consist of going through the motions. But just going through the motions will offer no true protection from a catastrophic process failure. In fact, aimlessly going through the motions often introduces more immediate hazards—ones that result in even more serious operating difficulties. It is important to factor-in how a thoughtful and truly professional failure investigation might make a proposed change unnecessary. Such thoughtful pursuits could avoid the calamitous situation where change begets change. In the rare event that a process change requiring an MOC is required, a hazard assessment should be allowed to reject the change if moving forward increases human dependence or involves uncertainties. Never would it be appropriate to use an MOC to validate a decision that has already been made. Additional lessons learned by examining this sequence of events behind the Bhopal disaster include:

- Casual compliance can be harmless at first, but ends in tragedy.
- When in doubt, err on the safe side and initiate an MOC.
- Never accept PPE as the only protection against a potentially severe process exposure hazard.
- Be cautious that solving a problem does not simply create another problem in exchange.
- Debris settling or forming in valves can cause process isolation failures.
- Accepting a recurring process fouling problem might require introducing potentially incompatible substances.
- When the opportunity to correct a problem arises, obtain and involve the necessary resources to solve it permanently.
- View every repair event as an opportunity to eliminate a chronic design defect.
- Industrial disasters can be expected to have multiple independent causes.
- Hiring more workers to attend to process performance issues is an unproductive approach to business stability.
- Repeat failures significantly increase maintenance requirements that can destroy production value.

REFERENCES

[1] T. D'Silva, The Black Box of Bhopal, 58, 2006, ISBN: 978-1-4120-8412-3.
[2] T. D'Silva, The Black Box of Bhopal, 37, 2006, ISBN: 978-1-4120-8412-3.
[3] T. D'Silva, The Black Box of Bhopal, 46, 2006, ISBN: 978-1-4120-8412-3.
[4] District Court of Bhopal, India, State of Madhya Pradesh Through CBI vs. Warren Anderson & Others, Criminal Case No. 8460 of 1996, 25, June 7, 2010.

[5] S. Diamond, The Disaster in Bhopal: Workers Recall Horror, The New York Times, January 30, 1985.
[6] T.R. Chouhan, Bhopal: The Inside Story, Carbide Workers Speak Out on the World's Worst Industrial Disaster, 37, 1994, ISBN: 0-945257-22-8.
[7] United States District Court Southern District of New York, Janki Bai Sahu, et al., Against Union Carbide Corporation and Warren Anderson, No. 04 civ. 8825 (JFK), 47, 2012.
[8] National Environmental Engineering Research Institute (NEERI), Assessment and Remediation of Hazardous Waste Contaminated Areas in and around M/s Union Carbide India Ltd., 4, 2010.
[9] I. Eckerman, The Bhopal Saga: Causes and Consequences of the World's Largest Industrial Disaster, 27, 2005, ISBN: 81-7371-515-7.
[10] A.S. Kalelkar, Investigation of large-magnitude incidents: Bhopal as a case study, in: IChemE: Symposium Series No. 110, Preventing Major Chemical and Related Process Accidents, May 10–12, 1988, 559.
[11] Supreme Court of India Criminal Appellate Jurisdiction, Application for Directions to Institute Charges U/S 302 (For Offence U/S 300 (4)) Read with S. 35 of the Indian Penal Code, 1860 Against the Respondents Herein, Curative Petition (Criminal) No. 39-42 of 2010 in Criminal Appeal No. 1672-75 of 1996 in the Matter of Central Bureau of Investigation Versus Keshub Mahindra & Ors., 50, 2011.
[12] American Petroleum Institute (API), API RP 574: Inspection Practices for Piping System Components, 9, 1998.
[13] District Court of Bhopal, India, State of Madhya Pradesh Through CBI vs. Warren Anderson & Others, Criminal Case No. 8460 of 1996, 52, June 7, 2010.
[14] Union Carbide Corporation, Bhopal Methyl Isocyanate Incident Investigation Team Report, March 1985, 5.
[15] T. D'Silva, The Black Box of Bhopal, 55, 2006, ISBN: 978-1-4120-8412-3.
[16] R. Van Mynen, Union Carbide Corporation Press Conference Transcript, March 20, 1985, 5.
[17] A. Agarwal, S. Narain, The Bhopal Disaster, State of India's Environment 1984–85: The Second Citizens' Report, 206, 1985.
[18] Union Carbide Corporation, Bhopal Methyl Isocyanate Incident Investigation Team Report, March 1985, 11.
[19] A. Agarwal, S. Narain, The Bhopal Disaster, State of India's Environment 1984–85: The Second Citizens' Report, 207, 1985.
[20] P. Shrivastava, Bhopal: Anatomy of a Crisis, 46, 1987, ISBN: 088730-084-7.
[21] I. Eckerman, The Bhopal Saga: Causes and Consequences of the World's Largest Industrial Disaster, 26, 2005, ISBN: 81-7371-515-7.
[22] T. D'Silva, The Black Box of Bhopal, 83, 2006, ISBN: 978-1-4120-8412-3.
[23] T.R. Chouhan, Bhopal: The Inside Story, Carbide Workers Speak Out on the World's Worst Industrial Disaster, 33, 1994, ISBN: 0-945257-22-8.
[24] K. Torzewski, Valves: chemical engineering facts at your fingertips, Chem. Eng. Mag. 115 (8) (2008) 49.
[25] Health & Safety Executive (HSE), The Fires and Explosion at BP Oil (Grangemouth) Refinery Ltd: A Report of the Investigations by the Health and Safety Executive into the Fires and Explosion at Grangemouth and Dalmeny, Scotland, 13 March, 22 March and 11 June 1987, 12, 1989.
[26] District Court of Bhopal, India, State of Madhya Pradesh Through CBI vs. Warren Anderson & Others, Criminal Case No. 8460 of 1996, 47, June 7, 2010.

[27] District Court of Bhopal, India, State of Madhya Pradesh Through CBI vs. Warren Anderson & Others, Criminal Case No. 8460 of 1996, 56, June 7, 2010.

[28] T. D'Silva, The Black Box of Bhopal, 43, 2006, ISBN: 978-1-4120-8412-3.

[29] T. D'Silva, The Black Box of Bhopal, 59, 2006, ISBN: 978-1-4120-8412-3.

CHAPTER

Process Isolation and Containment* 17

EVENT DESCRIPTION

As the end of 1981 approached, the Bhopal factory was set to finish the year on a good note. Abandoning the methyl isocyanate (MIC), transfer pumps had reduced the frequency of production interruptions. As a result, factory output peaked to its highest level since manufacturing operations got started in 1979 [1]. The productivity increase could be credited to a creative, new method of operating the MIC storage tanks. The solution involved simply closing the MIC storage tank vent valves to build up nitrogen pressure. It was that easy. The modification was made without having to ask permission for a capital project to reconfigure the process. In fact, the solution was already embedded in the process. Using pressure differential to replace the MIC transfer pumps reduced both process exposure incidents and invasive maintenance requirements. The solution therefore appeared to be an inherently safer design option—a bonus that made this particular change even more appealing.

Implementing this change required accepting a relatively minor maintenance inconvenience. Interrupting the nitrogen flow at the storage tanks allowed MIC trimer to start accumulating inside the process vent header (PVH). Eventually, these trimer deposits would restrict vapor passage through the PVH; thereby causing a process upset inside the MIC refining still (MRS). This problem was managed by regularly washing out the iron pipes and valves in MIC vapor service with water. Cleaning out the vapor line running between the MIC storage tanks and the vent gas scrubber helped keep MIC rundown quality on-spec for the duration of a production campaign. MIC disposal and reprocessing costs, and ultimately production constraints, were avoided by implementing this routine maintenance procedure. More workers had been hired to support the various additional operations and maintenance functions that were needed to maintain top production rates, which amounted to only 53% of what the factory was licensed for manufacturing. This production volume was sufficient for the factory to break even financially since becoming a manufacturing site [2]. Perhaps more improvements were expected in 1982, when the factory would surely be expected to start generating a profit.

But the prospect of remembering 1981 as an impressive year ended on December 24, 1981, when a maintenance worker was exposed to the lethal process [3]. Early that morning, a small maintenance crew consisting of three workers assembled at the MIC synthesis section to finish a repair that had started 2 days earlier [4].

* For timeline events corresponding to this chapter see pages 438–442 in Appendix.

Rethinking Bhopal.

Isolation blinds would first have to be removed. The process would then be bolted back together and restored to service.

Removing the isolation blind required breaking containment to access a removable insertion wedged between two pipe flanges. Since there was no way to verify positive isolation at this particular location, the workers had been instructed to wear fresh air–breathing masks [5]. This practice was required by policy in cases where bleeder valves were not available to check the pressure in isolated sections of piping before breaking containment (chapter: Licensing).

As one of the workers unbolted the flange to remove a slip blind, the process sprayed out, onto his sweater [6]. This meant that the block valve used to isolate the pipe section was leaking [4]. Hidden from the workers, the leaking valve had allowed a small amount of process to migrate into the cavity behind the blind. The process was waiting to be released by the unsuspecting person who would remove the blind.

Fortunately, the worker was wearing a fresh air–breathing mask in accordance with safety policies. With urgency, the exposed worker went to a nearby safety shower to wash the liquid process off his sweater. However, a residual amount of volatile process was still evaporating from his clothing when he removed his breathing mask [6]. Upon inhaling, some of the toxic vapor entered his lungs. He felt the burning sensation and was immediately taken to the on-site medical department for evaluation. The factory's medical staff examined the worker. He checked out okay and was later released.

While out with his family later that day, the worker who was directly exposed to the process started coughing up blood [7]. His wife called the factory, which immediately sent an ambulance to take him to a local hospital. Upon arrival, his condition rapidly deteriorated. The worker died on the evening of December 25, 1981, from complications of process exposure [5].

The grim reality of a fatal incident awakened the workers from their hypnotic state of acceptance. Ironically, the incident occurred while the worker was extracting a device that was intentionally installed to control a possible exposure hazard. In the worker's minds, an asset reliability problem had caused his death. A closed valve used to isolate the process had leaked. Chronic valve leaks were ordinary in the factory. If the valve had not leaked, then the process would not have been inside the pipe when containment was broken. If the process was not so maintenance-intensive, then the worker would not have been in the line of fire to begin with. To the workers, the problem could be narrowed down to the hazards of working with faulty valves [4].

However, site supervision disagreed. Their assessment of the situation was quite different. Their impression was that the worker had brought this unfortunate tragedy upon himself [8]. According to policy, the worker was wearing fresh air–breathing equipment. But in violation of policy, he was not wearing the required rubber apron intended to shield anyone who might come into contact with the process during maintenance. Although he went to the safety shower to rinse the chemical off his clothing, he removed his breathing mask too early. As a result, he inhaled some of the process still evaporating from his sweater. He failed to execute the decontamination procedure properly. The situation was completely under his control. Therefore the supervisors held the deceased worker accountable for this tragic incident that had tarnished the factory's enviable safety record [21]. While supervisors and workers alike both grieved over what had happened, the hazard was known and the policy was clear [9]. The fact that the valve used to isolate the process had leaked was irrelevant. In fact, knowledge of chronic leaking valves should have convinced the worker not to stand in the line of fire while loosening bolts. Therefore the supervisors considered the matter closed.

Pointing the finger back at the deceased worker did not exactly resonate well with the line organization. The incident made the line organization increasingly paranoid over a process that seemed to be getting more difficult to contain [4]. Leaking valves had become a common occurrence inside the Bhopal factory [10]. Supervision's indifferent attitude was not comforting to the workers who were affected by the hazards of leaking valves. However, both sides could agree that no matter what its cause, a similar incident could never happen again [9]. But since supervision did not think leaking valves were a problem, a solution would have to come from the line organization. Accordingly, the workers started to avoid installing blinds unless absolutely required. A fatality had occurred while removing a blind. Implementing this workaround solution would reduce the probability of another incident that might affect more workers at risk.

While this approach to improving process containment addressed one concern, other recurring process containment issues were still not resolved. Feelings of concern were transformed into requests for resolution. Any situation where the workers had reason to believe that there might be a potential exposure hazard was brought up for discussion. In the case of the unreliable and maintenance-intensive MIC circulation pumps, the workers' union insisted on specific design modifications on behalf of the workers who were now fearful that the MIC might not be as harmless as they wanted to believe [11]. After all, an exposure incident had just ended in a fatality. If no improvements were made in other areas where process releases were expected, could this incident perhaps be a warning sign foreshadowing problems yet to come? Under the circumstances, requests to apply more protection than personal protective equipment (PPE) alone could provide certainly seemed to be justified.

On January 7, 1982, an MIC circulation pump that had been undergoing maintenance was returned to service [12]. The MIC circulation pump had been shut down to replace a defective mechanical seal. This maintenance event provided an opportunity to upgrade the seal to a less-reactive material—perhaps a nonmetallic alternative. However, central engineering did not authorize changing the mechanical seal design for several reasons:

1. The specified, approved seal for MIC service was metallic.
2. A metallic seal had worked fine in the domestic factory.
3. A force-related failure mechanism (not time, reactive environment, or temperature) seemed to be involved. A nonmetallic seal would only address a potential reactivity problem (not force-related). In the opinion of central engineering, this was not the problem.
4. A nonmetallic seal design might not have the same strength as the specified metallic seal. Thus, switching to an alternative, weaker, nonmetallic seal design might make the problem worse. If the suspected force-related failure mechanism was indeed involved, then the seal might not last as long and could fail even more catastrophically than before.

After only 2 days of continuous operation the newly repaired circulation pump seal blew apart in an unprecedented catastrophic failure [13]. The seal's life cycle was far less than its anticipated mean time between failure of 24 days. The massive release of cold MIC that followed covered a widespread area. When the situation was brought under control, about 25 workers had suffered serious inhalation injuries and were immediately sent to the hospital for intense medical treatment. None of the exposure victims were wearing fresh-air breathing masks when they responded to the incident [14]. It took 5 days before all of the victims were released from the hospital [15]. Fortunately, none of their injuries were fatal.

Supervisors were deeply concerned over the sudden turn of events. A massive leak such as this was completely out of the ordinary compared to the relatively minor (and manageable) releases that had occurred many times before [22]. Another difference was that the new mechanical seal had worked for only two days before it catastrophically failed whereas previous seals had lasted about 24 days on average. Its destruction produced a leak so widespread that multiple workers had been injured.

A preliminary investigation determined that a ceramic seal had been installed instead of the recommended and approved metallic seal [15]. Site supervision was not amused by this discovery and considered it an act of sabotage [16], orchestrated perhaps by workers who had previously expressed dissatisfaction over allowing the problem to persist by not trying out a different seal. If this *was* a deliberate action, then it was a rebellious act of misconduct that had only made matters worse. At that point, supervisors took over the investigation to determine who was responsible for installing the improper seal.

The attempt to prove a deliberate act of sabotage was not substantiated. That is not to say that supervision's claim was wrong. In the minds of those in charge of factory performance, this is what happened although they could not prove it. Under the circumstances, it was hard for them to accept that the sequence of events was merely a coincidence. Immediately after the fatality, the workers' union had requested specific design changes to control repeat process releases and exposure hazards [23]. A ceramic, fouling-resistant mechanical seal was then installed, in the very next MIC circulation pump to be put back into service. This timing of events was suspicious. But the investigation could not clearly establish that a deliberate action was involved. The maintenance workers who installed the incorrect type of mechanical seal denied any participation in a subversive activity. It was perhaps an honest mistake made by the procurement department or the seal manufacturer. The maintenance workers completed the repair and safely restored the system to service exactly as they had done numerous times before.

Concurrent with the internal investigation, the catastrophic pump seal failure triggered a third-party investigation, led by the state factory inspector and the chief inspector of factories [15]. The independent investigation finalized two recommendations:

1. Regularly test the pump seals.
2. Regularly maintain the pump seals.

As is customary in many instances involving a serious process safety incident, immediate corrective actions are normally applied as a condition for restart [17]. The catastrophic circulation pump seal failure proved beyond a shadow of a doubt that MIC releases could represent a serious hazard. With this incident, there was instant alignment that repeating MIC pump seal failures could not be tolerated. Therefore mandatory changes would have to be applied to prevent further incidents.

Accordingly, on January 12, 1982 (3 days after the catastrophic process release), the factory issued a technical bulletin, stating that the MIC circulation pumps would be shut down immediately [12]. The circulation pumps would be switched on only when absolutely needed, which was when the MIC was being produced [18]. In other words, the MIC circulation pumps would be allowed to operate only when MIC from the MRS was flowing into the rundown tanks. That was also the only time that vent valves were allowed to be open.

TECHNICAL ASSESSMENT

By way of recap, note that this chapter includes recurring patterns that have been described in previous events. Improvisation and hazard exchange are the two most prominent patterns. In this chapter, we also see the return of casual compliance. We find it in a workaround procedure to reduce the potential for an exposure incident while inserting or removing blinds. These patterns were observed in the preceding chapter as well. But the sequence of events that develops in this chapter also contains three new patterns common to all industrial disasters, which are as follows:

- Warning signals,
- Hazard amplification, and
- Tribal knowledge.

A careful examination of the background to these two new patterns defines the source of decisions and actions that, individually, make no sense at all. Beneath this description, we learn valuable lessons about how a workplace fatality can increase the potential for more serious incidents to follow. This is one of the primary reasons why attention must always center on eliminating hazards that could lead to a serious injury or fatality.

As was described previously, basic inherently safer design practices were incorporated into the Bhopal factory's process configuration. For example, weld joints (Fig. 17.1) are often maximized in processes where leaks cannot be tolerated. Compared to flange joints, weld joints provide no discontinuities where a leak might be more likely to appear. However, flange joints provide a convenient location to insert solation blinds if maintenance should be required while another part of the process continues to operate. Mounting flanges face-to-face to seal a joint, however, makes inserting blinds difficult; especially if the pipe is held in place by stationary supports. Installing a spacer in the joint between two flange faces provides a physical gap where a blind can easily be slipped in and out. These provisions are often included where routine online maintenance is expected. However, installing a spacer doubles the discontinuities in pipe systems where the process is more likely to leak. Compared to flange joints fastened with spacers, weld joints are the inherently safer design option.

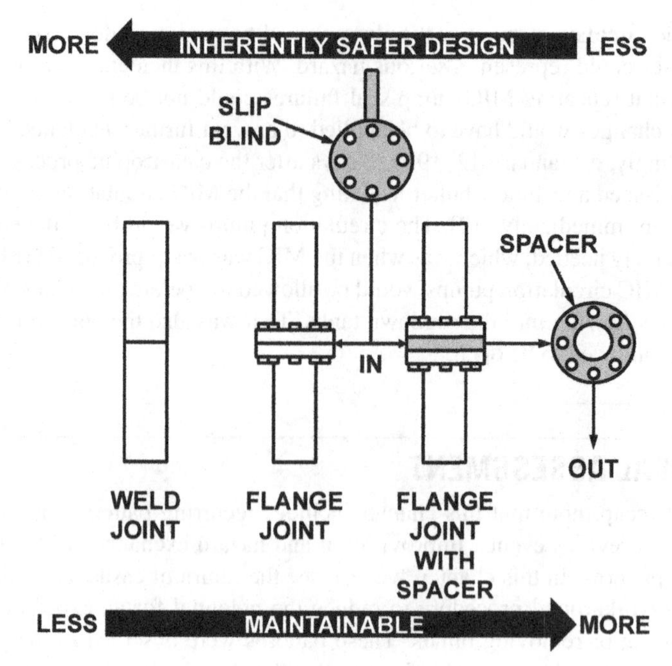

FIGURE 17.1

The relationship between safe and maintainable process piping design.

By viewing this relationship, it is evident that there is a trade-off between inherently safer process design and maintainability. Inherently safer processes must be supported by asset reliability. In cases where asset performance does not support reliable operation, invasive maintenance can become extraordinarily difficult and thus *unsafe* on processes built according to inherently safer design principles. For example, removing corroded sections of piping with weld joints requires cutting out sections of piping, perhaps by performing hot work on a potentially flammable process such as the one found in the Bhopal factory. However, flanged pipe spools can be swapped in and out with little difficulty, without introducing the potential for hot work. The trade-off for convenient (and thus safer) maintenance is a process that is more likely to leak.

Flange connections are not nearly as leak-restrictive as weld connections—unless of course corrosion is interfering with process containment. In that case, the inherently safer design option becomes significantly more hazardous because maintenance must be performed on a process not designed for this type of activity. Under these circumstances, pipe clamps or weld overlays might be used as an interim measure that allows working the process into a convenient turnaround cycle when damaged piping can be completely exchanged with the process offline. Otherwise, the process might have to be shut down and completely drained and decontaminated for corrective maintenance. The financial implications for making this type of unplanned repair are significant.

FIGURE 17.2

Bhopal factory relief valve vent header piping configuration.

Fig. 17.2 shows how inherently safer design principles were incorporated into a section of the Bhopal factory's relief valve vent header. In this process section, we see a wedge gate valve flanged into piping with multiple sections connected by weld joints. The upper and lower horizontal pipe runs are supported by stationary beams to stabilize the pipe. Inserting an isolation blind in either flange connection appearing in the photo would be extraordinarily difficult. The beams holding the pipes in place resist pipe movement needed to spread the flange connections apart. Considerable effort would be needed to install a flange in either of these locations. Isolating the process with blinds might even require modifying the pipe configuration. Modifying the joint to include a spacer would provide a more practical option, if this joint was expected to be blinded periodically for maintenance on one side of the valve but not the other. However, a spacer would not be installed here unless online maintenance requiring process isolation was intended by design. Since corrective online maintenance was not expected, there is no spacer. Performing isolation at this location would be manipulating the process in a way that was never intended. As explained earlier, it could, therefore, not be done safely—or as safe as it could if the design was configured appropriately for maintenance and not inherent safety.

A more practical blinding option in this particular location with the process configured as shown would be to completely remove a spool piece, such as the block valve, from the example piping (Fig. 17.3). Upon removing the valve, both open ends of the remaining pipe could then be blinded and closed with little difficulty. This approach, however, represents the least inherently safe option for process isolation. During blind insertion and removal, there is no way to isolate the process. Therefore this option would create the highest potential for a process release upon extracting the valve. The safest option in this process configuration, in terms of avoiding an accidental process release, would be to simply close the valve. Although the valve

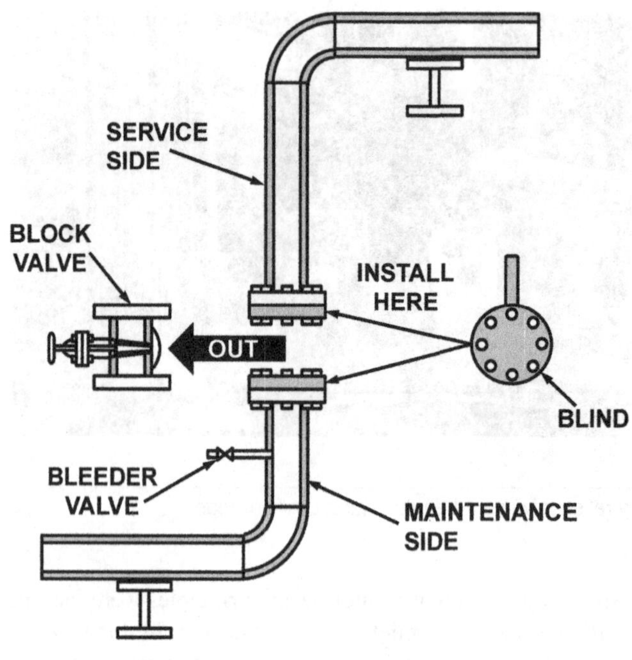

FIGURE 17.3

Alternative isolation method for rigid pipe structures, involving spool piece removal.

might leak, workers would be protected from any possible release scenario if the decision had been to break containment to insert a blind or remove the valve.

This appears to have been the workers' response to the fatal incident that occurred while removing a blind. Installing a blind for maintenance is not usually as hazardous as removing a blind in preparation for restart. The reason for this is that usually there is a bleed valve or some other means to confirm that isolation has been achieved before breaking containment to *insert* a blind. On the other hand, provisions are less likely to exist for confirming the integrity of isolation when containment is again broken to remove a blind.

For example, the photograph in Fig. 17.2 shows a block valve configuration that includes a horizontal pipe extension immediately beneath the block valve. It is unknown whether this small-diameter pipe served as a pressure tap or a bleeder valve extension. As photographed, there is neither a gauge nor a valve attached to the end of the pipe. Regardless, for this example we can assume that a valve (bleed valve) can be screwed onto the end (Fig. 17.3). A pressure gauge can then be fastened to the bleed valve. Upon closing the block valve, it would be possible to test the pressure to see if the "maintenance side" of the block valve is isolated from the "service side." If the pressure on the "maintenance side" is the same as the pressure on the "service side," then the valve is leaking and proper PPE must be worn before attempting to separate the flange, or performing maintenance if only the valve is to be closed for isolation. Depending on process conditions such as the severity of the process or the actual pressure reading on the gauge, it might not be safe to open the joint even when protected by PPE. In some cases, maintenance would need to be deferred until a time that the system is completely relieved of all pressure, which would again relate to a major production shutdown.

However, verification of isolation is not possible upon removing a blind at this location. Notice that inserting a blind in the flange between the valve and the pipe extension isolates the bleed valve from the block valve. Any process that might have leaked into the cavity between the valve disc and the blind would, therefore, be released upon spreading the joint to remove the blind. From the description of the incident involving the worker fatality in the Bhopal factory, this is exactly what happened. Upon returning to an isolated pipe section to remove a blind, a valve close to the blind had leaked [4]. Although the worker was protected by a fresh-air breathing mask, the process sprayed on his sweater. The worker acted in accordance with his training and immediately entered a safety shower to wash the process off his clothes. However, he did not successfully complete the decontamination procedure before removing his mask. As a result, he inhaled some of the lethal process and died the next day.

It is conceivable that a workplace fatality could have a positive impact on a defective organizational culture. Unfortunately, the opposite is more often true. This certainly was the case in the Bhopal factory where, immediately after the fatal incident, the local officers blamed the worker for bringing about his own fate. This attitude of indifference did not satisfy factory workers, whom were offended by the insensitivity shown toward the hazards created by chronic, severely leaking valves throughout the factory.

Prior to this incident, there appeared to be a cooperative effort between all departments to modify the process for improved performance. Indeed, in 1981 these efforts appeared to have begun paying-off, with the factory's production being raised to a new record—sufficient enough to break even for the first time since becoming a manufacturing location. The fatality event, however, drove a wall between site supervision and the line organization. After this incident that involved pinning the blame on the victim, the workers acted independently to protect themselves from further harm. They recognized that performing "normal" maintenance functions on the process put their lives in danger. The same hazard affected each of them equally. The way they saw the problem, they were all vulnerable to the same fate represented by the chronic hazard of leaking valves. Leaking valves were prevalent throughout the Bhopal factory.

Supervisors, however, disagreed with the workers and made perfectly clear their position that the hazard was limited to the mindless action of one single employee [13]. Therefore no changes would be made to prevent further incidents. Trained by their superiors to respond differently, the workers solved the problem exactly the same way that had been used previously, though improvisation. After the fatality that occurred when a worker was removing a blind, the workers resisted inserting blinds for routine maintenance.

This action followed the same precedent set forth by disabling the transfer pumps, which included reducing the failure probability by eliminating the task. Similarly, a new hazard was created by the action taken, which replaced the normal blinding practice in an effort to mitigate personal exposure hazards while executing routine maintenance. The fact is that the Bhopal factory was not designed to easily accommodate routine invasive maintenance. According to anticipated design conditions, invasive maintenance between scheduled turnaround cycles was not expected. Operating the process contrary to this design intention was, therefore, inherently more hazardous.

Within this incident we see the origination of tribal knowledge inside the Bhopal factory. Notice again that up to this point in the Bhopal factory's history the division between local supervision and the workers was transparent. Both sides were pulling in the same direction. The incident exposed a true hazard that remained hidden to supervisors. Thus, a separation between the two sides started to form. Since the workers sensed that supervision was unsympathetic to their concerns, they developed a workaround procedure to protect them from further harm. The fatal incident reinforced a developing concern that as time progressed and more changes were applied, the process was becoming more difficult to control and dangerous to operate. Acting independently, the workers introduced a practice to mitigate the hazard that was involved in the fatal incident.

It is not surprising to see union interests starting to surface during this part of the timeline as the workers begin acting independently. Likewise, after the fatal incident the workers' union continued to demand specific design modifications to make the process safer. It is important to note that the public record does not clearly state that supervisors blamed maintenance workers for inserting the inadequate or improper mechanical seal. But in the context of the uniformity in message represented by the consistent patterns repeating through the sequence of events, the discovery of an alternative, nonmetallic seal favored the continuum of creativity and innovation used to solve persistent process-related problems. If a deliberate attempt to improve process containment by inserting a less-reactive seal was involved, then the workers who felt empowered by previous experience to innovate would have expected to be rewarded by their action as they likely had been repeatedly in the past. This expectation was predicated on their certainty that installing an unauthorized mechanical seal would solve the problem. They would not, however, have contemplated what would happen if their actions should make the problem worse. They were not used to this kind of penalty following the performance improvement demonstrated by increasing production volume. By the end of the third year in the manufacturing business, the workers and factory supervision alike were conditioned for success and rewards by taking immediate action to solve problems; perhaps without fully understanding the problem to begin with.

The deliberate decision to disable the refrigeration system was sharply criticized after details behind the Bhopal disaster became public. However, looking at the scientific basis for this specific action, there seems to be a justification behind disabling the refrigeration system—or at least converting it to intermittent operation. Notice that according to industry guidelines covered in chapter "Process Selection" (Table 8.2), pressure vessels can be used to store highly volatile compounds *without* refrigeration. However, when atmospheric tanks are used, compounds with high vapor pressures must be refrigerated.

The modified storage tank operating procedure that was introduced on January 12, 1982, provided an immediate solution to the ongoing reliability problem associated with the MIC circulation pumps. The procedure was introduced to alternate the MIC storage tanks between atmospheric and pressurized conditions. Process operating manuals were later updated with specific instructions to comply with industry guidelines for refrigerating the product in the rundown tanks when they were being filled [19]:

> *Circulate MIC through the refrigeration unit and maintain tank temperature around 0°C. Once tank contents are chilled and no more MIC is being made into this then the circulation can be stopped…*

Notice that the procedure, although more complicated than the previous expectation for continuously running the refrigeration system, applied only to MIC *circulation*. The refrigeration function was provided by switching the MIC circulation pumps on. If MIC was not being "made," meaning that no MIC was running down from the MRS, then pressurized conditions would apply inside the tank. Under these conditions, the rundown tank vent valve would be closed and the MIC circulation pump would be shut off. Therefore the refrigeration assets (compressor, fluorocarbon loop, etc.) were left operating at all times. If refrigeration was needed, then the MIC circulation pump was turned on. However, most of the time the circulation pumps could remain off.

This discussion is not to suggest that switching the MIC circulation pumps off at certain times was a responsible practice. History would clearly suggest otherwise. However, it is important for readers to understand how similar circumstances apply within the manufacturing processes that we operate today. Unless the refrigeration system was specifically designed to provide a safety function, its usage was at the jurisdiction of the production team. In the case of the Bhopal factory, the refrigeration system was not designed to provide a safety function (chapter: Process Selection). Without this distinction, disabling it would simply allow the product to evaporate. Under these conditions, only economic penalties might result when the refrigeration system was turned off. Restoring the refrigeration function only when MIC vapor could escape the storage tanks, which was when a vent valve was opened, mitigated this consequence. Therefore it was a valid option that most industry professionals would be inclined to consider under similar circumstances today. Unfortunately, this specific change did not address the root cause for circulation pump failures. Neither was this change supported by the prevailing process design basis or configuration. Periodically shutting down the circulation pumps for extended periods of time introduced a voluntary common cause failure that disabled multiple safeguards (chapter: Process Hazard Awareness and Analysis). Therefore the change created additional hazards that would then have to be discovered and managed separately.

The lesson contained in this discussion, again, is that no change can be made without consequence. When considering a change—any change—ask first if you truly understand the problem. Under no circumstances should a change be made unless the problem can be solved on the basis of what is causing it. This might require exercising patience until an investigation can diagnose the root cause and a practical solution is recommended. Only then would it be safe to implement a change in full awareness of the consequences.

On the contrary, in this case the operation of the refrigeration system was modified to provide an immediate response to an unprecedented process release from the circulation pumps. Prior to the release on January 9, 1982, no other leak at the Bhopal factory had come close, in terms of its volume. Unlike the fatal incident which factory leaders were able to dismiss on the basis that it was an isolated event, the magnitude of the second release immediately put an end to the acceptability of repeat circulation pump seal failures. Based on the recent downturn in the factory's safety performance circulation pump failures that were previously tolerated were now unacceptable. The root cause for the seal failures had not been diagnosed. Therefore the alternative operating method was introduced to reduce the time at risk. If the MIC circulation pumps were not running, then they could not fail.

The action taken in response to the catastrophic MIC release was consistent with the two patterns that have started to define the partnership (improvisation and hazard exchange). Interestingly, we begin to see how the workers had fully adopted the qualities of the parent company. The exclusion of slip blinds, for example, was not really a shortcut. A shortcut is done to save time. In this specific case, avoiding the use of slip blinds appears to be a reaction for survival. Essentially no action was taken to protect the workers from further harm after a fatality resulted from the improper use of PPE. The workers and local supervision were not aligned on the primary cause for this incident. Therefore the workers took solving the problem into their own hands. This may also be one of the reasons why tampering with the process seemed logical after the catastrophic circulation pump mechanical seal failure. In the context of predictable patterns that were frequently observed within the Bhopal factory, employees that were not part of the line organization would likely have suspected worker involvement even if an honest mistake truly had been made. The net result of these events was a division that started growing inside the factory whereas up to this point, both sides had acted cohesively.

Before moving on, it is important to point out that in retrospect, the fatal incident on December 24, 1981 involving a minor process release was a warning signal. This incident led to a local change (modifying blinding practices) that solved one problem while creating another. Likewise, the catastrophic loss of MIC containment on January 9, 1982, was a second warning signal. A change immediately followed this event also. Intermittent usage of the refrigeration system further amplified the hazards related to operating the Bhopal factory. There was not a third discrete warning signal [20]. The next significant process release from the Bhopal factory would be recognized as history's worst industrial disaster.

After an industrial disaster, it is not uncommon to lament having not acted responsibly with respect to warning signals. By evaluating the sequence of events in the Bhopal factory, we notice how warning signals can be missed by acting to address a problem without first diagnosing its root cause. We also see how responding without this knowledge can actually make a process more dangerous to operate. It is, therefore, important to recognize that significant events in the processes we operate today represent warning signals. Resisting the temptation to act impulsively makes it possible to avoid responding in a way that creates more serious hazards.

LESSONS FOR US

A close look at the history behind two consecutive loss of primary containment incidents within the Bhopal factory demonstrates the importance of maintaining continuous process containment. Process release incidents create distractions that can make the process more dangerous to operate. It is important to avoid making hasty decisions after a serious asset failure, process release, personal injury, or fatality occurs. It is much easier to destroy than improve organizational culture in response to such events. The best solution is to prevent any process incident that could create misalignment between individual work groups. Also keep in mind that:

- Personal safety statistics do not accurately reflect the stability or safety of a process.
- Inherently safer design practices compete with the maintainability of a process.
- The example you set (good or bad) will rub off on the people you work with.
- Serious incidents always require an immediate correction, but the investigation must outline a long-term solution.
- The actions taken to reduce the chance for a personal injury can increase the potential for a more serious process safety incident.
- Concentrate on the things that you can control; not the things that you cannot control.
- Hazard alignment cannot be achieved unless everybody views risk the same way.
- All industrial disasters have clear warning signals.
- Incident response can never be as effective as incident prevention.

REFERENCES

[1] National Environmental Engineering Research Institute (NEERI), Assessment and Remediation of Hazardous Waste Contaminated Areas in and around M/s Union Carbide India Ltd., 4, 2010.
[2] T. D'Silva, The Black Box of Bhopal, 83, 2006, ISBN: 978-1-4120-8412-3.
[3] District Court of Bhopal, India, State of Madhya Pradesh through CBI vs. Warren Anderson & Others, Criminal Case No. 8460 of 1996, 18, June 7, 2010.
[4] T.R. Chouhan, Bhopal: The Inside Story, Carbide Workers Speak Out on the World's Worst Industrial Disaster, 34, 1994, ISBN: 0-945257-22-8.
[5] T. D'Silva, The Black Box of Bhopal, 72, 2006, ISBN: 978-1-4120-8412-3.
[6] D. Lapierre, J. Moro, Five Past Midnight in Bhopal, 174, 2002, ISBN: 0-446-53088-3.
[7] D. Lapierre, J. Moro, Five Past Midnight in Bhopal, 175, 2002, ISBN: 0-446-53088-3.
[8] I. Eckerman, The Bhopal Saga: Causes and Consequences of the World's Largest Industrial Disaster, 38, 2005, ISBN: 81-7371-515-7.
[9] D. Lapierre, J. Moro, Five Past Midnight in Bhopal, 176, 2002, ISBN: 0-446-53088-3.
[10] A. Agarwal, S. Narain, The Bhopal Disaster, State of India's Environment 1984–85: The Second Citizens' Report, 219, 1985.
[11] T.R. Chouhan, Bhopal: The Inside Story, Carbide Workers Speak Out on the World's Worst Industrial Disaster, 1994, ISBN: 0-945257-22-8, pp. 34–36.

[12] Supreme Court of India Criminal Appellate Jurisdiction, Application for Directions to Institute Charges U/S 302 (For Offence U/S 300(4)) Read With S. 35 of the Indian Penal Code, 1860 Against the Respondents Herein, Curative Petition (Criminal) No. 39-42 of 2010 in Criminal Appeal No. 1672-75 of 1996 in the Matter of Central Bureau of Investigation Versus Keshub Mahindra & Ors., 15, 2011.
[13] Supreme Court of India Criminal Appellate Jurisdiction, Application for Directions to Institute Charges U/S 302 (For Offence U/S 300(4)) Read With S. 35 of the Indian Penal Code, 1860 Against the Respondents Herein, Curative Petition (Criminal) No. 39-42 of 2010 in Criminal Appeal No. 1672-75 of 1996 in the Matter of Central Bureau of Investigation Versus Keshub Mahindra & Ors., 16, 2011.
[14] D. Lapierre, J. Moro, Five Past Midnight in Bhopal, 177, 2002, ISBN: 0-446-53088-3.
[15] T. D'Silva, The Black Box of Bhopal, 73, 2006, ISBN: 978-1-4120-8412-3.
[16] A.S. Kalelkar, Investigation of large-magnitude incidents: Bhopal as a case study, in: IChemE: Symposium Series No. 110, Preventing Major Chemical and Related Process Accidents, May 10–12, 1988, 567.
[17] Center for Chemical Process Safety (CCPS), Guidelines for Investigating Chemical Process Incidents, second ed., 262, 2003, ISBN: 0-8169-0897-4.
[18] T.R. Chouhan, Bhopal: The Inside Story, Carbide Workers Speak Out on the World's Worst Industrial Disaster, 58, 1994, ISBN: 0-945257-22-8.
[19] Supreme Court of India Criminal Appellate Jurisdiction, Application for Directions to Institute Charges U/S 302 (For Offence U/S 300(4)) Read With S. 35 of the Indian Penal Code, 1860 Against the Respondents Herein, Curative Petition (Criminal) No. 39-42 of 2010 in Criminal Appeal No. 1672-75 of 1996 in the Matter of Central Bureau of Investigation Versus Keshub Mahindra & Ors., 21, 2011.
[20] United States District Court for the Southern District of New York, In re Union Carbide Corporation Gas Plant Disaster at Bhopal, India in December, 1984, MDL No. 626; Misc. No. 21-38, 634 F. Supp. 842, 2A, 1986.
[21] D. Lapierre, J. Moro, Five Past Midnight in Bhopal, 184, 2002, ISBN: 0-446-53088-3.
[22] Supreme Court of India Criminal Appellate Jurisdiction, Application for Directions to Institute Charges U/S 302 (For Offence U/S 300(4)) Read With S. 35 of the Indian Penal Code, 1860 Against the Respondents Herein, Curative Petition (Criminal) No. 39-42 of 2010 in Criminal Appeal No. 1672-75 of 1996 in the Matter of Central Bureau of Investigation Versus Keshub Mahindra & Ors., 6, 2011.
[23] T.R. Chouhan, Bhopal: The Inside Story, Carbide Workers Speak Out on the World's Worst Industrial Disaster, 36, 1994, ISBN: 0-945257-22-8.

CHAPTER

Employee Participation* 18

EVENT DESCRIPTION

Exactly 1 month after the catastrophic methyl isocyanate (MIC) release that sent two dozen workers to the hospital, the union representing factory workers handed a formal letter of protest over to site supervision [1]. Among the Union's top complaints were the multiple injuries and fatality occurred during the two most recent process exposure incidents. The union took exception to supervision's refusal to own up to the asset failures behind both of these incidents. Instead, supervision had readily blamed these incidents on the mindless actions of factory workers. The letter of protest put supervision on formal notice that these insults were being taken seriously. Moving forward, aggressive action to contain the process was expected.

Supervision's official response to the union's letter of protest was that no additional actions were needed to improve process safety; nor would any action be taken. The process was already safe, and aside from the human error involved in both serious exposure incidents, there was no legitimate basis for any concern. In fact, the innovative design solutions that were implemented to address production issues had actually made the process inherently safer as the factory matured. These advances were responsible for the Bhopal factory's safety performance – which was among the finest, if not the best, in the enterprise [2]. Furthermore, the MIC circulation pumps had just been converted over from continuous to intermittent operation. Less asset utilization meant fewer circulation pump failures that the factory workers had complained about. Collectively, these actions were considered tangible proof that site supervision had always taken the workers' concerns seriously. In view of these developments, the supervisors considered the matter closed. Therefore, workers were encouraged to put their minds back into the business of making technical grade product and end this unnecessary distraction.

The official response to the workers' grievance only widened the growing division between the line organization and supervisors. The response to their letter of protest completely convinced factory workers that their supervisors had no interest in addressing their concerns. The process exposure incidents had occurred during normal operations; not some unusual operating condition that was new to them. The fact was that routine maintenance and operating tasks were creating unacceptable hazards.

*For timeline events corresponding to this chapter see pages 443–444 in Appendix.

Over the relatively brief period that the factory had been in the business of manufacturing technical grade product, the process appeared to have gotten progressively more dangerous. This simple fact was easy for the workers to understand, but supervisors saw matters differently. Had the various changes made to improve factory safety and performance put the process in a compromised position? On the basis of recent events, it certainly appeared to be so. Operating the process was definitely more complicated than it was in the beginning, when all that really bothered the supervisors and workers alike were MIC pump failures. Now there seemed to be a dazzling array of other operating difficulties and reliability incidents that made the MIC pump failures seem insignificant.

For example, process block valves were leaking profusely. Trimer was constantly choking off vapor flow through the process vent header (PVH). Water washing was frequently implemented to remove trimer debris collecting in the PVH. Intense corrosion problems seemed to be dissolving the PVH from the inside out making it necessary to constantly replace damaged pipe sections [12]. The rundown tanks could not be filled when PVH repairs were needed. This introduced another production constraint, similar to what was experienced with overlapping MIC rundown tank maintenance (chapter: Commissioning Phase). The workers had to remember when to run the circulation pumps and when to turn them off. The vent valves that were originally operated in automatic mode were now manually opened and closed at specific times. Operating the process required remembering various local procedure modifications that the workers had developed to protect themselves from process release hazards. Apart from all this, they could no longer "see" what the process was doing. The high-temperature alarm was disabled early in the life of the process [3]. The temperature gauge needles remained pegged-out against high range for extended periods of time [4]. MIC storage tank pressure gauges meant nothing to the workers since normal operation could be anything between 2 and 25 psig. Time flew by but despite the significant energy needed from the workers to control the wayward process, each year they were reminded that production was not good enough. More was expected from them. The factory was rated for 5000 MT annual production, but nothing more than about half that rate was realistically possible. Unfortunately, the production volume outlook in 1982 was not looking that great either. Something had to be done about the chronic PVH leaks and the production constraint that came along with them.

However, site supervision appeared to have complete confidence in the process and the adjustments that had been made to improve performance [5]. If anything, the workers just needed to work harder to prevent incidents that were bringing unnecessary attention to the business. The line organization disagreed and insisted that chaos dominated factory operation. The fact that the factory appeared to be operating well to the supervisors was a reflection of all the industrial choreography and gymnastics that experienced workers had learned to contribute just to keep things from coming apart. Their biggest concern was that the two incidents that affected them personally might be precursors of what to expect the moment they let their guard down.

Feeling insulted by supervision's demonstration of indifference to their complaint, factory workers abandoned their internal appeal to address their concerns.

Instead, they took their complaints to the public, beginning with a propaganda campaign to expose health hazards inside the factory. Pamphlets declaring "Worker's Protest for Better Safety Mechanisms" were distributed throughout the surrounding community in the hopes of getting the attention that the union failed to attract from the supervisors. The pamphlets warned [6]:

- Beware of fatal accidents,
- Lives of thousands of workers and citizens in danger because of poisonous gas,
- Spurt of accidents in the factory, and deficient safety measures.

Rallies and public meetings were held outside the factory gates to engage public support for the workers [7]. Factory supervisors remained tactfully silent during the uprising [2]. This seemed to infuriate the workers even more. The supervisors ended their silence by terminating the employment of the most prominent factory workers who were leading the union uprising [8]. This action was later regarded as the beginning of the campaign's end [9].

Later, a local journalist wrote a visionary article that presented a disturbing image of what Bhopal might look like in the future if the workers' concerns were not taken seriously [10]. His article described an apocalyptic scene where dead bodies would pile up in the streets of Bhopal over a period of about an hour and a half. People asleep would never wake up the next morning after a process release. He also envisioned a lack of community interest until the calamity fell upon them personally, which would then be too late (chapter: Commissioning Phase).

The journalist was right—the public did indeed show little interest in the workers' initiative. The workers also recognized the public's indifference to their message [7]. A growing concern among many of the workers was that the factory might close if through their public demonstrations the factory was to shut down [7]. Under these circumstances, Armageddon would come early by imposing unemployment upon them all. Thus the protest fizzled-out and the workers returned to work.

After the campaign ended, factory supervisors continued demonstrating interest in responding to the workers concerns but on site supervision's terms only. Accordingly, a joint health and safety committee was formed between the line organization and supervision. Maintaining an adequate supply of fresh air breathing masks, gloves, and aprons within the factory's seven operating sections was the purpose for the meeting [9]. System safety performance, operation, and process configuration relative to factory design were not discussed. That discussion was reserved for an independent safety audit that was scheduled for May 1982.

Factory supervisors continued to blame the deceased worker for his own fatality due to circumstances under his control. The debate over the acceptableness of chronic block valve leaks was rarely stoked again. Supervisors still believed that sabotage was somehow involved in the catastrophic MIC release. Dismissing the seriousness of process reliability within the factory was not helping to comfort the workers, who were still grieving the death of a coworker. The gulf between factory workers and supervisors could not have been any larger.

TECHNICAL ASSESSMENT

The sudden occurrence of two consecutive safety incidents involving process releases got the attention of the factory workers. The line organization was involuntarily committed to the inclination or direction in which the process was now moving. In other words, the process was controlling them; they were not in control of the process. However, they sensed this and were therefore devoted themselves to protecting each other from similar harm in any way possible. Herein exists a trap that concerns all workers in the manufacturing industry. Catastrophic failures instill fear into the workforce. When fear begins dominating judgment, it becomes very difficult to communicate effectively. Conversations tend to be led by emotion instead of sound technical knowledge or judgment. Under these circumstances, a bad situation can be made worse when a manufacturing process is involved.

Historical industrial disasters share a pattern similar to the one observed here. Unmitigated hazards are easily recognized by the group of individuals working closest to the problem. Others far removed usually find relating to their concerns extraordinarily difficult—at least initially.

Recall that a similar situation was documented in the timeline leading up to the space shuttle Challenger explosion (chapter: Role Statement and Fulfillment). Similarly, the workers at the Bhopal factory were in direct contact with the process—both contained and potentially uncontained. They were the ones who were closest to the problem. They had intimate awareness about the hazards involved in controlling the process. They had modified their decisions and actions to account for the difficulties involved with operating the uncooperative and unpredictable process. They had important information that needed to be shared with, and fully understood by, site supervisors who had decision rights over how to manage the process. However, the line organization broke communication off with site supervision upon receiving an unfavorable reply to their letter of protest. That is when the workers took their grievance to the public where their complaint was received with similar contempt. Shortly afterward, the protest against unsafe working conditions was organized.

It is important to note at this point in our discussion that the public record regarding the Bhopal disaster does not speak of a "strike" in the context of an action that a union might take against a factory. But the timeline related to these events closely matches that specific description. For example, the public record contains information referring to:

1. A walkout,
2. Distribution of propaganda (pamphlets),
3. Public meetings at the factory gate,
4. Use of media to increase public exposure,
5. Fear of unemployment, and
6. Loss of negotiating power.

This is somewhat significant since strikes are still used by unions today to gain negotiating power. However, strikes represent a risk for union members, who might walk away with less when the strike is over depending on how the employer responds and can maintain productivity during the strike. Whatever term might apply to this event in the Bhopal factory's history, supervision was able to break the morale of the workers, so that they came back willingly to retain employment.After this rebellious action, those with the power and authority to make the necessary adjustments were even less inclined to listen. They had a business to run. Circumstances before the walkout had already raised tensions between site supervision and the line organization. The next step taken by the workers did not help improve communication between the two workgroups that were now seriously divided. A reliable source of information that the supervisors could rely on for making responsible decisions was thereby silenced. The direct knowledge of the workers became inaccessible, and therefore of no value, to site supervision, who desperately needed to know what was happening in reality on the front line.

Technical details concerning risk calibration and alignment are not necessarily easy to understand. In the Bhopal factory, literal language differences might even have contributed to communication problems [11]. Individuals in possession of important information must be determined to make their case convincingly to those who are authorized to take action. That did not occur at the Bhopal factory. Neither did it occur inside NASA the evening before the space shuttle Challenger launch. In both examples, an internal debate over valid risk calibration concerns ended without obtaining the necessary commitments to make changes—up to and including the missions' suspension. On the contrary, those who had the information supervisors needed to make a responsible decision in both cases abandoned their argument upon encountering sufficient resistance.

The purpose for this discussion is not to establish blame for the communication breakdown that occurred between factory workers and supervisors in the Bhopal factory. Constructive use of this information makes it possible to avoid communication breakdowns that could lead to ineffective hazard control. Depending on the consequences of any decision made when a communication breakdown is encountered, all sides bear the burden of guilt—those who walked away from the argument, for allowing a poor decision to be made, and those who made the poor decision, for observing the side they defeated walk away in disagreement. The lesson that results from these examples and others that follow the same pattern is that nobody wins when communication over a technical matter governing process safety, factory operations, or asset reliability breaks down. Abandoning or avoiding a scientific debate that needs to be resolved is a certain prescription for a catastrophic failure. Nobody can make a responsible decision when the support function they depend on walks away or invokes the silent treatment. Those in possession of the information others might need to safely operate the process must therefore patiently, incessantly, and diplomatically work to address their grievances. They must reconcile their differences internally and do so responsibly. Reconciliation of differences is an indispensable ingredient of safe and profitable operation. It is never a mere option.

A careful analysis of events leading up to the Bhopal disaster is evidence enough that bringing the public into such situations is not an appropriate answer. The general public is given neither the power nor authority to act in behalf of the workers. Neither does the public have the process familiarity needed to understand the technical aspects of industrial hazards, until of course those hazards affect them personally. Only the factory supervisors who are charged with the responsibility for keeping the process stable can map out or approve an appropriate course of technical action. Nevertheless, many interests combine to influence or create an environment which addresses the various legitimate needs of all involved parties.

Abandoning the push for reform to address hazards known by some but not all fortifies tribal knowledge. Tribal knowledge describes a situation where those with the information continue to hold on to it. *Holding* is very closely related to *withholding* and both can be dangerous. Because information is held or withheld, unconventional systems are developed to control the hazards by alternative means. Productivity is maintained to the fullest extent by the workers who develop practices to protect themselves from the enemy they fear the most—an unstable process. Unless the needs of production (profits) and protection (safety) are suitably aligned and equitably satisfied, site supervision can potentially be left with misleading or erroneous impressions. Chances are they will stick to the belief that their process meets safety expectations. The company supervisors can then become unaware of the hazards which, for the time being, are managed or kept in check by improvised solutions.

For example, the line organization might implement a local workaround procedure to control an exposure hazard that, by design, should not exist. To the supervisor that does not have to manage this hazard, the hazard is being mitigated by design. The supervisor likely has no knowledge about the unconventional, home-made practice that is covering the hazard. These types of scenarios, again, are sustained by tribal knowledge. Only two things can expose forbidden practices hidden by tribal knowledge:

1. An independent compliance audit or
2. An unacceptable incident that involves a catastrophic process release.

Productive discussions between supervisors and the line organization are needed to avoid tribal knowledge that can create multiple operating levels inside a production site. The purpose for this type of interaction is not to keep track of *what* work is getting done but more importantly *how* the work is getting done. When workers freely admit to practices contrary to documented policies and procedures that they believe are needed to control an unmitigated hazard, a supervisor would do well to address the hazard to the workers' satisfaction. By doing this the incentive for deviating from expected operating methods is removed. Keeping the workers apprised of your efforts is the first big step toward retaining the workers' trust, confidence, and collaboration. The focus must always be on creating a dialog that maintains process operation according to design expectations. When deviations from design expectations are institutionalized into routine operation, anything is possible and bad things can happen.

LESSONS FOR US

Manufacturing organizations maintain stable process control by implementing a productive employee participation program (29 CFR 1910.119). Protection from a catastrophic process incident is lost when an adversarial relationship develops between multiple internal work groups that must cooperate to achieve their employer's mission. Avoiding disappointing results requires cultivating a spirit of trust, patience, and transparency between multiple entities that cannot successfully operate independently.

Focus should always be on having a meaningful technical exchange between members at all levels within a manufacturing organization. A meaningful technical exchange would prohibit superficial topics of discussion that typically hijack many employee participation programs, including:

- Labor relations,
- Nonprocess-related worker grievances,
- Company benefits,
- The distribution and availability of personal protective equipment, and
- Any other topics not directly related to *how* the work in production units is getting done.

A productive employee participation program concentrates strictly on understanding how the process is both operating and being operated. In these terms, company supervisors can usually provide the support needed to address hazards or deviations. Without employee participation, site supervision might not know or might be unable to fully appreciate significant issues. Being informed of modified blinding practices taking root (chapter: Process Isolation and Containment) is just one example that could represent valuable input to site supervision. Ignoring unexpected hazards endangers everyone. As was brought out and underscored many times in this text: Catastrophic process release incidents can be avoided. Avoiding incidents requires reconciling factory performance, verifying intended operation, and ascertaining conformance with original design requirements. Also remember that:

- Divisions can form instantly between people and groups that have long worked constructively together.
- A good technical debate must not end until one side wins the argument and the other side is genuinely comfortable with why they have lost the argument.
- Putting your concerns in writing after you have had the verbal and respectful technical discourse is not disloyalty. It is called professionalism and ethical conduct.
- Tribal knowledge results in communication breakdowns and is not safe.
- A manufacturing business stabilized through tribal knowledge is certain to fail when those who know how to control the process leave, or when new employees who do not know how to control the process are hired.

REFERENCES

[1] Supreme Court of India Criminal Appellate Jurisdiction, Application for Directions to Institute Charges U/S 302 (For Offence U/S 300 (4)) Read With S. 35 of the Indian Penal Code, 1860 Against the Respondents Herein, Curative Petition (Criminal) No. 39-42 of 2010 in Criminal Appeal No. 1672-75 of 1996 in the Matter of Central Bureau of Investigation Versus Keshub Mahindra & Ors., 16, 2011.

[2] D. Lapierre, J. Moro, Five Past Midnight in Bhopal, 184-185, 2002, ISBN: 0-446-53088-3.

[3] T.R. Chouhan, Bhopal: The Inside Story, Carbide Workers Speak Out on the World's Worst Industrial Disaster, 58, 1994, ISBN: 0-945257-22-8.

[4] Supreme Court of India, Criminal Appeal Nos. 1672, 1673, 1674 and 1675 of 1996, in the Matter of Keshub Mahindra vs. State of Madhya Pradesh, September 13, 1996. 16.

[5] D. Lapierre, J. Moro, Five Past Midnight in Bhopal, 184, 2002, ISBN: 0-446-53088-3.

[6] Supreme Court of India Criminal Appellate Jurisdiction, Application for Directions to Institute Charges U/S 302 (For Offence U/S 300 (4)) Read With S. 35 of the Indian Penal Code, 1860 Against the Respondents Herein, Curative Petition (Criminal) No. 39-42 of 2010 in Criminal Appeal No. 1672-75 of 1996 in the Matter of Central Bureau of Investigation Versus Keshub Mahindra & Ors., 18, 2011.

[7] T.R. Chouhan, Bhopal: The Inside Story, Carbide Workers Speak Out on the World's Worst Industrial Disaster, 35, 1994, ISBN: 0-945257-22-8.

[8] T. D'Silva, The Black Box of Bhopal, 76, 2006, ISBN: 978-1-4120-8412-3.

[9] T.R. Chouhan, Bhopal: The Inside Story, Carbide Workers Speak Out on the World's Worst Industrial Disaster, 36, 1994, ISBN: 0-945257-22-8.

[10] R. Keswani, Bhopal Sitting at the Edge of a Volcano, Rapat Weekly, October 1, 1982.

[11] District Court of Bhopal, India, State of Madhya Pradesh Through CBI vs. Warren Anderson & Others, Criminal Case No. 8460 of 1996, 24-25, June 7, 2010.

[12] A.S. Kalelkar, Investigation of large-magnitude incidents: Bhopal as a case study, in: IChemE: Symposium Series No. 110, Preventing Major Chemical and Related Process Accidents, May 10–12, 1988, 561.

CHAPTER

Compliance Audits* 19

EVENT DESCRIPTION

An independent process safety audit took place at the Bhopal factory in May 1982 [1]. At site supervision's request, the parent company sent a team of corporate auditors to address concerns over exposure hazards. These issues were the catalyst for the protest that had recently ended. The audit's specific purpose was to assess [2].

> *...possibilities for exposure of both maintenance and operating personnel to the wide variety of toxic materials handled or processed.*

The audit consisted of an extensive review of factory design documents and safety procedures. Factory personnel were interviewed during the audit. In addition, a thorough walk down of factory processes was performed [3]. The survey was to be followed by an action plan outlining the recommendations needed to address any concerns [10].

During the audit, the team recognized that numerous alternative operating approaches had been incorporated into the manufacturing process. For example, the audit team observed that differential pressure had replaced the methyl isocyanate (MIC) transfer pumps. Although this alternative operating method was promoted as an inherently safer design option, its implementation had interrupted nitrogen flow through the process vent header (PVH). The resulting trimer buildup was being managed by washing-out the iron pipes and valves with water. Also the MIC circulation pumps were now being activated intermittently, whereas by design they were intended to operate continuously. The most recent change involved interrupting MIC refrigeration for extended periods of time by switching off the circulation pumps when the tanks were not receiving MIC rundown.

In the final report issued in July 1982, the audit team praised the Bhopal factory for its constructive use of workaround solutions to address major asset performance issues [4]. In the context of how these actions had mitigated serious safety hazards, these changes constituted a major step in the right direction [5]. Accordingly, the action plan communicated no immediate concerns that required urgent attention [6].

* For timeline events corresponding to this chapter see page 445 in Appendix.

However, several long-term improvements were recommended. Among the more noteworthy recommendations were to:

1. Retrofit a nitrogen purge system at the 1-ton MIC charge pot in the derivatives section, complete with a low-flow pressure transmitter and alarm [7]. The MIC charge pot was equipped with an MIC feed pump that was used to deliver MIC into the reactors to make technical grade product. Unlike the MIC pumps at the storage tanks, the MIC charge pot pump location was functioning according to reliability design expectations. Therefore installing a nitrogen purge at this location would restore continuous inert gas protection in the PVH, all the way through the vent gas scrubber. The result would be eventual control over the trimer deposition problem, thus avoiding the constant practice of flushing trimer out of the MIC vapor transfer headers.
2. Equip the circulation pumps with dual seals. The notion was that although seal failures would still occur, primary seal leaks would be contained. Detecting a primary seal leak would allow an effective repair plan to be coordinated before the secondary seal also failed.
3. Install a vapor suppression system in the MIC pump area [8]. Numerous MIC leaks had been responded to by the time that the Bhopal factory's compliance audit was performed. One of the leaks was so massive that over two dozen employees were sent to the hospital to recover from inhalation injuries. Recognizing the limitations of fresh air breathing protection, a water spray system might suppress MIC and phosgene vapors more effectively upon future process releases. This addition would help reduce the potential for inhalation injuries regardless if the fresh air breathing program was effective or not.
4. Failure to use slip blinds while flushing out trimer with water. For example, exception was taken with the failure to install slip blinds while cleaning process filters [8]. The action plan documented that ineffective process isolation during routine maintenance might lead to serious exposure incidents [9].
5. Replace leaking block valves. The audit team acknowledged the hazard that workers complained about regarding chronic process block valve leaks. These asset performance limitations were viewed as serious impediments to effective process isolation. In fact, one observation was recorded about being unable to demonstrate effective isolation upon the auditor's request [8].

Leaking valves reportedly have been fairly common, compounding problems… A considerable number of valves were replaced in March 1982, but the problem still exists…Team members observed one case in which an MIC shutoff valve was leaking so severely that even evacuation of the line above the valve was not adequate to prevent MIC release when a blind flange was removed. Valve leakage would appear to continue to be a situation that requires continuing attention and prompt correction.

To address the concerns expressed over leaking valves, the report documented a recommendation to replace them [10].

No concern was expressed over supervision's decision to convert from continuous to intermittent MIC refrigeration [9]. This operating procedure modification had been implemented several months in advance of the audit.

TECHNICAL ASSESSMENT

Local production sites are expected to comply with corporate standards, policies, and administrative controls (SPAC) that specify how to maintain a satisfactory level of safety, environmental, reliability, and economic performance. Local sites might decide to incorporate more restrictive SPAC to increase the protection offered by complying with the fundamental expectations. Local sites are not, however, authorized to implement guidelines that contradict corporate SPAC. Regardless of where SPAC originates from, local sites must still comply with any that apply to maintain an acceptable level of standard performance. Compliance audits (29 CFR 1910.119) are needed on a regular basis to verify that factory operation is consistent with applicable SPAC.

A cold-eye review is another type of audit, not to be confused with a compliance audit. A cold-eye review taps into an auditor's personal knowledge and experience to identify improvement opportunities. Usually, a cold-eye review is performed after a compliance audit verifies that a problematic system is operating in accordance with mandatory SPAC. A cold-eye review generates recommendations that are discretionary, whereas compliance audit recommendations are mandatory.

Remaining in compliance means that local sites must reject any conduct that conflicts with corporate SPAC. The importance of this commitment cannot be overstated. It helps to avoid distractions that make it difficult to recognize the root cause for a chronic problem. Compliance audits verify that the foundation required for acceptable process control is properly set. Unless the basic expectations are fully implemented, advanced forms of protection cannot realistically be successful. As has been demonstrated numerous times in this case study thus far, advanced solutions are not able to recover performance lost by not satisfying design requirements. In fact, satisfying these fundamental design requirements makes experimenting with advanced solutions unnecessary. The process becomes optimized and the system meets design performance criteria.

In this sense the Bhopal factory's safety audit was a pivotal development in the factory's history. Multiple recognizable deviations had occurred to overcome problems related to:

1. Failure to construct the factory according to documented design specifications, and
2. Operating the process contrary to SPAC (to overcome problems related to deviating from design specifications).

The audit team was an independent resource that had not been blinded by the rationalization process responsible for SPAC deviations before their arrival. The team was therefore in a good position to detect numerous deviations from specified corporate and local SPAC. Independence would drive the audit team back to the

Table 19.1 Sample Bhopal Factory Independent Audit Protocol

Bhopal Factory Compliance Audit				
Sample Protocol				
System: MIC Storage Area				
Date:				
Auditor(s):				
Item	**Reference**	**Requirement**	**Comply?**	**Action Taken/Date Corrected**
1	UCC. 1976. Methyl Isocyanate. F-41443A.	All components in MIC service are stainless steel. Iron, copper, tin, and zinc are not allowed.	☐ Pass ☐ Fail	
2	UCC. 1976. Methyl Isocyanate. F-41443A.	The MIC storage tanks are operating under a slight pressure held by a dry nitrogen source.	☐ Pass ☐ Fail	
3	UCC. 1976. Methyl Isocyanate. F-41443A.	The MIC storage tank vent lines have no obstructions that could prevent free nitrogen flow into the vent gas scrubber.	☐ Pass ☐ Fail	
4	UCIL. 1979. Operating Manual Part-II: Methyl Isocyanate Unit.	MIC product in the storage tanks shall be continuously refrigerated so that the temperature never exceeds 5°C.	☐ Pass ☐ Fail	
5	UCIL. 1979. Operating Manual Part-II: Methyl Isocyanate Unit.	A maximum high temperature alarm is set at 11°C on all MIC storage tanks.	☐ Pass ☐ Fail	
6	UCC. 1976. Methyl Isocyanate. F-41443A.	All MIC storage tanks shall have dual temperature indicators to sound an alarm and flash lights if the temperature inside the tank rises above 11°C.	☐ Pass ☐ Fail	
7	UCIL. 1978. Unit Safety Procedures Manual.	The vent gas scrubber shall operate continuously (24/7) as long as MIC is on site.	☐ Pass ☐ Fail	

common cause for visible process malfunctions and SPAC deviations - the decision to pursue what appeared to be an inherently safer design option by transferring MIC into the derivatives section using differential pressure.

An audit protocol must be developed in advance to remain objective throughout an independent compliance audit. This is especially important when an internal audit team is being formed, where the auditors and those being audited are both on "the same team." Table 19.1 represents a sample audit template that might have served a useful purpose for evaluating the Bhopal factory's chemical storage area and adjunct processes. Proper use of this methodology would have eliminated subjective judgments from the evaluation. Not being focused and objective could allow an audit panel to enter into the same realm of rationalization that had directed the factory into justifiable noncompliance. In other words, the independent audit team could easily be persuaded to accept improvised practices that were not in agreement with design specifications and were therefore inherently dangerous.

The validity of the Bhopal factory's corporate SPAC was unquestionable. This SPAC was behind several decades of acceptable performance at the domestic factory that served as the Bhopal factory's blueprint. During the period leading up to the safety audit, the SPAC had been fine-tuned according to actual operating experience. This maturing process had increased both its valve and effectiveness. If the audit team could remain independent for the entirety of their visit, then the corporate SPAC would direct the factory out of the dangerous and uncompetitive operating territory it had drifted into. Otherwise, factory performance could not realistically be expected to improve.

During a compliance audit, the audit team can show no sympathy when scientific principles are provided as an excuse for deviating from SPAC. Deviating from corporate and local policies amounts to noncompliance. Items on the audit protocol must either pass or fail. Meeting the intention (which could involve improvisation or a workaround solution) is counted as a deviation. Failure to remedy any deviation in accordance with the audit team's recommendation would be grounds for discipline. Recognizing the potential legal implications involved in industrial noncompliance, failure to address deviations according to the action plan would be communicated to corporate management. This level of oversight is needed to give counsel, guidance, and direction to an operating division if formalized audit recommendations are not taken seriously.

Recall in previous discussions that changes in the Bhopal factory were made in accordance with scientific, industrial principles. These changes did not, however, comply with corporate policy. Examples include the selection of iron pipes and valves in the MIC vapor transfer headers (see chapter: Licensing) and intermittent refrigeration system usage (see chapter: Process Isolation and Containment). Any failure to implement the corporate SPAC as prescribed should generate a recommendation for closure and assigned to a specific owner along with a firm due date. One might therefore have expected the audit team to note both of these practices as deviations.

Unfortunately, the Bhopal factory audit team's composition created a dependency that influenced the audit's effectiveness. Because the audit team consisted exclusively of corporate staff members, they were likely already familiar with decisions that had previously been approved by central engineering [11], an entity operating from corporate headquarters with decision rights on process changes [12]. With this complication in mind, the corporate auditors undoubtedly knew much about what they could expect to find upon arriving in Bhopal. This technicality made it impossible for corporate staff representatives to act independently, unless they were guided by an objective audit protocol as described earlier (Table 19.1).

In reality, however, an objective audit protocol could have been implemented by anyone. The fact that experienced visitors from the USA were involved only promoted the outward appearance of an independent audit being carried out by process experts.

Without an objective audit protocol to direct the team, the safety audit was governed by hidden dependencies. Since an audit protocol was not used to guide the assessment, the auditors were really involved in a cold-eye review. The fact that intelligent professionals completely familiar with the manufacturing process were auditing the process was of no value here. Consider again the fact that most of the changes they observed had earlier been approved through the proper corporate channels. Therefore, the auditors' recommendations failed to point out fundamental compliance opportunities that were required to safely operate the assets.

The audit team was, no doubt, put in an uncomfortable position. On the one hand, taking exception to the corporate approvals in favor of the requirements defined in corporate and local SPAC would perhaps have been viewed as rebellious. At this point in the factory's history, operations were still under the direct control of veterans from the parent company, not the subsidiary [13]. This fact alone reinforces the point that the process designer had approved various SPAC deviations in place when the auditors arrived [14]. On the other hand, the audit team could not possibly ignore some of the observations that were not accounted for when these approvals were made. There were simply too many obvious hazards that needed to be corrected, regardless of possibly representing preapproved deviations—perhaps covered by exceptions. Some of these findings involved sensitive unresolved issues that the workers brought to the audit team's attention, which were then verified as legitimate concerns in the action plan. At a minimum, some of workers' concerns confirmed by the audit team (such as leaking block valves) definitely needed to be resolved. Therefore the audit team attempted to serve both interests by first acknowledging the preexisting changes as "improvements" in the factory's approach to managing ongoing performance issues. Then the audit team recommended additional corrective actions to achieve the *intent* of corporate and local SPAC without strictly complying with them as written; thereby acquitting those responsible for implementing corporate SPAC by accepting the scientific grounds for approved deviations, including cycling the circulation pumps off when MIC was not in production. Unfortunately, without adhering to the basic requirements as detailed in corporate and local SPAC, any miscellaneous recommendations were strictly experimental and could not reasonably be expected to promote stable operating conditions. Again, the basic framework needed to expect a satisfactory level of process performance had been stripped away by the changes that were made. This included loss of the refrigeration function and process sensing and monitoring capabilities. Interestingly, although the Bhopal factory was complimented for its creative approach to mitigating process hazards, the audit team did not recommend transferring any of that local knowledge over to the domestic factory as best operating practices.

For example, since the decision to interrupt continuous MIC refrigeration had previously been approved at the corporate level, the audit team did not mark this down as a deviation. If this SPAC deviation had been considered to be the "best practice" for operating the refrigeration unit, then it should also have been incorporated at the domestic facility. However, the domestic facility continued operating their refrigeration unit continuously [15]. Neither did the domestic factory ever adopt the practice of closing the MIC storage tank vent valves as an inherently safer design solution to reduce the probability for a mechanical seal failure [16]. Each of these examples introduced complexities that violated specific and previously documented acceptable operating requirements. In this manner the safety audit indirectly confirmed that it was more acceptable to adhere to the corporate SPAC rather than to deviate from it on the basis of legitimate scientific principle. It is neither productive nor safe to grasp for solutions through online process experimentation and improvisation that puts the process at risk.

The audit report called much attention to the failure rate of shutoff (block) valves in process service. Trimer choking in pipes and valves contributed much to this high failure rate. There was nothing fundamentally wrong with the valves other than the fact that they were incompatible with the highly reactive process. In most cases a valve malfunction was only temporary. Returning the valves to proper service required simply attaching a hose to an available bleeder valve or modified pressure tap and establishing water flow. Indeed, if a valve was known to be leaking, the first course of action would be to try purging the process with water or steam. Only then would it have been practical to consider a more invasive approach to restoring a valve's original function if necessary.

With the system operating as originally designed, the trimer would not accumulate inside the pipes and valves. Under design conditions there was neither the need to flush the pipes with water nor to blind the flanges in order to manage the hazard created by leaking valves during online maintenance. Note once again that this is the way that the process was *designed* to operate. Accommodating online maintenance involving the attachment of hoses to flush the pipes and valves with water was not a design provision. Since the connection between trimer deposits and malfunctioning valves was missed, the audit recommended addressing the two related issues independently:

1. Improving blinding practices to prevent contaminating the process with water.
2. Replacing leaking valves with new valves that did not leak (even though valves replaced three months earlier were again leaking.) [10]

The audit team's recommendations concentrated exclusively upon managing the multiple effects of the underlying failure mechanism (rust-catalyzing trimer accumulation in pipes and valves) without considering what might be done to address its cause. Removing the failure mechanism was the only way to restore acceptable operation to the factory. However, this would involve operating the process according to prescribed corporate and local requirements. Without a formal, independent incident investigation (see chapter: Incident Investigation), this was not an option. The effort to figure out what was wrong with the MIC pumps had been abandoned long before the audit team arrived. Unfortunately, the action plan did not recommend revisiting the matter.

The Bhopal factory was not adequately equipped for major online equipment repairs between scheduled maintenance outages. This was the basis for adopting new, unconventional practices to reduce the frequency of pump seal failures. However, the Bhopal factory was even less prepared for the functional losses in response to the alternative operating methods. These alternative operating methods obscured cause from effect and left no room for human error. To make matters even worse, human error was now much more possible due to the loss of multiple protective layers, and hazard exchange:

- The process monitoring gauges were of no value.
- The temperature alarms had been disabled.
- The refrigeration unit was not in use most of the time.
- Chronic valve leaks replaced chronic pump seal failures.
- Leaking pump seals, although less frequent than before, were still a periodic exposure hazard.
- Water was being introduced regularly into process piping sections to flush out copious amounts of trimer deposits.
- Sections of iron pipe in the PVH and RVVH circuits were corroding from the inside out and were in constant need of repair.

Instead of rejecting these developments in favor of documented operating requirements, the safety audit attempted to accommodate them by supplementing the factory's emergency response and online maintenance systems. Corrective actions were therefore centered on how to better manage recurring incidents without taking steps to prevent them. This only added more complexity to a factory already operating dangerously close to a disaster. In this way, the OSS became part of the growing list of missed opportunities to direct the factory into a safer operating condition before something bad could happen.

LESSONS FOR US

Audits are essential for stable process operation. However, audits can be sidetracked if they are not completely independent. Complete independence applies directly to the team members who are knowledgeable in SPAC that document mandatory operating requirements for local sites. A less rigid audit, perhaps one based solely on technicalities or science but lacking relevant procedural/corporate compliance puts the validity of actual factory operations into question. Chances are that it cannot offer adequate long-term protection. Other lessons directed by this evaluation include:

- Compliance audits are an integral part of a local site's ability to remain competitive.
- Cold-eye reviews can offer design and operating improvements but are not a substitute for the foundation preserved by compliance audits.
- Independent auditors use a predefined audit protocol to detect deviations.
- Items appearing on a compliance audit protocol must either pass of fail. Rationalizing deviations must not be allowed during an independent audit of any type.

- Local sites are obligated to comply with corporate policies unless an exception is written. However, exceptions are not preferred and should be used with caution (see chapter: Industry Compliance).
- Observations not related to the audit protocol are not findings. They are good ideas that might improve process performance.
- Improvised or workaround solutions do not satisfy audit objectives even if they meet the intent of the requirement.
- Local sites must own their MOC process so that corporate oversight can remain independent.
- Audit recommendations must be directed at eliminating the underlying causes of observed deviations.
- Experience alone does not make a good auditor. Persistence and audit specialization are more important qualities.

REFERENCES

[1] United States District Court for the Southern District of New York, In re Union Carbide Corporation Gas Plant Disaster at Bhopal, India in December, 1984, MDL No. 626; Misc. No. 21-38, 634 F. Supp. 842, 11, 1986.
[2] T. D'Silva, The Black Box of Bhopal, 79, 2006, ISBN: 978-1-4120-8412-3.
[3] Supreme Court of India Criminal Appellate Jurisdiction, Application for Directions to Institute Charges U/S 302 (For Offence U/S 300 (4)) Read With S. 35 of the Indian Penal Code, 1860 Against the Respondents Herein, Curative Petition (Criminal) No. 39-42 of 2010 in Criminal Appeal No. 1672-75 of 1996 in the Matter of Central Bureau of Investigation Versus Keshub Mahindra & Ors., 16, 2011.
[4] Supreme Court of India, Criminal Appeal Nos. 1672, 1673, 1674 and 1675 of 1996, in the Matter of Keshub Mahindra vs. State of Madhya Pradesh, September 13, 1996, 14.
[5] District Court of Bhopal, India, State of Madhya Pradesh Through CBI vs. Warren Anderson & Others, Criminal Case No. 8460 of 1996, 45, June 7, 2010.
[6] P. Shrivastava, Bhopal: Anatomy of a Crisis, 52-53, 1987, ISBN: 088730-084-7.
[7] T. D'Silva, The Black Box of Bhopal, 80, 2006, ISBN: 978-1-4120-8412-3.
[8] Union Carbide Corporation, Operational Safety Survey, 6, 1982.
[9] Supreme Court of India Criminal Appellate Jurisdiction, Application for Directions to Institute Charges U/S 302 (For Offence U/S 300 (4)) Read With S. 35 of the Indian Penal Code, 1860 Against the Respondents Herein, Curative Petition (Criminal) No. 39-42 of 2010 in Criminal Appeal No. 1672-75 of 1996 in the Matter of Central Bureau of Investigation Versus Keshub Mahindra & Ors., 17, 2011.
[10] T. D'Silva, The Black Box of Bhopal, 81, 2006, ISBN: 978-1-4120-8412-3.
[11] Supreme Court of India Criminal Appellate Jurisdiction, Application for Directions to Institute Charges U/S 302 (For Offence U/S 300 (4)) Read With S. 35 of the Indian Penal Code, 1860 Against the Respondents Herein, Curative Petition (Criminal) No. 39-42 of 2010 in Criminal Appeal No. 1672-75 of 1996 in the Matter of Central Bureau of Investigation Versus Keshub Mahindra & Ors., 15, 2011.
[12] W. Morehouse, M.A. Subramaniam, The Bhopal Tragedy: What Really Happened and What It Means for American Workers and Communities at Risk, 14-15, 1986, ISBN: 0-936876-47-6.

[13] Supreme Court of India, Criminal Appeal Nos. 1672, 1673, 1674 and 1675 of 1996, in the Matter of Keshub Mahindra vs. State of Madhya Pradesh, September 13, 1996, 14.
[14] District Court of Bhopal, India, State of Madhya Pradesh Through CBI vs. Warren Anderson & Others, Criminal Case No. 8460 of 1996, 8, June 7, 2010.
[15] District Court of Bhopal, India, State of Madhya Pradesh Through CBI vs. Warren Anderson & Others, Criminal Case No. 8460 of 1996, 38, June 7, 2010.
[16] W. Worthy, Methyl isocyanate: the chemistry of a hazard, Chem. Eng. News 63 (6) (1985) 29.

CHAPTER

Process Optimization* 20

EVENT DESCRIPTION

Hopes for continued product manufacturing growth at the Bhopal factory faded in 1982. Annual production levels again slipped below 50% of the factory's desired output capacity as the year ended. Excessive maintenance costs made it impossible for the factory to generate a profit. An aggressive corrosion mechanism had resulted in chronic process vent header (PVH) leaks. Unfortunately, the methyl isocyanate (MIC) rundown tanks could not be filled any time there was a process vapor leak between the MIC storage tanks and the vent gas scrubber (VGS). Frequent, costly, and lengthy PVH repairs allowed the MIC inventory in the rundown tanks to drop to their minimum storage level of 20% [1]. At this point production would be cut off until the PVH could again safely contain the process, if only for a brief time before another leak occurred. Therefore the MIC synthesis and derivatives sections sat idle many times while the PVH was being repaired.

The financial impact associated with the loss of the PVH's function made the thought of installing an independent, spare PVH very appealing [2]. A spare PVH could be used to route MIC vapors into the VGS any time that unfinished PVH repairs were a production constraint. Augmenting the process with this capability would eliminate or at least curtail production losses; it would allow MIC to fill the rundown tanks even when the PVH was under repair.

In the end, however, an improvised alternative solution was selected. The design change involved connecting the PVH and relief valve vent header (RVVH). together with a flex hose [3] (Fig. 20.1). This provision made it possible for MIC production to continue at times when the damaged PVH was unavailable. With the flex hose installed, MIC vapors in the rundown tanks could bypass the PVH. These vapors would be routed directly into the VGS through the RVVH. This workaround solution allowed MIC production to continue whereas previously it had to stop. Central engineering approved the change and the jumper line was installed in May 1983 [4].

Shortly after the PVH bypass solution was implemented, the RVVH also started to corrode rapidly. The problem could be traced back to the same trimer deposition phenomenon that was occurring in the PVH. Similarly, flash rust on the internal surface of the iron RVVH was sufficient to initiate trimer accumulation during normal process operation. Regularly washing the fouled RVVH out with water intensified the corrosion problem. Not before long, process containment was also lost in the

*For timeline events corresponding to this chapter see pages 446–448 in Appendix.

Rethinking Bhopal.

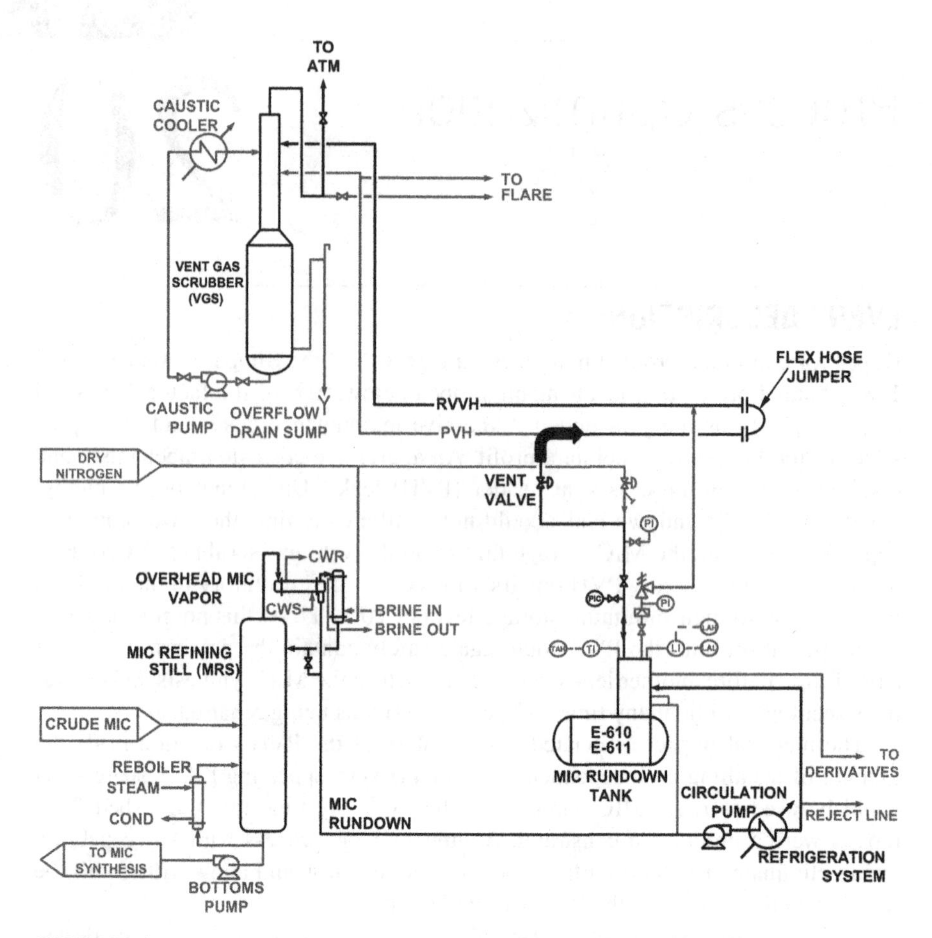

FIGURE 20.1

Installing a jumper made it possible to bypass the damaged PVH.

RVVH. When the PVH and RVVH were both down for repairs, the clock was ticking before production would again cease. MIC could not safely be routed into the rundown tanks when neither the PVH nor RVVH were available.

After installing the jumper, both the PVH and RVVH were operating under the same service conditions. Trimer was building up inside both of them. Therefore installing the jumper had simply spread the problem to a different location that was previously isolated from routine MIC vapor traffic. Both the PVH and RVVH were now rapidly corroding. Neither the PVH nor RVVH could adequately contain the process. RVVH leaks introduced a more immediate safety hazard as well, since a header leak during a pressure relief scenario could result in the catastrophic release of process gas. Both vapor transfer headers were in a perpetual state of repair. To keep up with the repairs, the factory discontinued the practice of replacing damaged sections of pipe. Instead, clamps and weld overlays were applied to expedite a quick fix to keep at least one way of accessing the VGS available [15].

The investors lost their patience at the end of 1983 when annual production again dropped for the second consecutive year; this time to only 35% of the factory's designated output capacity [5]. Maintenance costs were spinning out of control. The production constraint made it impossible to compete with local competitors who were stripping business away from the industry leader [6]. The rundown tanks could not contain enough MIC inventory to support manufacturing operations when both the PVH and RVVH were leaking. On January 26, 1984 the executives speaking on behalf of the investors issued a formal warning to the parent company that they had lost all confidence in the struggling business, and that they expected the "bleeding" to stop [7].

For as long as manufacturing operations had been conducted at the Bhopal factory, changes were applied to gain control over the chronic production constraint, a constraint wholly attributable to asset reliability. The collective result of these flaws was excess routine maintenance at enormous expense. Factory operation could not continue at this level of financial and production performance. Breaking even was out of the question. Profit generation was a fantasy. It had never happened. In hindsight, all the adjustments made to support safety and production had amounted to nothing. No net improvement had occurred since the factory was commissioned in 1979. In other words, methods used to address recurring MIC pump failures had not improved factory performance. These process modifications had simply transferred the original problem to other places or locations. The final result after 6 years of intense effort was no measurable improvement. The investors were acutely aware of this fact. They were the ones suffering financially. Their investment had never returned a profit. Despite all the promises that had been made over the years to end the constant capital drain by implementing cost-effective solutions, there had been no tangible improvement—only excuses and promises of the latest, and greatest, solution. Many solutions had been applied, but nothing seemed to work. Based on the constant deterioration of the factory, it might have been possible to say that things were getting worse as time progressed. For sure the factory had run out of inventive solutions that previously seemed to offer some hope. A total loss of $7.5 MM (US) had resulted during the operating period between 1978 and 1983 [8]. Financially the manufacturing experiment was a complete disaster.

Recognizing that something had to be done to regain the investors' confidence, an optimization and improvement program (OIP) was immediately implemented. Its purpose was to cut wasteful expenditures and eliminate the chronic production constraint [1]. Site supervision knew that the factory would have to return a profit to offset the financial losses cited by the investors, or the process would be shut down. In response to low production volume and high maintenance costs, there was no hope for the factory breaking even in its current condition [9]. Adding insult to injury was the local manufacturer that continued acquiring business at their expense. Out-competing the Bhopal factory meant that an alternative supplier figured out the secret to producing technical grade product at lower cost. Perhaps the competition might be willing to share their knowledge with the ones that created both the technology and market [10].

In one fell swoop, the OIP resulted in hundreds of layoffs for a total savings of $1.25 MM (US) by reducing employee compensation and benefits [6]. Unfortunately, the workforce reduction did not correspond with a drop in process maintenance requirements. Therefore the maintenance backlog grew as significantly fewer workers were available to support the ongoing maintenance commitment that showed no signs of easing.

Continuous VGS operation was also scrutinized. In application of the same principles that were used to modify circulation and vent valve operation, electricity was saved by running the VGS caustic circulation pumps only when needed, which was when the MIC rundown tanks were being loaded [11]. This is when MIC rundown tank and process vapors in the PVH or RVVH needed to be scrubbed for safe atmospheric discharge. Essentially the VGS caustic circulation pumps would operate only when the MIC circulation pumps were running. This would be when an MIC rundown tank vent valve was opened so that the tank could be filled with product from the MRS.

As the OIP neared its end, money in the short term was saved. At the same time, the investors noted that future savings would be more difficult [6]. Worker layoffs and modifying the VGS operating procedures were both "one-time deals"; neither of which addressed the chronic production constraint related to PVH and RVVH corrosion. In recognition of the excessive PVH and RVVH maintenance demands, either the availability of the MIC vapor transfer headers needed to improve or more storage tank volume needed to be made available. Otherwise the predictable frequency of PVH and RVVH failures and mandatory repairs would continue to limit production far beyond what the factory's investors were willing to accept. With factory performance where it was, break even production was not conceivable. The factory was on the verge of a financial collapse.

In one final attempt to eliminate the production constraint that had gripped the factory since MIC started entering the rundown tanks in 1979, the maximum level inside the rundown tanks was raised from 60% to 80% [11]. This adjustment was especially important in view of the layoffs that had increased the maintenance backlog. Raising the amount of available MIC in the rundown tanks provided more time for PVH and RVVH maintenance to overlap before running out of available MIC. However, storing more MIC in the rundown tanks did nothing to reduce the rate at which trimer was depositing inside the PVH and RVVH. Maintaining even the current level of production still required an enormous amount of personal attention. By keeping more MIC in reserve, then perhaps the anticipated, ongoing MIC production interruptions would not continue affecting factory output [12].

TECHNICAL ASSESSMENT

The sequence of events described in this chapter consistently follow patterns observed earlier. Improvisation was again demonstrated to address three optimization opportunities:

1. *Intermittent VGS operation.* Energy costs were reduced by running the VGS only when MIC was being manufactured. By design, the VGS was supposed to operate continuously regardless if MIC was being manufactured or not.

However, this design basis assumed that a nitrogen purge would always be carrying MIC vapor into the VGS. After the MIC transfer pumps were abandoned by closing the vent valves, this was no longer the case. Therefore, running the VGS caustic circulation pumps full time became optional. Available energy savings were captured by switching the VGS pumps off when the rundown tanks were not being filled.

2. *Increasing the MIC rundown tank levels.* More MIC was made available by increasing the maximum MIC rundown tank storage levels. Since there was no production constraint within the MIC synthesis section, the MIC rundown tanks were the bottleneck. When PVH and RVVH repairs were progressing, the vent valve was closed. This operating mode made it possible to move MIC into the derivatives section for final production. However, if the available supply of MIC in the rundown tanks was depleted before the repairs were finished, production would stop until the vent valve could safely be opened again. At that point the VGS would be restarted, the vent valve would be opened, the circulation pump would be restarted, and the MIC refining still (MRS) could begin delivering MIC product into the rundown tanks once again.
3. *Installing a flex hose jumper between the PVH and RVVH.* The prospect of running a spare vapor transfer header between the MIC storage tanks and the VGS had been considered. Utilizing the jumper hose satisfied the intent of installing a spare vapor transfer header between the MIC storage tanks and the VGS. This additional capability was needed to accommodate repeat PVH failures. However, the factory was already losing money. Installing a new asset to support more maintenance was probably not considered an acceptable investment. Therefore an improvised solution was implemented by simply connecting the PVH and the RVVH together. This zero-cost workaround solution allowed the rundown tanks to be filled even when the PVH was unavailable. In the final analysis, the solution simply transferred the problem over to the RVVH, which was also made of iron. The RVVH was therefore equally affected by the operating conditions responsible for frequent PVH failures. Trimer formation followed by water washing was needed to clear trimer out of the vent headers, so that MIC rundown quality would not be spoiled. Therefore installing the flex hose to perform corrective maintenance while avoiding production delays accelerated and spread damage within the factory's vapor management system. This included the pathways connecting the MIC storage tanks to the VGS and flare.

The OIP resulted in a series of actions that did, indeed, save money. However, most of the economic savings came from worker layoffs. Laying off about one-third of the workforce without reducing the factory's routine maintenance demand created a more immediate problem. Recall that the factory was overstaffed to keep up with the amount of maintenance needed to keep the process running, albeit at a reduced capacity (see chapter: Commissioning Phase). The OIP saved about \$1.25 MM (US) through headcount reductions that included eliminating 333 jobs and dropping the maintenance department headcount from six to two positions [13]. The resulting imbalance between the process' maintenance demand and the available workforce created a maintenance backlog. If the deferred repairs could not be completed in time, then the process would soon become unfit for service—making production impossible.

Short-sighted changes that could not offer long-term relief (including forced headcount reductions) are another pattern that returns in this sequence of events. None of the adjustments made during the OIP addressed the primary failure mechanism responsible for the factory's persistent operating, maintenance, and production difficulties. Aside from addressing the chronic pump failure mechanism at the center of all these problems and operating the factory as it was designed, there were only two practical options whereby the factory's safety and economic performance problems could be solved:

1. *Address causal factor 1* (see chapter: Factory Construction) by replacing the iron PVH and RVVH with stainless steel in accordance with design specifications. This would not solve the elusive asset reliability problem but would adequately control it. Replacing the MIC vapor transfer headers with stainless steel would prevent trimer formation inside the pipes and valves. It would eliminate the need for periodic water flushing, curtail aggressive system corrosion, significantly reduce maintenance costs, and possibly reduce the insurmountable maintenance backlog to a more practical level.
2. *Address causal factor 2* (see chapter: Management of Change) by opening the storage tank vent valves to reestablish the continuous flow of nitrogen from the MIC storage tanks to the VGS. This solution would require diagnosing and correcting the cause for repeat MIC pump seal failures. The decision to abandon the MIC transfer pumps was responsible for this failure mechanism. Resuming nitrogen flow through the PVH at all times would therefore involve recommissioning the MIC transfer pumps. Alternatively, following through on the safety audit team's recommendation to connect a new nitrogen source to the MIC charge pot in the derivatives section might also have solved the problem. Prior to abandoning the transfer pumps, the charge pot was protected by the nitrogen purge running between the MIC storage tanks and the VGS. Upon closing the vent valves, the MIC in the charge pot was unprotected and therefore became an ignition hazard. Regardless, a nitrogen purge by design was expected to be present at all times. Interrupting nitrogen flow early in the process' life created both direct (ignition) and indirect (contamination) process safety hazards.

With the factory in the middle of a financial crisis, deciding to invest in its future survival was not easy. Superficial changes were perhaps more attractive at this point in the factory's history. With the investors losing interest in the struggling project, convincing them to fund the installation of stainless steel PVH and RVVH pipes and valves would have been extraordinarily difficult. Still, it was the most direct way to satisfy the request for improvement that they had demanded.

An important lesson learned from this example is to step through the door of opportunity immediately when it opens for you. In this case, the investors clearly articulated their expectation to see the "bleeding stop." Installing stainless steel MIC vapor transfer headers would have stopped *the bleeding* in the least amount of time. That was the conversation that needed to occur when the directive was issued. Other problems would still have to be solved to safely return the transfer pumps and circulation pumps to continuous service. However, upgrading the metallurgy at these locations would at least have made the factory able to compete in the market that it had created and once dominated. The resulting corrosion resistance would also have significantly reduced the factory's maintenance demand, which was no longer under control following the loss of manpower that had kept it in balance. In the end, stainless steel MIC vapor transfer headers would also have eliminated the water washing dependency that threatened to contaminate the process. Eventually this hazard led to an incident that nobody could afford. Remember that specific lesson the next time you are struggling to conjure up the courage to have a difficult conversation with a business authority (supervisor) that has the purchasing power you need to get the job done. Spending money to equip the process to perform according to design expectations generates income in the short term while preventing incidents that can cause the loss of an operating license in the long term.

In the final analysis, the adjustments made during the OIP followed a third pattern observed since 1979, when the Bhopal factory was integrated and entered the commissioning phase (see chapter: Commissioning Phase). When the OIP ended the process was less profitable, less stable, and less safe than ever before. Competitors in the same market were able to generate a profit whereas the factory represented by the company that created the market could not stay in business. This type of comparative advantage could only have been achieved by operating a more reliable process. It was an example that corporate executives acknowledged they could learn from [14].

The line organization was the only stabilizing presence inside the factory after the OIP. These workers had developed local processes to control, and were skilled and practiced at managing, the dangerous process. Tribal knowledge, however, is an inferior alternative to engineered solutions applied on the basis of accurately diagnosing and removing the root causes for persistent operating difficulties.

LESSONS FOR US

Wasteful operating expenses can be reduced by optimizing the process. A successful process optimization effort might identify changes that could make a business more profitable and competitive. However, always remember that changes can, and often do, have unintended consequences. The value of stable system performance must therefore be weighed against the available savings opportunity before committing to any changes. Changes made when a process is unstable might withdraw necessary system functions when they are most needed. The results of these changes can be disastrous. Therefore, the best time to optimize the process is when system performance meets expectations and controlled adjustments can be made to determine additional savings that might be available. Other lessons to take away from this analysis include:

- Reliability is the secret to competitive production.
- Solving a problem early generates more income and stability in the future. It also makes a process more competitive.

- Make sure that the basics are adequately covered before resorting to stand-alone, add-on reliability solutions.
- Inexpensive solutions cost more money in the long run if they do not address the problem's root cause.
- Step through the door of opportunity immediately when it opens, even if it means mustering up the courage to have a difficult, direct conversation.
- Justify the cost for implementing effective solutions early when the resources and interests are available. This will prevent committing vital resources to endless work on superficial solutions later on.
- Manpower can be reduced safely only by removing the demand for human intervention, which is a function of asset reliability and process complexity.
- Reliability problems reward your competitors.
- Successfully contending with chronic process performance difficulties promotes a false sense of security that makes irrational decisions more likely.
- Failure to provide the necessary resources to address a chronic asset reliability problem institutes tribal knowledge while inviting the use of work-around solutions.
- Process optimization should never take the place of incident investigation.

REFERENCES

[1] S. Diamond, The Bhopal Disaster: How It Happened, The New York Times, January 28, 1985.

[2] P. Shrivastava, Bhopal: Anatomy of a Crisis, 7, 1987, ISBN: 088730-084-7.

[3] Supreme Court of India, Criminal Appeal Nos. 1672, 1673, 1674 and 1675 of 1996, in the Matter of Keshub Mahindra vs. State of Madhya Pradesh, September 13, 1996, 17.

[4] United States District Court Southern District of New York, Sajida Bano, Haseena Bi, Sunil Kumar, Dr. Stanley Norton, Asad Khan, Shiv Narayan Maithil, Devendra Kumar Yadav, Bhopal Gas Peedit Mahila Udyog Sangathan (Bgpmus), Gas Peedit Nirashrit Pension Bhogi Sangharsh Morcha (GPNPBSM), Bhopal Gas Peedit Mahila Stationery Karmachari Sangh (BGPMSKS), Bhopal Gas Peedit Sangharsh Sahayog Samiti (BGPSSS), and Bhopal Group for Information and Action (BGIA), on Behalf of Themselves and All Others Similarly Situated Against Union Carbide Corporation and Warren Anderson, Index No. 99 civ. 11329 (JFK), 14, 2000.

[5] National Environmental Engineering Research Institute (NEERI), Assessment and Remediation of Hazardous Waste Contaminated Areas in and around M/s Union Carbide India Ltd., 4, 2010.

[6] R. Natarajan, Letter to Mr. J. B. Law, Sub.: A Review of U.C. India Ltd.'s Ag Products Business, February 24, 1984, 2.

[7] T. D'Silva, The Black Box of Bhopal, 85, 2006, ISBN: 978-1-4120-8412-3.

[8] R. Natarajan, Letter to Mr. J. B. Law, Sub.: A Review of U.C. India Ltd.'s Ag Products Business, February 24, 1984, 1.

[9] T. D'Silva, The Black Box of Bhopal, 83, 2006, ISBN: 978-1-4120-8412-3.

[10] T. D'Silva, The Black Box of Bhopal, 2006, ISBN: 978-1-4120-8412-3, pp. 217–218.

[11] Supreme Court of India Criminal Appellate Jurisdiction, Application for Directions to Institute Charges U/S 302 (For Offence U/S 300 (4)) Read With S. 35 of the Indian Penal Code, 1860 Against the Respondents Herein, Curative Petition (Criminal) No. 39-42 of 2010 in Criminal Appeal No. 1672-75 of 1996 in the Matter of Central Bureau of Investigation Versus Keshub Mahindra & Ors., 14, 2011.
[12] W. Worthy, Methyl isocyanate: the chemistry of a hazard, Chem. Eng. News 63 (6) (1985) 32.
[13] Supreme Court of India Criminal Appellate Jurisdiction, Application for Directions to Institute Charges U/S 302 (For Offence U/S 300 (4)) Read With S. 35 of the Indian Penal Code, 1860 Against the Respondents Herein, Curative Petition (Criminal) No. 39-42 of 2010 in Criminal Appeal No. 1672-75 of 1996 in the Matter of Central Bureau of Investigation Versus Keshub Mahindra & Ors., 20, 2011.
[14] T. D'Silva, The Black Box of Bhopal, 218, 2006, ISBN: 978-1-4120-8412-3.
[15] District Court of Bhopal, India, State of Madhya Pradesh Through CBI vs. Warren Anderson & Others, Criminal Case No. 8460 of 1996, 56, June 7, 2010.

CHAPTER

Maintenance Failure*

21

EVENT DESCRIPTION

The Bhopal factory shut down in June 1984. By this time the process was no longer maintainable. All the coolant in the methyl isocyanate (MIC) refrigeration system had leaked out [1]. Even the most basic system functions needed to operate were unavailable. The process vent header (PVH) and relief valve vent header (RVVH) were both badly corroded and choked with trimer [2]. Valves were leaking, and washing them out only breifly solved the problem. The MIC rundown tanks E-610 and E-611 sat idle with 6 and 19 tons of MIC remaining in them, respectively [3]. Their fate was undetermined. There was no way to convert their contents into technical grade product.

The investors could not justify sinking any more money into a factory that had never returned a profit. After five years of trying, the project was a complete economic disaster [8]. Even if production could be restored, the factory could never compete with local suppliers that had found a way to manufacture the technical grade product without losing money. Therefore, the Bhopal factory was abandoned in place, in its damaged state [4]. The process was unable to operate and was completely unfit for service.

The optimization and improvement program (OIP) had resulted in a total maintenance failure. Not a single maintenance issue responsible for the factory's weak economic position had been addressed. Instead of generating fast income, the latest round of process adjustments had accelerated the Bhopal factory's destruction. At the start, recurring MIC circulation pump mechanical seal failures were an expensive maintenance nuisance and safety hazard that curtailed production rates. Later, frequent PVH and RVVH leaks became a source of constant downtime that allowed a production constraint to persist and perhaps even made it worse. The continual buildup of timer on the inside walls of the MIC vapor vent headers was difficult to control and became a perpetual drain on maintenance resources. Parts and labor for repairing corroded sections of the PVH and RVVH lines were expensive, and the situation had become progressively worse over time. Layoffs at the beginning of 1983 did nothing to ease the factory's oppressive maintenance demand. Since the work was no longer being done fast enough the maintenance backlog started to grow [5]. There was simply too much work to be done and not enough workers to do it—especially after the layoffs occurred. Piling superficial fixes and multiple deviations upon each other had undermined factory performance. Considering the excessive

* For timeline events corresponding to this chapter see pages 449–450 in Appendix.

maintenance costs and frequent process downtime events, even the faintest hope for breakeven production was now unrealistic. Adding insult to injury, the local competition was thriving on the Bhopal factory's quite obvious breakdown.

The factory's abrupt closure left the workers insecure about the future. They were not prepared for production to be aborted so suddenly. For weeks after the process shut down, the workers wandered aimlessly inside the factory grounds, not knowing exactly what to expect [6]. As the factory's closure became more real to them, many of them left. The factory had gone out of business. Future manufacturing activity was questionable at best, and not considered likely.

In August 1984 a meeting was held to coordinate the Bhopal factory's liquidation plan [7]. The factory, no longer working, was now officially for sale [8]. However, potential buyers showed little interest in purchasing the asset in the condition it was in. Specific concerns were expressed over the extensive corrosion damage they might inherit upon purchasing the MIC process. Unanswered questions about who would be responsible for funding the necessary repairs caused purchasing negotiations to stagnate [9]. Nobody seemed interested in buying an industrial process incapable of providing even the least of its required containment functions.

Another obstacle was the fact that MIC was not the only product remaining in substantial inventory when the factory was abandoned. Also in great supply were indigenous raw ingredients (chlorine and monomethylamine) that had been purchased in anticipation of continued production through the end of 1984. Regardless of whether or not the factory could be sold, the raw materials would have to be consumed before January 1, 1985, when the foreign collaboration agreement expired. The industrial license would not be renewed beyond that point in time. Therefore, factory operations would no longer be an option the turn of the New Year. Any products not consumed by that time would have to be disposed of at considerable cost. It was not reasonable to expect that any potential buyer would be willing to pay for the disposal or restocking of excess raw materials that were delivered to the factory by its previous owners.

The meeting ended with a sense of urgency since the factory was in no physical condition to resume production [10]. Action had to be taken quickly to restore the basic process functions needed to resume production. Immediately, the liquidation plan was set into motion. The objective was to deplete remaining chemical inventory levels inside the factory in the limited time remaining [11]. The three-phase plan involved:

1. Getting the production assets back into safe, working order, so that a final production run could be completed without incident.
2. Reducing raw material inventories in the storage area as much as possible while filling the rundown tanks with as much MIC as possible.
3. Transferring the entire MIC batch into the derivatives section for final conversion into technical grade product, by no later than December 31, 1984 [12].

Upon the final MIC production campaign's end (Step 2), the synthesis section would be isolated from the MIC storage tanks [13]. The process would then be dismantled [14].

Operations would take place continuously over three 8-h shifts until the entire MIC batch had been converted into technical grade products. Since many of the experienced workers were gone, any open positions would be filled by factory workers transferred-in from one of the subsidiary's other locations [15]. That way, workers already on the payroll who were familiar with company policies would be utilized instead of substitute workers with no background in the manufacturing industry.

A strict maintenance expense approval system was instituted while returning manufacturing capabilities to the process. Extensive PVH and RVVH repairs were needed. These assets were heavily choked and corroded. All maintenance purchases, down to the gaskets required to properly seal flange connections, were sharply scrutinized [16]. The goal was to get the factory into good enough shape for a final production run. Any cost beyond that was not acceptable. There was no longer any reason to think about the factory's future. The project had run its course and failed.

TECHNICAL ASSESSMENT

As the Bhopal factory nears the end of its life, we are introduced to a very important principle that governs all industrial processes. The message is that changes, including process adjustments, must be made slowly to be successful. Process control is more likely to be lost when sudden changes are applied. Changes must be well planned and coordinated between all interested parties to avoid serious complications that might otherwise develop. Communication, the conveying of relevant information, will take on an ever more important role as we dive deeper into the sequence of events described here.

Unfortunately, we find quite the opposite taking place in this specific sequence of events in the Bhopal factory's timeline. Notice that the manner in which the Bhopal factory's damaged process was abandoned amplified the rift that developed earlier between the factory workers and officers (chapter: Process Isolation and Containment). This division started when factory officers firmly pinned the death of a maintenance worker on his own personal actions. The situation became more contentious as time went on. Finally, the division tore the investors away from the parent company that held a slight controlling interest in the partnership. When these developments occurred, the two sides divided were no longer cooperating as partners. This fact is written into the warning issued by the investors, who were frustrated by the factory's poor economic performance. In the letter sent to corporate executives on February 24, 1984, shortly before the factory ceased production and was abandoned, investors expressed their interest to know what the parent company, "was going to do about the problem" [17].

A partnership divided to the extent where one side ceases cooperating with the other is destined for failure. Multiple partners in a business venture must continue cooperating as a cohesive unit to address system performance difficulties. In the context of these basic principles, the independent attitude expressed by the investors represents a partnership failure that contributed to a greater problem. The investors had already given up on the business by the time curiosity was being expressed about what the parent company planned to do about the situation. The collapse of a productive working relationship foreshadowed the next failure that developed when the factory heaved its last breath; suffocated under the factory's enormous maintenance demand that could no longer be satisfied. By June 1984, multiple process systems were no longer in service. In its nonfunctional condition, the factory was viewed as dead weight and was therefore abandoned [19].

Referring back to a reliability principle discussed in chapter "Equipment Reliability Principles," a functional failure occurs when an asset's deterioration margin is completely consumed. When this deterioration margin has been consumed, it can no longer meet the process's demand. It then becomes necessary to either restore the asset's lost capability or reduce its functional expectation by making necessary process adjustments—all at some cost.

We might consider a factory's maintenance function to be an "asset" that serves an important purpose. By these terms, a functional failure would occur if the maintenance system's performance was allowed to deteriorate beyond the point that its remaining capacity can no longer keep up with the demand. This is an accurate description of the situation that developed in response to the OIP (chapter: Process Optimization). Therefore, the useless condition of the Bhopal factory's assets in June 1984 provides evidence of a *maintenance failure*. The maintenance function was no longer adequate to support the factory's needs.

Abandoning the factory in a state of disrepair left the parent company in a somewhat awkward position. With very little time to spare before the Bhopal factory's industrial license would expire, the fate of the factory in its devastated condition became the parent company's problem. If the investors' decision was to walk away from the asset, then a graceful exit was deserved by all with a vested interest in its fate. This, however, did not happen. The factory was brought down suddenly in a damaged condition; unable to operate any longer. Although a more coordinated effort is generally needed to organize a safe shutdown, the factory could do no harm in its current condition. Production had stopped. The factory was stable—perhaps more than it ever had been.

The focus, however, now shifted to restarting production to draw down remaining product inventories, which included 25 tons of MIC still remaining in the rundown tanks. Accomplishing this objective required developing a liquidation plan. This plan concentrated on consuming remaining product inventories while factory operation was still legally permitted. Production was therefore scheduled to resume as soon as sufficient repairs could be made to make a final production run possible. Unlike previous campaigns, the process would be prepared to operate long just enough to convert the remaining product inventories to technical grade product. Since many workers had left when the factory ceased production, substitute workers with at least *some* industrial experience would be transferred-in to fill vacancies where needed during the last production run. Many of the better workers who found themselves unemployed when the factory closed down likely had no problem finding employment elsewhere. These workers were the ones that had figured out how to prevent exposure incidents by manipulating the process. Their jobs had been complicated by the fact that adjustments made to improve productivity had compromised vital alarms and gauge readings that would indicate a process upset. Although they had figured out ways to work around these issues, they could not sustain factory production indefinitely. A maintenance failure is what ultimately brought the process down. There was nothing they could do to reduce the factory's maintenance expenses with so many failures occurring so frequently.

Information regarding these lost functions would have been especially important to any substitute workers who might be familiar with the way things were intended to operate based on their personal experience in other locations. For example, a worker transferring into the Bhopal factory from the domestic factory would likely find it difficult to assimilate into the new location. By the time that the Bhopal factory was abandoned, so many changes had been made and so many deviations were normalized that the Bhopal factory looked nothing like the domestic factory that served as its blueprint [18]. While the process was identical, different knowledge was applied in order to maintain safe, continuous operation. Loss of that knowledge resulted when the factory closed abruptly and experienced workers were allowed to leave. This loss of operating knowledge put any other worker, no matter how experienced he or she was, at a serious and unfair disadvantage. Through trial and error, they might learn with time how to make the process operate safely so that they too could protect one another from potential process release incidents. The problem with this approach was that time had expired long before they would even set foot in the process area for the first time. The replacement workers would not be afforded the time required to learn how to manage the unforgiving process. The knowledge which the replacement workers and the parent company desperately needed walked out of the factory the moment that it was abandoned. A more effective plan would have made provisions to retain that knowledge long enough to safely decommission the process—perhaps by offering the displaced workers bonus pay until the liquidation plan was finished.

Despite the fact that its operation had been very dangerous, the workers had essentially made running the process look relatively easy. The best thing that could have happened at this point was to dismantle the factory. That was, however, not in the plan—or at least not until later. Thus, the plan was flawed in that it required restoring the process to a state of operability it could not reach with a workforce that was unfamiliar with its unconventional operating complexities – if even for just one last production campaign. With the liquidation plan's execution, the last opportunity to avoid a disaster was missed by those who considered the process safe enough to operate. Remember that these corporate advisors were far-removed from the local site practices that were developed to maintain acceptable process control (chapter: Tribal Knowledge). The safety audit (chapter: Compliance Audits) accepted these workaround solutions and therefore perpetuated this misconception. To anyone other than the line organization, the impression was that the factory was operating according to design expectations; similar to the domestic factory. Sadly, these impressions were wrong and proved unbearably costly. What followed next is an almost inescapable consequence of the many missed opportunities, of not examining and anticipating what will happen when we are blinded by optimism and faith resulting from superficial solutions.

The liquidation plan put a figurative noose around the Bhopal factory's neck. As the date approached when production would no longer be permitted that noose began to tighten. Looking back on the preceding sequence of events, we must revisit the overarching principle that governs all industrial processes. Changes must be introduced slowly to be successful. Process control is more likely to be lost when sudden changes are made. Changes must be well-planned and coordinated with all parties to be successful. Communication, the conveying of relevant information, is of huge importance to all parties involved in all industrial processes. All job functions are involved in the achievement of safety and reliability.

Become familiar with the merits of understanding that even the smallest of events has a cause. Investigating process failures is an ingredient of industrial survival and stability. Process reliability equals process safety. Process safety, not after-the-fact remedial steps, is an economical approach to doing business responsibly. Implementing workaround solutions, improvising with the process, and piling one deviation onto the next one is a palliative treatment that leads to regret. It certainly does not qualify as a value-adding preventive measure.

LESSONS FOR US

We want to amplify the message: manufacturing processes do not typically respond well to abrupt changes. Changing any aspect of process operation requires a carefully constructed and articulated plan. The plan must effectively be communicated to all involved parties prior to its execution. This is especially true when significant changes are involved in the life of a process. Abandoning a process without coordinating a realistic plan is a direct path toward a disaster. Even in the midst of tension between divided partners when an industrial project fails, a sense of duty must prevail to protect each other and those less intimately involved from serious harm that can result from a disorganized plan. The same principle applies to individuals when, for various reasons, they decide to separate from their employer. Especially during shift turnover between individual workers, a serious handoff must occur whereby the leaving worker provides a full status update to the oncoming worker. Maintaining safe business continuity requires a mature attitude and communicating freely, setting each other, as well as those who follow, up for success.

Other lessons received from this analysis include:

- Never abandon a live process for any reason, especially if it is complex, unstable, or not responding as expected.
- Maintenance failures are human failures.
- Tolerating unreliability by accepting flaws in assets as the new normal makes production unprofitable, unstable, and more hazardous.

- Shutting down a manufacturing process might be the safest option for managing an unstable asset.
- Shutting down an unreliable process might result in the loss of knowledge needed to safely restart the process—careful planning is required.
- A useless asset can be converted into a productive asset by shifting the focus to reliability.
- It is easier to sell a reliable asset than an unreliable asset.
- There is no better investment than a reliable asset, but reliable assets are rarely put up for sale.

REFERENCES

[1] Union Carbide Corporation, Bhopal Methyl Isocyanate Incident Investigation Team Report, March 1985, 23.
[2] District Court of Bhopal, India, State of Madhya Pradesh Through CBI vs. Warren Anderson & Others, Criminal Case No. 8460 of 1996, 57, June 7, 2010.
[3] S. Varadarajan, et al., Report on Scientific Studies on the Factors Related to Bhopal Toxic Gas Leakage, 19, 1985.
[4] D. Lapierre, J. Moro, Five Past Midnight in Bhopal, 228, 2002, ISBN: 0-446-53088-3.
[5] T. D'Silva, The Black Box of Bhopal, 89, 2006, ISBN: 978-1-4120-8412-3.
[6] D. Lapierre, J. Moro, Five Past Midnight in Bhopal, 226, 2002, ISBN: 0-446-53088-3.
[7] D. Lapierre, J. Moro, Five Past Midnight in Bhopal, 226–227, 2002, ISBN: 0-446-53088-3.
[8] D. Lapierre, J. Moro, Five Past Midnight in Bhopal, 227, 2002.
[9] T. D'Silva, The Black Box of Bhopal, 219, 2006, ISBN: 978-1-4120-8412-3.
[10] T. D'Silva, The Black Box of Bhopal, 88, 2006, ISBN: 978-1-4120-8412-3.
[11] W. Morehouse, M.A. Subramaniam, The Bhopal Tragedy: What Really Happened and What It Means for American Workers and Communities at Risk, 4, 1986, ISBN: 0-936876-47-6.
[12] T. D'Silva, The Black Box of Bhopal, 82, 2006, ISBN: 978-1-4120-8412-3.
[13] T. D'Silva, The Black Box of Bhopal, 90, 2006, ISBN: 978-1-4120-8412-3.
[14] P. Shrivastava, Bhopal: Anatomy of a Crisis, 45, 1987, ISBN: 088730-084-7.
[15] W. Morehouse, M.A. Subramaniam, The Bhopal Tragedy: What Really Happened and What It Means for American Workers and Communities at Risk, 16, 1986, ISBN: 0-936876-47-6.
[16] Supreme Court of India Criminal Appellate Jurisdiction, Application for Directions to Institute Charges U/S 302 (For Offence U/S 300 (4)) Read With S. 35 of the Indian Penal Code, 1860 Against the Respondents Herein, Curative Petition (Criminal) No. 39–42 of 2010 in Criminal Appeal No. 1672-75 of 1996 in the Matter of Central Bureau of Investigation Versus Keshub Mahindra & Ors., 20, 2011.
[17] R. Natarajan, Letter to Mr. J. B. Law, Sub.: A Review of U. C. India Ltd.'s Ag Products Business, February 24, 1984.
[18] District Court of Bhopal, India, State of Madhya Pradesh Through CBI vs. Warren Anderson & Others, Criminal Case No. 8460 of 1996, 24, June 7, 2010.
[19] Supreme Court of India, Criminal Appeal Nos. 1672, 1673, 1674 and 1675 of 1996, in the Matter of Keshub Mahindra vs. State of Madhya Pradesh, September 13, 1996. 14.

CHAPTER

Human Error*

22

EVENT DESCRIPTION

On October 7, 1984 [1], the Bhopal factory was in good enough condition to begin the final production run, in accordance with the liquidation plan. The purpose for restarting the process was to consume as much remaining chemical inventory as possible [2]. Upon resuming methyl isocyanate (MIC) synthesis, all of the MIC produced would be distilled into tank E-610 [3]. Tank E-610 was designated as the receiver tank for the entire operation. This decision was made to take advantage of the only circulation pump that was still available. Directing the entire MIC batch into tank E-610 would allow the necessary quality assurance/quality control (QA/QC) checks to continue as long as MIC circulation was possible. Upon filling tank E-610 to its maximum capacity (80%), the circulation pump would also be used to transfer MIC into tank E-611 at the reject distribution manifold. Then MIC rundown would continue into tank E-610 until both rundown tanks were filled to their maximum 80% limit.

If everything went as planned, MIC synthesis would last about 15 days before both rundown tanks were fully loaded with 42 tons of MIC each. With an average mean time between failure of 24 days, there was a 54% chance that the MIC production run could be completed without losing MIC circulation (chapter: Equipment Reliability Principles). The exclusive use of tank E-610 as the intermediary tank between the MIC refining still (MRS) and rundown tank E-611 made it possible to restore MIC production without first repairing tank E-611's circulation pump. This was the most economical approach to restarting the process. After the production run was completed, the process would be blinded from the storage tanks to be dismantled [4].

The fact that the refrigeration unit was useless with no coolant was irrelevant. MIC cooling was technically not required during this particular production run. The purpose for restarting the process was to reduce remaining raw ingredients (chlorine and MMA) and MIC inventories. Under these circumstances, losing more MIC through evaporation was of little consequence. In fact, depending on how the liquidation plan's purpose was viewed, excessive loss of MIC inventory through evaporation might have been more desirable. More evaporation into the vent gas scrubber (VGS) would result in consuming more of the remaining raw ingredients before the rundown tanks reached maximum capacity. This might assist in reducing disposal cost if any raw ingredients remained after the rundown tanks were fully charged.

*For timeline events corresponding to this chapter see pages 451–466 in Appendix.

Temporary workers employed by the company were transferred in from another site to fill vacancies created when the factory was abandoned. In accordance with operating procedures the VGS was restarted to control environmental emissions. Then MIC production resumed and commercial grade MIC from the MRS began distilling into rundown tank E-610. Warm MIC vapor engulfed the process vent header (PVH) as MIC entered the rundown tank. Although this was expected, trimer began immediately coating the walls anywhere that MIC vapor touched the iron pipes and valves.

On October 18, 1984, 11 days of continuous MIC production had occurred and rundown tank E-610 was nearly full. Unknown to the workers staffing the MIC process, the chloroform content in the MIC rundown stream started to increase that day [5]. With QA/QC checks being regularly performed on the contents in tank E-610, the initial contamination incident was not detected within the first 24 h. The process configuration did not allow MIC samples to be drawn directly from the rundown line. Only bulk MIC samples could be taken from the rundown tanks [6].

The undetected appearance of excess chloroform in the MIC rundown hid the fact that a process upset had begun. Indeed, the temperature in the MRS had started to creep up in response to the copious amount of trimer now inhibiting MIC vapor flow into the VGS [7]. However, the slight increase in back pressure through the PVH went unnoticed as MIC production continued.

Later that day, the MIC level in tank E-610 reached its maximum limit. At this point, 23 tons of MIC in tank E-610 was transferred into tank E-611 [3]. Everything still seemed to be going as planned. When the MIC transfer was complete, tank E-611 held about 42 tons of MIC and tank E-610 contained about 19 tons. Production then continued into tank E-610 to consume more of the remaining raw ingredients that were still in storage.

The circulation pump's mechanical seal failed on October 19, 1984. The loss of MIC circulation disabled both tank mixing [8] and sampling [5] at the same time. The chloroform level in the MRS rundown stream was still increasing; however, QA/QC checks were no longer possible. The contents of tank E-610 could not be sampled unless the MIC circulation pump was working. Therefore, the unintended operating condition persisted, undetected.

The circulation pump failure had serious safety implications depending on the circumstances that the factory might encounter before during the liquidation plan's implementation (chapter: Process Hazard Awareness and Analysis). No longer having access to the reject distribution manifold meant that contaminated MIC in tank E-610 could not be transferred into the VGS accumulator section for destruction. Neither could the reject MIC be routed to the flare tower. Tank E-610 was essentially isolated from both tanks E-619 and E-611. The failure to have access to tank E-611 would be extraordinarily significant later in the sequence of events. Without reject line access, reprocessing the contaminated product was also out of the question. A deadline extension might be required even if it was possible, since the liquidation plan did not account for potential off-spec product generation.

MIC production continued for another 6 days until tank E-610 was fully loaded with 42 tons of MIC. On October 22, 1984, the process was shut down [8] and the VGS was turned off [9]. The MIC rundown line from the MRS was then permanently isolated from the chemical storage area. Preparations were then made to dismantle the

MIC process [10]. During production, however, the chloroform level in the MIC rundown stream reached as high as 16% [5]—well above the established 0.5% upper limit. By the end of the production run, the PVH and associated valves between the MIC storage tanks and VGS were severely choked with timer and no longer fit for service.

Rundown tank E-611 was lined up to the MIC charge pot in the derivatives section for final processing into technical grade product. Production started on November 24, 1984, [11] with just over 1 month left to convert 84 tons of MIC into final product. There was a problem at tank E-610 that made it impossible to transfer its contents into the derivatives section. Completing the forerunning MIC production campaign with tank E-610's vent valve wide open for 15 days had resulted in excessive trimer buildup inside the iron valve. The vent valve was now leaking [12] due to the accumulation of solids within the valve body while MIC was running down into the tank. This physical blockage made it impossible to pressurize the tank when the MIC production campaign ended [16]. Any nitrogen fed into the tank readily escaped through the vent valve, held open by trimer, lodged between the valve's seat and diaphragm (Fig. 22.1). With neither tank E-610's transfer or circulation pump available, the only way to send MIC into the derivatives section was by pressure differential. This, however, would require removing the temporary obstruction from tank E-610's vent valve.

On November 25, 1984 [13], the jumper line was installed, which connected the PVH to the relief valve vent header (RVVH) in preparation for maintenance. Failure to establish nitrogen pressure inside tank E-610 at the end of the MIC production run meant that the vent valve was stuck open by trimer deposits. Water washing was the method that the factory had adopted to remove trimer deposits from process valves and pipes when necessary, after disabling the transfer pumps.

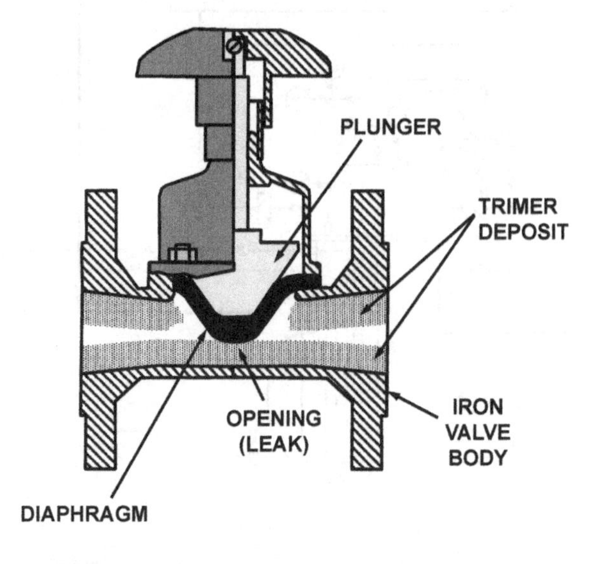

FIGURE 22.1

Cross-sectional view of diaphragm motor valve (vent valve) operation with physical obstruction (trimer).

On November 29, 1984 [15], water again started flowing into the MIC vapor transfer headers to free the vent valve from debris that had caused a temporary malfunction. Tank E-610's common valve was firmly closed so as to avoid any potential contamination issue with a leaking vent valve unable to securely isolate the rundown tank from the PVH. On November 30, 1984 [5], the pressure at tank E-610 was again tested, with no success after washing the vent lines for 24 h. Neither was the pressure test successful after 48 h of continuous water washing on December 1, 1984 [16]. The water was simply not reaching the vent valve, where trimer had deposited.

On December 2, 1983, the MIC vapor line flushing procedure was modified to target cleaning tank E-610's vent valve more directly. Time was running out to completely convert the chemical intermediates into technical grade product. Within less than a month, the factory's production privileges were to be suspended. Tank E-610's transfer function needed to be restored well before then. Otherwise there could be excess unreacted MIC inventory remaining on factory grounds when the industrial license expired.

The modified cleaning plan involved attaching a water hose directly to the pressure tap close to the common valve (Fig. 22.2). Upon verifying that the common valve was tightly closed, water could then be added to flush trimer out of the malfunctioning vent valve. Twenty-four hours later, the pressure inside tank E-610 would again be tested to confirm that the procedure was successful.

With that goal in mind a hose adapter was fitted onto the threaded end of the pressure tap, after removing the pressure gauge [27]. A utility hose was then connected and

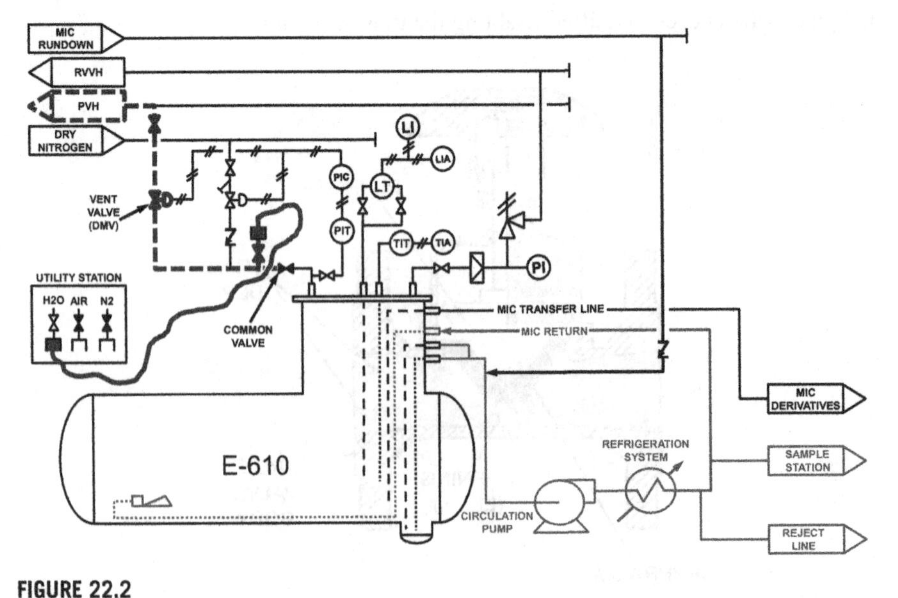

FIGURE 22.2

Intended water-washing route to remove trimer from the malfunctioning diaphragm motor valve (DMV) (vent valve).

water was introduced directly into the vent line. Water flow was established through the malfunctioning diaphragm motor valve at about 9:30 pm [17]. A maintenance tag would normally be hung on the valve undergoing water flushing operations to designate it as part of a maintenance routine [15]. However, no tags were attached to the valve being used to implement the modified cleaning plan [18]. The pressure gauge read 2 psig at the start of the water flushing operation [19]—no change from where it had been since October 22, 1984, when the MIC production run ended. Normal pressure inside a rundown tank could be anywhere between 2 and 25 psig depending on the tank's operating mode [20]. Upon closing the vent valve and adding nitrogen, pressure above 2 psig was expected. Thus, it was known that the valve was leaking and needed to be cleaned.

Unseen and unknown to the workers at the time was that the common valve, also constructed of iron, was leaking as well. Trimer had accumulated inside the valve body during the final production run just as it had in the vent valve. The temporary deposit inside the common valve was also creating a gap that resulted in a leak (Fig. 16.6). The resistance felt upon turning the valve handle to close the valve promoted a sense of confidence that tank E-610 was properly isolated from the water-washing activity. However, the trimer plug dissolved upon contacting water. This left a gaping entryway for water to drain down into the tank containing 41 tons [20] of MIC (Fig. 22.3), instead of through the vent valve where it was expected to travel. The hidden failure went undetected until shift change began at about 10:45 pm. During the last routine surveillance round at 10:20 pm, the pressure on the MIC control

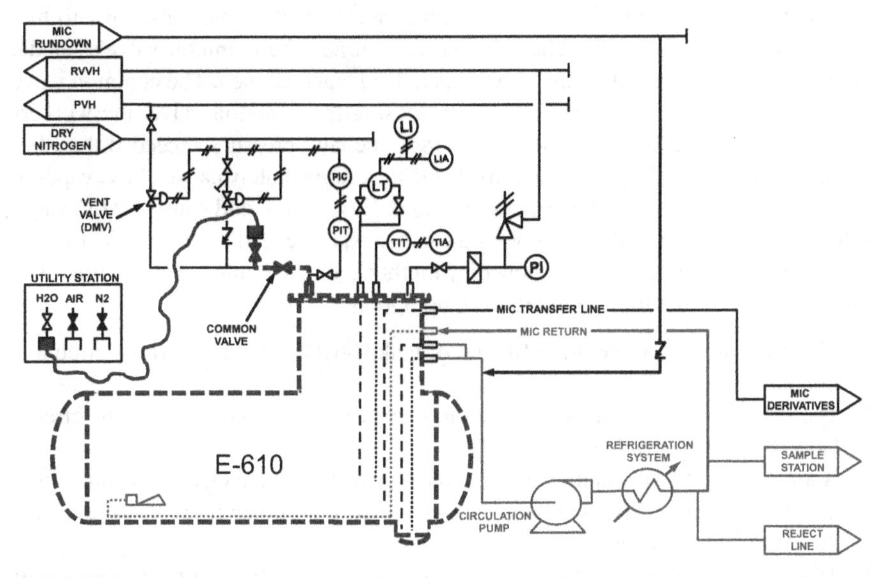

FIGURE 22.3

Actual water-washing route, through the leaking common valve and into tank E-610. *DMV*, diaphragm motor valve.

room tank gauge was still reading 2 psig [20]. However, an exothermic reaction was now in progress. The heat produced was increasing the reaction rate inside the contaminated tank that continued to be fed by the water source.

The pressure inside tank E-610 was 10 psig at 11:00 pm when the new control room operator took over on December 2, 1984 [21]. At about the same time, reports were received about MIC odors and eye irritations inside the factory [22]. To the workers, this meant that an MIC leak existed somewhere inside the complex. Accordingly, a leak patrol was organized to figure out where the problem was. About 30 min later, the leak patrol reported back to the control room that they had found MIC gas and water leaking from a disconnected relief valve (RV) line close to the VGS [23]. This discovery seemed a bit odd, since MIC was not in production at the time [24] and the problem had appeared independent from any known changes. Neither had there been any unusual alarms or gauge readings. Up to this part of the liquidation plan's execution, everything looked normal.

Back inside the control room, workers again checked the pressure gauge and this time noticed that the pressure inside tank E-610 had risen to 30 psig [20] in the course of only 30 min. This pressure reading was now outside of the tank's normal operating range. The unusual tank pressure level was the first indication of any abnormality. Upon discovering higher than normal pressure inside the MIC rundown tank, workers donned fresh air breathing supplies [25] and immediately assembled at the tank to investigate the unexpected gauge pressure [26].

When the workers arrived at the MIC storage tanks a few moments later they found that the second shift had fastened a water hose to the pressure tap at tank E-610, to clean the vent valve. Fearing the worst, they removed the water hose only to find that water was flowing [27]. Then the workers turned the common valve's handle clockwise and confirmed that the valve was indeed cracked open. The common valve was now clean and no longer leaking. By closing the common valve, the workers had blocked the tank's vapor exit. In response, the tank pressure exceeded the RV's 40 psig set point [28]. High pressure inside the tank immediately caused MIC vapor to penetrate the rupture disc; thereby opening tank E-610's RV and causing MIC vapor to flow directly into the RVVH through the expected route leading into the VGS [29].

The workers then considered how they might be able to control the situation [30], but soon realized that they were left with no option:

1. There was no way to get the MIC out of tank E-610 without a working circulation pump.
2. The refrigeration system was useless to them without coolant, even if the circulation pump was available.
3. Adding solvent to the tank would not have provided an effective heat sink, since the level in tank E-610 made it impossible to accommodate more liquid even if its contents were stable [31].
4. Transferring a portion of the unstable process into tank E-611 or E-619 was also not an option since the loss of circulation flow made it impossible to access the reject distribution manifold.

5. Without being able to access the reject distribution manifold, the unstable process could not be dumped into the VGS accumulator or flare.
6. The flare was not accessible even if the circulation pump was still available due to leaks the aggressive PVH and RVVH corrosion problem [32] that had made production so difficult before the factory shut down.

Rather than just wait, hoping that the reaction would subside now that the water source was removed, the workers improvised a possible solution by transferring 2300 pounds of contaminated MIC into the charge pot [33]. Contaminated MIC was propelled into the charge pot under sufficient pressure generated by the exothermic chemical reaction [34]. However, this action did not lower the tank pressure. Instead, the situation continued to deteriorate. The VGS was restarted shortly after midnight [21] to help bury the process' release from the atmospheric vent stack. However, quenching hot MIC vapor in the VGS did nothing to curtail the local release of MIC vapor through various RVVH leaks [35]. The factory became progressively more enveloped in MIC fog.

The thermal runaway reaction occurred at about 12:30 am [31]. The thermal runaway reaction corresponded with a sharp pressure spike, above tank E-610's maximum gauge range of 45 psig [36]. Disturbing screeches were then heard at tank E-610 [37] as the volume of vaporizing MIC rushing through the pressure relief system hit sonic velocity. Under severe stress, the concrete deck surrounding the MIC storage tanks cracked open a few moments later [38]. Choked vapor flow through the pressure relief nozzle created about 180 psig back pressure inside tank E-610 [39] compared to the tank's design pressure rating of 40 psig [40]. The temperature inside the tank was estimated to have exceeded 200°C, whereas it was only rated for 121°C, at design pressure. Remarkably, the tank did not burst under these extreme operating conditions [41].

The exponential increase of MIC flowing into the RVVH exceeded the scrubber's functional capability. Consequently, a thick plume of MIC vapor could be seen exiting 120 ft above the factory through the VGS' atmospheric vent stack [23]. The VGS had been sized to control the maximum amount of vapor traffic that could be expected during routine operation [42], possibly upon the loss of refrigeration as a credible worst-case scenario. Suppressing an MIC release during a full-scale thermal runaway reaction was not factored into the VGS' design basis [43]. As a result, the sky over the city of Bhopal, India, was filled with about 28 tons of toxic vapor over the course of the next 2 h.

After the thermal runaway reaction was underway, an attempt was made to encapsulate the fugitive vapor release in a water curtain [35]. The directive was issued to spray water from fire monitors at the VGS' atmospheric vent stack and MIC process area in order to control the release. Workers responded by activating a series of fire monitors as requested, but found that the water spray could not reach the tip of the atmospheric vent stack where the release was occurring. Unfortunately, the firewater header pressure was too low due to the number of fire monitors that were activated [35]. Had fewer fire monitors been activated it might have been possible to contain at least a portion of the process vapor emitted from the atmospheric stack.

Ambient temperatures were below the hot vapor's 39°C boiling point during the process release [8]. The wind initially carried the airborne process by to the west [44].

As the process vapor cooled, it condensed and returned to the ground as a white fog [45]. The fugitive dispersed process then drifted to the south, inflicting serious harm upon human, animal, and plant life in the more populated areas of the city [46].

At about 2:15 am [47] on Sunday, December 3, 1984, the pressure inside tank E-610 dropped below 40 psig. As the reaction subsided, the pressure inside tank E-610 dropped below the RV's set point. At this point, the valve reseated and isolated the tank from the RVVH. In the days to come, the remaining MIC inventory was safely converted into technical grade product under very strict control. The remaining conversion operation was finished on December 22, 1984 [20].

The Bhopal factory's Industrial License expired on January 5, 1985. Production stopped and the factory never reopened. The incident imposed significant residual damage upon personal health, safety, and the environment [48]. These long-term effects have persisted beyond three decades following the process release. The businesses and people directly affected by the disaster suffered losses that can never be replaced. In the final analysis, it is no doubt appropriate that the Bhopal disaster was recorded as history's worst industrial accident [49].

TECHNICAL ASSESSMENT

Three recurring themes appear in the final sequence of events leading up to the Bhopal disaster:

1. Repeat failures
2. Improvisation
3. Tribal knowledge.

In the context of normal operations, the activities observed within the Bhopal factory during its last days of operation were nothing out of the ordinary. The segment of time examined more closely in this chapter is simply an extension of the history that preceded it. Even though the liquidation plan changed the way the process was operated and configured, changes were constantly being made inside the factory. The liquidation plan was just another link in the continuous chain of process operating and configuration changes. In fact, looking back over the entire sequence of events, an abnormal situation could be defined as any day that things stayed exactly the same.

The temporary factory workers that were transferred in from a different location appeared to be right at home under these working conditions. Perhaps the temporary workers were already conditioned for this type of operating environment. This would make sense since the parent company's corporate philosophy was to capture value through the creativity of its employees. A spirit of change was embraced. It was the formula for decades of unprecedented success.

To reiterate, change was a critical part of that company's success. In fact, change is what led to inventing the product that justified the Bhopal factory's construction. Ironically, making changes to the operation and configuration of the Bhopal factory is not only what ruined the factory, but also what ultimately brought down the entire enterprise and stripped it of its fine reputation.

Stepping back for a moment to view the big picture, the question might be asked about how these events relate to the industrial incidents that have appeared in more recent decades? When we take a comprehensive look at the entire sequence of events underlying the Bhopal disaster, it is evident that matters need to be viewed with a greater depth of vision. Considering an industrial disaster strictly on the basis of the events closest to the actual incident speaks nothing about the depth of the problem. Taking into account the final sequence of events covered in this chapter alone would lead to a more superficial and therefore improper analysis. This level of understanding would satisfy only those who are willing to make drastic changes based on incomplete knowledge. A shortcut like this could hijack what might otherwise be an effective investigation and valuable learning opportunity. If the details discussed in this chapter were all that was known about the disaster, we might be fooled into thinking that a downturn in market performance is what made the Bhopal factory so unprofitable. We might also be led to believe that human error is what caused the incident. Nothing could be farther from the truth—at least not in the context of the human error that appears in the final days of the Bhopal factory's life cycle.

Human error causes all incidents. This comes as no real surprise. Even when asset malfunctions are directly involved, as they were in the Bhopal disaster, humans are directly involved in specifying, building, and operating the machines. A mechanical failure, therefore, is a reflection of our own imperfections. Machines function exactly according to how they are designed when exposed to certain operating conditions. But "designed" deserves to be better defined. Note that machines will function as they are *actually* designed, not as conceived, hoped for, or intended to be designed. Simply diagnosing human error as the cause for an industrial disaster does not provide adequate depth of knowledge to prevent future incidents of any size, order, or magnitude. A definitive analysis of the Bhopal disaster impresses upon the reader the absolute importance of considering human actions in the context of prior events. Doing so makes it possible to distinguish the root of the problem.

In this case history, the initial breakdown appeared several years prior to the disaster. More precisely, a very specific change addressed a problem at the MIC transfer pumps. The discretionary maintenance costs associated with this change were exchanged for specific productivity improvements. The intention for implementing a creative solution to poor transfer pump reliability was to maintain continuous MIC flow into the derivatives section. The decision to make the change had no recognized process safety implications. It was simply a discretionary maintenance choice that seemed like a good business decision, to offset the economic penalties associated with regular production losses and manufacturing delays.

But changing the way that MIC was transferred into the derivatives section led to closing the vent valves that were designed to remain open at all times. Closing the vent valves interrupted the nitrogen flow, which led to:

1. Trimer accumulating in the MIC vapor vent headers.
2. MRS upsets, which caused chloroform limits in the MIC rundown stream to be exceeded.

3. Washing iron pipes and valves with water.
4. Excluding blinds while isolating the process.
5. Intermittent MIC circulation pump operation.
6. Disabling the MIC storage tank high-temperature alarms.
7. Temperature and pressure gauge failures.
8. Loss of process containment due to pipe corrosion.
9. Hidden failures involving leaking valves.
10. Connecting the PVH to the RVVH with a jumper line.
11. Raising the maximum level in the MIC storage tanks.
12. An excessive maintenance backlog after indiscriminate factory worker layoffs.
13. Running the VGS only when MIC was being produced.
14. Cutting off flare tower access for safe MIC disposal.
15. Not being able to access the VGS accumulator for safe MIC disposal.
16. Not having access to the reject MIC tank, E-619.
17. Losing the ability to transfer MIC between the rundown tanks.
18. Removal of the option to dilute MIC in the rundown tanks with a heat sink.
19. Failure to access tank samples for QA/QC testing.
20. Loss of tank mixing capabilities.
21. Abandoning the factory when it was no longer functional.

Incremental changes continued to be applied until the process was completely out of control. The chaos that dominated the factory when it was abandoned in June 1984 made it impossible to determine what was causing all of these problems. Many of the chronic maintenance issues seemed independent and yet they were all related to a common asset performance difficulty. The entire manufacturing process was ultimately crushed by the force of an excessive maintenance appetite that could no longer be satisfied. Economically, an unsustainable maintenance demand started attacking the factory long before the catastrophic process release occurred. But when the imbalance between available maintenance functions and the factory's actual maintenance demand was severe enough, a process safety hazard developed. The problem was no longer simply a discretionary maintenance (economic) issue. Five years of struggling to control a chronic asset reliability problem transformed a discretionary maintenance decision into a mandatory process safety management (PSM) issue.

Prior events often reveal a precedent that explains how an organization can be expected to respond to specific situations. In the Bhopal factory, for example, outstanding consistency was demonstrated between decisions and actions that were taken in response to asset failures and process upsets as the timeline became more transparent. We also saw how factory supervisors responded to incidents that instilled fear within the workforce, and how the workforce resorted to inventing and implementing workaround solutions for their own protection, as they were trained by supervision's example. These precedents give strong clues as to how the workers would probably respond to a problem involving leaking iron valves in MIC vapor service. The problem in this case was that tank E-610's vent valve could not be sealed shut after the final production campaign ended. This left 41 tons of MIC stagnating inside tank E-610, the rundown tank, with nowhere to go. Unless the vent valve's sealing function could be returned, transferring the MIC into the derivatives section for final conversion was impossible.

It is important to note that the anticipated loss of tank E-610's circulation pump on October 19, 1984, was extraordinarily significant. Notice that an MIC transfer into tank E-611 had been made on October 18, 1984—prior to the circulation pump's failure. Although tank E-611's circulation pump was not in working condition, its vent valve had remained closed for most of the final production campaign. It was opened only for a brief period of time on October 18, 1984, before excessive trimer buildup inside the PVH started interfering with MRS performance. After the MIC transfer into tank E-611, the tank was still able to contain nitrogen pressure. Suppose tank E-610's circulation pump continued working past October 22, 1984. In that case, the MIC inventory in tank E-610 could have been transferred into tank E-611 when empty space was available. Tank E-611 could then have been used to feed the remaining MIC into the derivatives section for final processing. Unfortunately, this option was not available. Tank E-610's circulation pump stopped operating on October 19, 1984—with 3 days of production yet to go. Afterward, transferring tank E-610's inventory into tank E-611 was not possible. Consequently, attention focused on restoring the vent valve's tight shutoff function so that the MIC could be pushed into the derivatives section with nitrogen, as it had been ever since the transfer pumps were eliminated.

Taking into account prior actions that underscore a willingness, acceptance, expectation, and perhaps even a preference for improvising solutions to asset performance problems, an attempt was probably made to clean tank E-610's malfunctioning vent valve with water. Most readers with manufacturing industry experience would be familiar with this type of action. An initial attempt to restore normal operation would be anticipated before resorting to more aggressive maintenance measures that would require removing the valve. Cleaning the vent valve online would have been predicated on the suspicion that a trimer deposit was causing it to leak (Fig. 22.1). This scenario would certainly make sense, given the fact that the vent valve was working fine before the final MIC production run. Borrowing from the objective investigation principles described in chapter "Incident Investigation," however, facts and evidence must be located to have confidence in any proposed explanation.

A pattern matching exercise (Table 22.1) provides the objective framework needed to screen possible tank E-610 contamination causes against collected facts and evidence. This problem-solving approach ranks the proposed scenarios according to how well they match the pattern created by the available information (chapter: Incident Investigation). This analysis will evaluate three possible scenarios that might explain how water managed to enter the tank:

1. The common valve leaked while attempting to clean tank E-610's leaking vent valve online.
2. Cleaning water added at a different part of the process migrated into the tank through the RVVH and jumper line [50].
3. Mischief (sabotage) [41].

Table 22.1 Pattern Matching Worksheet for Rundown Tank E-610 Contamination Event

Pattern Matching Worksheet

Define the Purpose
Determine the most probable cause for water to have entered Tank E-610 on December 2, 1984

List the Facts and Evidence (I, L, T, and S)	
ID	Facts and Evidence
1	Tank E-610's vent valve was leaking at the time of the incident.
2	The incident occurred 42 days after a vent valve leak at Tank E-610 was detected.
3	The incident occurred the day after two consecutive pressurization tests at Tank E-610 failed.
4	The incident was detected at the beginning of the third shift.
5	Cleaning was taking place at the other side of the process when the incident occurred.
6	A thermal runaway reaction occurred.
7	A water-washing routine had been developed to remove trimer from iron pipes and valves.
8	Improvised solutions were common at the Bhopal Factory.
9	Leaking valves were common at the Bhopal Factory.
10	

Propose Possible Scenarios	
ID	Possible Scenarios
A	The common valve leaked while attempting to clean Tank E-610's leaking vent valve online.
B	Cleaning water added at a different part of the process migrated into the tank through the RVVH and jumper line.
C	Mischief (sabotage).
D	

Fact-Hypothesis Matrix										
ID	1	2	3	4	5	6	7	8	9	10
A	+	NA	+	NA	-	NA	+	+	+	
B	+	NA	NA	NA	+	NA	NA	NA	+	
C	NA	NA	NA	NA	NA	NA	NA	NA	NA	
D										

Create a Logic Problem	
ID	If-then statement
A4	If cleaning was taking place at the other side of the process when the incident occurred, then the vent valve was also being cleaned when the incident occurred.
B	No contradictions detected.
C	No contradictions detected.
A4 (2)	If the vent valve was being cleaned, then a reference to valve cleaning on the night of the incident can be found.

Solve the Logic Problem	
ID	Results
A4 (1)	Pass: The prior (second) shift drew a diagram showing how to configure the process for cleaning the vent valve. [Kalelkar, 565]
A4 (2)	Pass: Valve cleaning was recognized as one of the two possible causes in court proceedings. [Supreme Court of India, 13Sep1996, 17]

Document conclusions
Scenario A (the common valve leaked while attempting to clean Tank E-610's leaking vent valve online) reconciles five items and ranks as the most probable cause. Neither Scenario B nor Scenario C can be ruled-out and create no logic problems. Scenario C (sabotage) is the least probable cause based on its failure to reconcile facts and evidence.

The fact–hypothesis matrix in Table 22.1 documents the screening results. A failed attempt to clean the leaking vent valve using the pressure tap on the tank's vent line (Scenario A) provides the best match, with a ranking of +5. Water migration from another location tied into the RVVH (Scenario B) also fits, but not as well, with a ranking of +3. Mischief (Scenario C) is also a plausible explanation, but finishes last with a ranking of 0.

Since neither Scenario B nor Scenario C creates a logical contradiction with the facts and evidence, we might conclude that Scenario B is a better fit than Scenario C strictly based on the information provided (+3 vs 0). Mischief seems to run independent from facts and evidence that are not ambiguous and thus require no further interpretation.

Even though Scenario A is the best fit (+5) corresponding with the assessment criteria, note that a contradiction is generated by evidence that water was being used to clean a section of pipe at the other side of the process (A5). If the common valve leaked when the vent valve was being cleaned, the fact that water cleaning was being performed elsewhere has nothing to do with the intent to flush trimer out of the leaking vent valve. Therefore, we place a minus sign in the cell at A5.

Unfortunately, we have no evidence that any shift attempted to wash the vent valve—at least not yet. Neither have we hit a dead end in our search for this evidence. All we know is that the water was being used to clean a section of pipe at the other side of the process. Therefore, we must create a logic problem in the form of an "if–then" statement:

If cleaning was taking place at the other side of the process when the incident occurred, then the vent valve was also being cleaned when the incident occurred.

The public record confirms that a diagram was found in the second shift's notes that might closely resemble Fig. 22.2 [51].

While reviewing the daily notes of the MIC unit for the period prior to the incidents, a sketch was found on the reverse side of one page, the first page available for writing. This sketch showed a hose connection to an instrument on a tank, and it appears to have been made to explain how the water entered the tank.

This information solves the logic problem and thus can be added to our rationale for concluding that Scenario A represents the most probable cause. A drawing detailing how the process was configured for vent valve cleaning is reported to have existed in documents generated on the night of the incident. This drawing was obtained from notes "for the period prior to the incident," which suggests that vent valve cleaning started during the second shift. This fact also corresponds with the timing of events - unless, of course, the drawing was made by a worker on the third shift who had access to the second shift's notes. In that case, the evidence would favor Scenario C (mischief/ sabotage.) Here we find another possible contradiction or inconsistency.

One of the principles of defining the *most probable cause* is that once it is found, no contradictions will result—only solutions. Therefore, a new logic problem must be solved to address the inconsistency created by the possibility that the diagram was drawn by the *third shift* upon finding that the process had been sabotaged. Ultimately, we must be able to find evidence that the vent valve was being cleaned the night of the incident. Therefore, we might phrase a new "if–then" statement this way:

If the vent valve was being cleaned, then a reference to valve cleaning on the night of the incident can also be found.

Now that we have an idea of what we are looking for, we once again turn to the public record. Previously we noted that the public record explains that workers used water to remove trimer deposits from process equipment [17]. That information by itself is insufficient to conclude that a failed attempt to clean the vent valve resulted in water entering the tank. Applying falsifiability principles to attempt to disprove something as much as we wish to prove it, we must find more specific evidence that water was being used to clean the vent valve on the night of the incident.

The information we are looking for is contained in a credible reference concerning the Bhopal disaster, which states [52]:

> *...there was possibility of ingress of water and other contaminants from the RVVH or during cleaning of the valve due to rupturing the disc valve which had resulted into this grim tragedy.*

The cited reference indicates that either Scenario A or Scenario B could explain how the water got into the tank. From our pattern matching exercise we know that the facts and evidence support Scenario A over Scenario B as the most probable cause. Moreover, two logic problems have been solved with additional evidence that clarifies important details about Scenario A. We thus have a sound basis to defend Scenario A as the most probable cause: the common valve leaked while attempting to clean tank E-610's leaking vent valve online, which contaminated the MIC rundown tank. This explanation also supports the organizational culture developed in this case history, which was equally responsible for the monumental success and failure of an industrial enterprise.

The public reference supporting valve cleaning corresponds nicely with some of the principles of process operation we covered in chapter "Management of Change." The statement made about "rupturing of the disc valve" refers to the common valve, which must have leaked for water to enter the tank. Closing this valve against a trimer deposit would have left an opening between the valve disc (wedge) and seat. Detecting that the valve was not closed properly would have been difficult. By the sense of touch that applies to operating manual block valves, the common valve would have felt as if it was completely closed. Once again we see how the practices that evolved within the Bhopal factory created a process that was very difficult to operate properly. Had the trimer deposit not been lodged inside the common valve, the cleaning operation would likely have been successful. Any worker transferring in from another site could only learn about the Bhopal factory's deceptive valve behavior by personal experience. Sometimes we are given a second chance to apply the knowledge we learn from an unexpected process experience. Other times, the knowledge we receive from experience is of little value to us, because the process has lost its tolerance for human imperfection. Under these circumstances there are no second chances. Over the course of 5 years, the Bhopal factory had lost its tolerance for human imperfection.

Obviously, responsible process operation requires a more reliable approach. Chronic asset failures of any sort are not to be tolerated. Long-term reliable process operation requires controlling the causes for repeat asset failures in accordance with design expectations. As was demonstrated here, our ability to acceptably control human error is also involved.

Before continuing, it must be recognized that the vent valve failure in this scenario was *revealed.* The failure to establish nitrogen pressure inside tank E-610 meant something very specific to anyone operating the process. Nitrogen being added to the tank was escaping at some unspecified rate through the leaking valve. On the other hand, the common valve failure was *hidden.* The only way to detect the common valve failure was through an incident with some varying degree of potential consequence.

There are many potential hidden failures waiting to be discovered in the factories that we operate today. In a case like this one, previous operating knowledge was needed to develop an effective margin of protection that could prevent a disaster. Under the circumstances, anyone assigned the responsibility of closing the valve would be lured into a false sense of security–perhaps thinking that the valve was indeed closed based on feel alone. Knowledge about previous incidents was needed for any newcomer to adequately protect against the hazards of leaking valves at the Bhopal factory. Unfortunately, much of that protective knowledge was lost when the factory was abandoned weeks before the last and final production run. The factory was staffed by substitute workers that had little time to learn the intricacies of operating this particular factory in its modified condition [53]. Had those intricacies not been present, the incident would not have occurred. However, replacement workers might not have been needed at all, if proper attention had been given to maintaining simple operation consistent with original design. Likewise, solving the original problem would have made the factory more profitable and possibly competitive down to this day. In other words, the incident could have been prevented by eliminating the intricacies involved with operating the factory. Had the factory's original design intent been preserved, it is likely that the Bhopal factory would still be operating today.

There is a very important lesson embedded in the discovery of this hidden failure. The misguided effort made to control the original hazard by disabling the transfer pumps created the maintenance hazard that resulted in the Bhopal disaster. The workers who encountered the contamination incident were no different from any other worker who passes through the turnstiles in today's factories. The same is also true for the workers on the previous shift who were responsible for administering the cleaning operation. Hidden failures were involved. Repeat failures were involved. Tribal knowledge was involved. Improvisation was involved. Everything that was normal at the Bhopal factory was involved. Nobody was acting as if anything was wrong before the incident. They did not know that the factory they were controlling was not in control. By the time that the liquidation plan was implemented, the factory training manuals had been revised. The training manuals then covered how to operate the process in its modified state [13]. Nobody recognized that the factory was unsafe in its modified condition. Think about how these lessons apply to the processes under your control.

Training is important in all situations. An adequate level of process control cannot be achieved without proper and effective training. However, no level of training could have adequately prepared the workers for what they would encounter on the night of the Bhopal disaster. The system was primed for the scenario that developed by the time that the last production run took place. It was out of the workers' hands by that time. This conclusion is supported by updating the hazard analysis we started in chapter "Process Hazard Awareness and Analysis."

The hypothetical contamination scenario analyzed in chapter "Process Hazard Awareness and Analysis" closely relates to the one present in the factory when the Bhopal disaster occurred. In the preincident analysis, we properly applied the rules of layer of protection analysis and double jeopardy to arrive at a one-in-one hundred year probability for the scenario to develop. In that analysis, however, we selected an initiating event frequency (IEF) of 1/10 (or 1×10^{-1}) years to for a circulation pump failure. Recall, however, that this credit corresponded to generic industry data that can be used in the absence of practical system performance data. This might be the case when performing an initial hazard and operability study on a new process installation or during the design phase of a covered process. In our case, however, we have determined that the circulation pumps failled at an average rate of every 24 days each. After some time went by, their operation was changed to intermittent operation. Upon implementing this change, the circulation pump would sit idle for long periods of time in a "failed" (off-line) condition. Regardless, it is not responsible to accept the generic 1×10^{-1} year IEF for a pump that fails at a much higher frequency than once in every 10 years.

Adjusting the IEF to correspond with *actual* circulation pump reliability lowers the final probability for a contamination incident to lead to a catastrophic process release to once every 10 years (Table 22.2). From reality we know that a very similar situation occurred 5 years into the life of the process. Furthermore, it occurred in the center of the calculated range of 10 years.

The way that the original hazard analysis was presented in chapter "Process Hazard Awareness and Analysis" implies that the incident would come about suddenly and perhaps unexpectedly. A rare combination of multiple failures would converge at once to cause an incident. Reality, however, was quite the opposite. The actual scenario was much more deliberate and slow, albeit sudden and unexpected.

We can learn a valuable lesson from understanding how the actual scenario transpired. The changes that were historically made to solve specific problems created different ones. Over the course of the factory's life, essentially every system capable of preventing a release or helping to provide early detection of an unstable process condition was disabled. At the same time, everything looked normal until the unexplained MIC leak was detected. By that time every condition described in the hypothetical contamination scenario was a reality. The contaminant was provided by a routine maintenance function that was created or imposed relatively early into the life of the process.

After the Bhopal disaster, corporate executives recognized that the factory had been plagued by "a whole litany of nonstandard operating procedures, omissions, and commissions" [54]. We might take the time to understand how this situation was able to develop. Recall that the company involved in the Bhopal disaster was an experienced chemical manufacturer. For decades they had introduced innovative products that served a useful purpose. Their formula for success was strict adherence to their compliance commitments while tapping into the creative minds of the talent they attracted. A chemical or mechanical engineer could only dream of being employed by a company with the prominence, stability, and track record of the one

Table 22.2 Asset Reliability-Adjusted Hazard Analysis for Contamination Scenario

Summary Sheet for LOPA Method

Scenario Number: 22.2	Equipment Number: E-610	Scenario Title: MIC storage tank overpressure upon process contamination incident following a Circulation Pump failure.	
Date:	Description	Probability	Frequency (per year)
Consequence Description/Category	Category-5 incident (MIC gas release): Potential for inhalation injuries and fatalities.		
Risk Tolerance Criteria (category or frequency)	Maximum tolerable risk for a fatal injury		1×10^{-6}
Initiating Event (typically a frequency)	(1) Entry of iron, water, or any other compound incompatible with MIC process		1×10^{-1}
	(2) Circulation pump failure		1
Enabling Event or Condition		N/A	
Conditional Modifiers (if applicable)			
	Probability of ignition	N/A	
	Probability of personnel in affected area	1	
	Probability of fatal injury	1	
	Others	N/A	
Frequency of unmitigated Consequence			1×10^{-1}
Independent Protection Layers			
	Automatic tank pressure relief system	1	
Safeguards (non-IPLs)			
	Refrigeration system		
	High temperature storage tank alarm, leading to dilution with inert liquid heat sink		
	Phosgene spike		
	Transfer into Reserve Tank (E-619)		
	Transfer into parallel MIC Rundown Tank		
	Route MIC into VGS Accumulator section		
Total PFD for all IPLs		1	
Frequency of Mitigated Consequence			1×10^{-1}
Risk Tolerance Criteria Met? No (1×10^{-1} is greater than 1×10^{-6})			
Actions Required to Meet Risk Tolerance Criteria: Safely shut down process unit until sufficient protection can be added to meet minimum risk tolerance, under corporate management's direction.			
Notes: Process design is five layers of protection short of specified Category-5 consequence risk tolerance. Corporate management notified before ending the PHA meeting. IEF credit taken for loss of MIC Circulation Pump creates dependency with listed Safeguards , so IPL credit for all preventive Safeguards was denied. Inadequate capacity does not allow credit to be taken for VGS/ flare system.			
References (links to originating hazard review, PFD, P&ID, etc.):			
LOPA analyst (and team members, if applicable):			

MIC, *methyl isocyanate;* VGS, *vent gas scrubber;* LOPA, *layer of protection analysis;* PHA, *process hazard analysis;* PFD, *probability of failure on demand;* P&ID, *piping and instrumentation diagram;* IEF, *initiating event frequency.*

whose name was imprinted on the factory's gate. Yet, that formula for success later became a recipe for disaster. Throughout this discussion we cannot help but notice this company's compliance commitment. For example:

1. Sacrificing production for reliability and safety. In two known cases, production was consistently suspended upon experiencing process leaks. Loss of primary containment was not tolerated by this company. When the transfer pumps failed, production was shut down until the necessary repairs could be completed. This was also the case for PVH and RVVH leaks. If MIC vapor could not be routed into the VGS without leaking into the atmosphere, then MIC production would stop.
2. Consuming remaining chemical inventories. During the liquidation plan, the MIC storage tanks were filled to 80% capacity [55]. Both tanks were filled to 80% (42 tons) for the purpose of reducing remaining chemical inventories. Although additional raw ingredients still remained in storage when both tanks were full, this was as far as the company was willing to go. A headspace of 2 ft 9 inches remained inside the tanks at the end of the procedure [20], but 80% was the approved limit and no further production was allowed. As discussed in chapter "Process Optimization," the decision to increase the permissible storage level from 60% to 80% was flawed by favoring production to offset economic reliability penalties. The point brought out in this discussion is that the company did not callously disregard established operating limits. If an operating limit was the subject of justified debate, then it might be changed. However, limits were never intentionally exceeded.
3. Maintaining minimum storage tank levels. The MIC storage tanks were protected by a 20% low-level limit (chapter: Factory Construction). All three tanks, including tank E-619, were always operated in full compliance with this limit. At the time of the process release, the level at tank E-619 was 22%. This reinforces the point made above; the company was commendably enforcing established policies which included operating according to prescribed limits.
4. Operating within the industrial license constraints. At no time was the thought of deviating from the terms of the industrial license or foreign collaboration agreement considered acceptable. Information in the public domain paints the portrait of a company unwilling to negotiate on compliance matters. This philosophy trickled down to the execution of the liquidation plan, which had to be finished by December 31, 1984, to decommission the factory in full compliance.

Paradoxically and quite the opposite from the discipline we observed with respect to compliance, there was no corresponding commitment to preserving asset reliability and integrity relative to design conditions. Instead, there was unlimited freedom to make discretionary changes to overcome any safety, environmental, reliability, and production difficulty that was encountered. Making these changes distanced the Bhopal factory from its original design intentions. Yet, all process design and operating changes were approved through the proper technical channels; again in strict adherence to internal compliance commitments. With these checks and balances in place, a disaster seemed remote at best in the eyes of a company whose success had been built by blending compliance with creativity. The Bhopal disaster represents an indispensable example of how things can go wrong when compliance creates an outward image of superiority that does not match one's internal dedication to process reliability.

LESSONS FOR US

The final sequence of events behind the Bhopal disaster explains how the unrestricted pursuit of discretionary changes can result in serious process safety incidents. In general, the manufacturing industry thrives on creative talent that invents new processes and pathways for improvement. However, incremental losses of available safety functions can result when numerous changes are applied without experiencing corresponding success. Maintaining safe process operation requires demonstrating as much interest in asset reliability as is being shown to compliance. Allowing a reliability problem to persist can result in the voluntary (self-inflicted) loss of process control. Protection from loss of process control results from making changes that allow production assets to function according to process design expectations and requirements.

Other important messages contained in this chapter include:

- Compliance integrity does not constitute mechanical integrity.
- Discretionary maintenance issues multiply to become a mandatory PSM issue.
- Every revealed failure has one or more hidden failures.
- Training is most effective when it focuses on proper process operation that is consistent with design.
- An incident investigation should correspond directly to the appropriate process hazard analysis (PHA).
- There is never enough time to recover what you have lost through previous actions when you need it the most.
- Even after eliminating a defect, the aftereffects can still harm you.
- Explaining human error requires going back further in time.
- Training drills are needed to define what defenses are available and how they can be used in the event of a crisis.
- The proper application of PHA concepts accurately simulates actual process performance.

REFERENCES

[1] P. Shrivastava, Bhopal: Anatomy of a Crisis, 45, 1987, ISBN: 088730-084-7.

[2] W. Morehouse, M.A. Subramaniam, The Bhopal Tragedy: What Really Happened and What It Means for American Workers and Communities at Risk, 4, 1986, ISBN: 0-936876-47-6.

[3] S. Varadarajan, et al., Report on Scientific Studies on the Factors Related to Bhopal Toxic Gas Leakage, 19, 1985.

[4] T. D'Silva, The Black Box of Bhopal, 89-90, 2006, ISBN: 978-1-4120-8412-3.

[5] Union Carbide Corporation, Bhopal Methyl Isocyanate Incident Investigation Team Report, 21, March 1985.
[6] District Court of Bhopal, India, State of Madhya Pradesh Through CBI vs. Warren Anderson & Others, Criminal Case No. 8460 of 1996, 17, June 7, 2010.
[7] District Court of Bhopal, India, State of Madhya Pradesh Through CBI vs. Warren Anderson & Others, Criminal Case No. 8460 of 1996, 61, June 7, 2010.
[8] Union Carbide Corporation, Bhopal Methyl Isocyanate Incident Investigation Team Report, 23, March 1985.
[9] S. Diamond, The Bhopal Disaster: How It Happened, The New York Times, January 28, 1985.
[10] S. Varadarajan, et al., Report on Scientific Studies on the Factors Related to Bhopal Toxic Gas Leakage, 80, 1985.
[11] Union Carbide Corporation, Bhopal Methyl Isocyanate Incident Investigation Team Report, 13, March 1985.
[12] R. Van Mynen, Union Carbide Corporation Press Conference Transcript, 7, March 20, 1985.
[13] Supreme Court of India Criminal Appellate Jurisdiction, Application for Directions to Institute Charges U/S 302 (For Offence U/S 300 (4)) Read With S. 35 of the Indian Penal Code, 1860 Against the Respondents Herein, Curative Petition (Criminal) No. 39-42 of 2010 in Criminal Appeal No. 1672-75 of 1996 in the Matter of Central Bureau of Investigation Versus Keshub Mahindra & Ors., 21, 2011.
[14] Supreme Court of India Criminal Appellate Jurisdiction, Application for Directions to Institute Charges U/S 302 (For Offence U/S 300 (4)) Read With S. 35 of the Indian Penal Code, 1860 Against the Respondents Herein, Curative Petition (Criminal) No. 39-42 of 2010 in Criminal Appeal No. 1672-75 of 1996 in the Matter of Central Bureau of Investigation Versus Keshub Mahindra & Ors., 23, 2011.
[15] A.S. Kalelkar, Investigation of large-magnitude incidents: Bhopal as a case study, in: IChemE: Symposium Series No. 110, Preventing Major Chemical and Related Process Accidents, May 10–12, 1988, 561, 1988.
[16] District Court of Bhopal, India, State of Madhya Pradesh Through CBI vs. Warren Anderson & Others, Criminal Case No. 8460 of 1996, 15, June 7, 2010.
[17] P. Shrivastava, Bhopal: Anatomy of a Crisis, 46, 1987, ISBN: 088730-084-7.
[18] District Court of Bhopal, India, State of Madhya Pradesh Through CBI vs. Warren Anderson & Others, Criminal Case No. 8460 of 1996, 48, June 7, 2010.
[19] District Court of Bhopal, India, State of Madhya Pradesh Through CBI vs. Warren Anderson & Others, Criminal Case No. 8460 of 1996, 49, June 7, 2010.
[20] Union Carbide Corporation, Bhopal Methyl Isocyanate Incident Investigation Team Report, 11, March 1985.
[21] R. Van Mynen, Union Carbide Corporation Press Conference Transcript, 5, March 20, 1985.
[22] T.R. Chouhan, Bhopal: The Inside Story, Carbide Workers Speak Out on the World's Worst Industrial Disaster, 48-49, 1994, ISBN: 0-945257-22-8.
[23] S. Varadarajan, et al., Report on Scientific Studies on the Factors Related to Bhopal Toxic Gas Leakage, 20, 1985.
[24] A. Agarwal, S. Narain, The Bhopal Disaster, State of India's Environment 1984–85: The Second Citizens' Report, 206, 1985.
[25] D. Lapierre, J. Moro, Five Past Midnight in Bhopal, 291, 2002, ISBN: 0-446-53088-3.

[26] Supreme Court of India Criminal Appellate Jurisdiction, Application for Directions to Institute Charges U/S 302 (For Offence U/S 300 (4)) Read With S. 35 of the Indian Penal Code, 1860 Against the Respondents Herein, Curative Petition (Criminal) No. 39-42 of 2010 in Criminal Appeal No. 1672-75 of 1996 in the Matter of Central Bureau of Investigation Versus Keshub Mahindra & Ors., 25-26, 2011.
[27] A.S. Kalelkar, Investigation of large-magnitude incidents: Bhopal as a case study, in: IChemE: Symposium Series No. 110, Preventing Major Chemical and Related Process Accidents, May 10–12, 1988, 566, 1988.
[28] W. Worthy, Methyl isocyanate: the chemistry of a hazard, Chem. Eng. News 63 (6) (1985) 28.
[29] District Court of Bhopal, India, State of Madhya Pradesh Through CBI vs. Warren Anderson & Others, Criminal Case No. 8460 of 1996, 14-15, June 7, 2010.
[30] District Court of Bhopal, India, State of Madhya Pradesh Through CBI vs. Warren Anderson & Others, Criminal Case No. 8460 of 1996, 13-14, June 7, 2010.
[31] T.R. Chouhan, Bhopal: The Inside Story, Carbide Workers Speak Out on the World's Worst Industrial Disaster, 49, 1994, ISBN: 0-945257-22-8.
[32] A. Agarwal, S. Narain, The Bhopal Disaster, State of India's Environment 1984–85: The Second Citizens' Report, 207, 1985.
[33] A.S. Kalelkar, Investigation of large-magnitude incidents: Bhopal as a case study, in: IChemE: Symposium Series No. 110, Preventing Major Chemical and Related Process Accidents, May 10–12, 1988, 563, 1988.
[34] A.S. Kalelkar, Investigation of large-magnitude incidents: Bhopal as a case study, in: IChemE: Symposium Series No. 110, Preventing Major Chemical and Related Process Accidents, May 10–12, 1988, 564, 1988.
[35] Union Carbide Corporation, Bhopal Methyl Isocyanate Incident Investigation Team Report, 12, March 1985.
[36] Supreme Court of India, Criminal Appeal Nos. 1672, 1673, 1674 and 1675 of 1996, in the Matter of Keshub Mahindra vs. State of Madhya Pradesh, 11, September 13, 1996.
[37] P. Shrivastava, Bhopal: Anatomy of a Crisis, 47, 1987, ISBN: 088730-084-7.
[38] A.S. Kalelkar, Investigation of large-magnitude incidents: Bhopal as a case study, in: IChemE: Symposium Series No. 110, Preventing Major Chemical and Related Process Accidents, May 10–12, 1988, 559, 1988.
[39] Union Carbide Corporation, Bhopal Methyl Isocyanate Incident Investigation Team Report, 24, March 1985.
[40] T. D'Silva, The Black Box of Bhopal, 211, 2006, ISBN: 978-1-4120-8412-3.
[41] A.S. Kalelkar, Investigation of large-magnitude incidents: Bhopal as a case study, in: IChemE: Symposium Series No. 110, Preventing Major Chemical and Related Process Accidents, May 10–12, 1988, 567, 1988.
[42] W. Morehouse, M.A. Subramaniam, The Bhopal Tragedy: What Really Happened and What It Means for American Workers and Communities at Risk, 15-16, 1986, ISBN: 0-936876-47-6.
[43] District Court of Bhopal, India, State of Madhya Pradesh Through CBI vs. Warren Anderson & Others, Criminal Case No. 8460 of 1996, 18, June 7, 2010.
[44] W. Morehouse, M.A. Subramaniam, The Bhopal Tragedy: What Really Happened and What It Means for American Workers and Communities at Risk, 22, 1986, ISBN: 0-936876-47-6.
[45] T.R. Chouhan, Bhopal: The Inside Story, Carbide Workers Speak Out on the World's Worst Industrial Disaster, 39, 1994, ISBN: 0-945257-22-8.

[46] P. Shrivastava, Bhopal: Anatomy of a Crisis, 74-75, 1987, ISBN: 088730-084-7.
[47] W. Morehouse, M.A. Subramaniam, The Bhopal Tragedy: What Really Happened and What It Means for American Workers and Communities at Risk, 21, 1986, ISBN: 0-936876-47-6.
[48] W. Morehouse, M.A. Subramaniam, The Bhopal Tragedy: What Really Happened and What It Means for American Workers and Communities at Risk, 36-38, 1986, ISBN: 0-936876-47-6.
[49] M.S. Mannan, A.Y. Chowdhury, O.J. Reyes-Valdes, A portrait of process safety: from its start to its present day, Hydrocarbon Process. 91 (7) (2012) 60.
[50] T.R. Chouhan, The unfolding of Bhopal disaster, J. Loss Prev. Proc. Ind. 18 (2005) 205.
[51] A.S. Kalelkar, Investigation of large-magnitude incidents: Bhopal as a case study, in: IChemE: Symposium Series No. 110, Preventing Major Chemical and Related Process Accidents, May 10–12, 1988, 565, 1988.
[52] Supreme Court of India, Criminal Appeal Nos. 1672, 1673, 1674 and 1675 of 1996, in the Matter of Keshub Mahindra vs. State of Madhya Pradesh, September 13, 1996. 17.
[53] P. Shrivastava, Bhopal: Anatomy of a Crisis, 50, 1987, ISBN: 088730-084-7.
[54] M. Wines, Firm Calls 'Deliberate' Act Possible in Bhopal Disaster, The Los Angeles Times, March 21, 1985.
[55] Supreme Court of India Criminal Appellate Jurisdiction, Application for Directions to Institute Charges U/S 302 (For Offence U/S 300 (4)) Read with S. 35 of the Indian Penal Code, 1860 Against the Respondents Herein, Curative Petition (Criminal) No. 39-42 of 2010 in Criminal Appeal No. 1672-75 of 1996 in the Matter of Central Bureau of Investigation Versus Keshub Mahindra & Ors., 32, 2011.

PART

Lessons Learned

5

After the Bhopal disaster, a considerable amount of attention was devoted to understanding how multiple emergency response systems broke down during the night of the process release. Indeed, the complete loss of preventive functions that resulted in the disaster carried over to ineffective control after the disaster. Many casualties could have been prevented if these systems had been available when the process was released. Various sources in the public domain accurately describe how lessons from the Bhopal disaster were used to improve emergency response capabilities. We continue to benefit from these improvements, today.

An analogy might be drawn to the sinking of the Titanic in 1912. Prior to that disaster, lifeboats were provided but not to the degree that they would be effective. More precisely, the Titanic was equipped with 20 lifeboats able to carry 1178 people in the event of an emergency. However, the ship was loaded with 2224 passengers when the incident occurred. By today's industrial standards, the lifeboats' capacity would have limited the number of passengers that could board the ship. If the ship's utilization was constrained by the capacity limit, then raising that limit would require first adding more lifeboats. Applying this principle to the Bhopal disaster, a common attitude was demonstrated between these two disasters. Again, there is a shared pattern that might easily be missed. In both cases, emergency response capabilities were available, but not enough to prevent a worst-case scenario. Neither the people that boarded the Titanic nor the ship's designers honestly believed that more capability was needed until a worst-case scenario occurred—resulting in the complete loss of asset and multiple fatalities. Improvements to address inadequate emergency response functions came later. Unfortunately, a worst-case scenario was needed for these changes to occur. We are now speaking of the same circumstances that apply to the Bhopal disaster. The same pattern is consistently observed in other industries. Before a disaster, overconfidence in incident prevention results in inadequate emergency response provisions. After a disaster, overconfidence in emergency response capabilities results in inadequate incident prevention. Both cases represent an imbalance.

A further analysis into the details of the catastrophic emergency response failure following the Bhopal disaster is outside the scope of this text. The reader is encouraged to personally research available information in the public domain if more insight into these matters is needed. The manufacturing industry's history following the Bhopal disaster provides compelling evidence that improvements are needed to *prevent* the incidents that could necessitate an effective emergency response. This includes the conditioning process that affects people living outside the factory, as much as it does people inside the factory. Therefore, the focus of our discussion remains on *preventing* industrial disasters and also on the lesser precursor events that almost always lead to far more significant consequences.

The remaining two chapters are provided with these thoughts in mind. Their intention is to instill a sense of personal accountability into the business of preventing disasters. Accepting the professional and ethical imperative of preventing industrial incidents is critical for industry's success. Acting on the knowledge we acquire through case studies may range from simply disseminating information to actively defending the reliability of our physical assets.

Regardless of where we rank in an organizational lineup, adopting the right mindset offers gratifying results to those who truly recognize their role. It is a role whereby people contribute and use their talents to solve problems while preserving an asset's functional capabilities. That manner of problem solving and its ultimate outcome makes it safer to operate the processes in which the asset is required to perform.

CHAPTER

Crude Unit Fire: A Case Study in Human Error 23

INTRODUCTION

Most people can recognize the pattern that dominates the instinctive response to a potentially embarrassing situation. A desperate attempt to rewind the past instead of finding a way to constructively deal with it is usually observed. This pattern translates into shredding paper evidence during the collapse of Enron (2001). The same principle also applies to embellishing the truth to downplay the seriousness of an oil slick that polluted the Gulf of Mexico after the Macondo Well blowout (2010). It even governs the denial of any safety concern in the nuclear reactor core meltdown at the Three Mile Island nuclear plant (1979).

These examples involve different owners and different industries. Each of them occurred at a different time scattered across a period of over three decades. Nevertheless, the people who were directly involved all acted similarly. Is this distinctive pattern merely a coincidence?

What determines a person's reaction to a catastrophic event? How does this reaction impact an investigation's effectiveness? Can this knowledge be used to prevent industrial disasters? Examining the details behind a fire that broke out in a live refinery crude unit answers these and many other important questions.

PROCESS OVERVIEW

Crude distillation is the first step in converting hydrocarbon mixtures into refined petroleum products. The crude distillation unit (CDU) is a two-stage process that starts by distilling crude oil at atmospheric pressure. In both the atmospheric process and the vacuum distillation process that follows, an incoming hydrocarbon mixture is fed into a furnace. Heating the incoming hydrocarbon mixture raises the vapor pressure of the individual organic compounds before they enter the distillation tower. Hydrocarbons with a higher vapor pressure than the pressure in the distillation tower vaporize out of the mixture and travel up the tower. These hydrocarbon fractions condense into their various components as they cool while moving up the distillation tower. As these components separate according to their boiling ranges, they are drawn out of the tower by pumps that distribute them into storage tanks or other processes.

The heavy hydrocarbon liquid that remains at the bottom of the atmospheric distillation tower is sent to a furnace that feeds the crude vacuum section (Fig. 23.1).

In this section of the CDU process, distillation takes place at about 6 mmHg (vacuum) pressure created by a series of vacuum ejectors. These conditions cause heavier organic fractions that do not vaporize in the atmospheric tower to flash from the incoming mixture so that they too can be separated and collected. Vacuum tower bottoms, otherwise known as "residuum," can then be sold as asphalt or cracked into smaller-chain hydrocarbon fractions, depending on what represents the highest economic value and what assets are available for further processing.

Light hydrocarbon fractions move further up the tower. The heaviest hydrocarbons do not vaporize under vacuum, but remain as a thick liquid at the bottom of the tower. The vacuum distillation tower vents into an overhead (elevated) vacuum system, consisting of three ejectors configured in series. The three ejectors are attached to independent overhead vacuum condensers, also operating in series. The three vacuum condensers are supplied with cooling water that runs from the first-stage inlet to the third-stage outlet. The pressure inside the vacuum tower drops as the steam condenses.

In the vacuum system, any loss of ejector efficiency raises the tower pressure. This causes vaporized components to slump down the tower, which under severe circumstances could cause the liquid levels on the internal collection trays to drop. Under these conditions the pumps could cavitate or lose suction. This situation can be addressed by raising the furnace temperature to further increase the vapor pressure

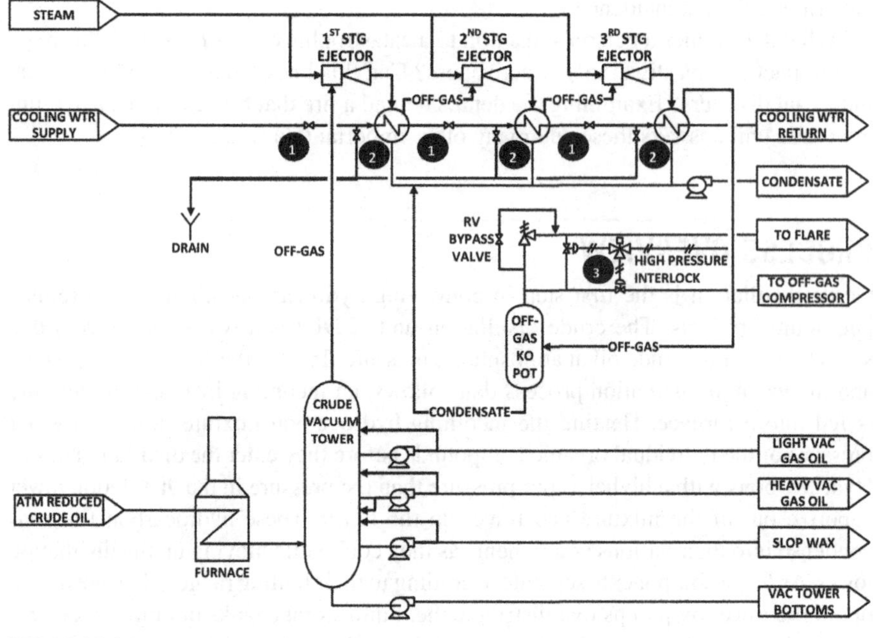

FIGURE 23.1

Vacuum distillation section.

of the reduced crude feeding the vacuum tower. However, this reactive approach increases energy costs and must be watched carefully to avoid exceeding a high temperature limit. Otherwise, tenacious coke deposits might form inside the furnace tubes and distillation column, which creates even more problems.

The uncondensed "off-gas" vapor leaving the third-stage condenser exits the process through a liquid knockout drum (KO pot). The volume of steam exiting the third stage with the off-gas increases as condenser efficiency deteriorates. Excess steam exiting the third-stage condenser raises the KO pot pressure. An automatic solenoid valve (high-pressure interlock) is programmed to switch residual off-gas flow to the flare if the pressure in the drum pressure exceeds 17.7 psia (about 3 psig). Alternatively, the console operator can switch the off-gas flow directly into the flare by operating the system in manual control mode. The off-gas KO pot is protected by a pressure relief valve (RV) configured to automatically reroute excess vapor traffic into the flare header at 50 psig.

EVENT DESCRIPTION

On a normal day at the refinery, CDU operators were performing routine maintenance inside the crude unit. Heat-exchanger backflushing was one of several items on the work list that day. The backflushing procedure involves momentarily interrupting cooling water flow by closing the condenser's inlet valve (designated as "1" on Fig. 23.1) then briefly opening the drain valve (designated as "2" on Fig. 23.1) on the cooling water supply line. This maneuver allows cooling water in the return header to flow through the condenser tubes in reverse. This procedure removes debris that could reduce condenser efficiency. Unless the routine is performed quickly, the loss of cooling water flow causes the steam to stop condensing. Under these circumstances, the pressure inside the vacuum tower will rise and excess steam begins flowing into the KO pot.

At about 8:00 am, a CDU outside operator notified the console operator that he would be backflushing heat exchangers, just in case any unusual readings should appear. At 8:12 am, the CDU console operator received a vacuum tower high-pressure alarm. At 8:13 am, the KO pot also transmitted a high-pressure alarm. These signals directly warned of lost steam condensing and excessive vapor traffic leaving the third-stage vacuum condenser. However, having been alerted by the outside operators earlier that condenser backflushing was being performed, these readings made sense. The console operator believed that everything was under control and temporarily ignored the alarms, thinking that everything would soon be normal.

A minute later the situation in the control room turned chaotic. Suddenly the control panel lit up with numerous alarms, crushing the console operator under the heavy load of meaningless information. Alarms were coming in faster than anyone could possibly process them. The KO pot pressure shot up to 17.7 psia. This caused the interlock (designated as "3" on Fig. 23.1) to automatically divert off-gas flow to the flare. When that happened, the off-gas KO pot pressure instantly dropped below

17.7 psia which automatically switched the high-pressure interlock back to its normal lineup—into the off-gas compressor. However, vapor traffic did not decrease so the pressure immediately climbed to 17.7 psia again, which automatically switched the solenoid valve back to the flare lineup. The alarm flood was caused by a critical "solenoid contact alarm" that was set to alert the console operator any time that the interlock automatically switched positions—to the flare or to the compressor. This unintended operating condition generated a critical alarm every few seconds. The result was serious confusion in the control room.

The console operator checked the condition of the vacuum tower. He noticed that the tower pressure was increasing and showed no sign of retreating (Fig. 23.2). Immediately, the console operator contacted the outside operators by radio. The outside operators had finished their routine maintenance tasks and were on their way back into the control room for a break. The console operator asked the outside operators to retrace their steps to determine what might be wrong. At first the outside operators requested patience from the console operator. They were confident that the situation would resolve itself in a few minutes. However, the console operator persisted and finally convinced the outside operators to start looking for a definitive answer. The outside operators returned to the process to start troubleshooting the instability that seemed to start about the same time that they were finishing their routine maintenance tasks.

Next, the console operator turned his attention to the nuisance alarms. At 8:15 am, he attempted to stop the alarm flood by switching open an off-gas circulation valve. This adjustment did nothing to stop or slow the nuisance alarms. At 8:22 am, the

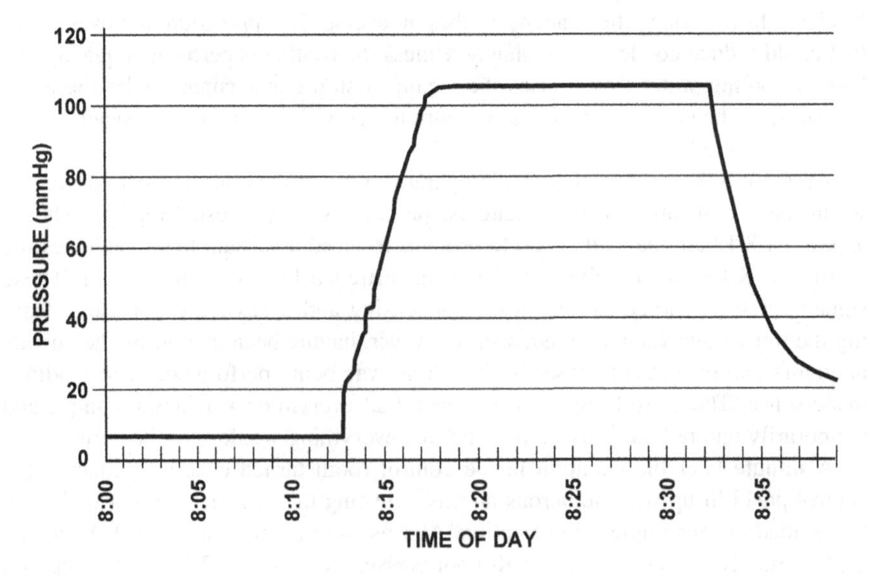

FIGURE 23.2

Vacuum distillation tower pressure trend.

console operator figured out that he could stop the alarm flood by forcing the interlock valve open in manual mode. Upon making this adjustment, all condenser vapor traffic was routed directly into the flare.

With this move, peace finally settled upon the control room that had been in a constant state of disarray for the past 10min. At this point, the vacuum tower pressure was operating above high-range (105mmHg) as indicated by flat-lining (Fig. 23.2). Nobody was sure exactly how high the vacuum tower pressure was, but suspicions were that loss of the condensing function created conditions similar to atmospheric distillation (0psig).

The next 10min passed with an eerie silence while troubleshooting continued outside in the production unit. Then suddenly at 8:32am, the vacuum tower pressure suddenly dropped without warning. To the console operator, this meant that the vacuum had been restored.

Four minutes later, an operator from an adjacent unit saw a fire ignite and dark smoke coming out of the CDU, near the base of the crude vacuum tower. He immediately alerted the CDU operators across the unit radio channel. The CDU operators immediately responded by activating a deluge system and area fire monitors and implemented emergency shutdown procedures. At 8:42am, the refinery fire department arrived at the scene and took over coordinating the emergency response. At 10:52am, the fire was extinguished. As per unit emergency response procedures, water continued flowing to completely cool down the unit. At 3:00 pm, the firefighters shut down the remaining water and announced that the emergency had ended. At that point the all-clear advisory was broadcast throughout the refinery.

When the immediate danger had passed, a profound sense of reality struck those who could see what had happened. Although a safety incident had been avoided, there was significant property damage. Repairs were required that would take over a month of unplanned downtime to complete. Perhaps even more sobering was the fact that flames had lashed into the primary pipe rack running through the CDU. Had these pipelines ruptured in the fire, the situation would have been much worse. Thankfully, an operator in the adjacent unit was at the right place at the right time. This individual saw the fire when it ignited because he was diligently attending to his outside duties, according to his expectations. His immediate response prevented an even more serious situation from developing.

INVESTIGATION

Although hours would pass before the CDU was safe for entry, an abundance of historical process information could be harvested. Historic process information (PI), the unit alarm summary, and the electronic console record were accessed before the "all-clear" was announced. The preliminary information suggested that the initial timeline should start around when tower pressure started increasing.

A process upset had taken place immediately after routine maintenance, which had included backflushing the vacuum condensers. That much was known. However,

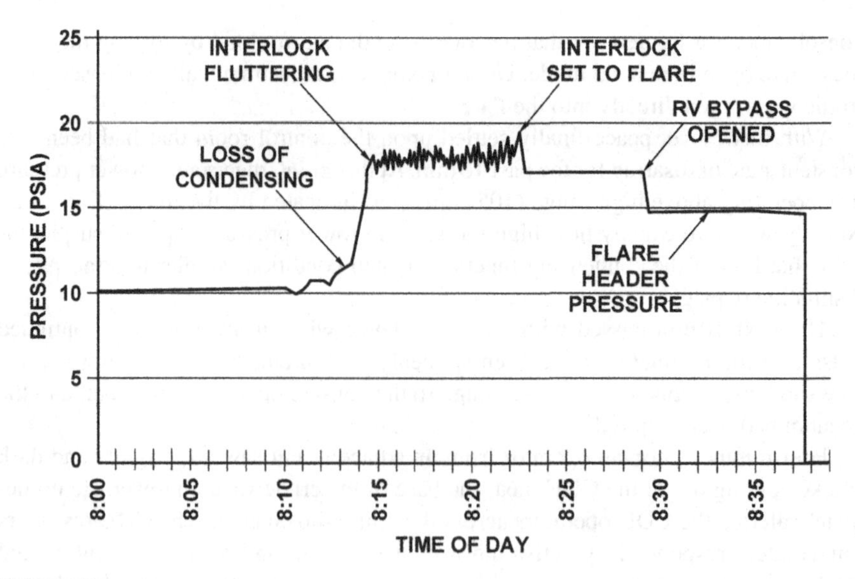

FIGURE 23.3

Off-gas knockout pot pressure trend.

the investigation team decided to let the facts and evidence determine the course of the investigation. After all, backflushing had occurred regularly in this unit for 8 years without an incident.

The first sign of the impending process upset was recorded at about 8:10 when the KO pot pressure increased sharply (Fig. 23.3). This unexpected pressure rise indicated that excessive gas flow was exiting from one specific location, the third-stage vacuum condenser (Fig. 23.1). The KO pot usually handles only trace amounts of process vapor (hydrocarbon) that travels through the vacuum condensers. Excessive vapor flow into the KO pot results from loss of steam condensing in the vacuum condensers.

At 8:13 am (3 min later), the pressure in the vacuum distillation column started to climb rapidly (Fig. 23.2). This activity not only followed the first indication of deteriorating condenser performance, but also would be expected upon the loss of steam condensing. Next, the light vacuum gas oil (LVGO) level on the collection tray at the top of the tower started to drop (Fig. 23.4). Afterward, the level on the heavy vacuum gas oil (HVGO) tray began dropping too. These indicators signaled that the vaporized hydrocarbons were moving down the tower. At the same time, the tower bottoms level started to increase as fewer hydrocarbons vaporized out of the process entering the vacuum distillation tower. This process behavior might be expected if the tower pressure is increasing. Based on the process indications, it was as if the vacuum system had been turned off.

At 8:14:01 am, the KO pot pressure reached 17.7 psia for the first time (Fig. 23.3). This pressure caused the interlock to flutter, which resulted in the observed pressure

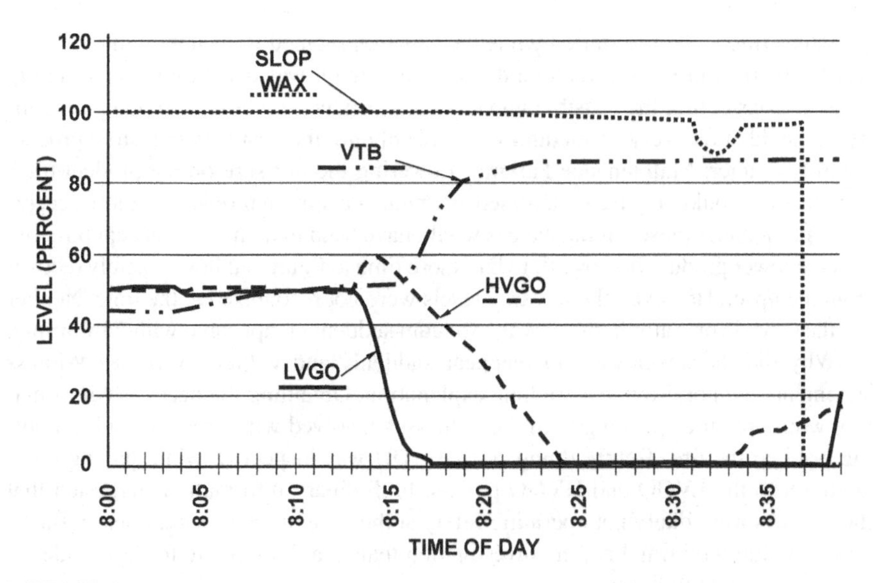

FIGURE 23.4

Vacuum distillation tower collection tray levels.

reading choppiness as well as the alarm flood that created a distraction inside the control room. This event was extremely significant because two critical alarms were buried in the nonsense created by the alarm flood:

1. Low (20%) LVGO tray level alarm at 8:14:24 am and
2. Low (20%) HVGO tray level alarm at 8:20:43 am.

The console operator was not aware that the LVGO tray had completely emptied and the HVGO tray was reaching a dangerously low level, because the two critical low-level alarms were hidden in the list of nuisance solenoid contact alarms. Unfortunately, no other warnings were issued to alert the console operator that the collection tray levels inside the distillation column were below their minimum levels.

The low tray level alarms were programmed to activate a response to an unstable condition that could result in a catastrophic pump seal failure. Pump cavitation that results from operating at low suction pressure may cause severe pump vibration. This vibration might lead to a catastrophic thrust bearing failure that can damage the pump's mechanical seal. It can also produce high-frequency vibration capable of loosening nuts that fasten flange joints. Either scenario might result in the involuntary loss of primary containment.

At 8:21 am, the console operator ended the alarm flood by manually opening the interlock solenoid valve to the flare. Previously that valve had been automatically diverting off-gas flow into the flare any time the KO pot pressure was 17.7 psia. At 8:24 am, the HVGO tray ran completely empty. At 8:31 am, the slop wax tray level started to drop, but remained above the minimum level alarm set point.

Interestingly, the fire started while the situation seemed to be improving. In other words, the pressure in the vacuum distillation column was decreasing when the fire ignited. Considering the possible impact of low product tray levels on pump reliability, the sudden recovery of vacuum was probably not the best way to restore process control (chapter: Maintenance Failure). Lowering the pressure on the product trays even further would only have increased any pump cavitation problem. A more acceptable option under these circumstances would have been to do nothing except possibly turn the tower product pumps off, to buy more time to figure out how to safely recover from the upset. However, the low tray levels were not recognized at the time. Neither did the console operator know why the vacuum suddenly reappeared without warning.

Why did the tower vacuum reappear suddenly and without warning? Witness statements did not provide a credible explanation. Obtaining the necessary information would require speaking with the witnesses involved with process troubleshooting prior to the fire. Before doing so, the CDU was inspected, while paying close attention to the LVGO and HVGO pumps. Preliminary information suggested that these assets were likely not operating safely at the time of the process release. Based on the preliminary timeline, the investigation team started looking for any evidence of excessive pump vibration.

INSPECTION RESULTS

Identifying the process release point was not difficult. There was no physical evidence of a thrust bearing or mechanical seal failure on either the primary or spare pump. However, removal of their insulation revealed that the HVGO pump discharge flange had separated (Fig. 23.5). Severe vibration likely resulting from pump cavitation had loosened all of the stud bolts that held the joint together to some degree. One stud completely dropped out of its hole while its nut remained wedged between

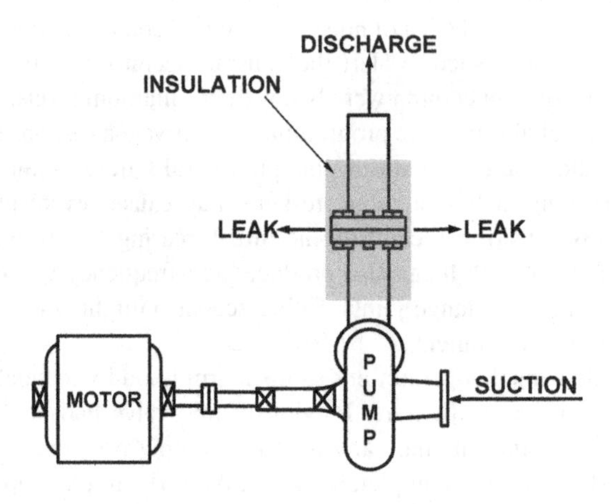

FIGURE 23.5

Heavy vacuum gas oil pump discharge flange configuration and leak location.

the flange and insulation. The joint failure opened up a pathway for hot HVGO to leak out. Upon contacting oxygen, the fire ignited as would be expected for an atmospheric hydrocarbon release above autoignition temperature.

INTERVIEWS

Next, the two workers performing routine maintenance were asked to describe their observations leading up to the fire and anything that might explain how the vacuum was restored. They stated that the vacuum loss seemed to correspond with backflushing the overhead vacuum condensers. This apparently had happened in other units before. Eventually they opened the manual bypass around the RV to purge any noncondensables from the KO pot. Afterward, the tower vacuum recovered. Shortly thereafter the fire was reported.

This input seemed reasonable—definitely something for the investigation team to pursue further. However, the information they provided directly contradicted the facts and evidence obtained in the initial phase of the investigation leading up to the interview. Experienced investigators live by the rule that "PI doesn't lie." Neither does process configuration act in mysterious ways. Processes do not exercise free will. They operate exactly according to design, depending on specific operating conditions. When the process acts contrary to expected performance, then there is a problem to solve.

Did you notice the contradiction? Looking back at the process configuration (Fig. 23.1), the manual bypass valve around the RV serves exactly the same purpose as the interlock that diverts the process into the flare when the KO pot pressure reaches 17.7 psia. Process data retrieved after the incident confirmed that the console operator ended the alarm flood by opening the interlock solenoid valve in manual control. This adjustment was made about 20 min before the first sign of tower vacuum recovery. However, the two operators involved in the backflushing procedure claimed that the tower vacuum was restored by opening the RV bypass valve to vent excess noncondensable vapor into the flare. What they did not know is that the console operator had already opened an alternative path directly into the flare about 10 min earlier to silence the alarm flood. While the true cause for the process upset remained unknown, it was obvious that the outside operators knew something that they were not telling. The cause for the loss of vacuum that initiated the entire sequence of events needed to be resolved. Therefore, the investigation continued.

INTEGRITY

It is impossible to predict where the most valuable information will come from at the start of an investigation. Many times, process data and maintenance records contain valuable information that the investigation needs for progress to be made. In this case, a third operator had been involved in troubleshooting the process several minutes before the fire started. This operator stepped forward as an anonymous source of information that filled the cavity surrounding the process upset that led to the fire.

According to the anonymous operator, the third-stage vacuum condenser was backflushed last. The vacuum condenser backflushing procedure had been followed by first closing the third-stage cooling water supply valve and then opening the inlet–drain valve. The drain valve was closed after several seconds of reverse cooling water flow through the exchanger tubes.

After completing the procedure, the operators left the elevated platform, but forgot to reopen the cooling water supply valve. This meant that no cooling water was passing through the vacuum condensers, which caused the upset. Confusion had dominated the troubleshooting effort. Nobody was really sure what to look for. It seemed to take forever to check the cooling water supply valves. This was a simple maintenance procedure that had been performed multiple times previously without incident. Nobody thought that the answer could be that simple. Instead of returning to the valves, they wasted time checking all kinds of things at the ejectors. Then finally one of the two operators who performed the procedure noticed the error and *immediately* corrected it. The cooling water supply valve was swung wide open, instantly restoring full condensing capability to the vacuum condensers. Feeling quite embarrassed, the workers descended the platform as if nothing was wrong and started walking back to the CDU control room. They were then notified about the fire in the CDU and started activating fire monitors.

This information was the missing link that brought together all the various sources of information. The loss of tower vacuum resulted from the extended closure of a critical cooling water supply valve. Since the cooling water supply operated in series, forgetting to reopen a condenser inlet valve interrupted flow through all three condensers. This error resulted in a total loss of steam condensing. The tower responded exactly as would be expected upon losing cooling water flow through the condensers. With no vacuum present, a smooth recovery could only occur by implementing a controlled process shutdown, then restarting the vacuum section from scratch according to normal startup procedure. Indeed, the hasty and uncontrolled move had made matters worse. By this time, the LVGO and HVGO product collection trays had emptied. Loss of pump suction head created a severe vibration event that likely separated the HVGO pump's discharge flange. The fire started about 5 min after level on the HVGO tray reappeared (Fig. 23.4).

Embarrassed over their error, the workers resisted admitting it. They wanted to believe that the problem had to do with noncondensables accumulating in the vacuum system, which they had vented by opening the RV bypass valve. They felt guilty that their failure to restore flow through the condensers, followed by an undisclosed and abrupt adjustment to correct it, might have caused the fire.

CLOSURE

This is where most industrial failure investigations would stop. There is complete clarity over the sequence of events that led to the fire. An act of omission, made worse by an act of commission, occurred during a routine maintenance task that had been implemented successfully hundreds of times before. The error resulted in a process upset that compromised asset integrity. Upon hitting air, conditions were right to autoignite the leaking process. Many lessons and improvements could be recommended based on this analysis, covering topics and practices such as:

- Insulating hot flanges,
- Interlock programming,
- Alarm rationalization,
- Process troubleshooting,
- Operator training,
- Emergency operating procedures,
- Disciplining the operators for their direct involvement, which included the error they made in (1) performing the routine maintenance procedure, (2) implementing an extreme adjustment to restore control, and (3) failing to communicate openly and honestly during the investigation.

In the final analysis, none of these items really digs deep enough to prevent a similar incident. That is not to say that it was inappropriate to look into resolving these more superficial and readily apparent issues. However, real value can only be generated by addressing the common pattern that underlies all human error in an industrial setting. We saw it in the Bhopal disaster as clearly as it is seen here. The problem is much deeper than what we have considered thus far. Digging further into the problem provides a satisfying and actionable conclusion that applies to this sequence of events. It directly relates to all other conceivable sequences of events leading up to industrial disasters.

The truth is that the timeline we have just finalized does not end here. Or perhaps it is more accurate to say that the sequence of events started much earlier than about 8:00 am on a normal day at the factory, when one worker advised another worker that he was about to start backflushing the vacuum condensers. Ending the timeline here barely even scratches the surface of the governing issue. Our investigation comes down to answering one simple question:

Why were the workers on the platform that day?

The answer to this question at first seems relatively obvious. The workers were conducting a normal, approved maintenance procedure as they had done successfully many times before. However, in reality the incident had nothing at all to do with backflushing steam condensers. Likewise, the Bhopal disaster had nothing to do with how water got into a tank full of a reactive product. The seeds for industrial disasters are planted years—perhaps decades—in advance. The common cause has relatively little to do with individuals who might be present on any given day.

Maybe the workers on the platform should never have been there. Or maybe they *should* have been there, but for more productive reasons. Say, for instance, that it would have been better for them to be there checking on the condition of critical assets in a low foot traffic area, off the beaten path. Examining the sequence of events prior to the day of the incident explains why:

Eight years before the incident, heat exchanger backflushing became a *routine* online maintenance procedure to keep the steam condensers clean. What was previously a seldom-executed, condition-based maintenance task was converted to a mandatory activity. The backflushing operation was thereafter carried out at least once a week. The procedure involved aggressively backflushing tenacious deposits out of the cooling water heat exchangers that, for some unknown reason, suddenly started to appear. In reality, the backflushing procedure did not solve the problem. Therefore, "air rattling" was added, which involved injecting plant air into the heat exchanger cooling water inlet prior to backflushing. This workaround procedure made it possible to limp the process into the spring turnaround season when the CDU could be shut down for mechanical cleaning. Due to the severity of the chronic problem, it was not possible to achieve the intended 5-year turnaround maintenance cycle. The CDU had to be shut down *at least* annually to physically remove thick layers of mineral scale at great economic expense, including production losses.

Without the routine maintenance procedure, degraded tower performance would force even more frequent shutdowns to avoid exceeding the furnace tube high temperature limit. However, implementing this workaround solution increased the system's human dependency. In other words, operators were then expected to compensate for the loss of an asset function that by design did not require routine human interaction. With this information in mind, the failure did not occur when human error caused a fire. It occurred 8 years earlier when the backflushing procedure was changed to manage the chronic problem that was causing repeat failures.

As might be expected, the asset performance issue was not isolated to the CDU. Starting eight years before the incident, booster pumps were retrofitted into the cooling circuit without significantly improving performance. Bleeder valve usage was modified to recover lost efficiency by injecting acid directly into the inlet of heat-exchange assets. This workaround solution caused a severe corrosion problem that resulted in numerous leaks, cooling water contamination incidents, and unplanned production suspensions. In response, many critical heat transfer assets were coated with an acid-resistant epoxy polymer to accommodate the unconventional, creative approach to removing cooling system deposits without shutting down. Unfortunately, the coating further reduced heat transfer efficiency and did nothing to curtail the rapid accumulation of debris inside the cooling system. In many cases, the coating flaked off; creating a macrofouling problem that further increased the dependency on heat exchanger backflushing.

Five years before the incident, the problem was so widespread and out of control that all heat exchange assets throughout multiple production units had to be replaced. The multimillion dollar cost for asset repairs (not including production losses) created sufficient financial losses to trigger a formal investigation to diagnose and address the cause for the chronic heat exchanger performance issue. The investigation report detailed an action plan based on inspections, analyses, and a timeline that associated historic system performance changes with specific events. Implementing the action plan eliminated the problem before the end of the next year.

When the fire occurred, the CDU was approaching a 5-year uninterrupted run without shutting down to clean and replace damaged cooling water assets. The aggressive backflushing campaign implemented when the problem first appeared, however, remained intact. Did this creative workaround solution set the workers up for success? The fire that followed 8 years later answers that question. The incident occurred long after the purpose for modifying heat exchanger backflushing practices was forgotten. Although the investigation solved the problem several years prior to the incident, the change that was introduced when the problem first appeared persisted.

The truth is that the workers should never have been performing the routine maintenance task that was normal to everyone on that day. Those responsible for preserving the integrity of the cooling water assets years before should never have been allowed to modify the heat exchanger backflushing procedure, install booster pumps, coat or rattle the heat exchangers, or supplement inadequate system performance by injecting acid into cooling water bleeder valves. They should have been held accountable for diagnosing and correcting the problem immediately when it appeared. Instead, they created more complexity that inserted a process safety hazard while making it very difficult for the business to compete.

In exchange for accepting repeat failures, operators inherited the responsibility of controlling the problem with brute force. Those responsible for preserving system performance and integrity were excused by those willing to accept creative work-around solutions that made normal process operation significantly more complex and hazardous. These practices were detrimental to the owner's financial performance. In making these adjustments, the operators took over process functions that were never intended by design. The operators were expected to perform better than the assets could. Avoiding the failure that occurred required executing the backflushing procedure with perfection every time. By process hazard analysis (PHA) standards, that type of performance is impossible. Anyone serving as an operator in that capacity would have eventually made the same mistake. In this case, it took 10 years to happen.

To illustrate this point, please consider how this scenario applies to a PHA before the fire occurred. In PHA meetings, great attention is given to predicting how equipment or human failures might create an intolerable risk. In all cases where cooling water heat exchangers are involved, the inclusion of a supply valve represents some probability for misoperation [1]. The potential to leave the valve shut, therefore, was an anticipated hazard validated by this scenario. The error rate is agreed to by all PHA meeting participants. In the case of improperly operating a manually operated block valve, the failure probability can never realistically be zero (chapter: Process Hazard Awareness and Analysis). An initiating event frequency of 0.1 would likely have been selected for the misoperated valve, based on its frequent usage.

The scenario can be eliminated by removing the manual block valve in the cooling water supply line, which workers might be tempted to close (Fig. 23.6). But that would make the backflushing procedure impossible. Once again, we find that reliable processes built for inherent safety are less able to accommodate routine maintenance that by design intentions should not occur (chapter: Licensing). Since we are dealing with a system where the valve exists, improving human performance in this case requires reducing the demand for closing the valve. Implementing a procedure to work around a chronic equipment performance issue does not achieve this purpose. A routine maintenance procedure only *increases* the demand on the valve, and therefore the incentive to use it. The only acceptable solution is to effectively address any issue that might suddenly increase the demand on the asset. This might require an

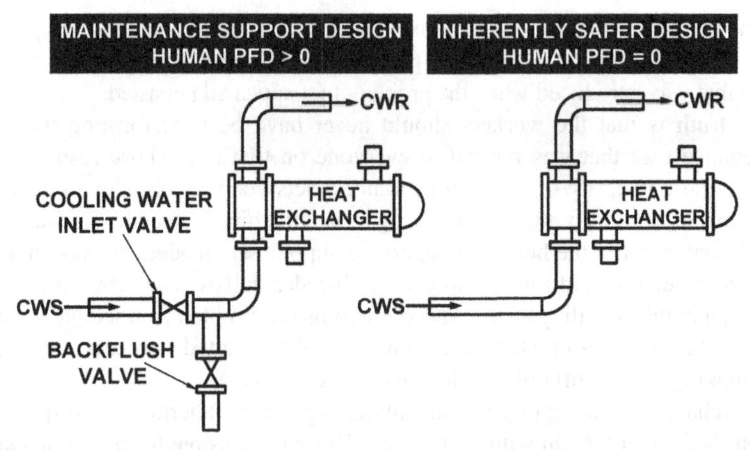

FIGURE 23.6

Inherently safer design reduces potential human failures but requires a greater reliability commitment. *CWR*, cooling water return; *CWS*, cooling water supply.

investigation to understand the phenomenon responsible for an observed change in process performance immediately when it appears.

Through this explanation we have derived the universal relationship that applies to all industrial disasters. By these terms, all industrial disasters are traceable to human performance difficulties. Although none of the case histories related in this text could ever equate to the Bhopal disaster's magnitude, they all follow a distinctive pattern. Deep beneath their superficial and distracting exterior they are all identical. The Bhopal disaster occurs daily in minor forms throughout the manufacturing industry today. Some of these incidents are consequential enough to be recorded as disasters. Preventing disasters and even precursor incidents of lesser consequence requires applying this knowledge.

Maintaining an acceptable level of system protection requires preserving mechanical functions. Using brute force to compensate for asset capability losses upsets the balance of human performance. This relationship is true in all cases where man must interact with machines.

INSTINCT

Regrettably, the case study does not end here. About a year after the fire, the third operator who provided the information needed to complete the investigation with certainty was personally involved in an incident. While taking a hot process sample from an unauthorized, alternative location, he attempted to poke through blockage in an open bleeder valve with a welding rod. The pressure behind the obstruction suddenly burst through the valve and splashed hot process onto his face. He was able to close the valve before being seriously injured but left the area streaked with hot oil. He immediately washed the process off his face, but fresh burn marks remained. When his fellow workers asked him if he was alright, he explained that he was fine except for an accident that occurred before leaving for work. This explanation created a contradiction since his coworkers knew quite well that he started the shift without fresh burn marks on his face. In an unfortunate twist of irony, he followed down the same road as those who were personally involved in the fire a year earlier. Regardless, the same question applies that revealed the true cause for human error in the CDU fire:

Why was the worker using this alternative sample point?

Here we have another, completely different, set of circumstances with essentially the same answer. Since the *authorized* sample station was not available, the workers were using an alternative sample location (hazard awareness). They had been doing so for quite a long period of time when the incident occurred (tribal knowledge). A welding rod was kept nearby to clear the valve when needed (improvisation). The personal protective equipment that was needed to safely implement the workaround procedure was not being worn (casual compliance). Here again we find a direct link back to an event in the Bhopal disaster that occurred on January 9, 1982 (chapter: Process Isolation and Containment). Also again, the worker who was splashed with hot oil should not have been involved in the workaround procedure that was normal to him on that day (uniformity in message).

The pattern we observe following human error is different from the pattern that causes human error. In all cases we find that individuals who sense any possible involvement after an incident tend to express denial [2]. Time is needed for individuals and perhaps even more so for organizations to work through "the five stages of grief" (denial, anger, bargaining, depression, and acceptance). At the end of this period, however, there is acceptance. An objective, well-written, fact-finding (not fault-finding) investigation report shortens the time needed for acceptance to arrive. The faster you get there, the sooner you can protect yourself and others with the lessons you learn.

LESSONS FOR US

Once the circumstances behind an industrial incident of significant consequence are known, the actions of individuals who might somehow be involved are always questioned. That is why it is more important for the investigation to explain *why* people were involved instead of simply exploring *what* they were doing. Drilling down to that level of comprehension usually requires going back in time; maybe back to the very beginning. Time governs all failures. Time also heals all wounds, except for the ones that run too deep to forget. It is therefore important to remember that your responsibility is to operate far enough away from an asset failure to prevent it. Doing so is the safest, most rewarding, and most economical approach to operating a business in the manufacturing industry. Other points to remember are:

- Hold individuals performing a service, reliability, or technical function accountable for what they are expected to do.
- When confronting a problem, look for ways to eliminate it before settling to control it.
- Workaround maintenance solutions introduce unmitigated hazards.

- Most disasters occur after workaround maintenance solutions become the routine.
- In the midst of chaos, contemplate your next move before making it. When things do not make sense, investigate to obtain the missing knowledge.
- Despite the circumstances and how abstract the causes may seem, someone usually knows what happened [3].
- Investigations create lasting value by digging beneath the surface of what others initially see.
- Contradictions result when misleading or partial statements are made during an investigation.
- Communicating openly and honestly avoids contradictions.
- If mishandled, a minor issue can lead to tragic circumstances.
- A complete timeline links together events that *independently* appear to be unrelated.
- Use an investigation to determine what you are looking for before trying to find it.
- Look for ways to simplify process design by improving process reliability.
- Asset reliability governs human PFD.
- Safe operation costs much less than unsafe operation.

REFERENCES

[1] Center for Chemical Process Safety (CCPS), Layer of Protection Analysis: Simplified Process Risk Assessment, 71, 2001, ISBN: 0-8169-0811-7.
[2] R. Noon, Kübler-Ross and root-cause evaluations, Maint. Technol. 25 (6) (2012) 39.
[3] A. Sofronas, Case 70: twenty rules for troubleshooting, Hydrocarbon Process. 91 (8) (2012) 35.

CHAPTER

Reliability Commitment 24

A comprehensive, updated look at history's most influential process safety failure provides compelling new insight into decisions and actions that concern us daily. As we consider lessons learned from this case study, we appreciate how the basic concepts of process safety management (PSM) create a solid foundation against catastrophic process release incidents. The basic lesson that applies to all of us, no matter where the future of the manufacturing industry might go, is that this foundation must exist before adequate protection can be reasonably expected.

LESSONS LEARNED

Had the Bhopal factory's design and operation been guided by the PSM concepts that were formalized after the disaster, then the incident would have been prevented. The same is true for industrial incidents of today; all of which fit the patterns developed fully by the case study. We can use this lesson to our advantage. Incident prevention is possible when we adhere to the standard principles that strengthen our resistance to failure. Recognizing the patterns that led to the Bhopal disaster in the processes that we operate today indicates a weakness deserving of our attention. Compliance audits, incident investigation, and process hazard analysis are the most effective ways to acquire the knowledge we need by examining our own processes. A corrective (not adaptive) approach to process performance that does not meet expectations helps to maintain a failure-restrictive organizational culture where deviance is not accepted. This is just one of many lessons that were demonstrated in the previous chapters.

Another lesson has to do with the Bhopal factory itself, relative to the factories that safely operate today. An objective review of this process introduced us to a truly impressive chemical manufacturing site for those who both designed and operated it. The factory was constructed to provide safe, reliable, and affordable production for infinite decades to come. Its simplified design concept incorporated various interwoven functions, which made the process both more compact and economical to build and operate. The design of the Bhopal factory took full advantage of optimization opportunities that made it more advanced than the one from which it was conceived. These upgrades eliminated unnecessary complexity. On a minor scale, the factory was a testimonial to industry's technological advancement. On a much larger scale, it represented an achievement in human creativity.

But despite how impressive the Bhopal factory looked both inside and out, the process was not perfect. An asset reliability problem soon developed. The problem was solved by reconfiguring the process. This change was considered a necessary improvement by those who designed, approved, and operated it. The change further simplified process design and made it inherently safer. It exemplified all the virtues that had made the company so successful for many years.

The only problem was that the factory was not designed to accommodate this change, which instantly became the preferred way to operate. Immediately, the process retaliated with more demanding maintenance. The additional maintenance, however, was considered a fair price to pay for a solution to a chronic production problem. At least, that was the prevailing impression when the decision was made.

As time went on, the control processes needed to responsibly operate the factory capitulated under the ever-increasing maintenance demand. Many of the process functions that were needed to maintain safe, reliable, and affordable production were lost. The importance of retaining these functions was obscured by the image of a process that was becoming inherently safer.

Unrecognized by many was that the factory was, in fact, becoming progressively less safe. To those who were responding to the constant needs of the demanding process, the changes to improve process performance were made responsibly and according to established engineering practices. No shortcuts were taken—all changes were approved through the proper technical channels prior to implementation, in conjunction with established policies.

Eventually, those who envisioned a career that involved being in control of a process, rather than being controlled by a process, left the factory. The experience that was needed to safely control what had become a very complex and dangerous process left with them. Soon afterward, the factory responded in a way that nobody previously thought was possible. On December 3, 1984, a tragic loss of process containment resulted in the death and injury of thousands of people in the course of only a few hours. The aftereffects of the industrial incident that occurred that night have persisted since then. The incident raised difficult questions about a trusted company whose name became connected with a tragic process incident. How could a company that was so successful and set the example in compliance that others only dreamed to accomplish have allowed this to happen? How could they have hidden their true colors for so long?

The true colors of the company were transparent both inside and outside the sites that they operated. Their colors were the same both before and after the incident. Their secret for success was ultimately a formula for destruction. The truth is that the company *did* set a fine example in compliance. Perfection was their goal. That expectation was firmly implanted and strictly enforced within the ranks at every level both internationally and at home. However, all restraints were off when discretionary matters such as risk calibration and asset reliability were involved. Controlling discretionary matters could stifle the creativity that had made them the most successful chemical manufacturing company on the planet earth. It was the unbalanced combination between mandatory and discretionary issues that made the company so very dangerous. The thinking prevailed that if more maintenance was all that was needed to upgrade process design, then it was a good business decision. Eventually, however, this philosophy created overwhelming economic and safety difficulties for all involved. Always remember that reliability is preferred over maintenance. If a maintenance demand exceeds design expectations, then it is only a matter of time before a human failure reminds you of this lesson.

The disaster proved beyond any doubt that the failure to firmly establish a *reliability commitment* can offset all of the progress that industry makes. The relentless pursuit of compliance at the expense of reliability created a hidden failure. Revealing that failure destroyed an enterprise and the future of many in the community that were affected by it. On the outside, the factory's owners had earned a well-deserved good reputation among their contemporaries. But inside, the same discipline was not applied to discretionary matters. Discretionary choices about managing hazards, risk, and maintenance caused a catastrophic process release that could only be understood in the context of the tools that were adopted in direct response to the disaster.

The false impression of internal compliance was created by indiscriminately changing internal policies. These changes allowed the implementation of improvised, creative workaround solutions. As is the case with all industrial disasters, the price we pay for learning these lessons may be unaffordable. This was certainly the case for one particular company whose commitment to compliance exceeded its reliability commitment. The defect was this imbalance. This transparency creates the ability to avoid operating under the false impression of safety that does not actually exist. No matter how industry's PSM regulations might change, the safest and most productive organizations always show *at least* as much interest in internal compliance as they do in regulatory compliance. This creates the uniformity in message that promotes a failure-restrictive organizational culture. There can be no contradictions between the way the business is managed internally and externally.

This analysis of the Bhopal disaster provides the knowledge you need to see the interwoven relationship between asset reliability, process design, human error, process safety, and regulatory compliance. In all cases involving the manufacturing industry, there is an unswerving commitment to compliance. Compliance matters are mandatory. They get done first. However, the processes we operate become more dangerous if our reliability commitment does not match our compliance commitment. Failure to continually demonstrate a proper respect for either of these commitments can damage our reputation. The Bhopal disaster demonstrates how nothing damages a fine reputation faster than losing control of an industrial process. When an industrial process becomes a menace to society, there is a limited time before the factory doors will be closed forever. We keep these doors open by responsibly controlling our creativity, to avoid manipulating the process in ways not supported by design.

Absorbing the lessons from this case study could reveal a reliability commitment that is not as firm as it needs to be. Take a lesson from the Bhopal disaster to understand how a workaround solution might conflict with process design. Take note of the process configuration. Leave your desk and walk the process down, now knowing what to look for. Where are the flanges and bleeders? Are there valves installed in various locations to support online maintenance or was the process designed to be inherently safer? Is it designed for maintenance or reliability? Recall that the process design will limit your flexibility on corrective maintenance. To truly be safe, your asset reliability must match the process design specification. Recall the example of the methyl isocyanate (MIC) transfer and circulation pumps to understand the basis for this requirement.

WHAT YOU CAN DO

Make inherently safer design your ultimate goal. Processes that are designed to be inherently safer are also more economical and more productive than more human-dependent processes. However, your process must be designed in parallel with your reliability. If you tolerate repeat failures on a process that was constructed on the principles of inherently safer design, then your maintenance cannot be safely or economically performed to your satisfaction. A choice must then be made. Will you raise process reliability to the standard that is reflected in process design, or will you reduce process design to the standard set by poor reliability? There are cost implications involved in this decision. But the decision must be made because you cannot afford the safety incident that will result from accepting an imbalance. This is a major theme that is supported by a definitive analysis of the Bhopal disaster. Your decision will define your reliability commitment.

Talk to your operators. Listen to what they tell you. Ask them to explain how they are operating the process. Is it in accordance with design or does the process only respond by doing things differently? What issues must they struggle with to make the process work properly? If they could solve one process-related problem that they must contend with to maintain acceptable performance what would it be? They will tell you before you need to perform an incident investigation. This is by far the most direct, cooperative, efficient, and preventive path to improvement.

Understand also that, as a case history, the Bhopal disaster is no different from any other industrial incident investigation. When an investigation report is read by two different people, their conclusions usually do not match. Rarely is there disagreement, but one person sees things that others cannot and vice versa. In this way, the Bhopal disaster represents an "ink blot" that everyone can benefit from. "Lessons for us" are appended to the end of each chapter. These lessons are by no means the only learnings offered by this deep analysis. The manufacturing industry is staffed by a diverse composition of different work groups all supporting the same objectives. The question you need to ask upon learning about how processes and management systems under your control malfunctioned is how you personally can make a difference. As was stated in chapter "Process Isolation and Containment," you can control what you can do but not what others can do. Concern yourself, therefore, primarily with what you can do to prevent an industrial disaster. Our processes run the safest when all are committed to this principle. At the same time, take responsibility for protecting others by pointing out to your coworkers what they must do to prevent an industrial disaster if you observe them about to make a mistake that relates to what you have learned. Advise them to read the book you are holding or the source that contains the lesson that applies. Only by doing this can they appreciate the importance of what you are speaking of, while you develop the uniformity in message that prevents more serious incidents.

In closing, never underestimate the importance of your reliability commitment. Remember the Bhopal disaster every day when you enter through the factory gate. See it all around you and listen to what the process is saying. Be sensitive to asset performance that does not meet reliability expectations. Confront these problems directly and reject any opportunity to implement a convenient workaround solution. Remember that safety really is your top priority. But recognize that process safety and productivity are measurements of your reliability commitment.

Appendix

CHAPTER 6

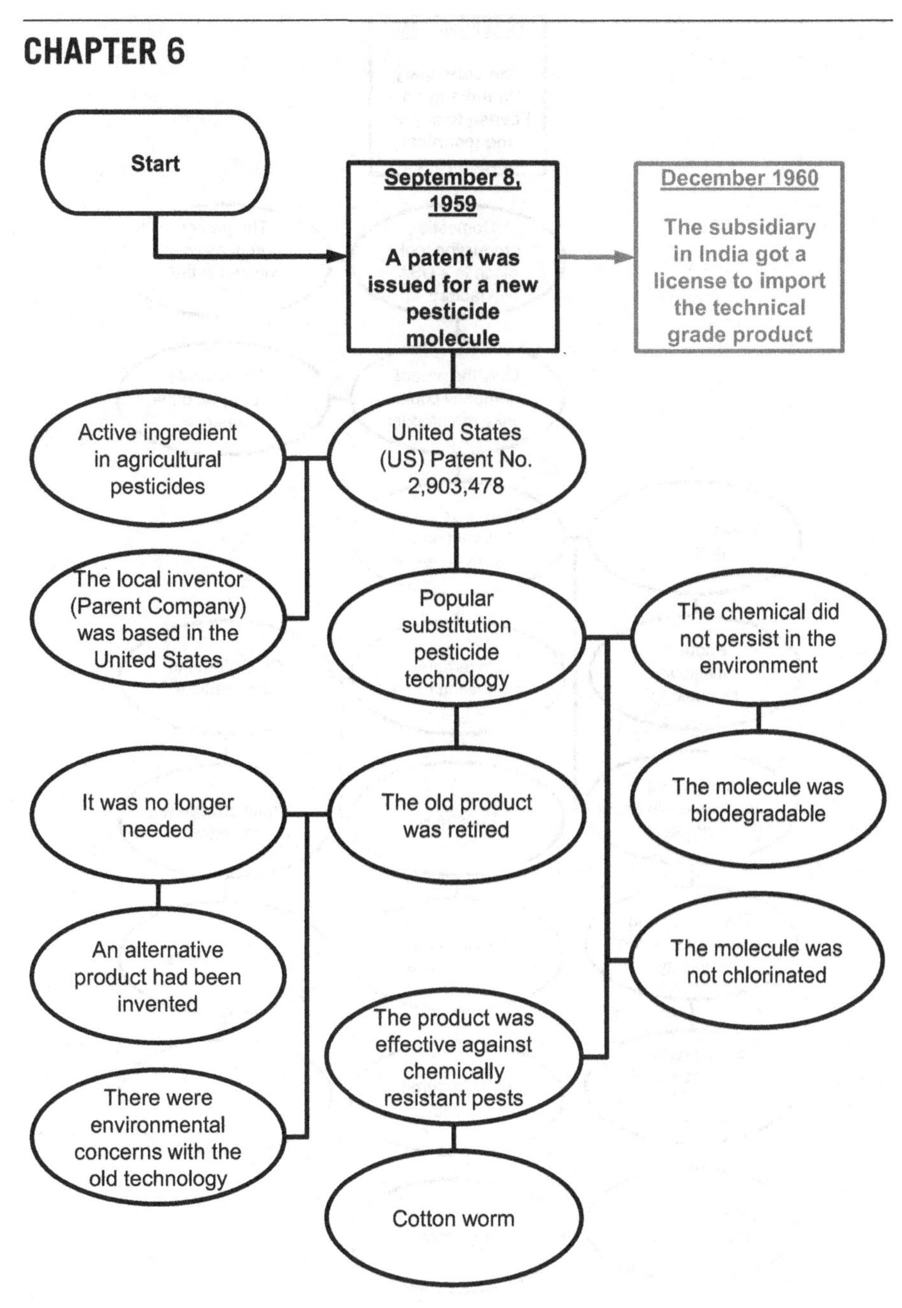

Start
September 8, 1959
A patent was issued for a new pesticide molecule
December 1960
The subsidiary in India got a license to import the technical grade product
Active ingredient in agricultural pesticides
United States (US) Patent No. 2,903,478
The local inventor (Parent Company) was based in the United States
Popular substitution pesticide technology
The chemical did not persist in the environment
The molecule was biodegradable
It was no longer needed
The old product was retired
An alternative product had been invented
The molecule was not chlorinated
The product was effective against chemically resistant pests
There were environmental concerns with the old technology
Cotton worm

Bhopal Disaster Timeline Part 01

CHAPTER 7

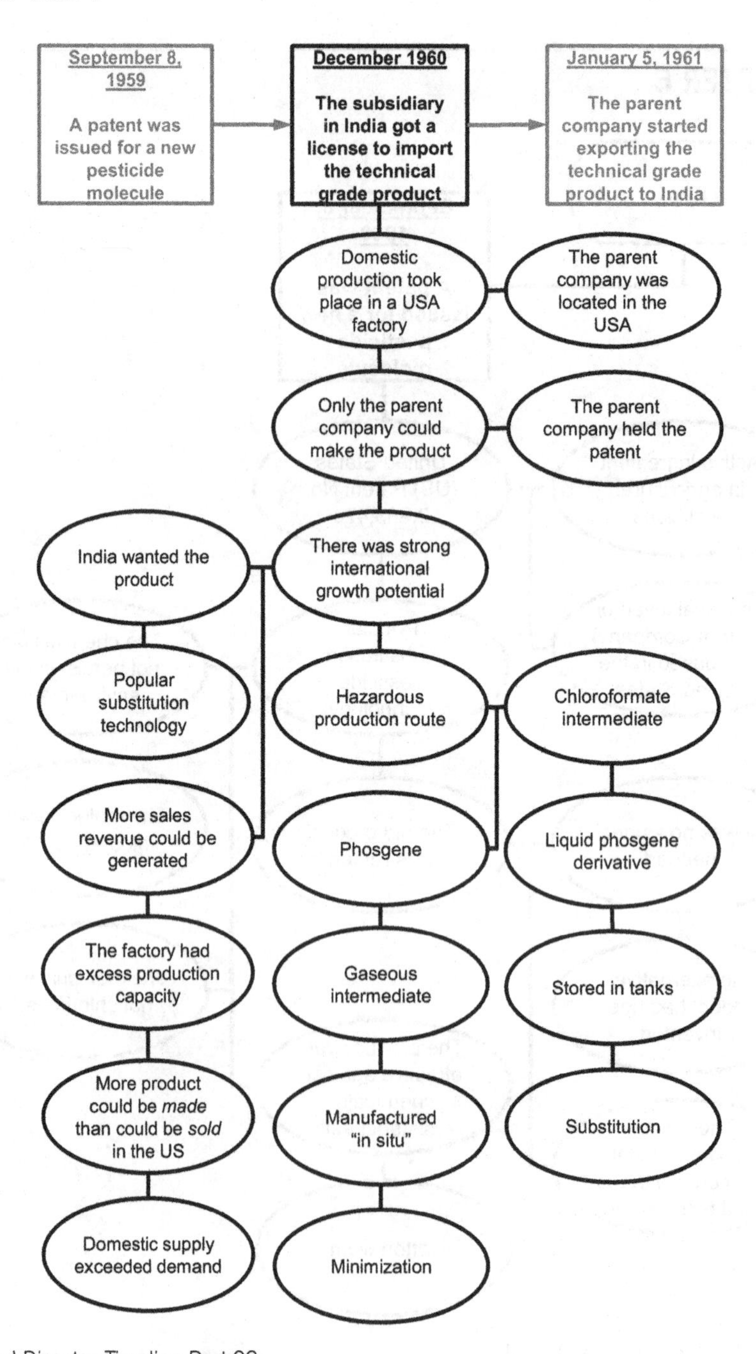

Bhopal Disaster Timeline Part 02

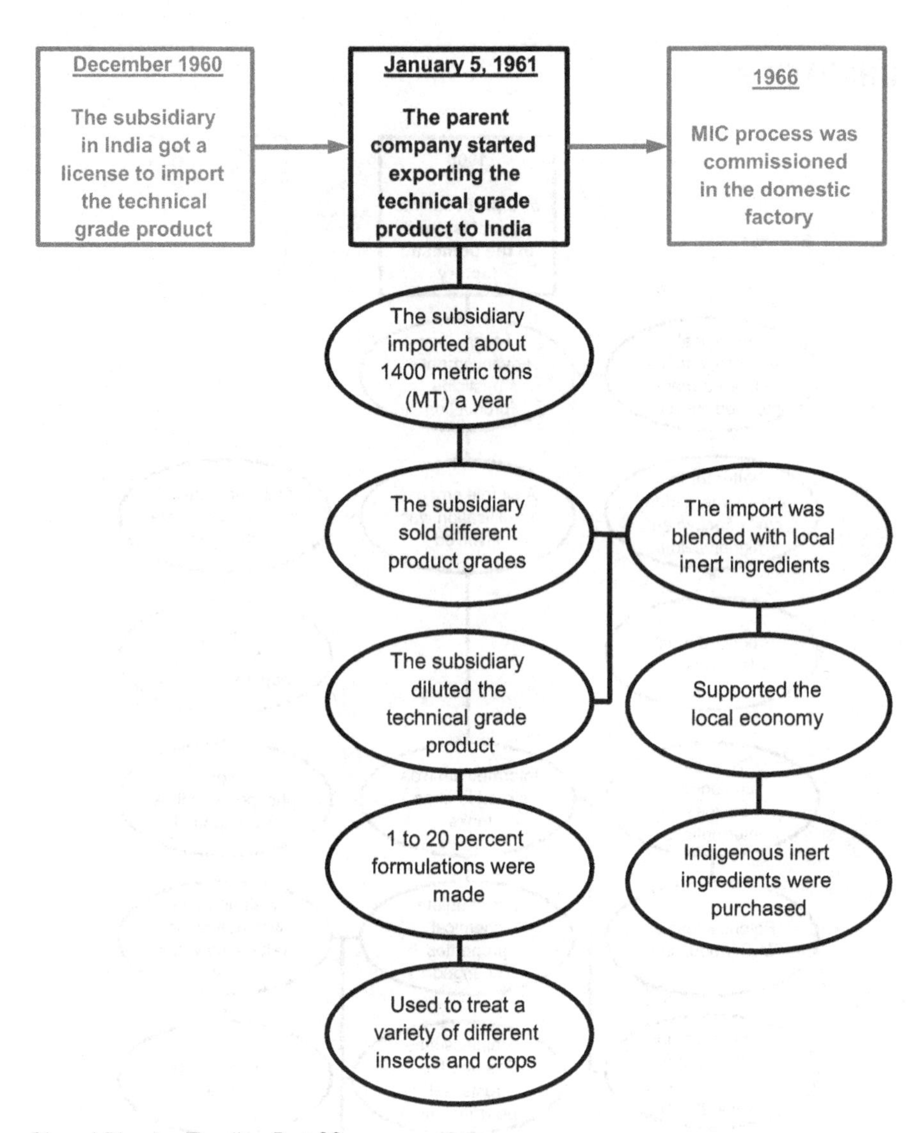
December 1960
The subsidiary in India got a license to import the technical grade product
January 5, 1961
The parent company started exporting the technical grade product to India
1966
MIC process was commissioned in the domestic factory
The subsidiary imported about 1400 metric tons (MT) a year
The subsidiary sold different product grades
The import was blended with local inert ingredients
The subsidiary diluted the technical grade product
Supported the local economy
1 to 20 percent formulations were made
Indigenous inert ingredients were purchased
Used to treat a variety of different insects and crops

Bhopal Disaster Timeline Part 03

CHAPTER 8

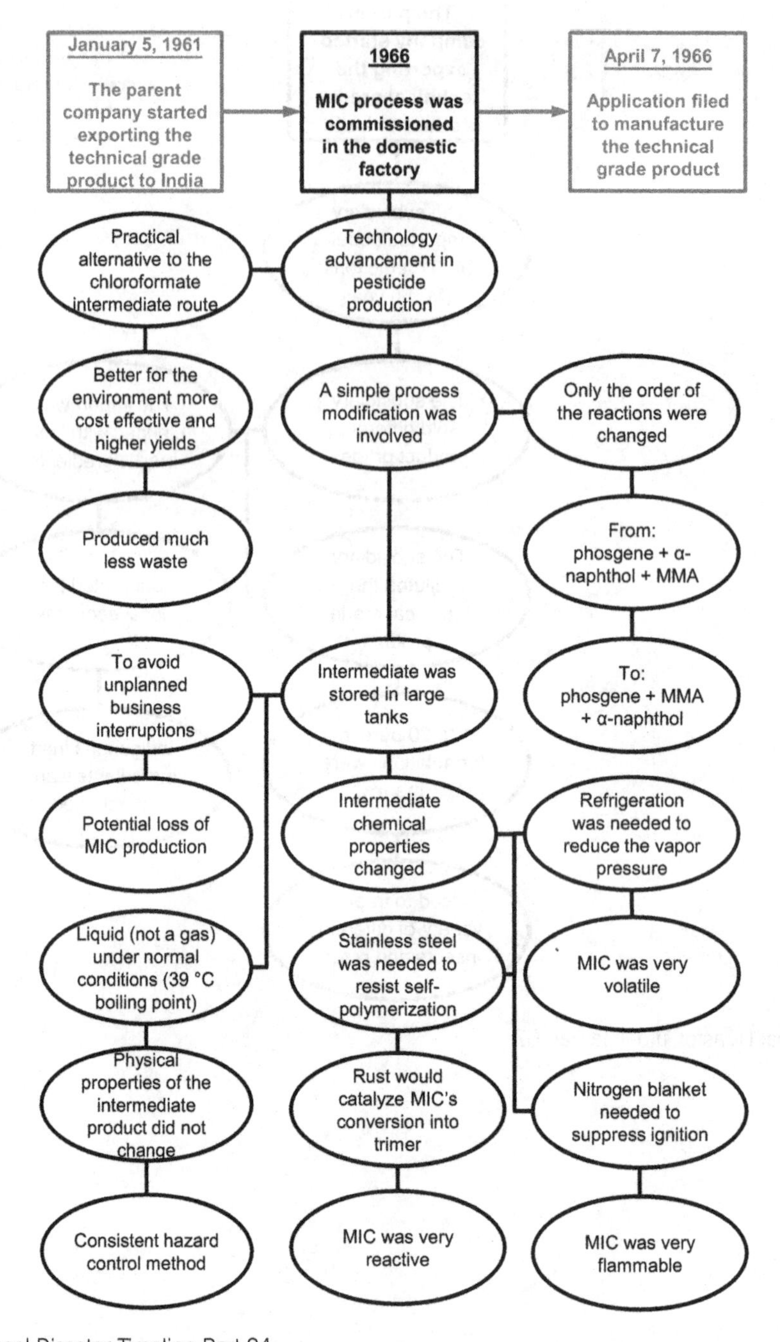

Bhopal Disaster Timeline Part 04

CHAPTER 9

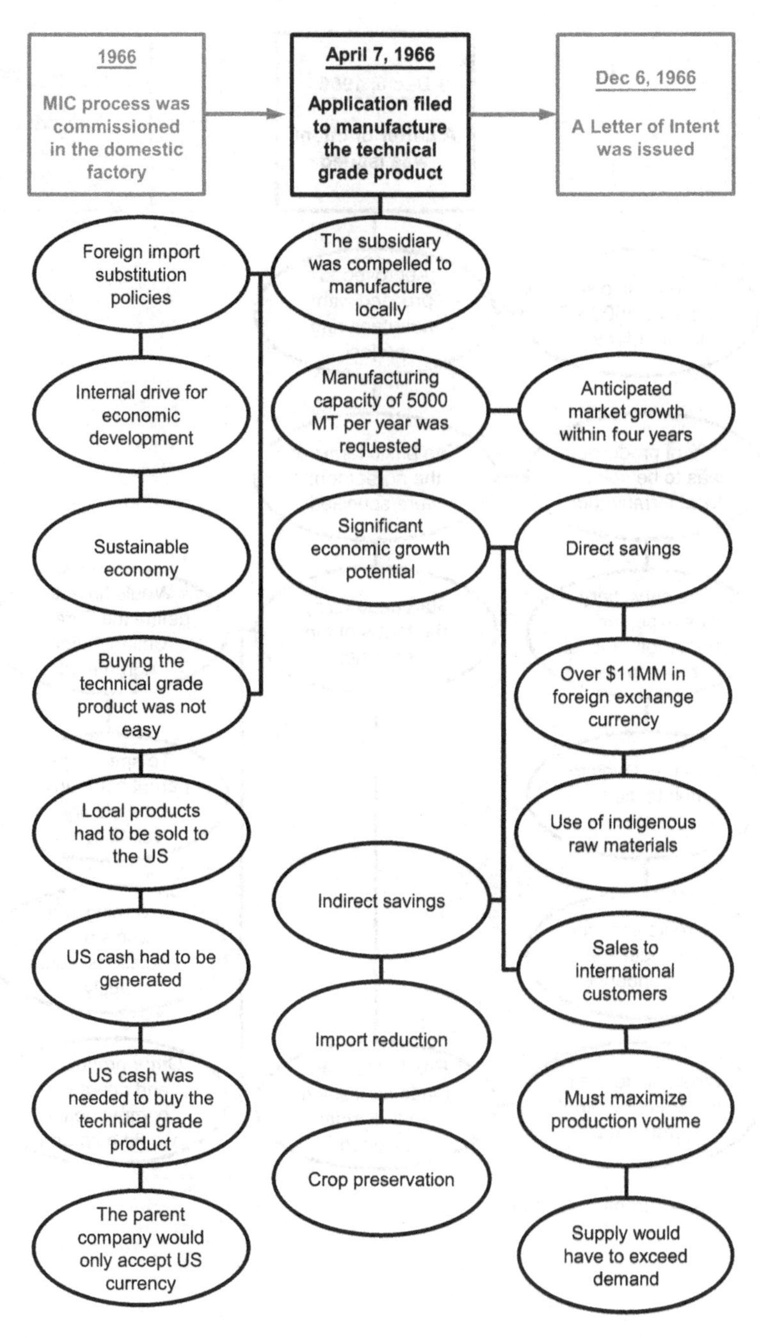

Bhopal Disaster Timeline Part 05

CHAPTER 10

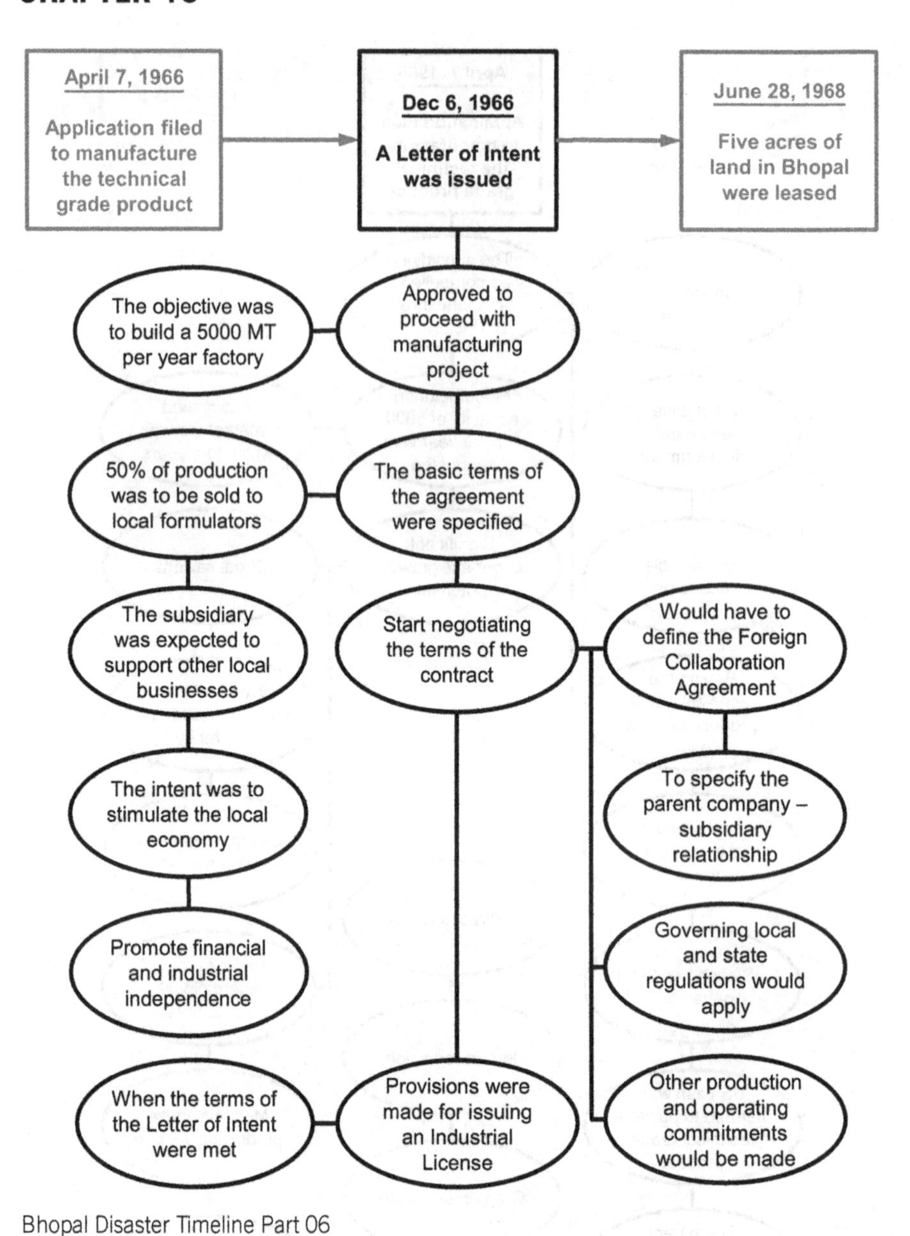

Bhopal Disaster Timeline Part 06

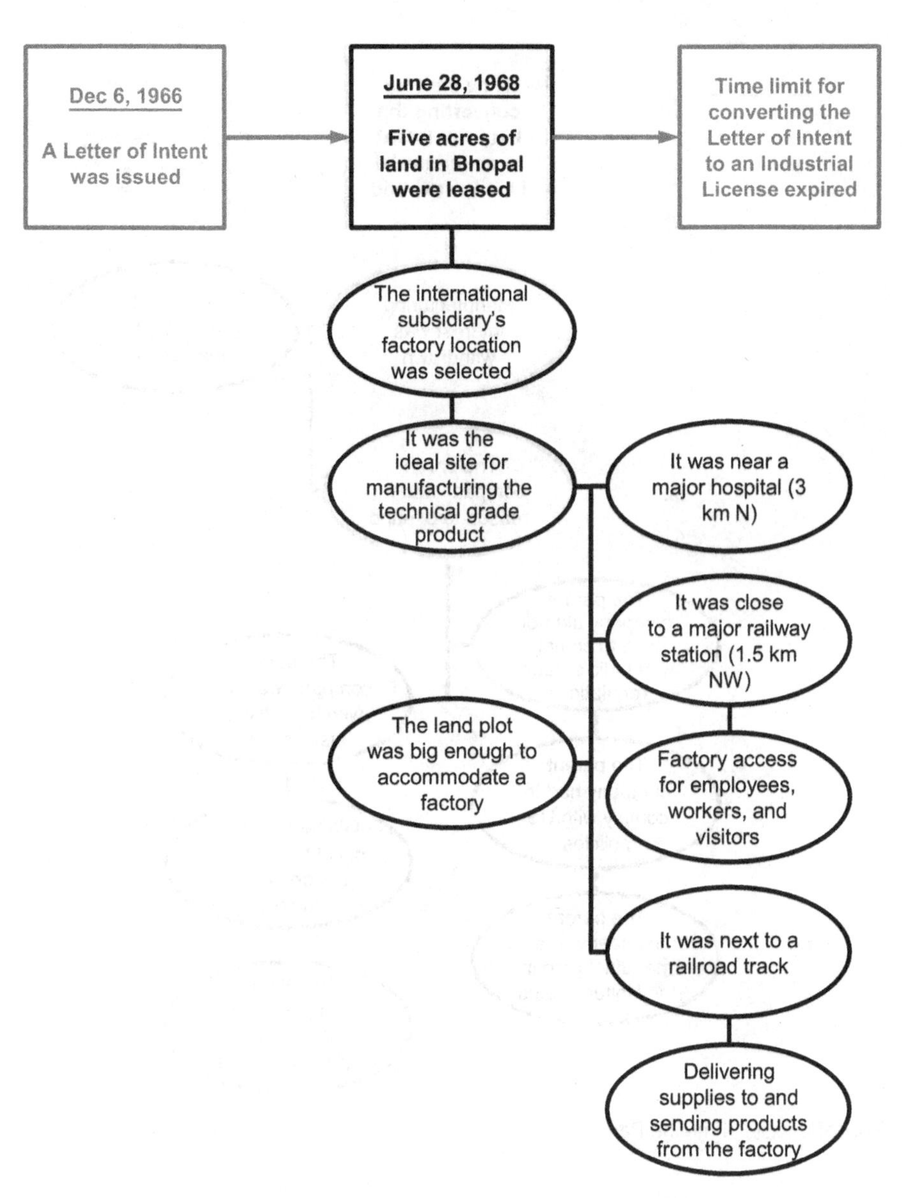

Bhopal Disaster Timeline Part 07

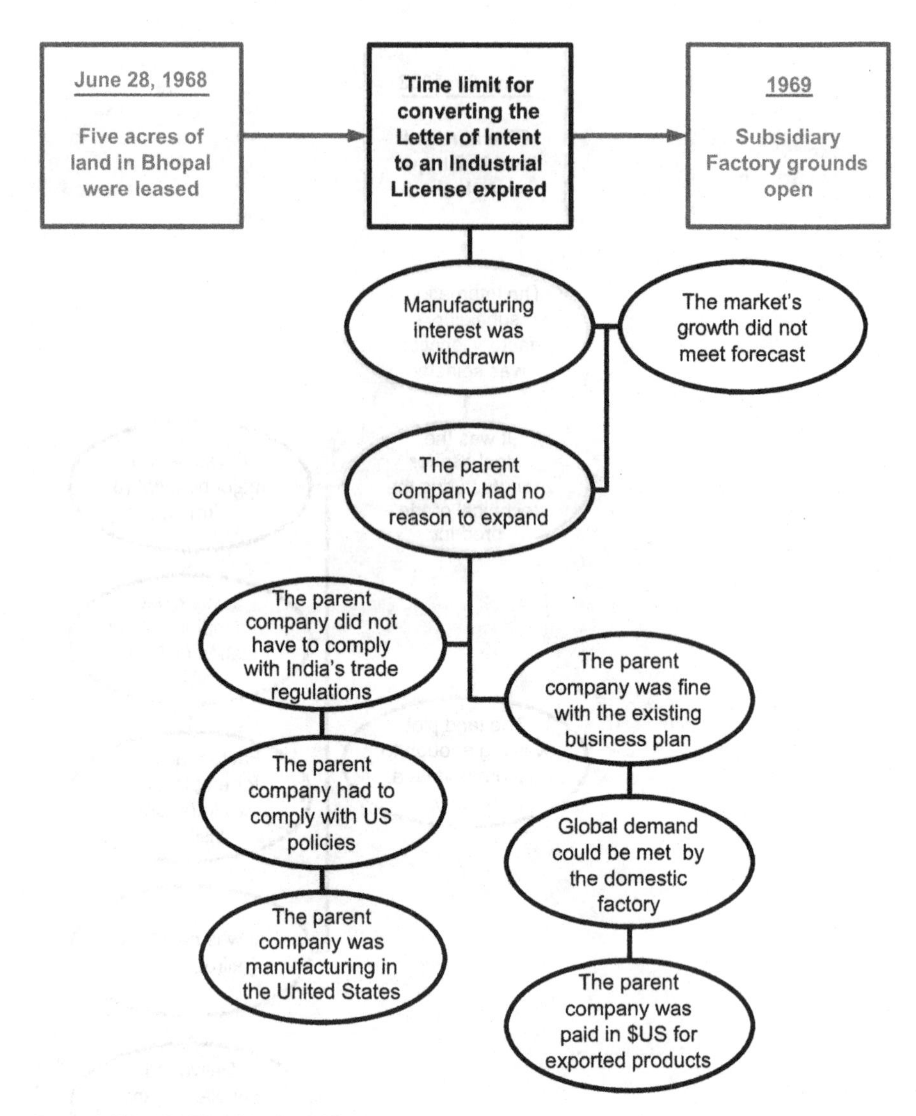

Bhopal Disaster Timeline Part 08

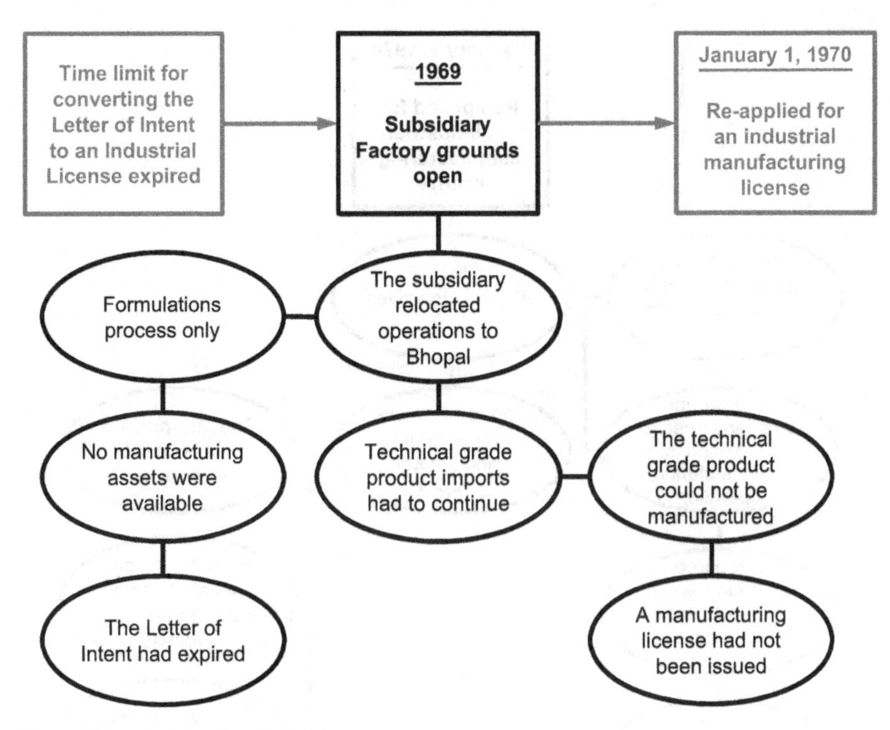

Bhopal Disaster Timeline Part 09

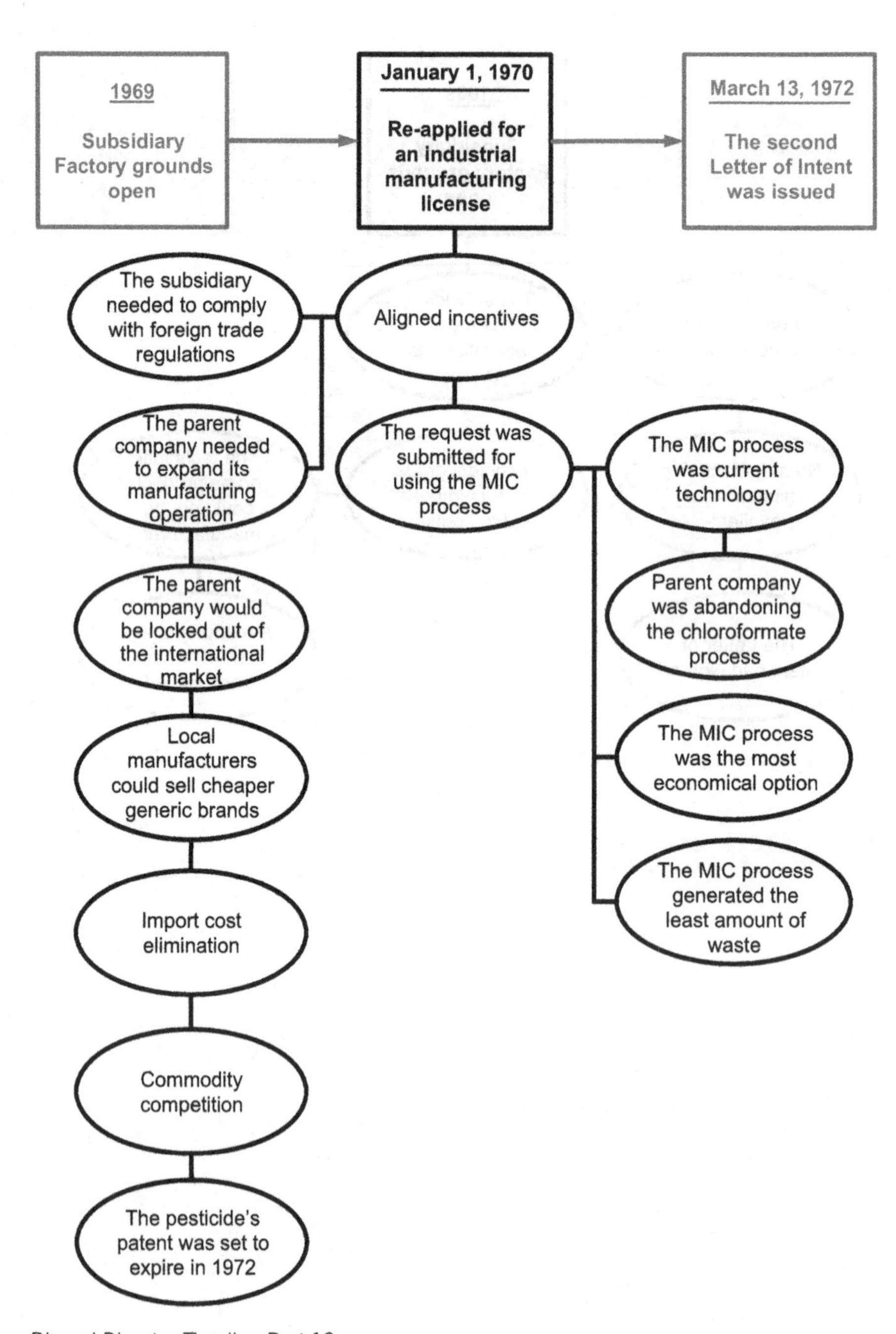

Bhopal Disaster Timeline Part 10

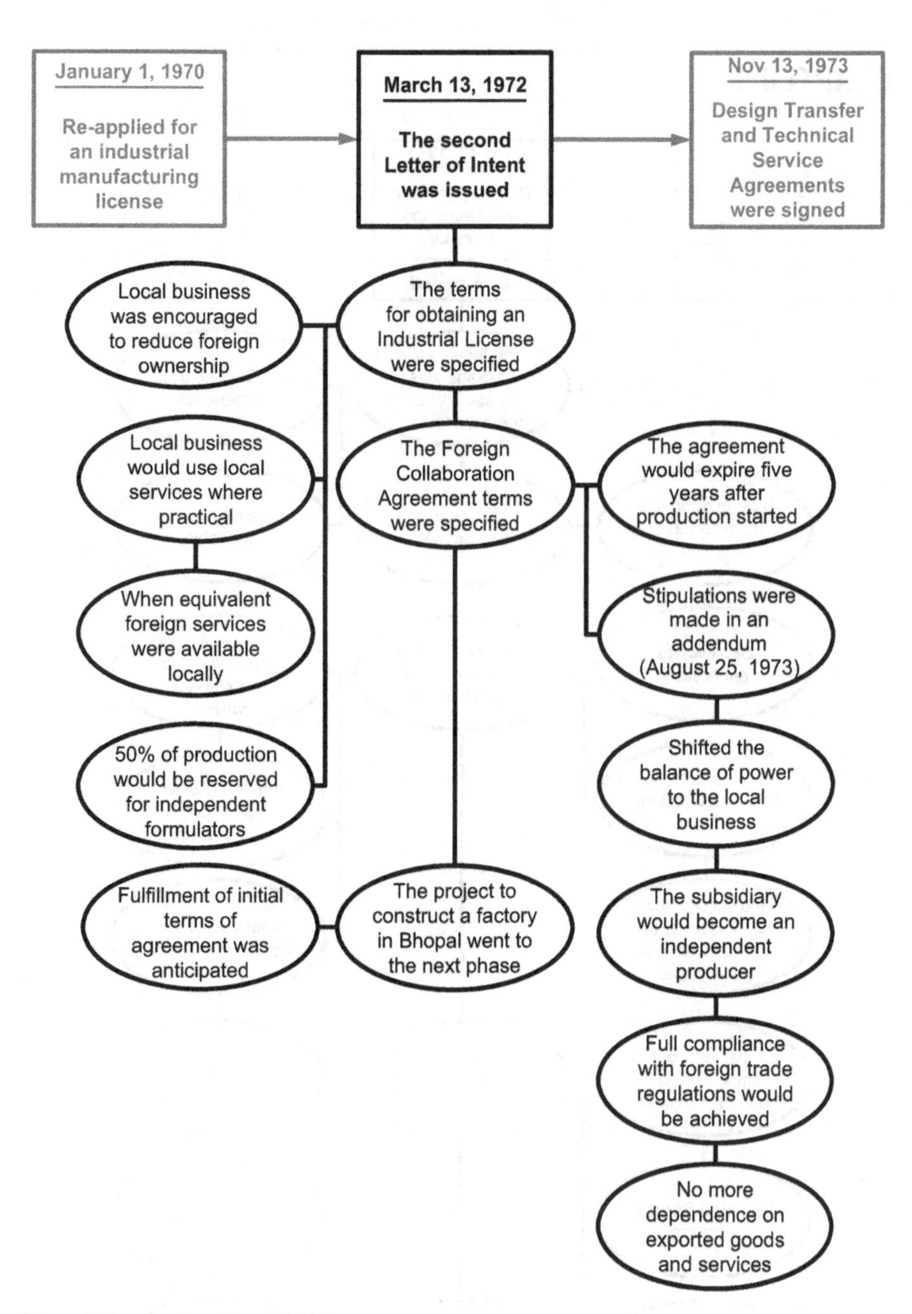

Bhopal Disaster Timeline Part 11

CHAPTER 11

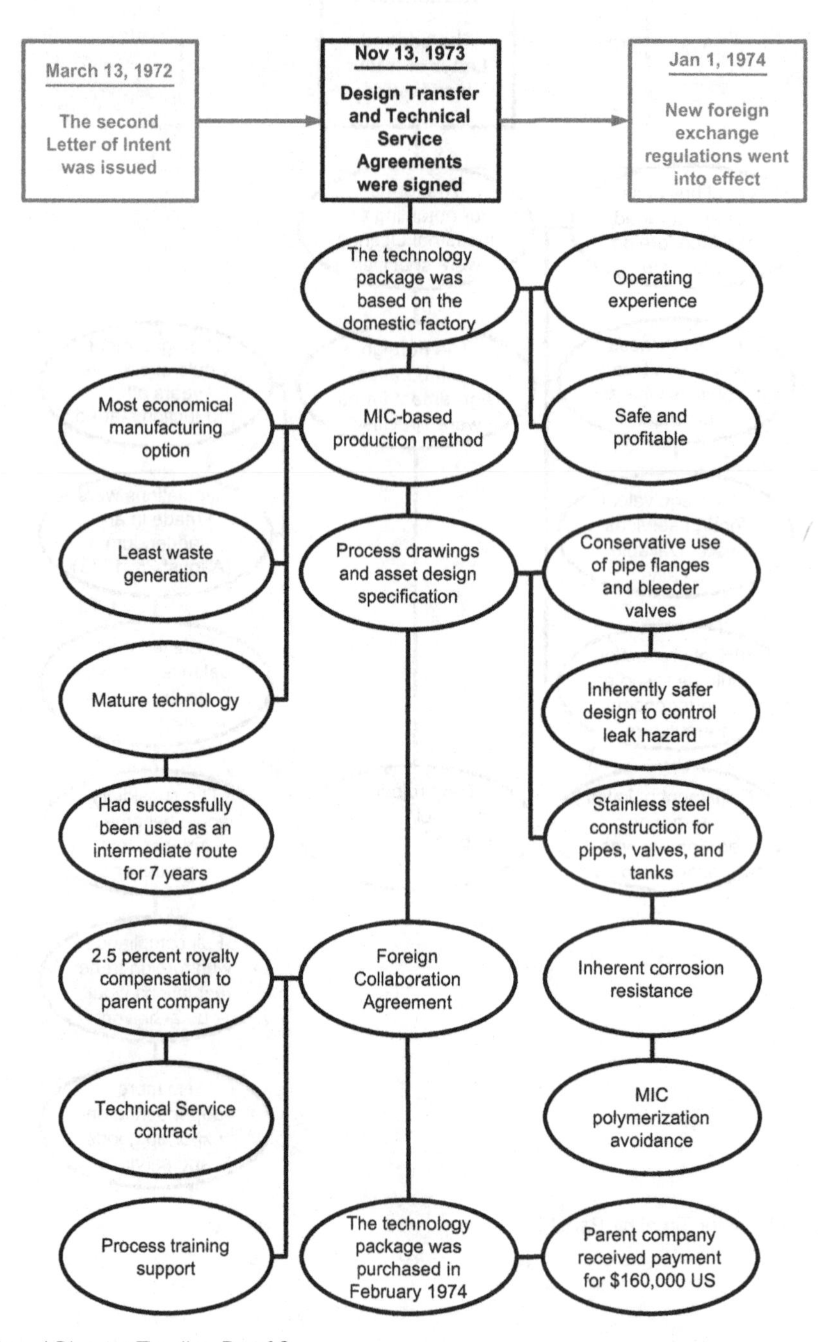

Bhopal Disaster Timeline Part 12

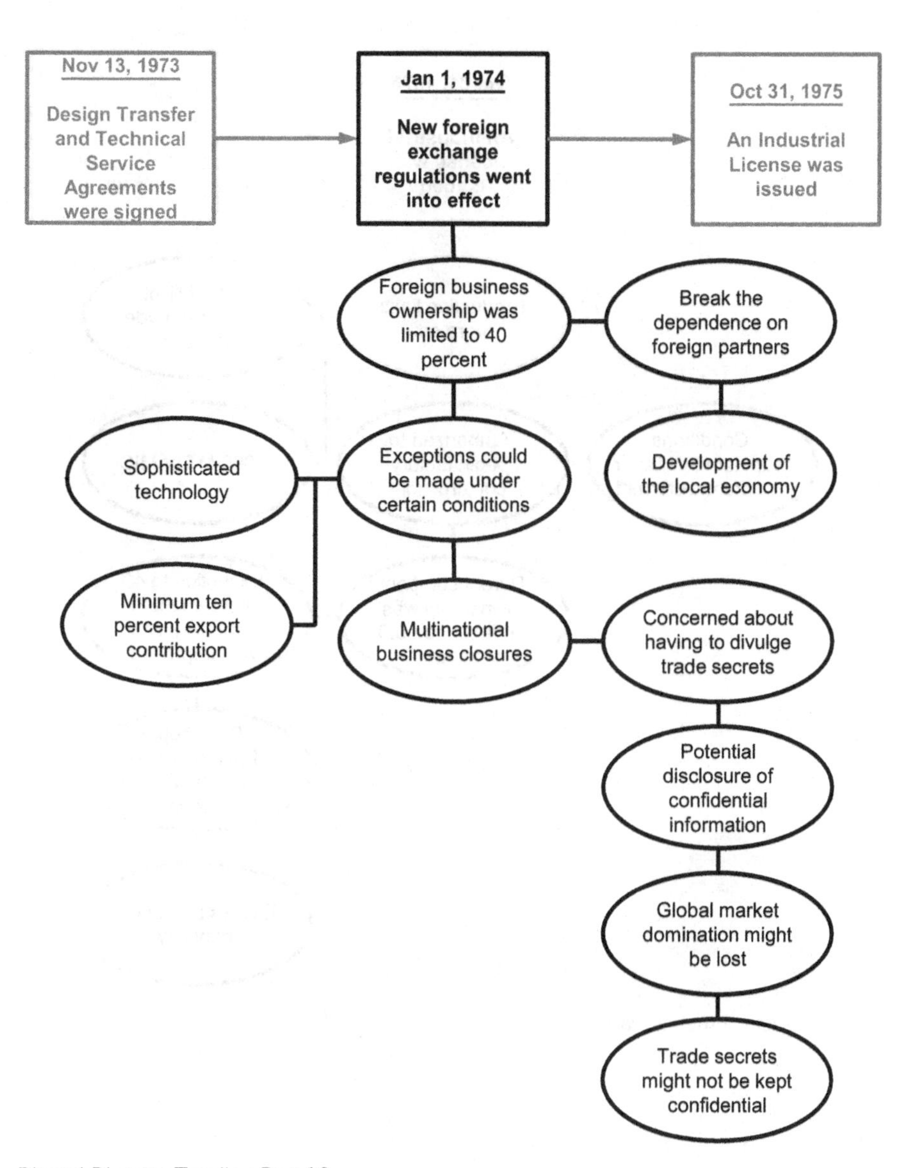

Bhopal Disaster Timeline Part 13

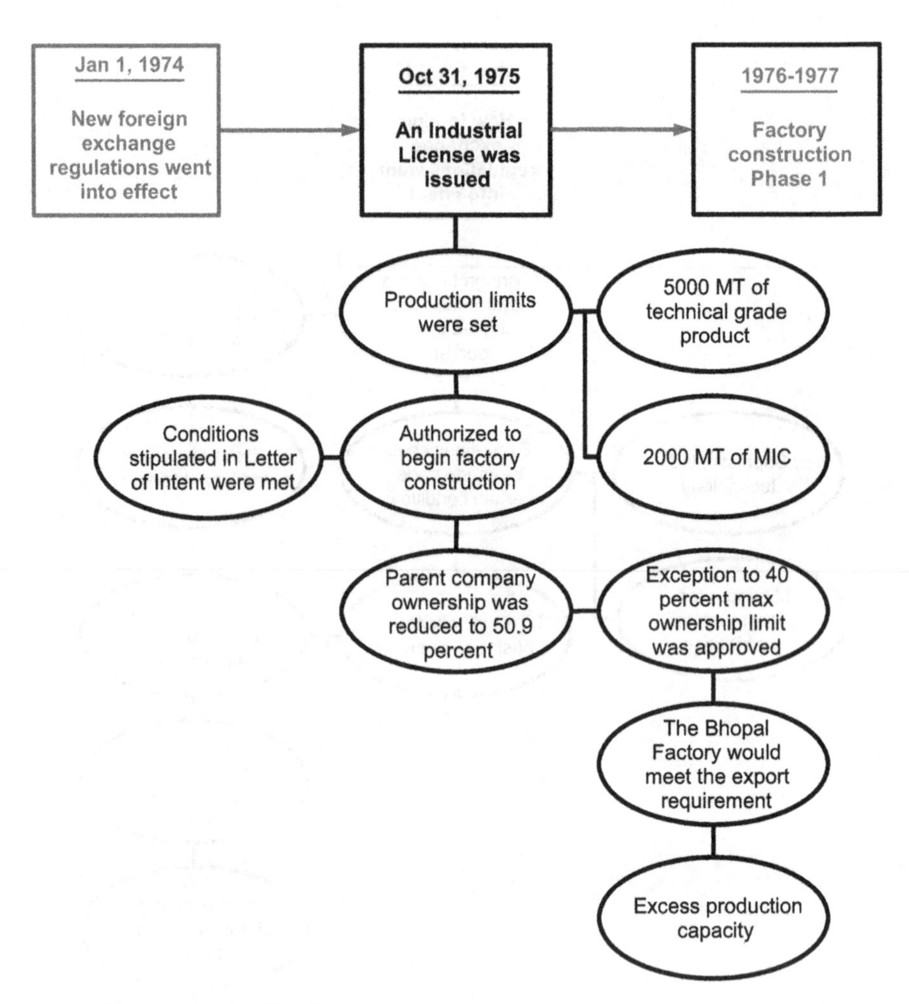

Bhopal Disaster Timeline Part 14

CHAPTER 12

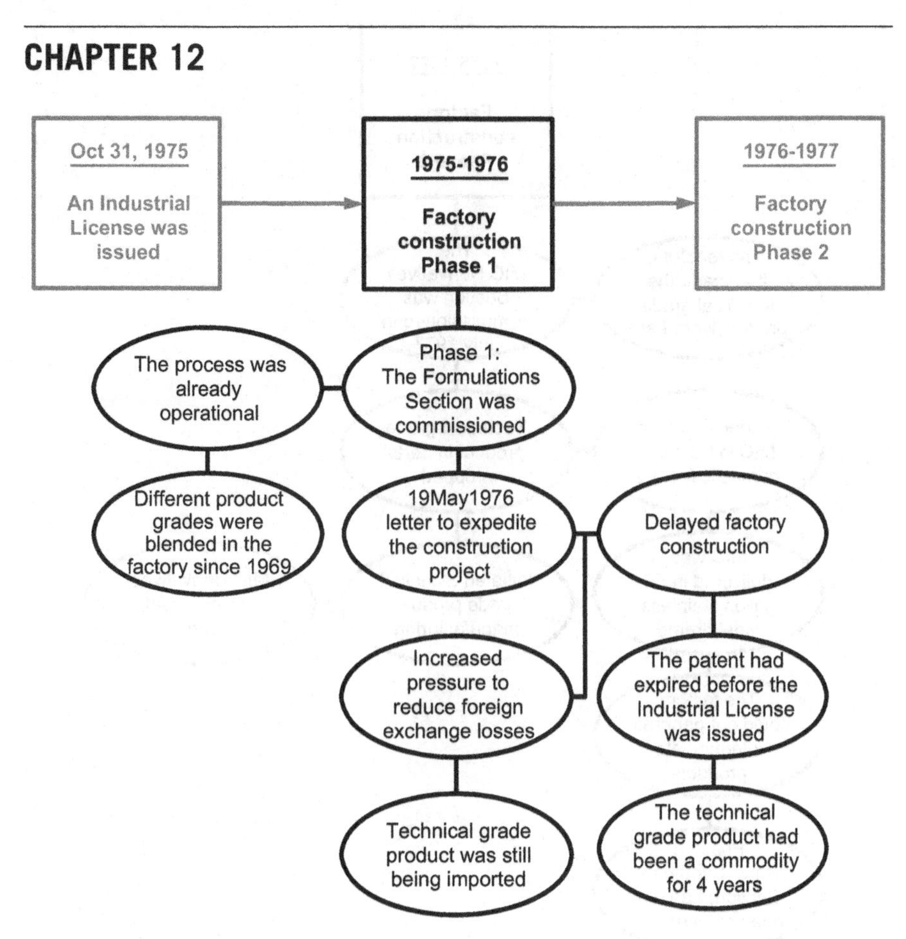

Bhopal Disaster Timeline Part 15

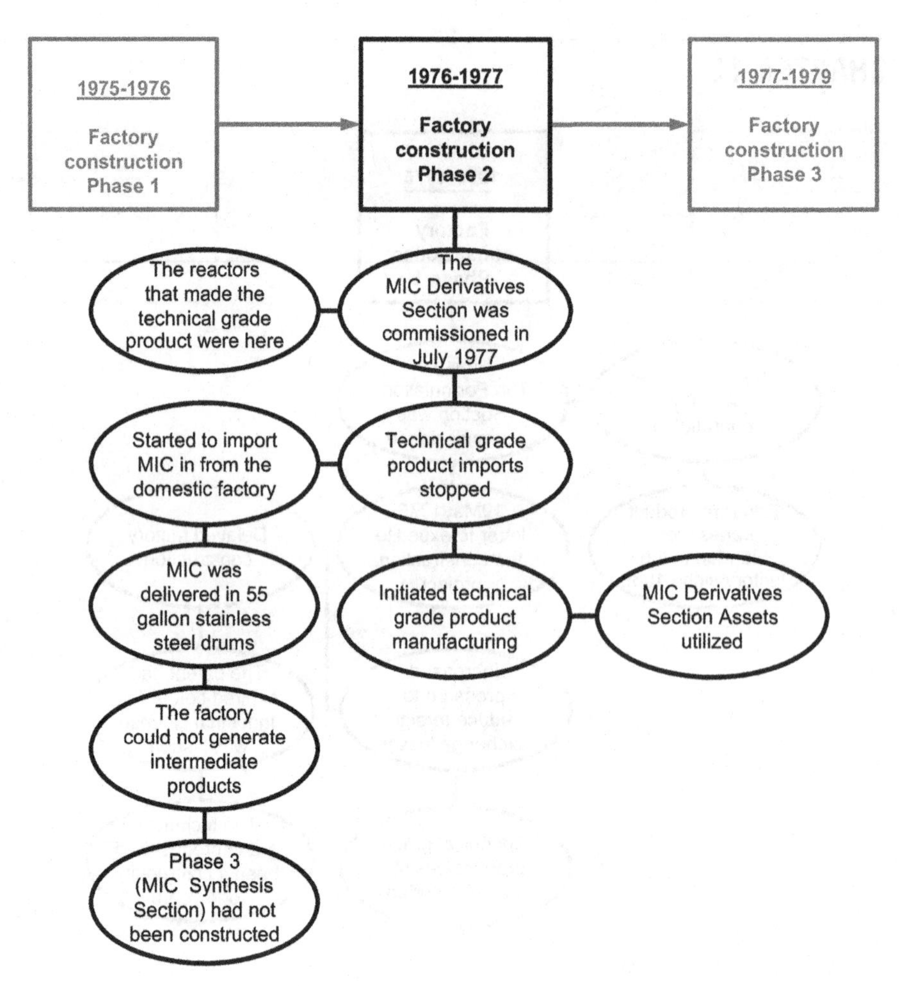

Bhopal Disaster Timeline Part 16

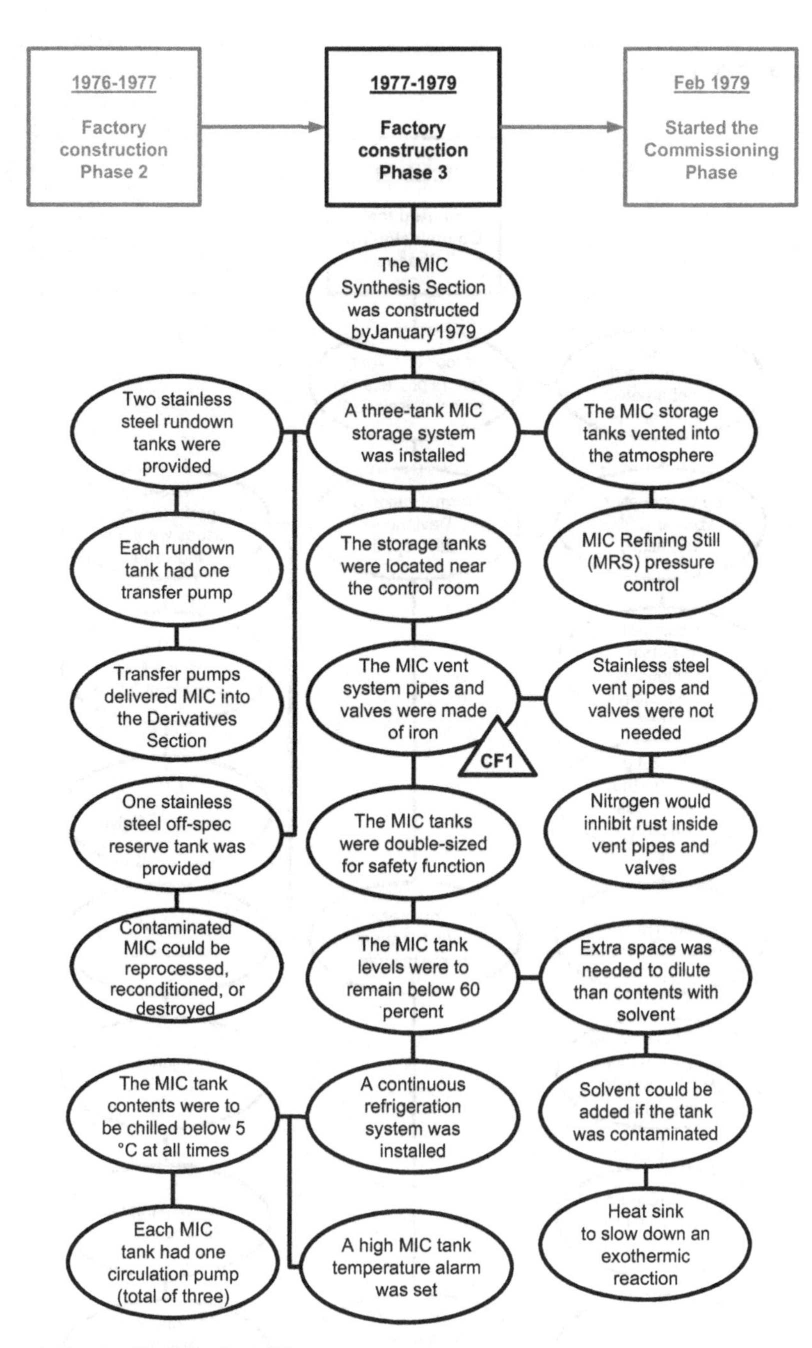

Bhopal Disaster Timeline Part 17

CHAPTER 14

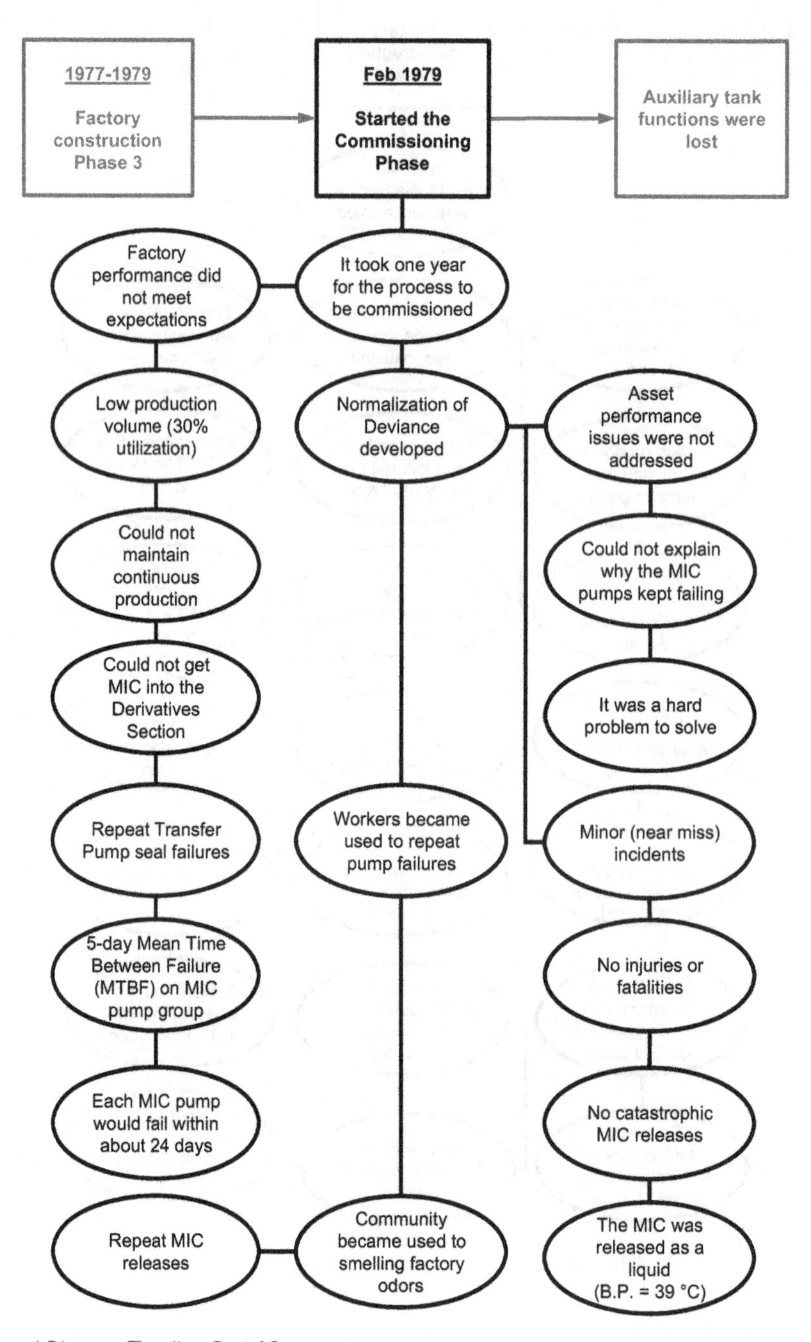

Bhopal Disaster Timeline Part 18

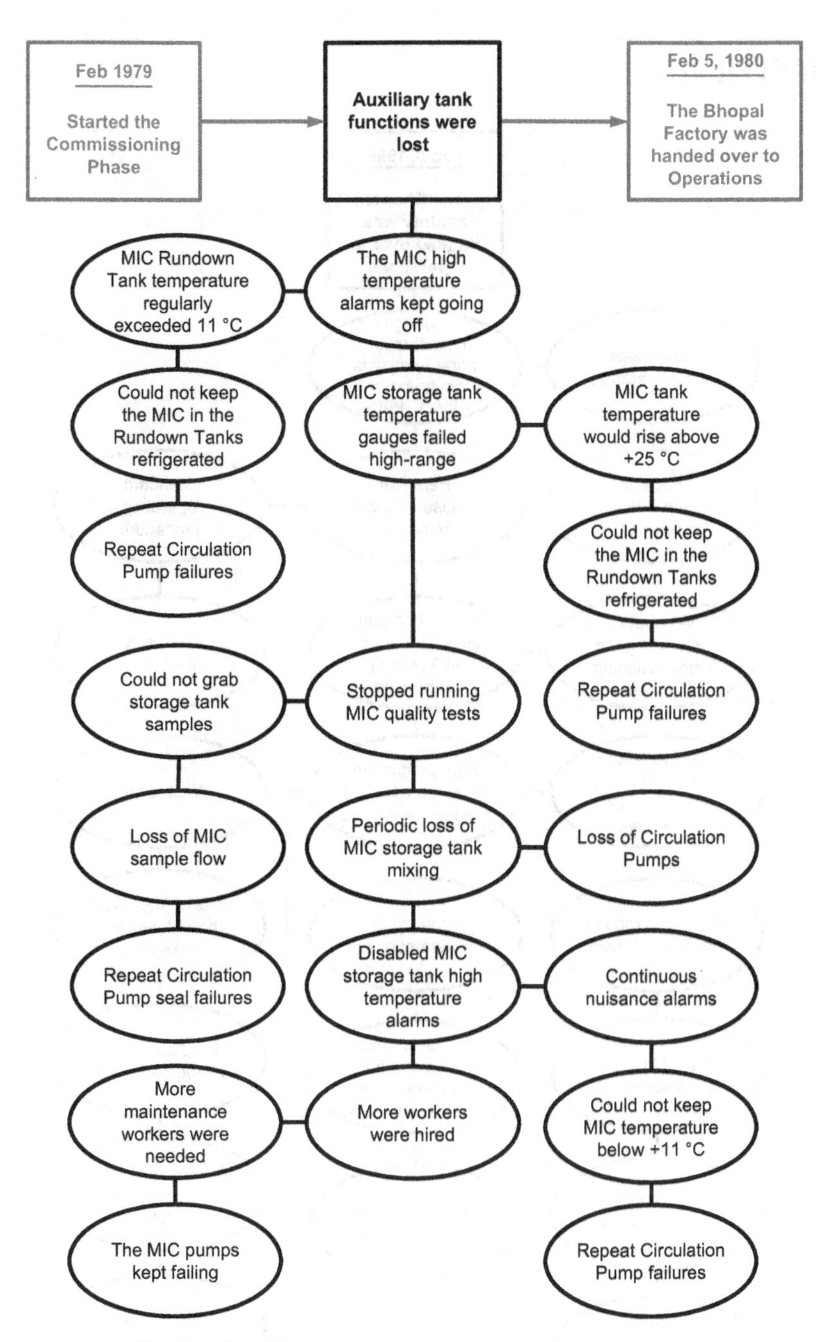

Bhopal Disaster Timeline Part 19

CHAPTER 16

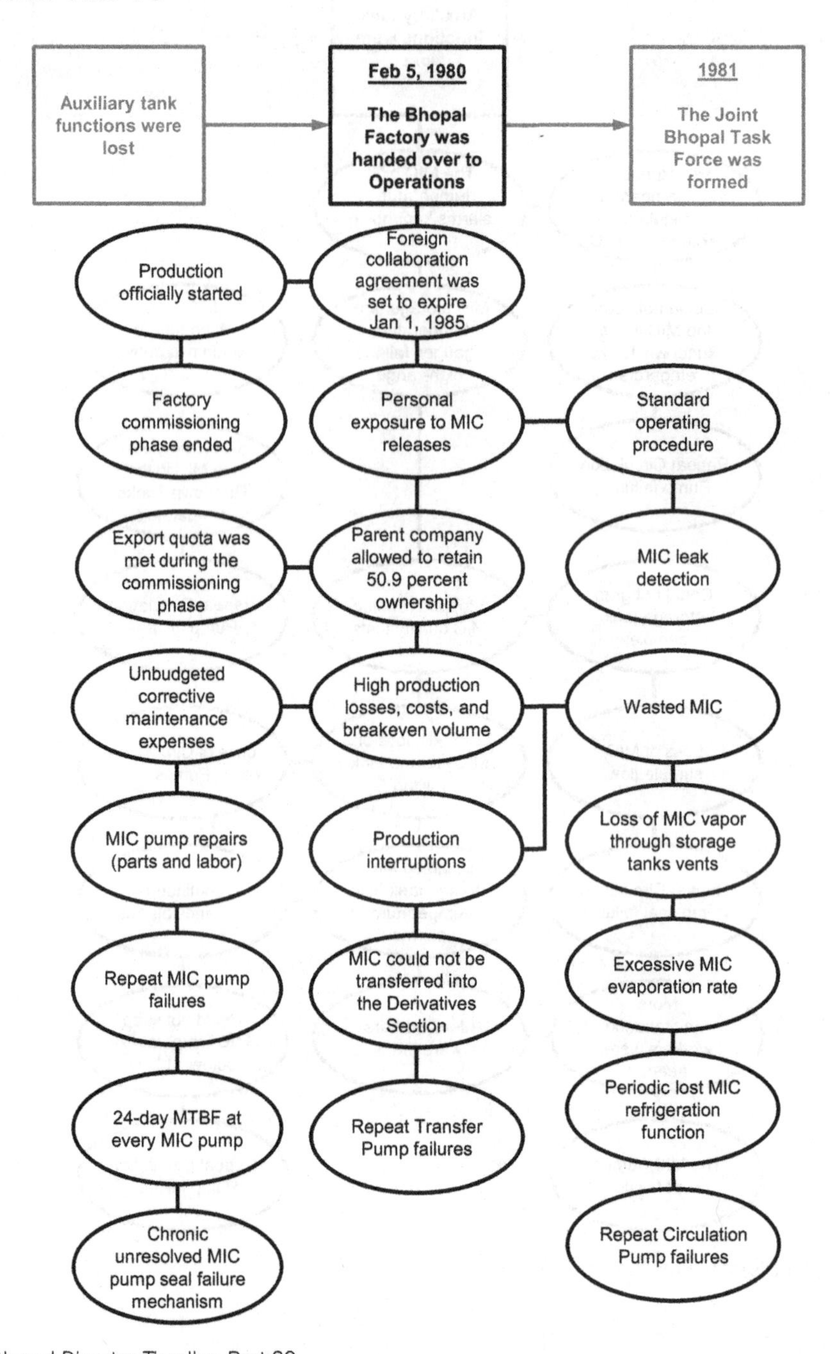

Bhopal Disaster Timeline Part 20

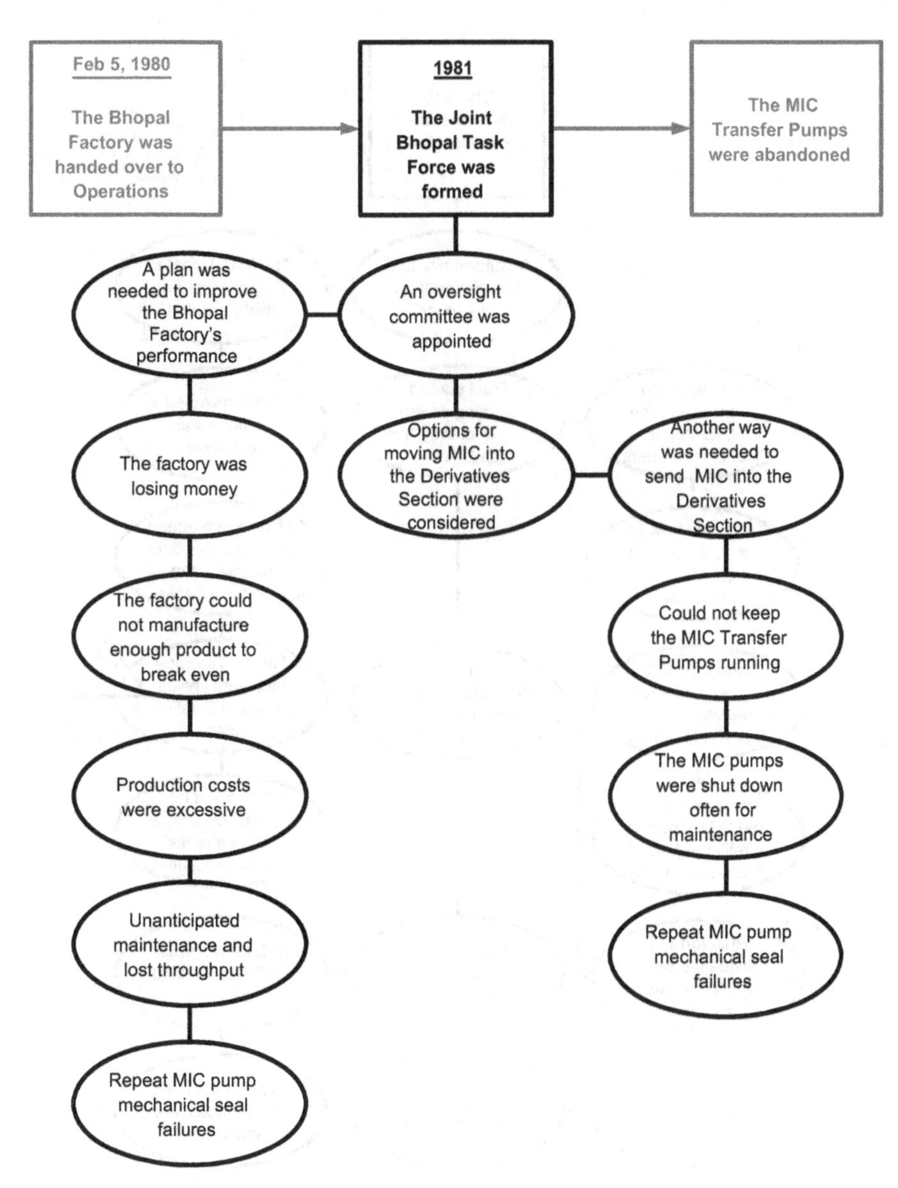

Bhopal Disaster Timeline Part 21

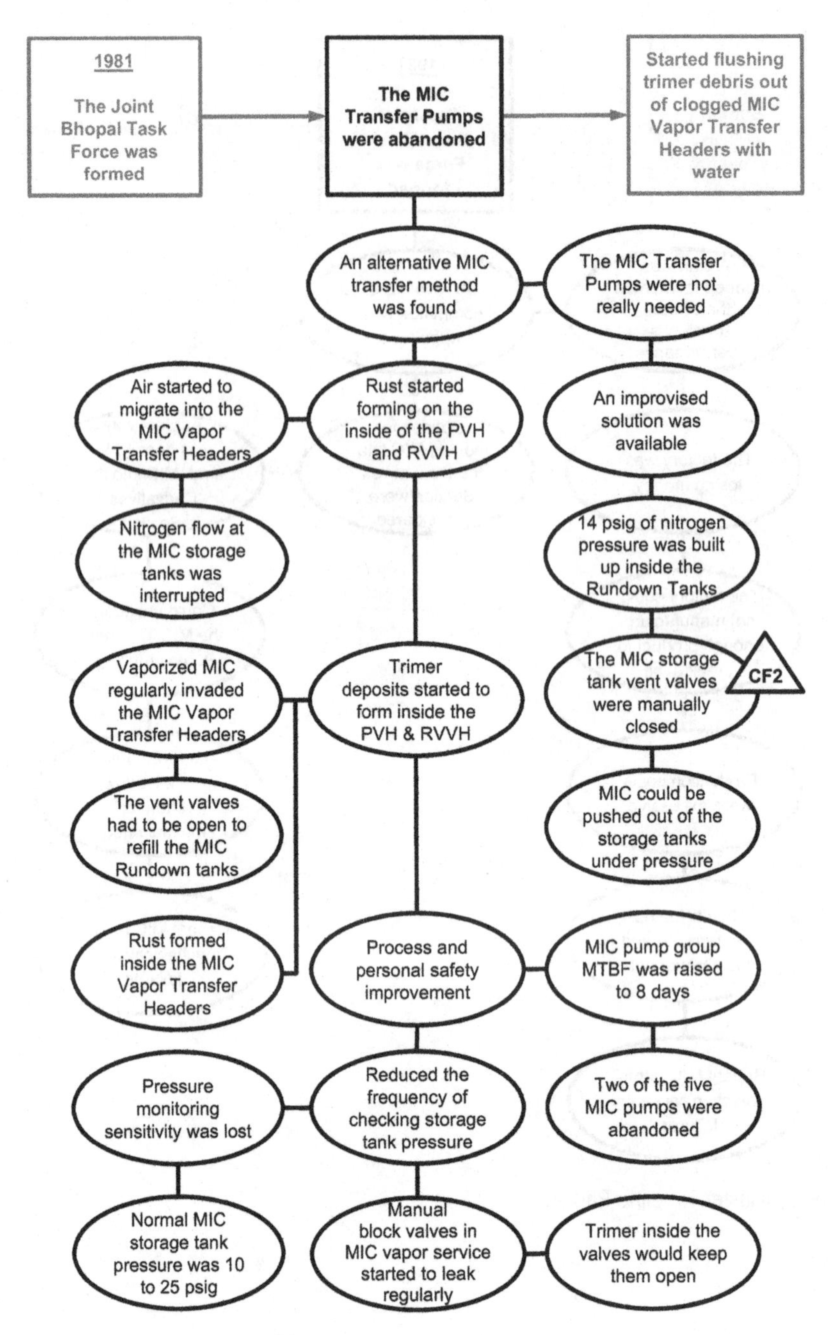

Bhopal Disaster Timeline Part 22

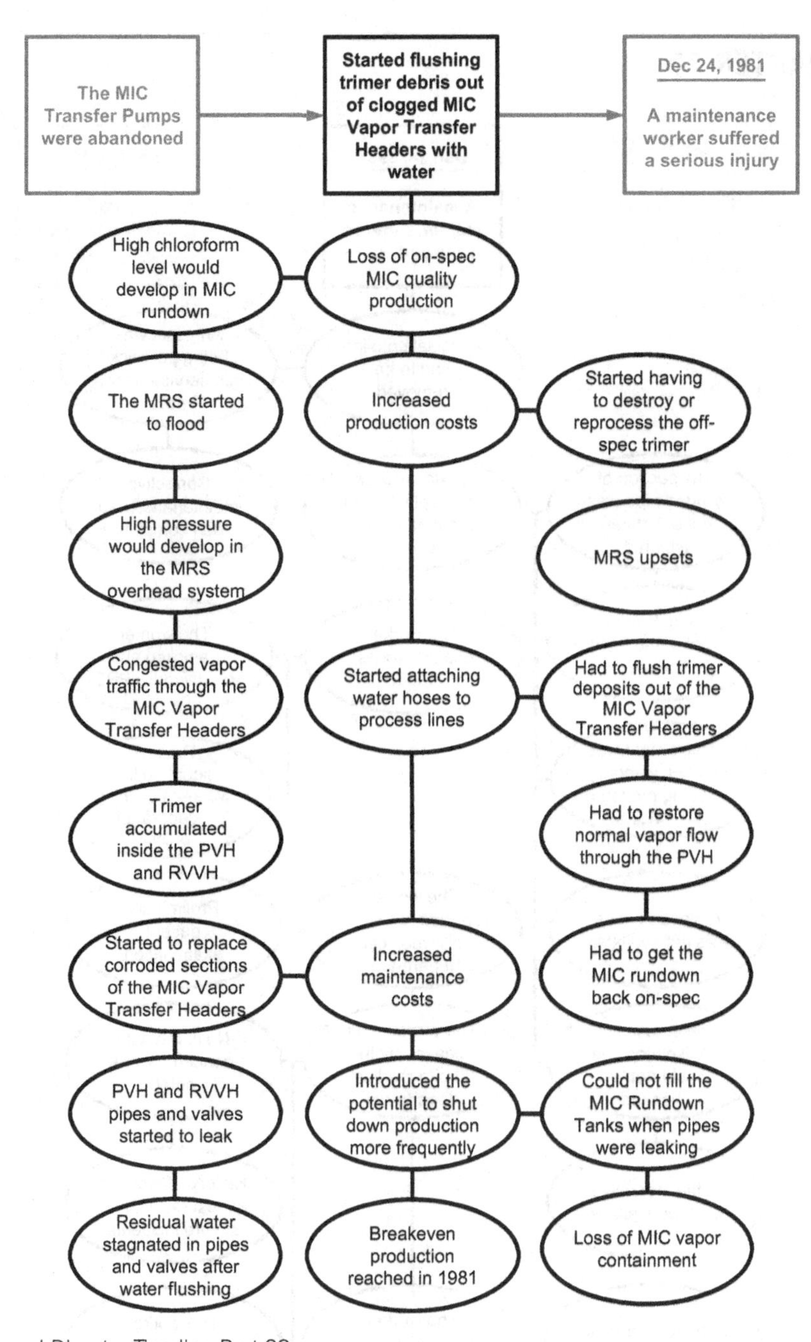

Bhopal Disaster Timeline Part 23

CHAPTER 17

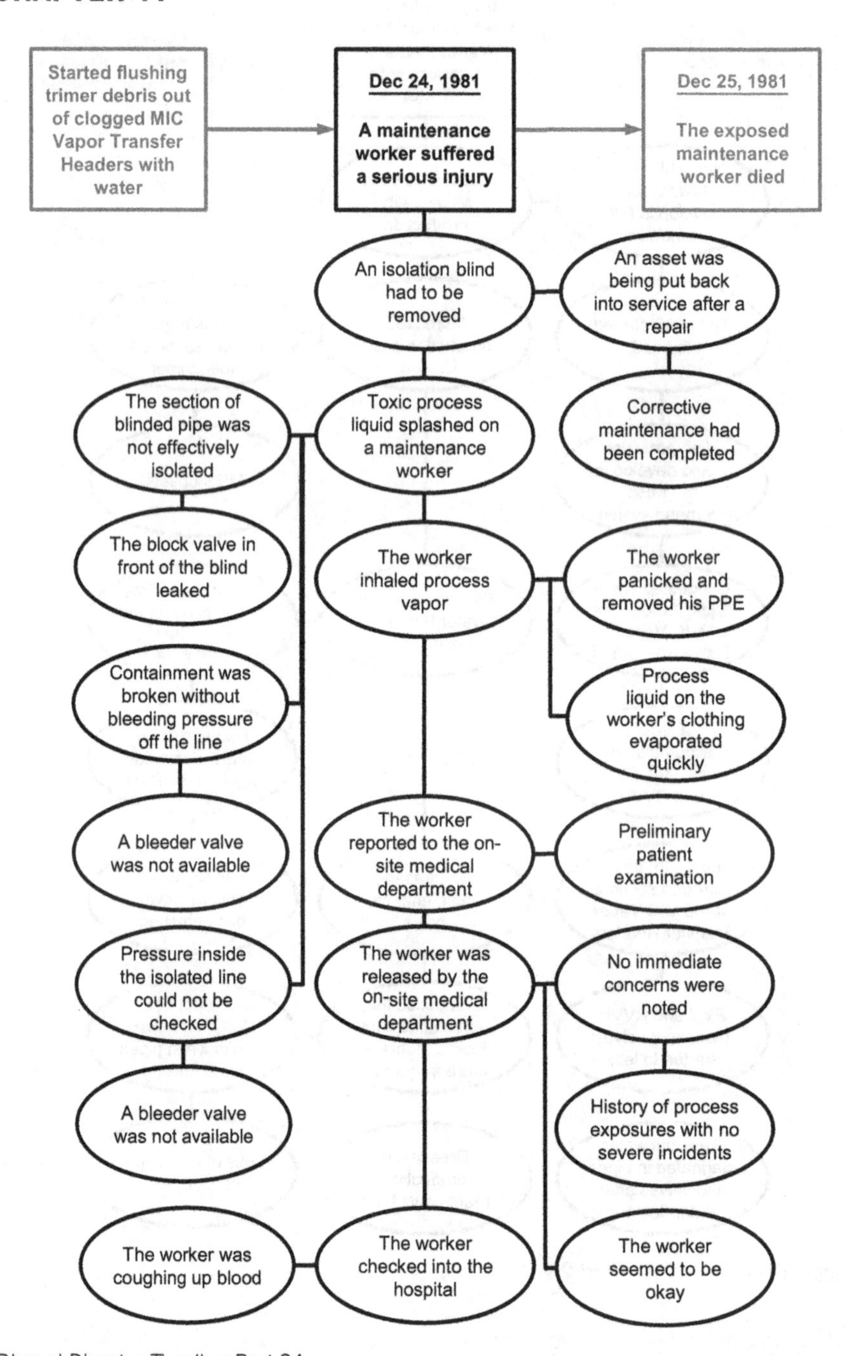

Bhopal Disaster Timeline Part 24

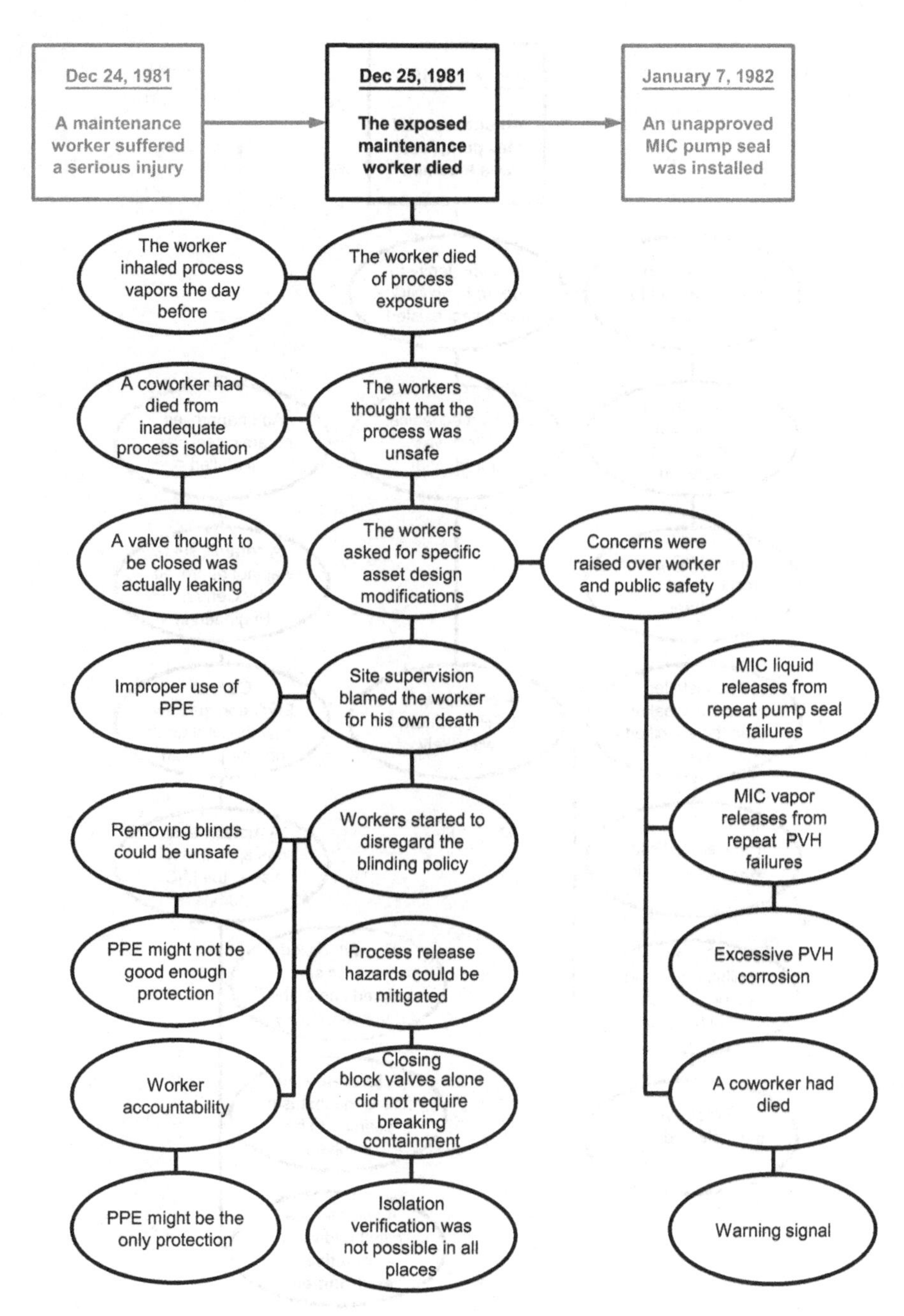

Bhopal Disaster Timeline Part 25

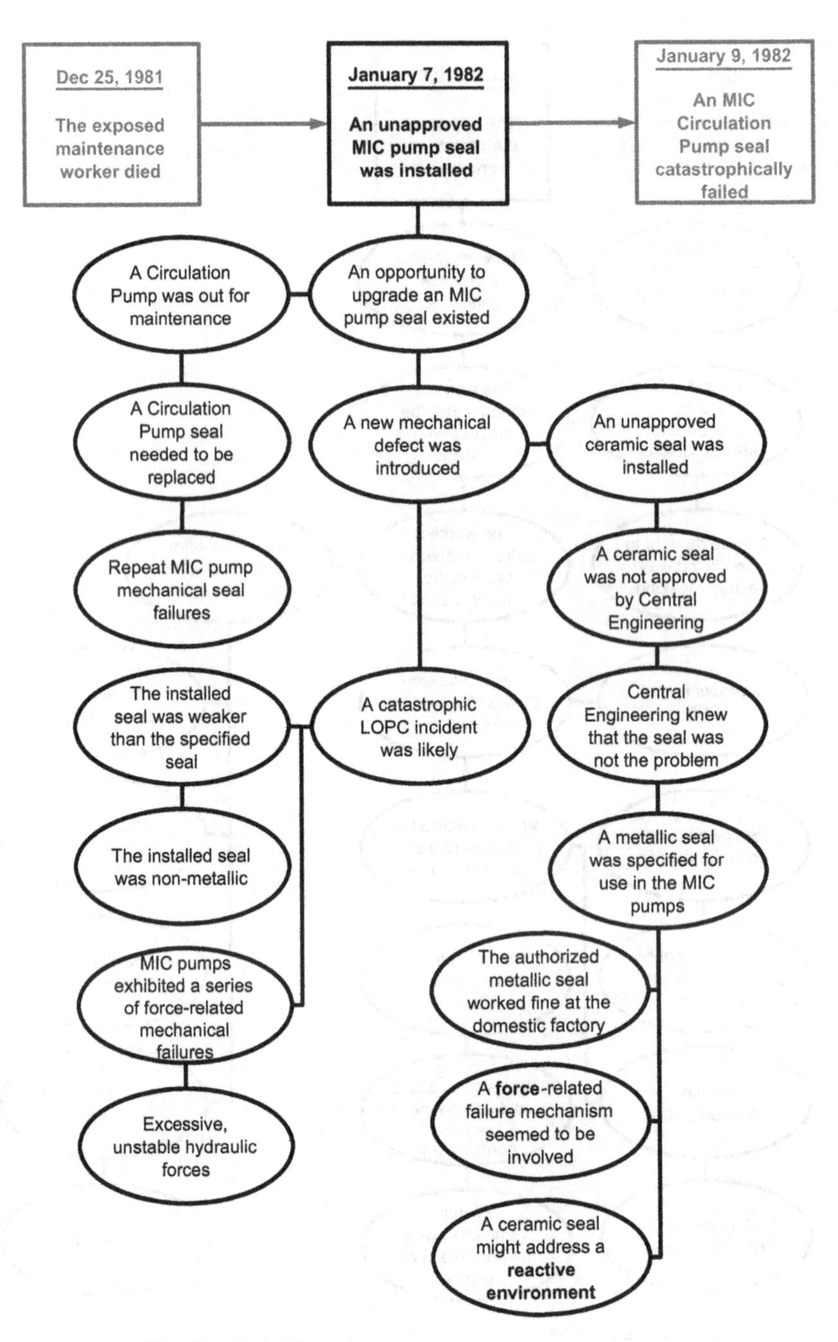

Bhopal Disaster Timeline Part 26

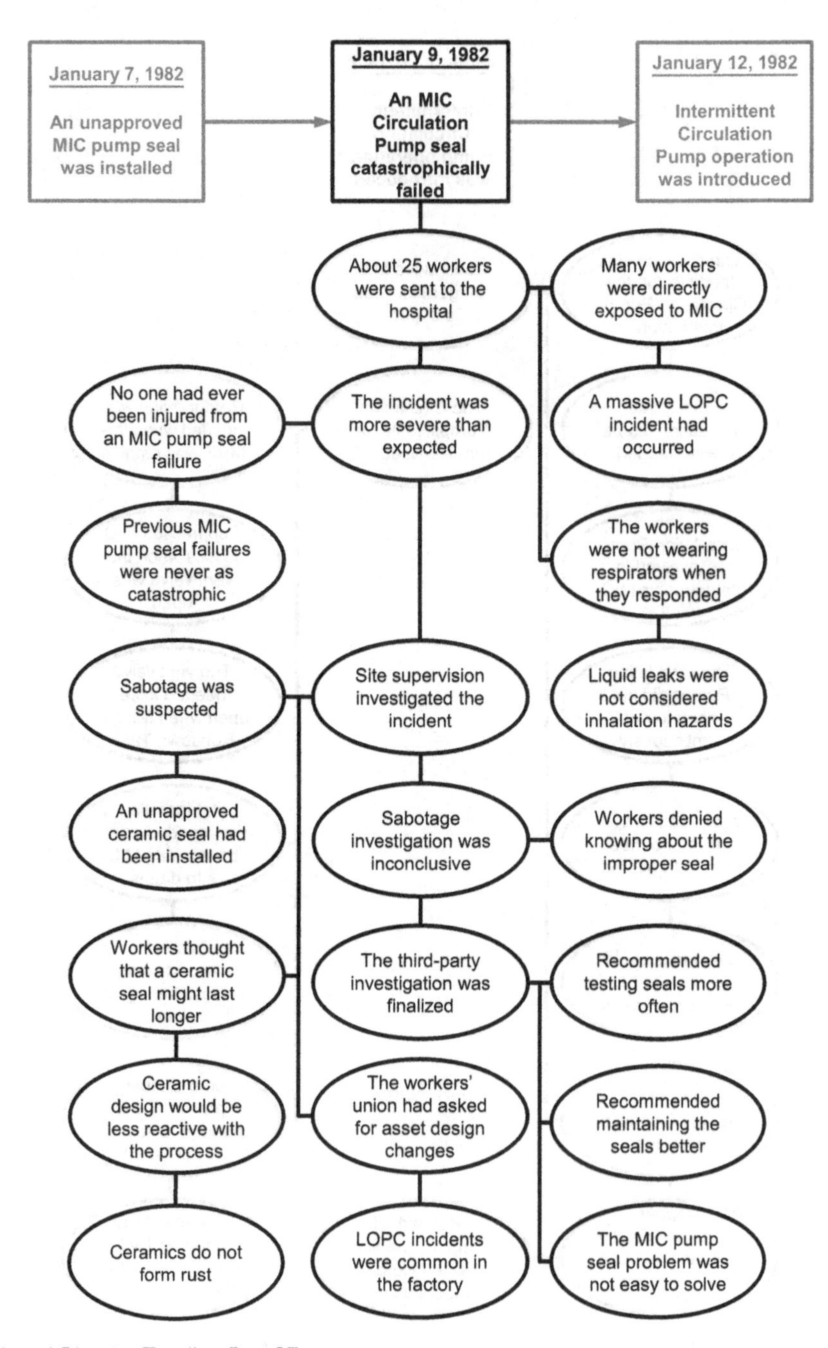

Bhopal Disaster Timeline Part 27

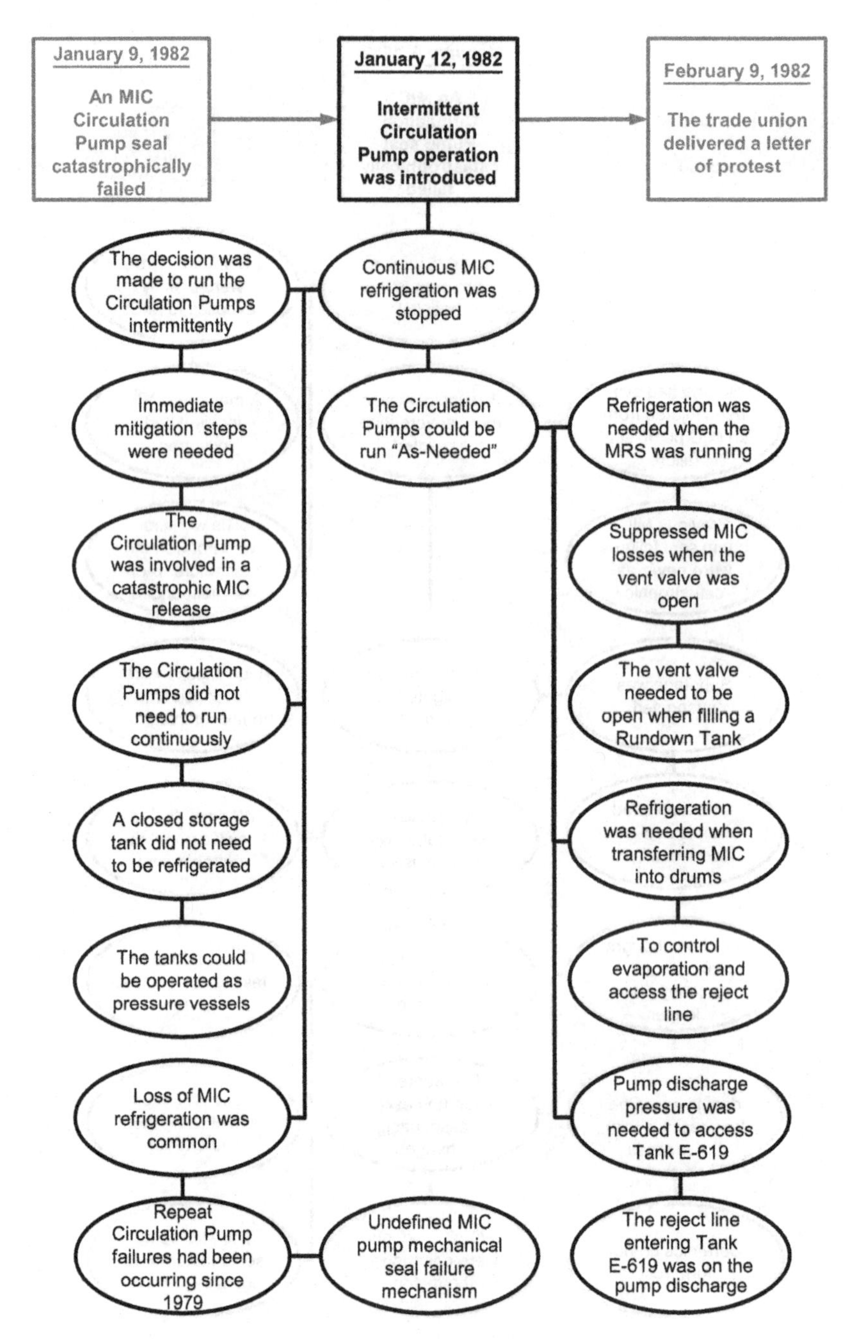

Bhopal Disaster Timeline Part 28

CHAPTER 18

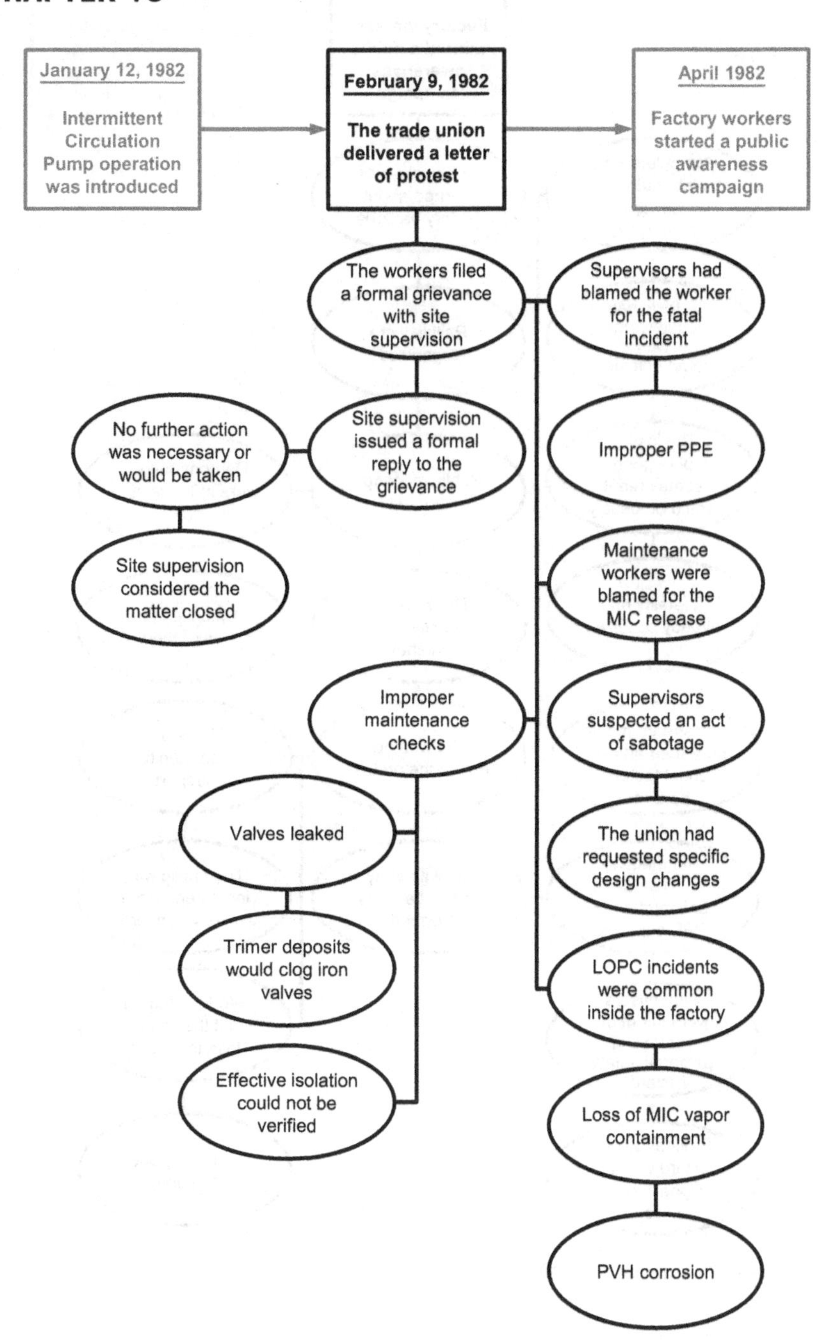

Bhopal Disaster Timeline Part 29

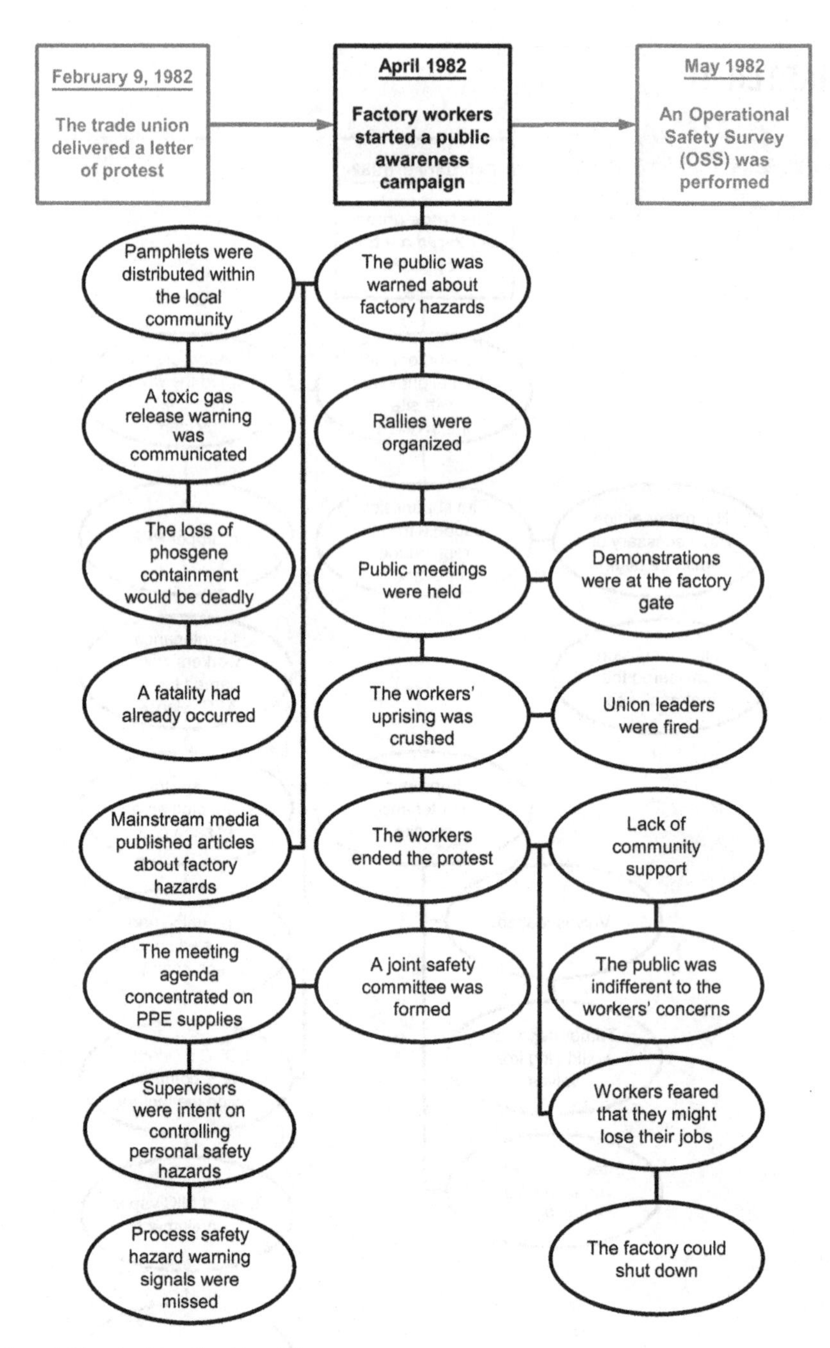

Bhopal Disaster Timeline Part 30

CHAPTER 19

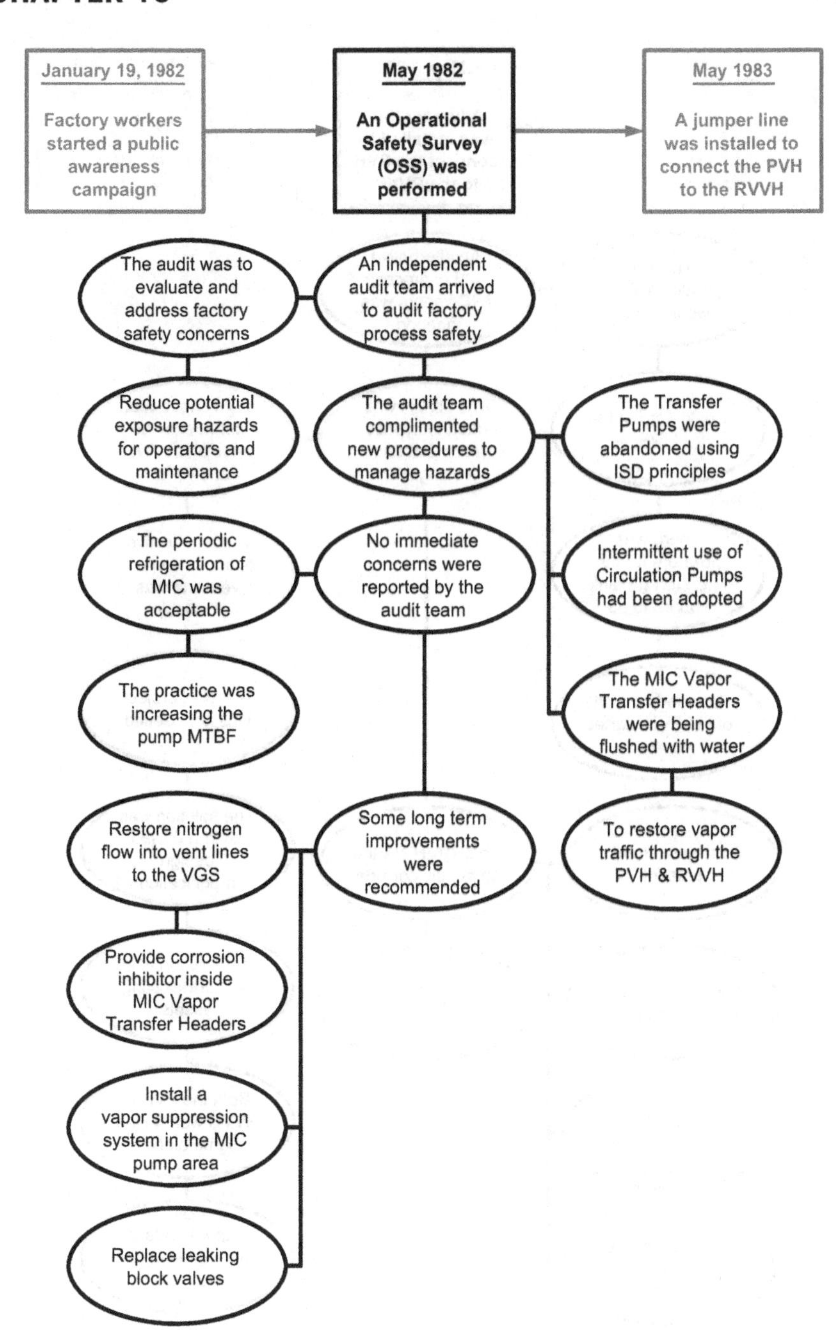

Bhopal Disaster Timeline Part 31

CHAPTER 20

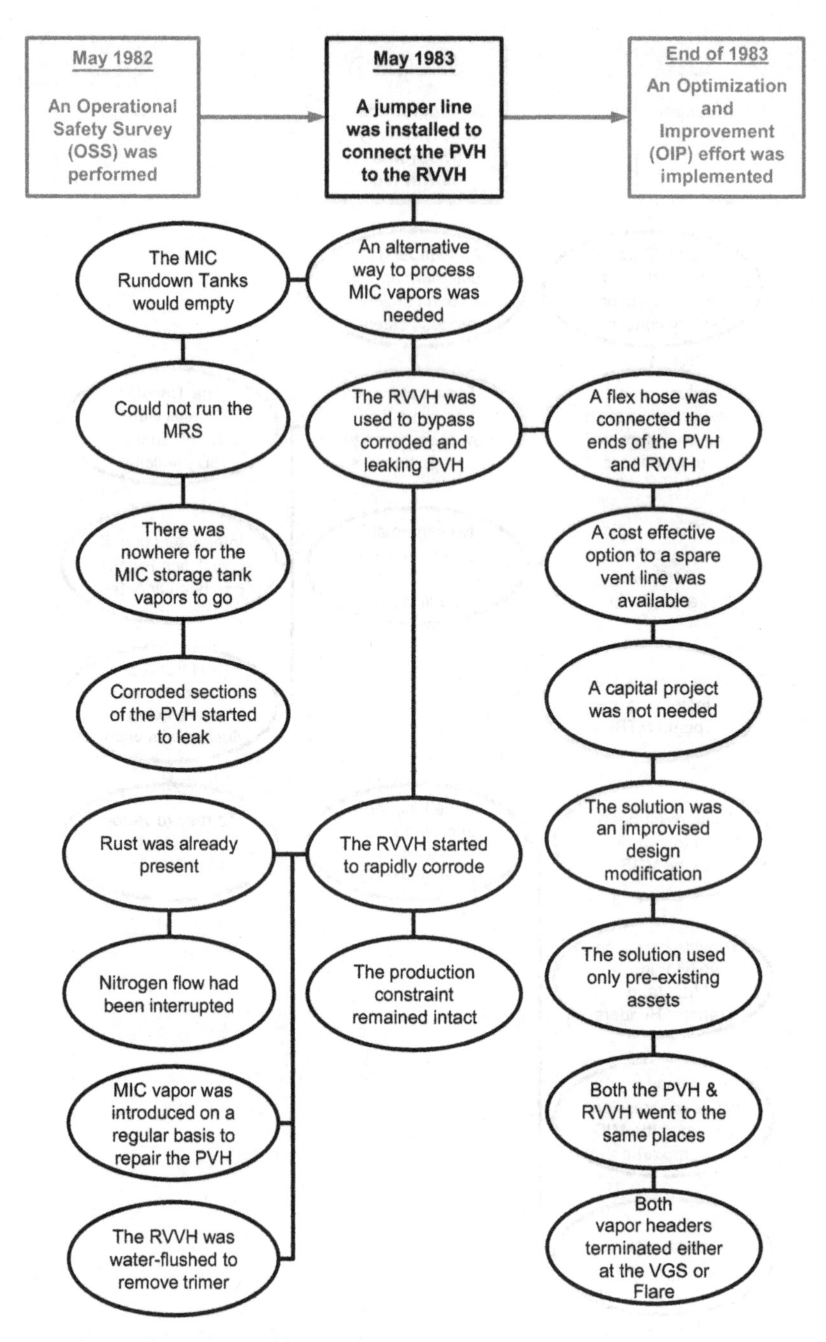

Bhopal Disaster Timeline Part 32

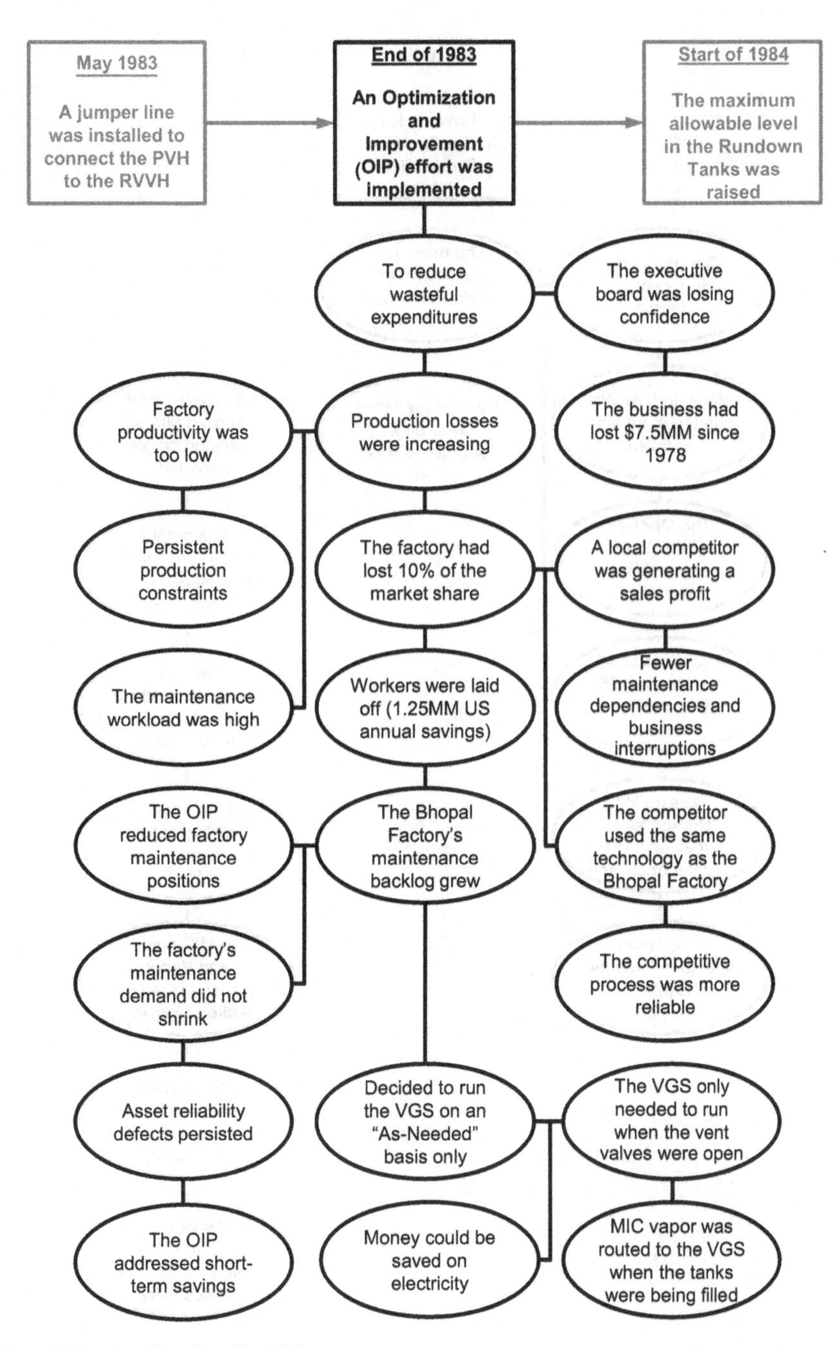

Bhopal Disaster Timeline Part 33

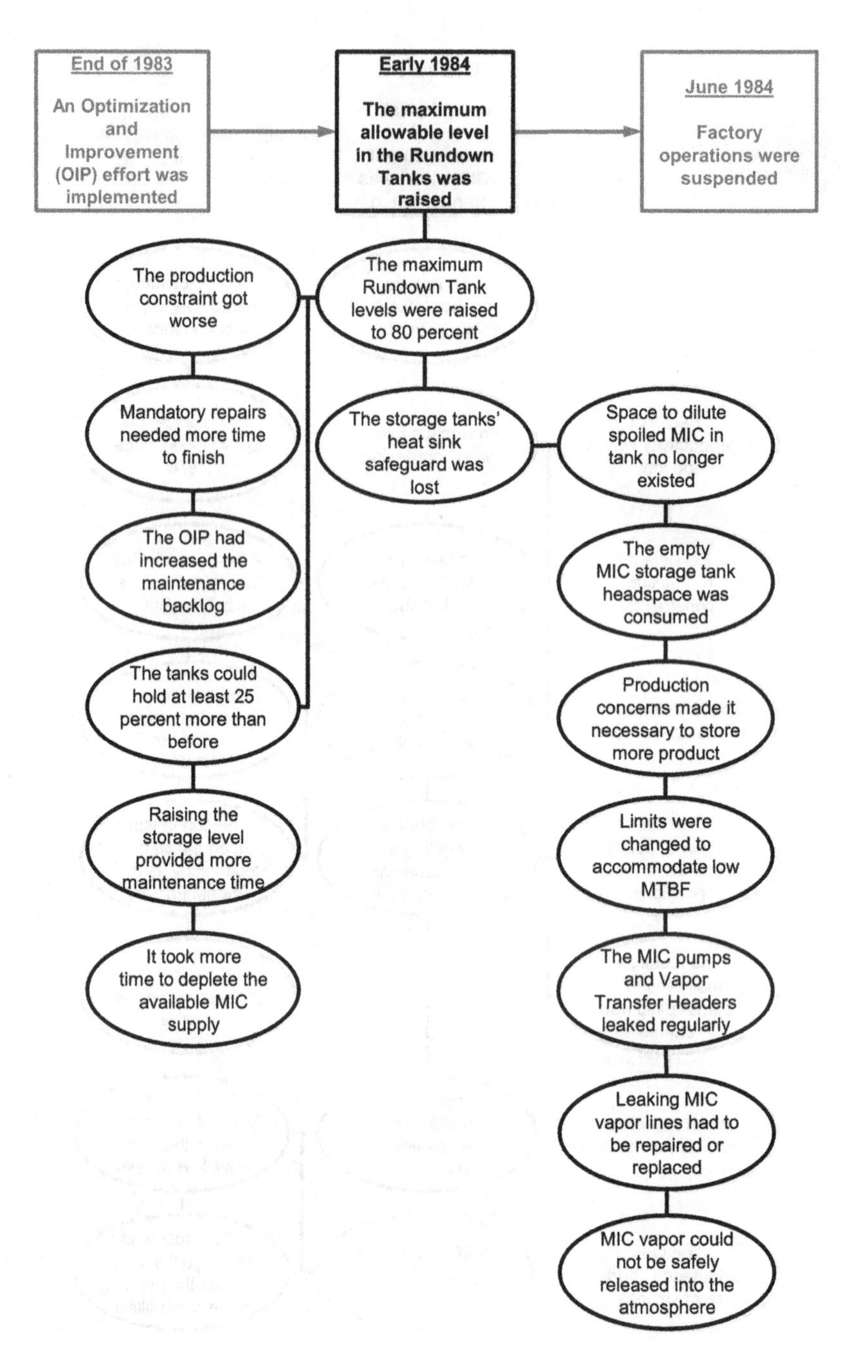

Bhopal Disaster Timeline Part 34

CHAPTER 21

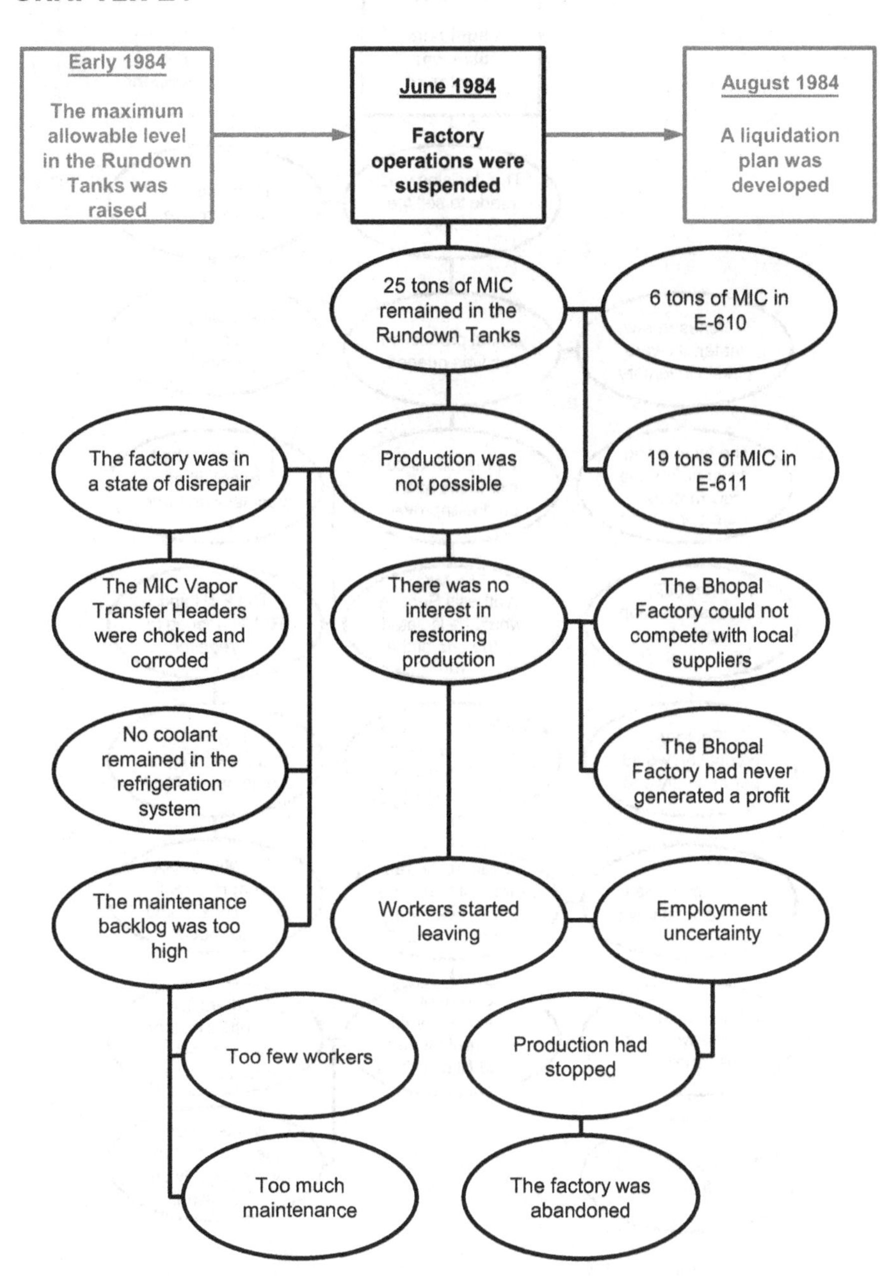

Bhopal Disaster Timeline Part 35

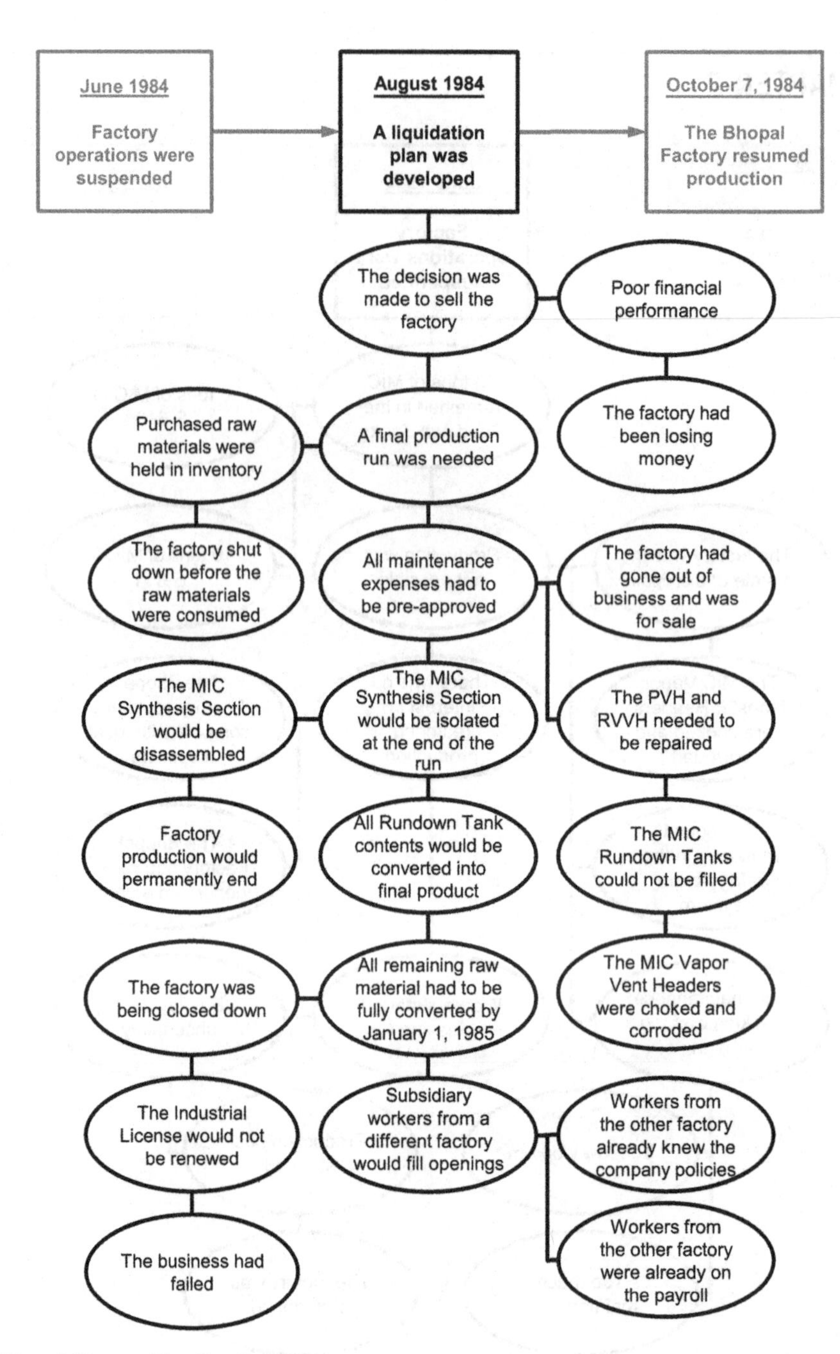

Bhopal Disaster Timeline Part 36

CHAPTER 22

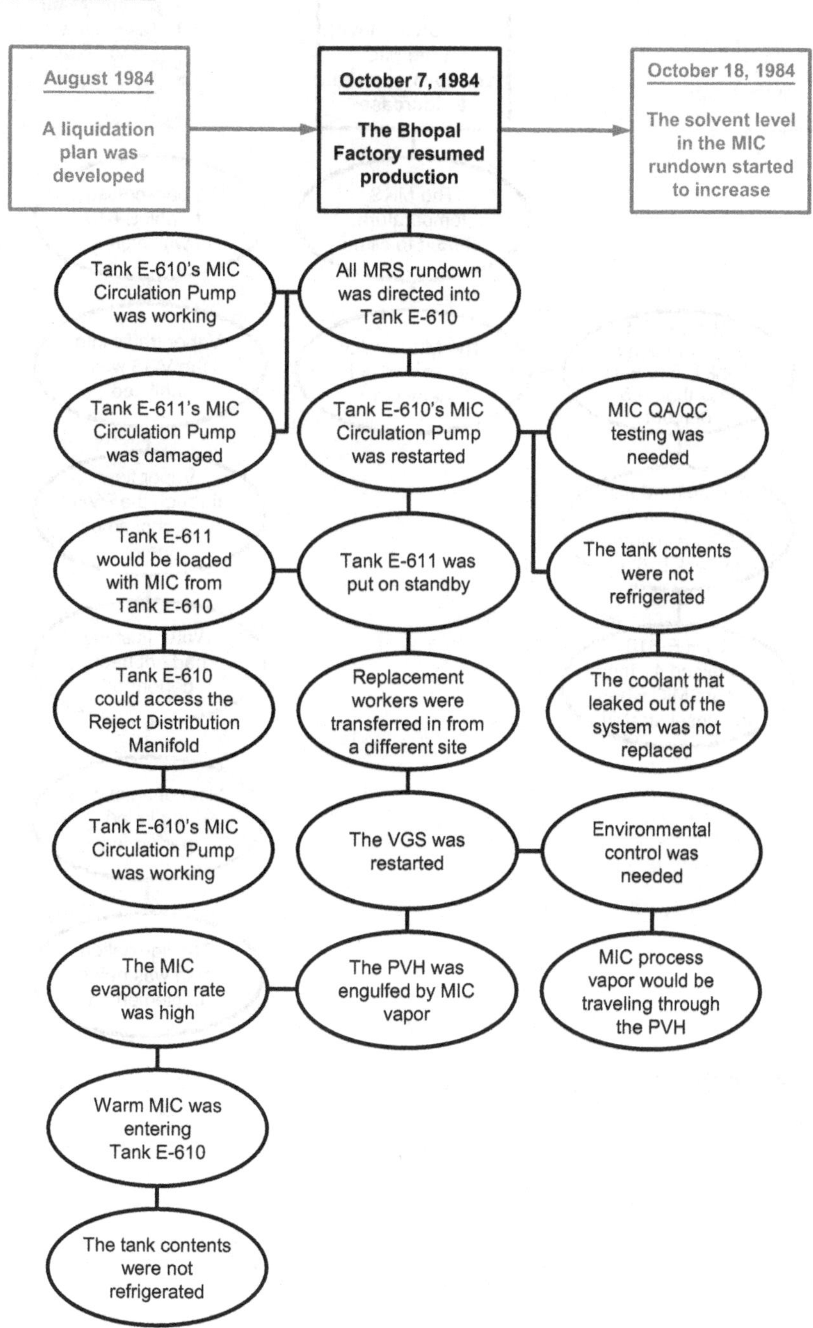
August 1984
A liquidation plan was developed
October 7, 1984
The Bhopal Factory resumed production
October 18, 1984
The solvent level in the MIC rundown started to increase
Tank E-610's MIC Circulation Pump was working
All MRS rundown was directed into Tank E-610
Tank E-611's MIC Circulation Pump was damaged
Tank E-610's MIC Circulation Pump was restarted
MIC QA/QC testing was needed
Tank E-611 would be loaded with MIC from Tank E-610
Tank E-611 was put on standby
The tank contents were not refrigerated
Tank E-610 could access the Reject Distribution Manifold
Replacement workers were transferred in from a different site
The coolant that leaked out of the system was not replaced
Tank E-610's MIC Circulation Pump was working
The VGS was restarted
Environmental control was needed
The MIC evaporation rate was high
The PVH was engulfed by MIC vapor
MIC process vapor would be traveling through the PVH
Warm MIC was entering Tank E-610
The tank contents were not refrigerated

Bhopal Disaster Timeline Part 37

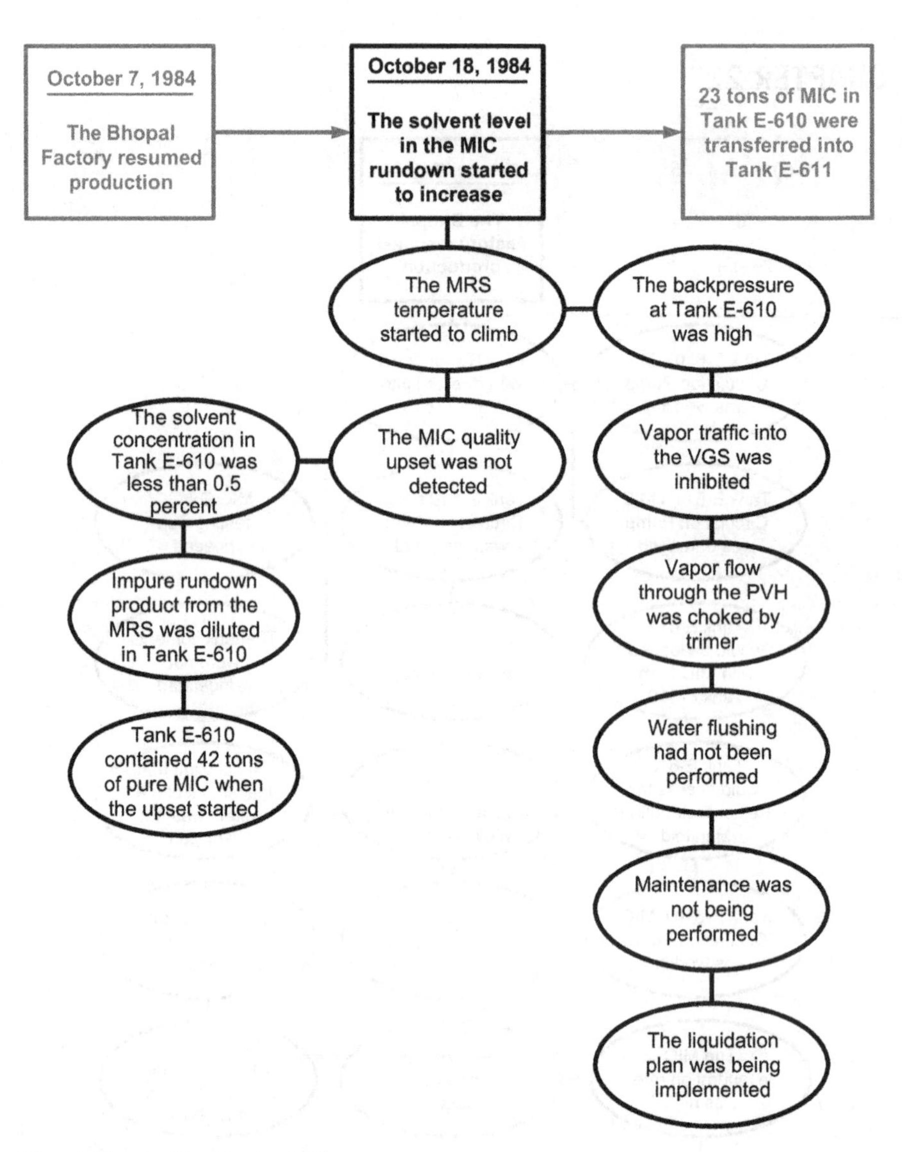

Bhopal Disaster Timeline Part 38

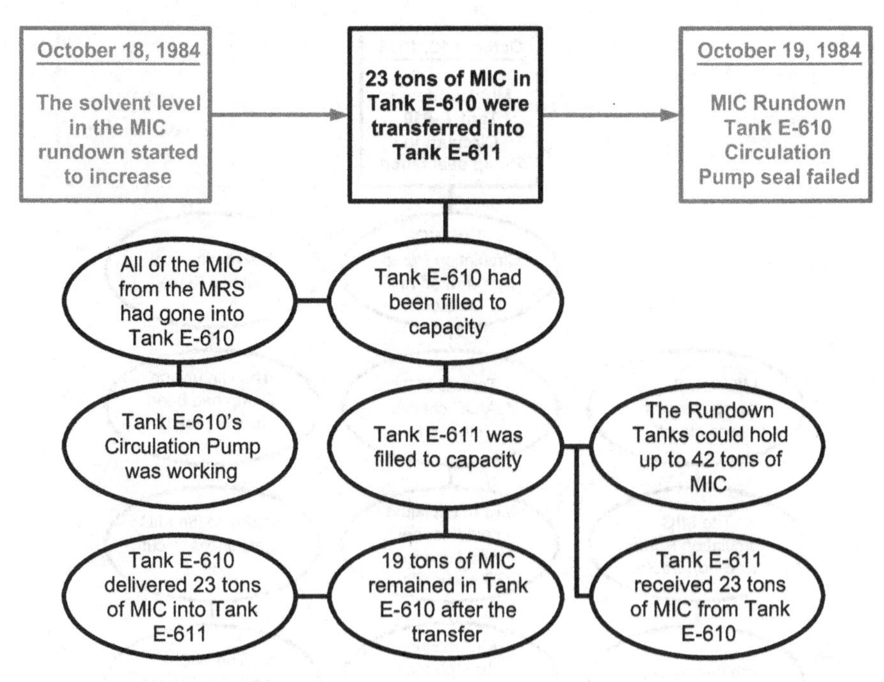

Bhopal Disaster Timeline Part 39

Bhopal Disaster Timeline Part 40

Bhopal Disaster Timeline Part 41

Bhopal Disaster Timeline Part 42

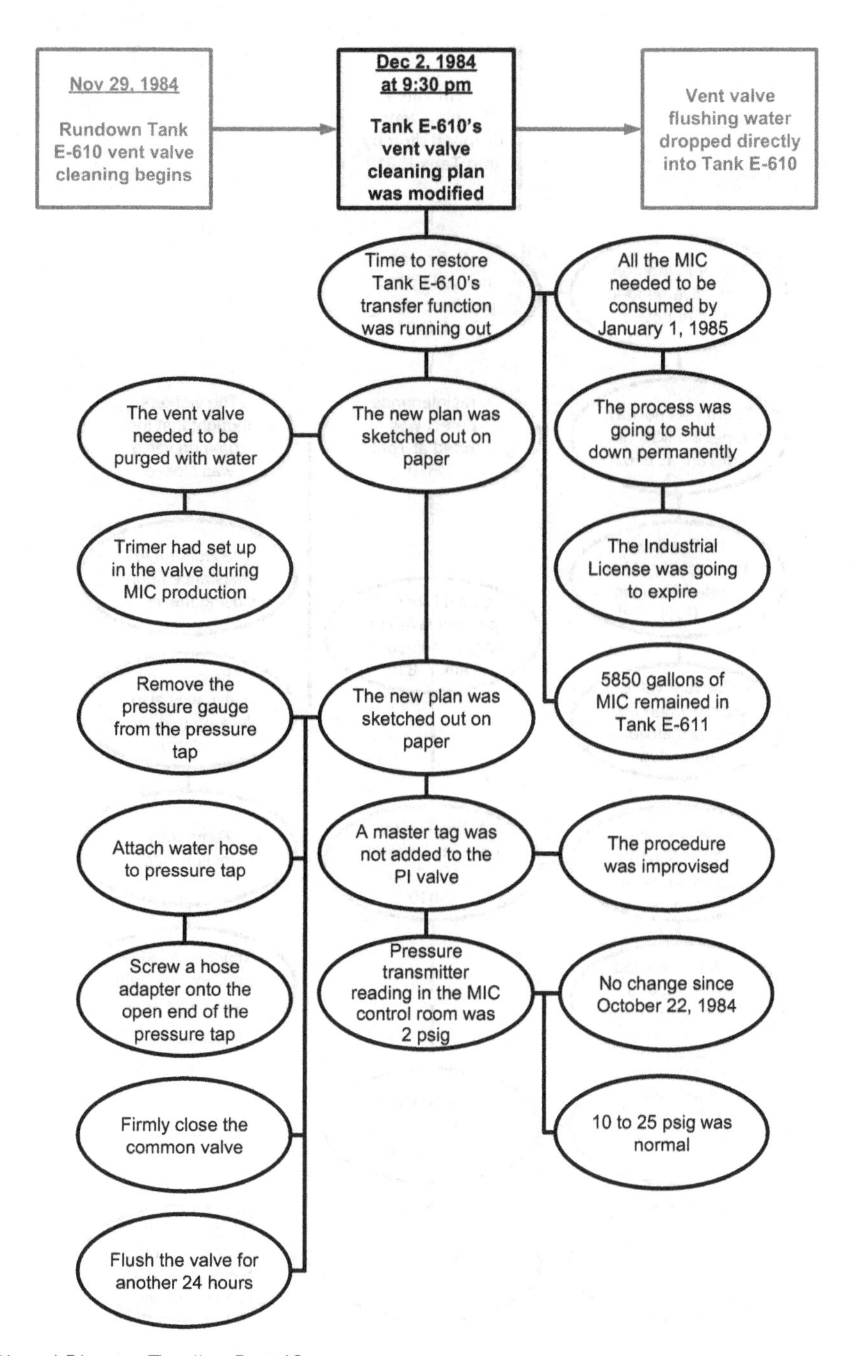

Bhopal Disaster Timeline Part 43

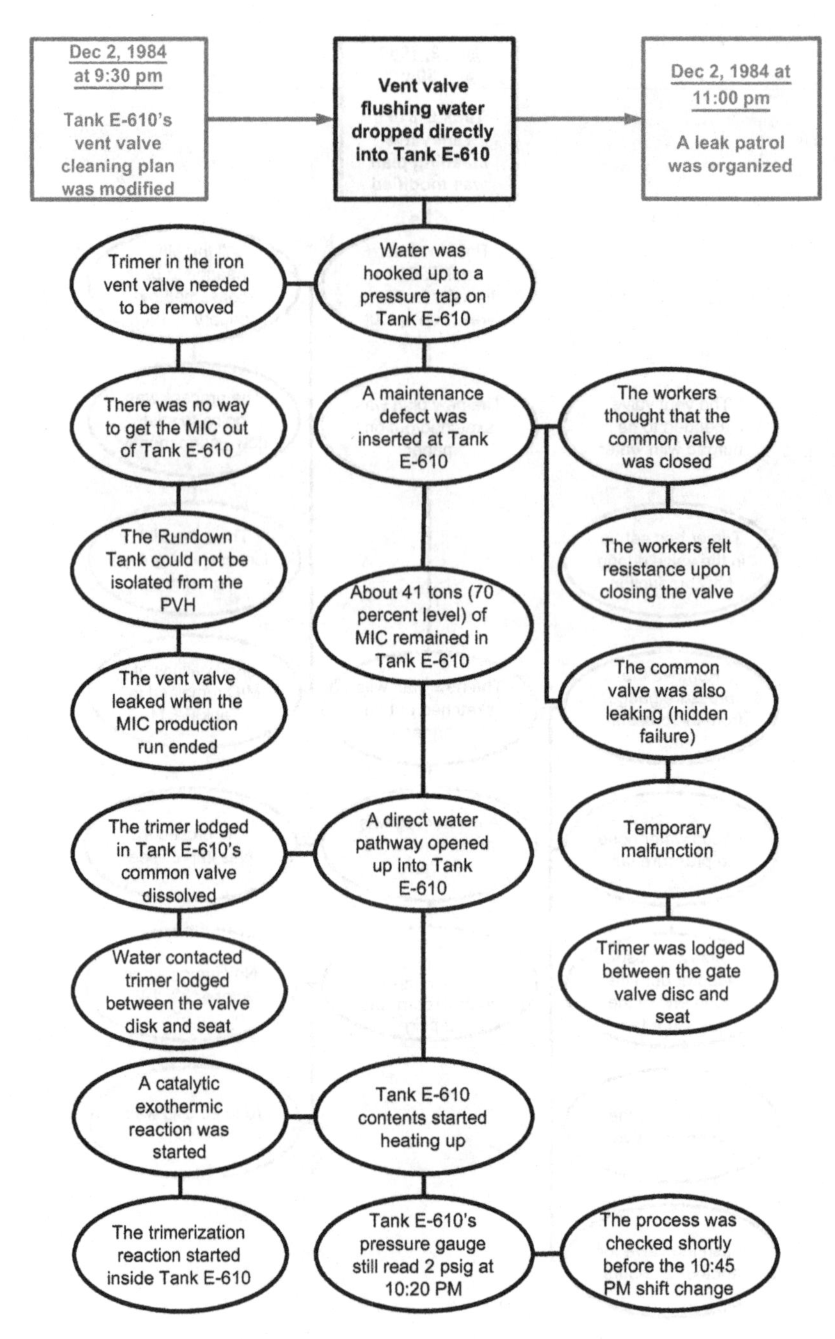

Bhopal Disaster Timeline Part 44

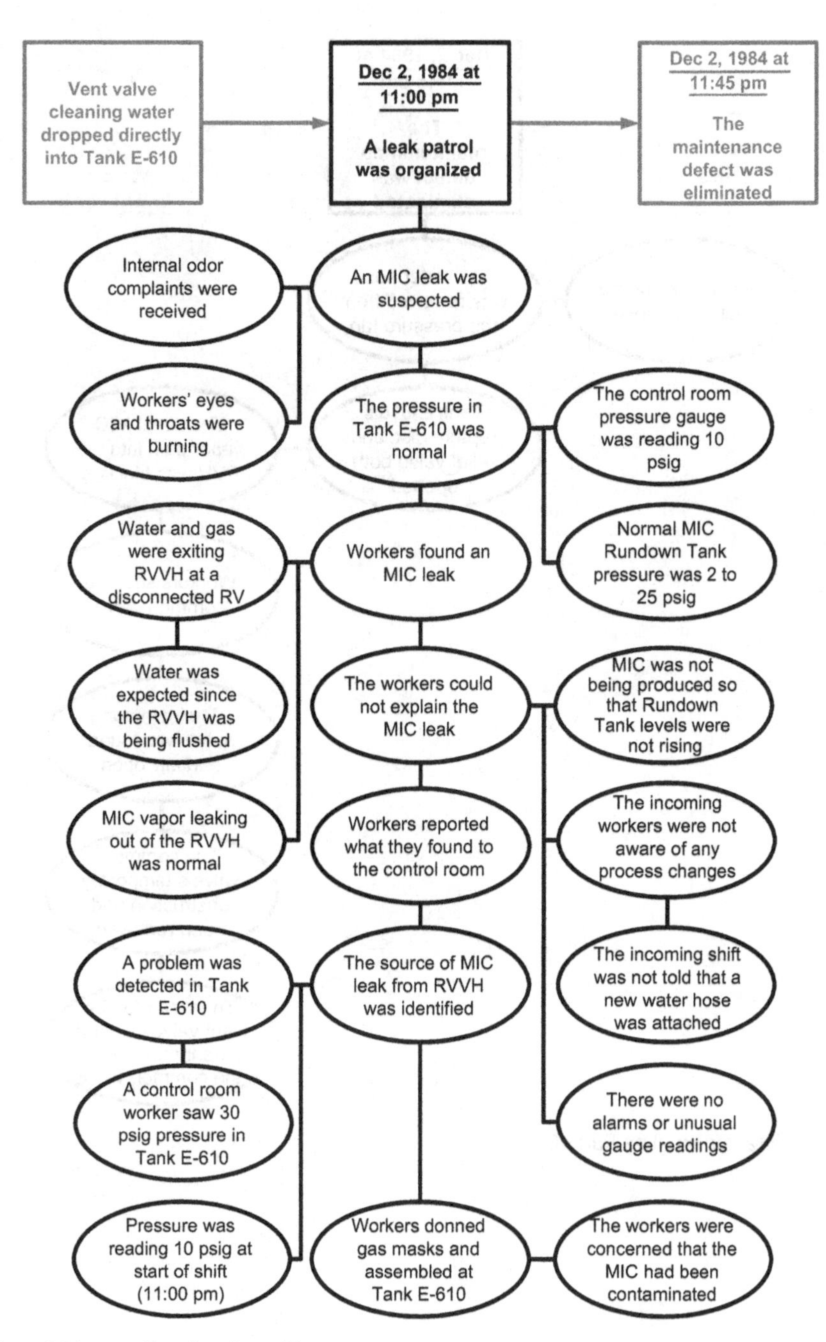

Bhopal Disaster Timeline Part 45

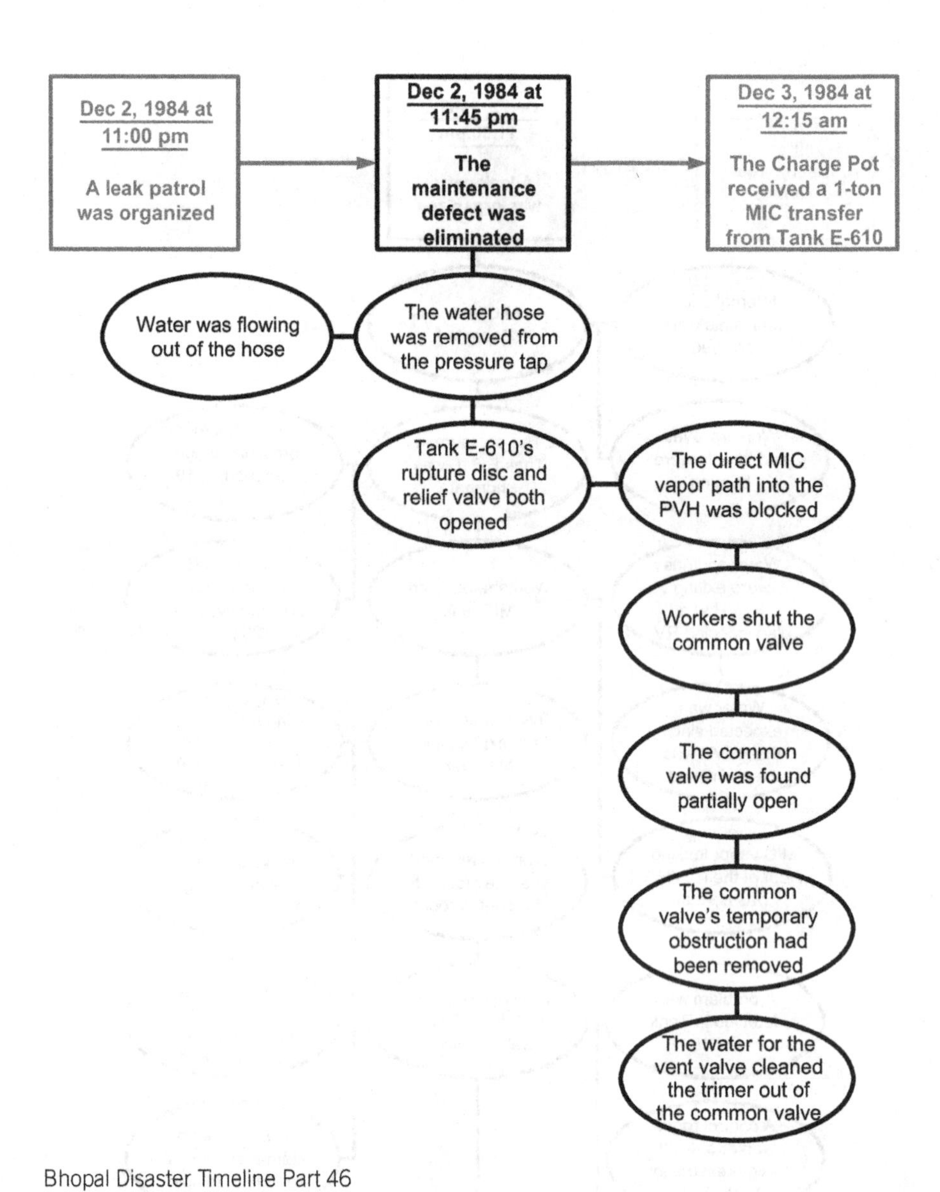

Bhopal Disaster Timeline Part 46

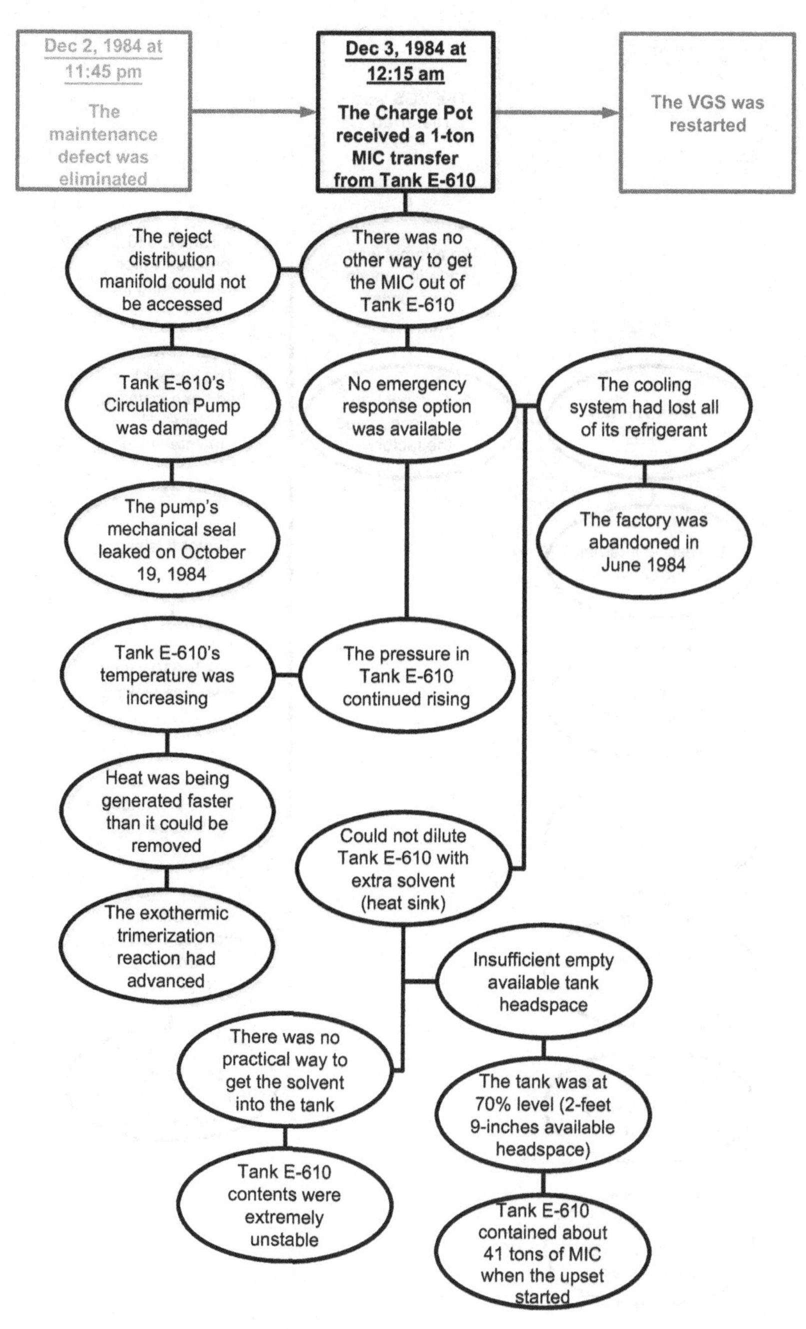

Bhopal Disaster Timeline Part 47

Bhopal Disaster Timeline Part 48

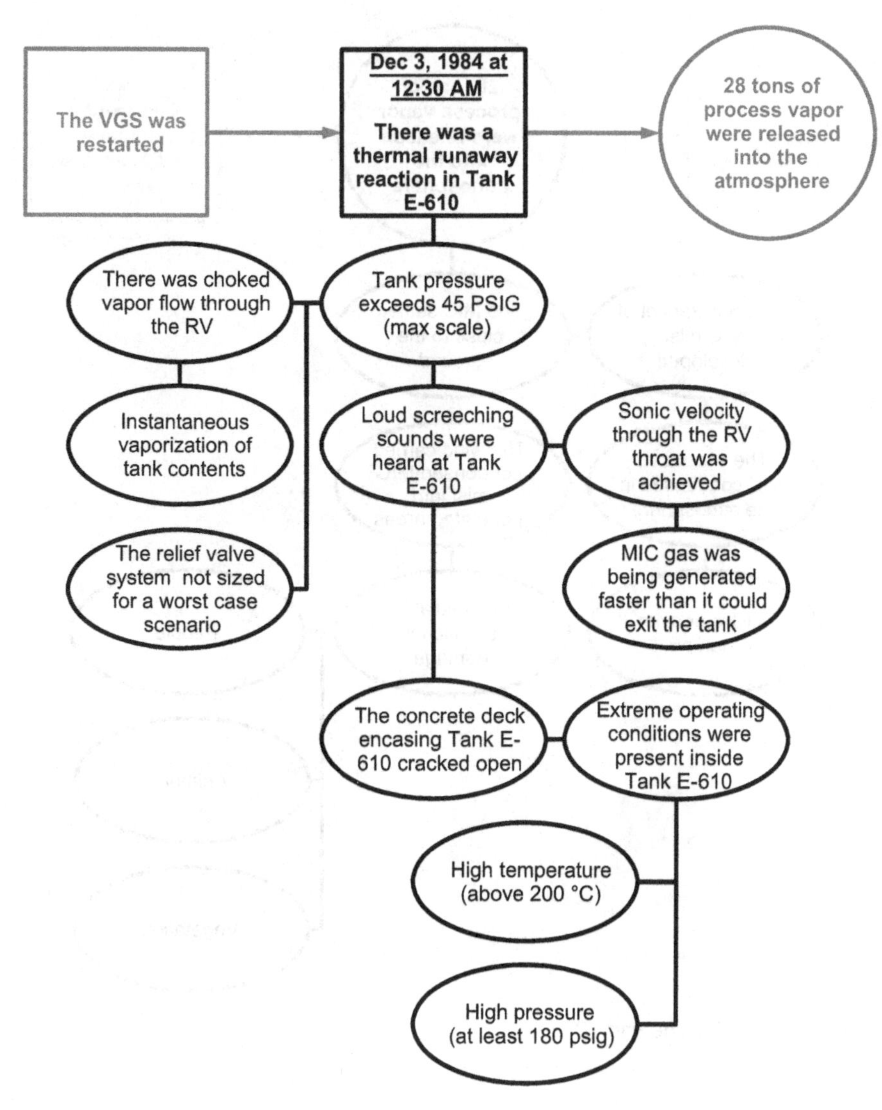

Bhopal Disaster Timeline Part 49

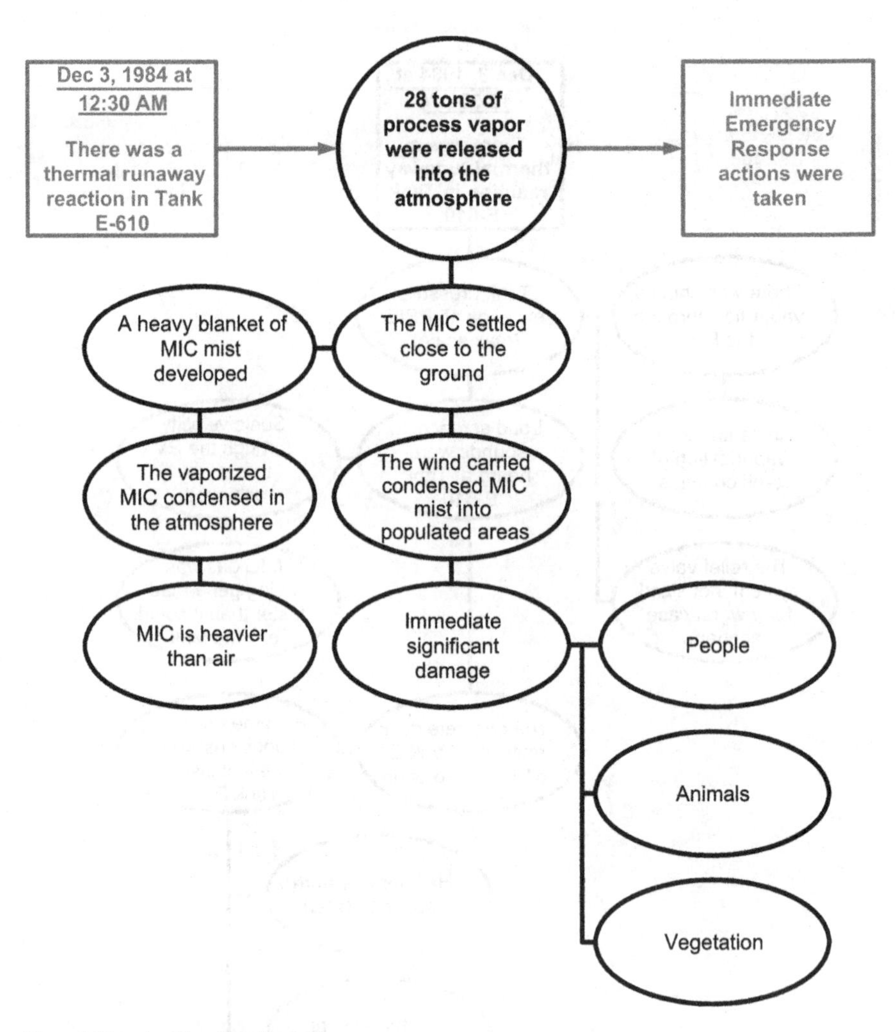

Bhopal Disaster Timeline Part 50

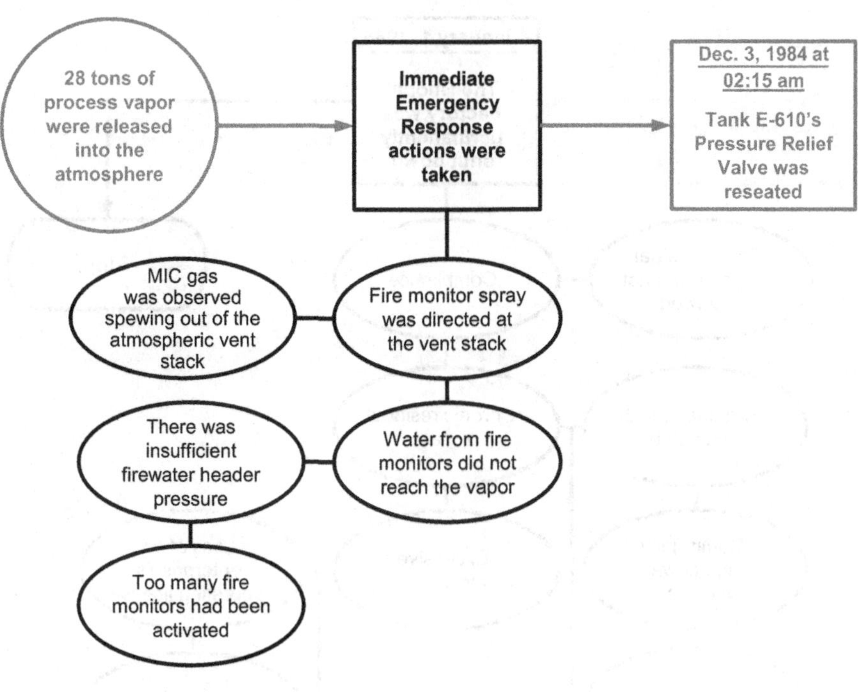

Bhopal Disaster Timeline Part 51

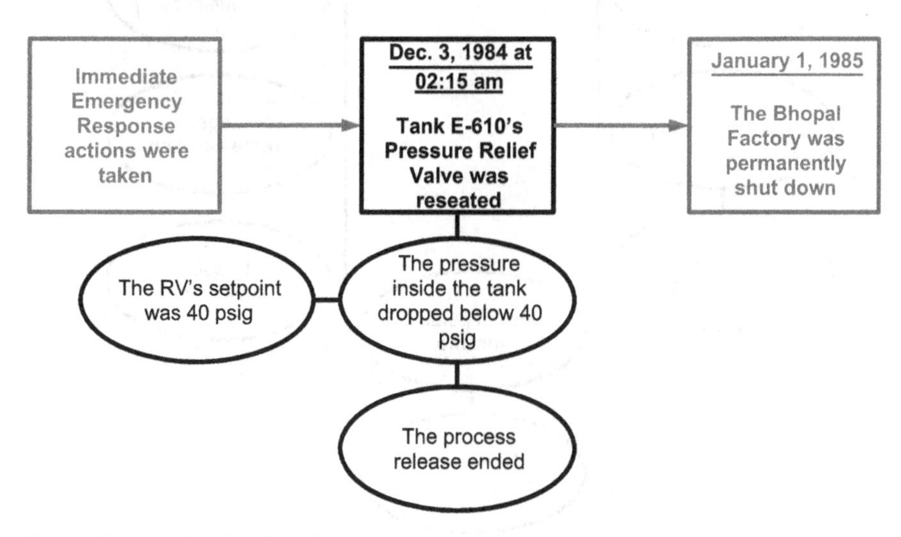

Bhopal Disaster Timeline Part 52

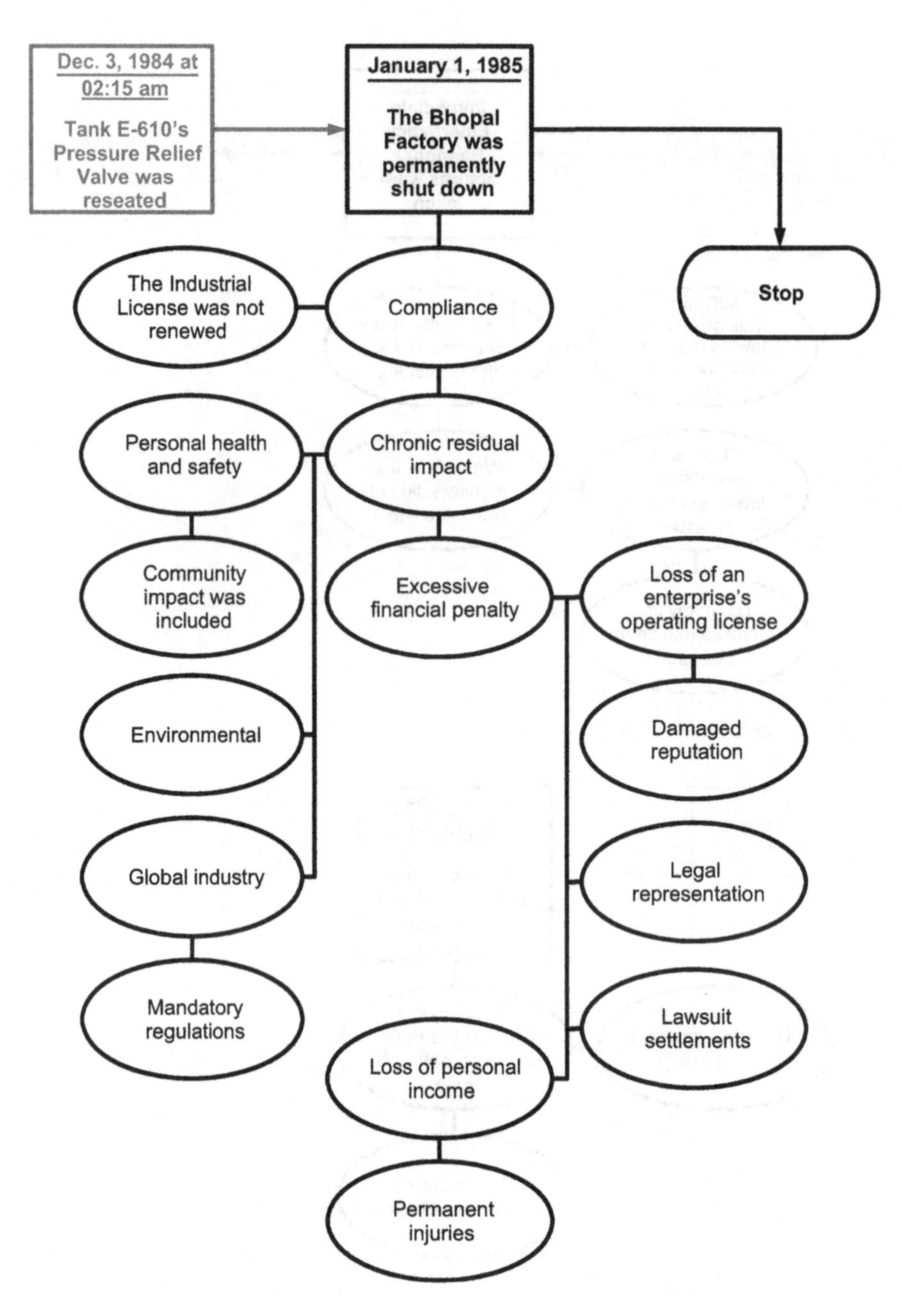

Bhopal Disaster Timeline Part 53

Glossary

Accelerometer A measuring device used in monitoring vibration-related movement of a body.
Act of commission Performing or allowing an irresponsible action.
Act of omission Failure to perform an intended action.
Active control An asset or function that provides protection by changing from one condition to another (see also passive control).
Autoignition temperature The temperature at which a flammable material ignites upon contacting oxygen.
Availability Percentage of time an asset is available for use in a production process.
Best operating practices The work procedures and work processes utilized by highly reliable and profitable industrial entities.
Board of directors Investor spokespeople or representatives that are ultimately responsible for the success of a manufacturing operation.
Carryover Involuntary transfer of a process to unintended location by spillage, entrainment, or overflow.
Catastrophic failure An event that creates enough damage to render an asset incapable of providing its intended design function.
Check valve A device inserted in a pipe that prevents process flow in the opposite direction.
Commodity A bulk or raw material with no manufacturing restrictions (see also patent, specialty product, and trade secret).
Common cause failure Unresolved process dependencies that disable multiple asset functions upon a single, usually unrelated, failure (see also independent protective layer).
Compliance Action in agreement with stipulated commitments.
Condition-based maintenance Maintenance that is performed on an "as-needed" basis and according to certain performance indicators.
Corrective maintenance Intervention and restoration of an asset after a functional failure occurs.
Customer loyalty Action by a customer or client who consistently supports a supplier.
Debottlenecking Removal of a weak link or constraint in a process.
Derivative A product that is that is produced with two or more intermediates.
Design function Asset behavior in accordance with as-intended or as built-in operating modes.
Discretionary A nonmandatory action or decision.
Executives Corporate directors including the CEO, president, and vice president.
Failure mechanism The specific way in which an asset becomes damaged.
Falsifiability A systematic approach to validating evidence by attempting to disprove it.
Findings Conclusions based on close examination.
Fire monitors Water spray cannons or similar nozzles which can be aimed at a fire.
Function The definition of what an asset is supposed to do in a manufacturing process.
Functional failure Inability to perform as intended.
Generic Function Basic description of an asset's primary purpose that uniformly applies to other similar assets.
Glut Supply in excess of demand.
Grounding and bonding Physical preventive steps used to suppress potential static ignition of flammable mixtures.

Hazard awareness The ability to detect a safety impediment.

Human factors Inclinations of personnel based on training, experience, intelligence, or imperfections.

Hydrocarbon entrainment Turbulent binding of a flammable mixture in an inert type of media.

Incident An unexpected event that carries with it some unwanted consequence.

Independent protective layer (IPL) A safety barrier that is not influenced by the performance or reliability of another safety barrier (see also common cause failure).

Inherently safer design (ISD) Assets with built-in failure-proof components, subassemblies, or protective means.

Intermediate A product that undergoes further reactions to enhance its value (see also derivative).

Invasive maintenance Action(s) that include entering into or dismantling an asset before it can be returned to as-intended service or performance.

Isolation Separating an asset from its surroundings.

Lachrymator An agent that produces watering of the eyes (a tear producer).

Line organization The business or department in a manufacturing organization that interfaces directly with the process.

Location A reliability term used to describe where or how an asset fits in a manufacturing process.

Macrofouling Large physical process obstructions that tend to inhibit flow.

Maintenance program Organized activities aimed at maintaining assets in as-designed condition.

Mandatory Compulsory actions or decisions that are consistent with business commitments.

Noncondensables Process that remains in the vapor phase instead of being converted into liquids.

Normalization of deviance Nonconformance that becomes an accepted routine.

Operational discipline A structured and adhered-to means of running a process.

Operating limits Maximum and/or minimum conditions or parameters required to maintain an acceptable level of process control.

Organizational culture Mindset, conduct, or behavior allowed to exist or purposely fostered.

P-F interval A measure of time between when signs of a developing failure can be detected and when the failure actually occurs.

Passive control A process or asset function that provides continuous protection in its natural form (see also active control).

Patent Time-limited legal coverage whereby an originator or inventor is protected from competition (see also commodity, specialty product, and trade secret).

Personal health and safety Efforts aimed at protecting individuals from endangerment.

Physical asset Human or material contributors needed to enable process performance.

Probable cause Explanation for an incident based on the totality of facts and evidence.

Process A collection of assets that makes manufacturing a product possible, or the material contained by that collection of assets.

Process configuration Sequence of assets involved in manufacturing a product.

Process flexibility Ability to reconfigure a process in response to changing market demands and asset performance, to meet specified production commitments.

Process optimization Imparting interacting features, work executions, specifications, implementation routines, etc., which combine to yield reliable, safe, and profitable product manufacturing.

Process reliability The discipline of continuously controlling physical and chemical manufacturing parameters in accordance with asset capabilities.
Production constraint Impediment to achieving intended output.
Raw materials Primary substances that enter a process before being converted to manufactured product.
Safeguards Controls that provide a barrier against undesirable incidents.
Secondary function A function other than the primary intended function.
Shortcut A quick way to achieve results, although usually by sacrificing safety or reliability.
Source documents Primary references that support an incident investigation with credible information that must be qualified on the basis of acceptable investigation practices.
Specialty product A unique or custom-designed product manufactured by a specific supplier (see also commodity, patent, and trade secret).
Standards, policies, and administrative controls (SPAC) Human-dependent management systems that define the workflow required to safely operate a process.
Time-at-risk The total amount of time that an asset is in an operating condition where its failure is possible.
Trade secret Knowledge safeguarded by owner's action and precautionary moves, usually supported by legally binding limited and controlled access to detail (see also commodity, patent, and specialty product).
Transparency Openness to examination by others; clarity.
Turnaround Preplanned downtime or shutdown action used to restore assets to as-designed condition; removal of wear and tear; upgrading and removal of production constraints.
Utilization Completeness or extent of actual available asset usage.
Witness statement A personal, documented recollection of observed events.
Workaround solution Purposeful and often flawed operational strategy that circumvents established rules (see shortcut).

Index

'*Note*: Page numbers followed by "f" indicate figures and "t" indicate tables.'

F

H

L

M

R